CALCULUS WITH FINITE MATHEMATICS FOR SOCIAL SCIENCES

MARY W. GRAY · American University · Washington, D.C.

CALCULUS WITH FINITE MATHEMATICS FOR SOCIAL SCIENCES

ADDISON-WESLEY PUBLISHING COMPANY
Reading, Massachusetts · Menlo Park, California · London · Don Mills, Ontario

This book is in the
ADDISON-WESLEY SERIES IN MATHEMATICS

LYNN H. LOOMIS
Consulting Editor

 Printed in the United States of America. Published simultaneously in Canada. Library of Congress Catalog Card No. 74-174335.

PREFACE

This book is intended to be a text for a year's course in mathematics for students of the social, biological, and management sciences. However, mathematics majors may also benefit from the application-oriented approach in their first encounter with calculus. This book includes both more and less than is covered in the traditional first-year calculus course. For example, very few of the functions with which social scientists are concerned are functions of one variable, hence, since most of these students will have only one year of mathematics, such things as partial differentiation must be included in that year. On the other hand, there is less emphasis on theory. A number of theorems are stated without proof and some concepts are introduced intuitively. A great deal of attention is given to the use of maximization, integration, and differential equations in social and biological sciences.

This book makes no attempt to turn anyone into a mathematical economist or biologist. Only the most basic material is presented. It is expected that the student will come out prepared to understand and use the elementary mathematical methods of his field and equipped to read current publications with a deeper understanding. Should he be interested, or should his specialization require it, he should be ready to take a course in advanced calculus or beginning analysis, an upper-division course in linear algebra, or an advanced course on mathematical methods in his own field.

The first two chapters are remedial in the sense that the material should have been covered in high school. Usually, however, the student without an early expressed interest in physical science or mathematics takes a minimal amount of mathematics in high school and consequently needs this preparatory work. It is also handy for those whose algebra may be rusty or whose high school training predates the introduction of "new math" at that level. For the same reasons the chapter on trigonometric functions assumes no prior knowledge of trigonometry. Should a class be unusually well prepared, the first two chapters and some of Chapter 8 can be omitted.

Chapters 10 and 11 depend only on Chapters 1 and 2 and Section 4.1, plus examples from later sections, so that Chapters 1, 2, and 3; Section 4.1, Chapters 10, and 11 could be used for a one-semester course in finite mathematics. A two-semester (three credits per semester) course can be based on Chapters 1, 2, 4, 5, 6, 7, 8, 9, and 10, with parts of Chapters 3, 10, 11, 12, and 13 included as desired. If the first two chapters are omitted, Chapter 14 can be included.

Chapters 4 through 8 can serve for a one-semester course embracing material usually covered in the first semester of calculus, easing the difficulties of the student who decides to transfer to a traditional sequence. Nothing in the calculus portion of this book depends on Chapter 3.

An unusual feature of this book is that it points out common errors and warns against them. In the "warnings" scattered throughout, the student is shown what he may be doing wrong.

My thanks to Alfred Gray, Joyce Samuels, Judith Sunley, my students at American University, and the staff and consultants of Addison-Wesley.

Washington, D.C. M.W.G.
January 1972

CONTENTS

1 / TOOLS AND RULES

This chapter is a very brief survey of those concepts from high school algebra which we need for the subsequent material in this book. The emphasis is on mechanics and the material may be omitted by those with an adequate background.

1.1 SETS

The notion of set is described rather than defined. We can say that a set is a collection, a family, an aggregate; but the best way to familiarize oneself with the idea is to observe many examples: the set of all chairs in a room, the set of all registered students at a particular university, the set of all integers greater than zero and less than five, and so on. The distinguishing property of sets is that one should be able to determine whether or not any individual is a member of a particular set. In the first example above, one can decide whether a given object is a chair and whether it is in a room, and hence one can determine whether it is an element of the set under consideration. Similarly, the last set described above has 1, 2, 3, and 4 as its only members.

Members of a set are called *elements* and are said to be *contained* in the set. If S is a set and s an element of S we write $s \in S$. If, as in the last example above, there are only a few elements in the set, we may write them out as

$$S = \{1, 2, 3, 4\}.$$

For the set of all even integers greater than zero and less than 50 we may write

$$\{2, 4, \ldots, 48\}.$$

Suppose we want to represent the set of all integers greater than 0. We could write

$$\{1, 2, 3, \ldots\}$$

to indicate the infinite extent of the set; but usually we write

$$\{x \mid x \text{ is an integer} \quad \text{and} \quad x > 0\}.$$

It is at this point that panic frequently sets in. Suddenly a *variable*, the letter x, has appeared in place of the old familiar numbers. Take heart, x represents a number, any number. Do not wish for a revelation regarding what x *really* stands for, as the whole point of introducing letters as variables is that they do vary; that is, they may represent different numbers.

The notation

$$\{x \mid \ldots\}$$

is used to indicate all objects having the property given after the vertical bar and is read "the set of all x such that . . . ." Any other letter may be used instead of x, but it is customary to use letters toward the end of the alphabet as variables. An equivalent way of representing $\{1, 2, 3, 4\}$ is

$$\{x \mid x \text{ an integer} \quad \text{and} \quad 0 < x < 5\}.$$

As a useful convention we talk about a set with no elements, called the *empty set*, denoted by $\varnothing$. For example,

$$\{x \mid x < 5 \quad \text{and} \quad x > 6\} = \varnothing.$$

We are familiar with various operations on numbers such as addition and multiplication. There are also operations on sets which will be useful. If A and B are sets, we define

$$A \cup B,$$

read "A union B," to be the set of all elements in A or in B, that is

$$A \cup B = \{x \mid x \in A \quad \text{or} \quad x \in B\};$$

and

$$A \cap B,$$

read "A intersect B," to be the set of all elements in both A and B, that is,

$$A \cap B = \{x \mid x \in A \quad \text{and} \quad x \in B\}.$$

It needs to be emphasized that "or" is always used in mathematics in the inclusive sense: "P or Q" means "P or Q or both."

Example. If

$$A = \{x \mid x \text{ an integer} \quad \text{and} \quad x > 0\}$$
$$B = \{x \mid x \text{ an integer} \quad \text{and} \quad x < 5\},$$

what are $A \cap B$ and $A \cup B$?

Solution

$$A \cap B = \{x \mid x \text{ an integer} \quad \text{and} \quad 0 < x < 5\} = \{1, 2, 3, 4\}$$
$$A \cup B = \{x \mid x \text{ an integer} \quad \text{and} \quad (x > 0 \quad \text{or} \quad x < 5)\}$$
$$= \text{the set of all integers.}$$

PROBLEMS

1. Write the following in set notation.
   a) The set of all states bordering the Pacific Ocean.
   b) The set of all living ex-Presidents of the United States.
   c) The set of all negative integers greater than -6.
   d) The set of U.S. Senators from New York.
   e) The set of all integers larger than 10.
   f) The set of all outcomes from throwing a die.
   g) The set of all outcomes of tossing a coin twice. (Let H = heads, T = tails.)
   h) The set of all English words beginning with qh.
   i) The set of all negative integers greater than 5.
2. In each of the following, find $A \cup B$ and $A \cap B$.
   a) $A = \{a, b, c, d\}$, $B = \{c, d, e\}$.
   b) A = {pink, blue, green}, B = {red, white}.
   c) $A = \{1, 2, 3, 4, 5\}$, $B = \{2, 4, 5\}$.
   d) $A = \{x \mid x < 6\}$, $B = \{x \mid x \geq 5\}$.
   e) $A = \{x \mid x > 5\}$, $B = \{x \mid x < 10\}$.
   f) A = set of all red cars, B = set of all Fords.
3. We say that S is a *subset* of T (written $S \subset T$) if every element of S is an element of T. Find all the subsets of each of the following:
   a) {red, pink, gold}. b) $\{1, 2, 3, 4\}$.
4. If A is a set, we define the *complement* of A to be the set of all elements not in A and write
$$\bar{A} = \{x \mid x \notin A\}.$$
   Clearly the complement must be relative to some universe of discussion. For example, the complement of the rational numbers is the set of real numbers which are not rational, i.e., the irrational numbers, and does not include such things as apples and oranges. Find the complement of each of the following sets:
   a) all positive real numbers.
   b) $\{1, 2, 5\}$ relative to the universe $\{1, 2, 3, 4, 5\}$.
   c) $\{x \mid x > 5\}$. d) $\{x \mid 6 < x < 8\}$.
   e) $\{x \mid a < x < b\}$. f) $\bar{A}$.
   g) the real numbers relative to the real numbers.
5. If A and B are as stated, write A and B in set notation and find $A \cap B$, $A \cup B$ and the indicated complements.
   a) A is the set of all real numbers greater than 5, B the set of all real numbers less than 10 and greater than -3. Find the complements of A and of B with respect to the set of all real numbers.
   b) A is the set of all integers greater than 2 and less than 4, B the set of all positive integers less than 5. Find the complements of A and of B with respect to the set $\{-1, 0, 2, 3, 4, 5\}$.
   c) A is the set of all real numbers less than or equal to -3, B the set of all real numbers greater than -10. Find the complements of A and of B with respect to the set of all real numbers.

d) A is the set of all letters following T in the English alphabet, B the set of all letters between P and W. Find the complements of A and of B with respect to the alphabet.

6. One means of illustrating the concepts of intersection and union is by means of Venn diagrams:

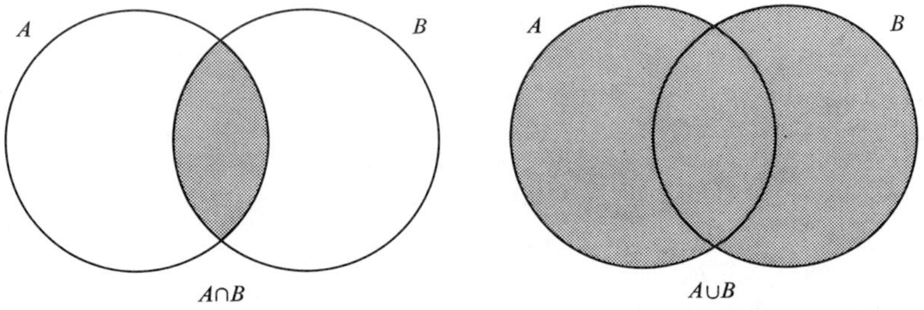

where the shading indicates the set under consideration. Draw a Venn diagram using three sets to illustrate the following:

a) $A \cap B \cap C$. b) $(A \cap B) \cup C$. c) $A \cap (B \cup C)$. d) $A \cap \bar{B} \cap C$.

e) $\bar{A} \cup (B \cap \bar{C})$. f) $A \cap \overline{(B \cup C)}$. g) $\bar{A} \cup (B \cap C)$.

7. In a certain city 300 salesmen sell life insurance; 520 sell fire insurance and 400 sell auto insurance. One hundred and fifty sell both life and fire insurance, 300 sell fire and auto insurance, 90 sell life and auto insurance and 30 sell all three. Use a Venn diagram to determine how many insurance salesmen there are in the city.

8. In a precinct with 1434 registered voters, 94 are under 25 years old, 213 are black and 512 are "poor," i.e., have incomes under \$5000 per year. If 197 of the blacks and 87 of those under 25 have incomes under \$5000, and of the 47 blacks under 25, two have incomes over \$5000, how many voters are there in this precinct who are neither young, black, nor poor?

9. Of the 42 students studying French, 6 are studying only French, 19 are studying French and German, and 21 are studying French and Spanish. If 10 students are studying only German, 20 only Spanish, and 13 both German and Spanish, how many students are studying at least one language out of French, German, and Spanish? How many are studying all three languages?

10. In a returning planeload of 200 passengers, 90 have been in France; 75 have been in Germany but not in France or Russia; 25 have been in Russia but not in France or Germany; 10 have been in Russia, France, and Germany; 23 have been in France and Germany; and 18 have been in France and Russia. If every passenger visited at least one country out of France, Germany, or Russia, how many visited Germany?

1.2 REAL NUMBERS

Even primitive societies have some notion of numbers used for counting. However, their number system may extend no farther than "one, two, many." The numbers we use for counting are called "natural numbers" or *positive integers*:

$$1, 2, 3, \ldots.$$

The set of positive integers is infinite, as is illustrated by the fact that no matter what integer one chooses, say n, there is always another larger integer, say $n + 1$. A somewhat more sophisticated culture may add *negative integers* to the positive integers. The concept of -3 is 3 coins, sheep, or whatever, *owed*. Still more sophisticated is the concept of *zero*. Its significance in the notation of our number system is as a place holder. The decimal system of notation depends on distinguishing between 12 (one ten and two one's) and 102 (one hundred and two one's). Computation is greatly simplified by the use of place holders; multiplication via Roman numerals soon gets out of hand. The set of positive and negative integers together with zero is called the *integers* and usually is denoted by Z.

A need develops, however, for something more than the integers. A *rational number* is the quotient of integers

$$\frac{a}{b}, \qquad b \neq 0.$$

Here there is a slight problem. We know what equality is in the case of integers. For n, m integers,

$$n = m,$$

precisely when n and m are the same integer. We surely want $\frac{1}{2}$ and $\frac{2}{4}$ to represent the same rational number even though they don't look the same. Hence for any two rational numbers

$$\frac{a}{b} = \frac{c}{d}$$

precisely when $ad = bc$. We are back to the letter problem, but it can be easily solved by remembering that a, b, c, and d stand for any integers with the restriction that $b \neq 0$ and $d \neq 0$.

Warning. There is nothing to prevent any two of the letters in the equation

$$\frac{a}{b} = \frac{c}{d}$$

from representing the *same* number. For example,

$$\frac{2}{2} = \frac{3}{3} = \frac{1000}{1000} \quad \text{and} \quad \frac{4}{8} = \frac{8}{16} = \frac{16}{32}.$$

In general the use of different letters for variables indicates possibly (but not necessarily) different values. Note that integers can be considered as special cases of rationals—4 as $\frac{4}{1}$ (or $\frac{8}{2}$ or $\frac{12}{3}$, etc.). It is also important to note that rationals can always be reduced to lowest terms: all common factors can be divided out of the numerator and denominator. For example, $\frac{1}{3}$ is in lowest terms, but $\frac{2}{6}$ is not. We frequently denote the set of all rational numbers by Q.

A word needs to be said about the addition and multiplication of rational numbers. By definition,

$$\frac{a}{b} + \frac{c}{d} = \frac{ad + bc}{bd},$$

$$\frac{a}{b} \cdot \frac{c}{d} = \frac{ac}{bd}.$$

It is *not* the case that for all a and b

$$\frac{1}{a + b} = \frac{1}{a} + \frac{1}{b}$$

nor does

$$\frac{a + b}{c + d} = \frac{a}{c} + \frac{b}{d} \qquad \text{for all } a, b, c, \text{ and } d.$$

Examples

1. Find

$$\frac{2}{3} + \frac{1}{2}.$$

Solution

$$\frac{2}{3} + \frac{1}{2} = \frac{4 + 3}{6} = \frac{7}{6}.$$

2. Find

$$\frac{2}{3} \cdot \frac{1}{2}.$$

Solution

$$\frac{2}{3} \cdot \frac{1}{2} = \frac{2}{6} = \frac{1}{3}.$$

There are numbers which are not rational; that is, they cannot be written as the quotient of integers. It is not always easy to prove that a particular number is not rational. For our purposes we simply accept the fact that they exist. The numbers $\sqrt{2}$ and π, which represents the ratio of the circumference of a circle to its diameter, are examples. They are called *irrational numbers.* It is possible to approximate irrational numbers by rational numbers; for example

$$\frac{3}{1} = 3, \quad \frac{31}{10} = 3.1, \quad \frac{314}{100} = 3.14, \quad \frac{3141}{1000} = 3.141, \quad \frac{31{,}416}{10{,}000} = 3.1416$$

are successive approximations to π.

The set of all rational and irrational numbers is called the *real numbers,* and is usually denoted by R. What properties do real numbers have? For our purposes the most important is that there are no gaps. That is, between integers there are nonintegers and between rationals there are irrationals, but there is nothing else between real numbers. This property is called *completeness*. Consequently, we represent real numbers as an unbroken line (Fig. 1.1) extending infinitely in both directions, called the *real line*. We choose a reference point for zero and a unit of measurement for one and set up a correspondence (see Fig. 1.2), in which each real number corresponds to exactly one point and, conversely, each point corresponds to exactly one real number.

Fig. 1.1

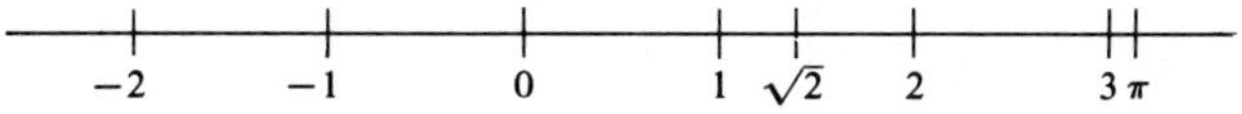

Fig. 1.2

For the set

$$\{x \mid 0 < x < 2\}$$

we write (0, 2). Note that "x a real number" is omitted from the description of the set. We shall adopt this procedure throughout most of this book since our *universe* (the set from which the objects under discussion come) will be the real numbers. In general, for any real numbers $a < b$,

$$\begin{aligned}(a, b) &= \{x \mid a < x < b\}\\ [a, b] &= \{x \mid a \leq x \leq b\}\\ (a, b] &= \{x \mid a < x \leq b\}\\ [a, b) &= \{x \mid a \leq x < b\}.\end{aligned}$$

This scheme is referred to as *interval notation*: (a, b) is an *open interval*, $[a, b]$ is a *closed interval*, $(a, b]$ and $[a, b)$ are *half-open* (or *half-closed*) *intervals*. Note that each of these sets has an infinite number of elements, since it includes all real numbers between a and b. In particular, the interval (a, b) should not be confused with the 2-element set $\{a, b\}$. If x is a real number such that $a < x < b$, then we can write $x \in (a, b)$ and similarly for the other intervals. If (a, b) is an open interval such that $x \in (a, b)$, (a, b) is said to be a *neighborhood* of x.

Examples. $1 \in (0, 2)$, $3/2 \in [0, 2)$; $2 \in [0, 2]$, but $2 \notin (0, 2)$. $(-1, 1)$ is a neighborhood of 0 and also a neighborhood of $-1/2$, $1/3$, $\pi/4$, etc.

PROBLEMS

1. Perform the indicated operations and reduce to lowest terms:

a) $\frac{1}{3} + \frac{1}{4}$. b) $\frac{2}{5} - \frac{5}{6}$. c) $\frac{7}{9} + \frac{1}{3}$. d) $\dfrac{a^3}{b} + \dfrac{3a^3}{2b}$.

e) $\frac{2}{5} \cdot \frac{3}{4}$. f) $\frac{1}{3} \cdot \frac{7}{15}$. g) $\dfrac{xy}{z} \cdot \dfrac{x}{z^2}$. h) $\frac{3}{7}(\frac{1}{5} + \frac{5}{9})$.

i) $\frac{2}{5} + \frac{3}{4}$. j) $\frac{1}{6} - \frac{3}{7}$. k) $\frac{2}{9} \cdot \frac{1}{8}$. l) $\frac{8}{9}(\frac{3}{4} - \frac{1}{6})$.

2. Write the following sets in interval notation:
   a) $\{x \mid -3 \le x \le 2\}$. b) $\{x \mid 0 < x \le \frac{7}{3}\}$. c) $\{x \mid -3 \le x < 1\}$.
   d) $\{x \mid \sqrt{2} < x < 2\sqrt{2}\}$. e) $\{x \mid \pi < x < 3\}$. f) $\{x \mid 2 < x < 6\}$.
   g) $\{x \mid -1 < x \le 5\}$.
3. Write the following intervals in set notation:
   a) $(2, 3)$. b) $(-1, 0]$. c) $[-10, -5)$. d) $[c, d]$.
   e) $(-2, 3)$. f) $[-1, 10)$. g) $(-4, -3]$.
4. In each of the following, write $A \cap B$ and $A \cup B$ in interval notation if possible:
   a) $A = \{x \mid 1 \le x \le 2\}$, $B = \{x \mid x < \frac{3}{2}\}$.
   b) $A = \{x \mid x > 5\}$, $B = \{x \mid x < 4\}$.
   c) $A = \{x \mid x < -1\}$, $B = \{x \mid x > -2\}$.
   d) $A = \{x \mid 1 < x < 5\}$, $B = \{x \mid -1 \le x \le 3\}$.
   e) $A = \{x \mid -2 < x \le 0\}$, $B = \{x \mid -1 \le x < 2\}$.
5. Find a neighborhood of each of the following:
   a) 0. b) -1. c) 2. d) $\sqrt{\frac{2}{3}}$.
6. If one third of the voters are registered Republicans and two fifths are registered Democrats, what fraction are Independents and those affiliated with minor parties?
7. If two fifths of the population have incomes under \$10,000 and one third of those have incomes under \$5000, what fraction of the population have incomes between \$5000 and \$10,000?
8. If 60 percent of the eligible voters are registered and two thirds of the registered voters voted in a recent election, what percentage of the eligible voters voted? (Note: 60 percent $= \frac{60}{100} = 0.6$).

1.3 OPERATIONS AND ORDER

We can perform two operations on real numbers, addition $(+)$ and multiplication $(\cdot)$, which satisfy the following axioms:

Commutativity. For any real numbers a and b

$$a + b = b + a$$

$$ab = ba$$

(we indicate multiplication by juxtaposition where no confusion is possible).

Associativity. For any real numbers a, b, and c

$$a + (b + c) = (a + b) + c$$

$$a(bc) = (ab)c.$$

Identity. There are real numbers 0 and 1 such that for any real number a

$$0 + a = a$$

$$1 \cdot a = a.$$

Inverse. For each real number a, there is a real number, denoted by $-a$, such that

$$a + (-a) = 0.$$

For each real number $a \neq 0$, there is a real number, denoted by $1/a$, such that

$$a \cdot \frac{1}{a} = 1.$$

Distributivity. For any real numbers a, b, and c

$$a(b + c) = ab + ac.$$

These axioms are called the *field axioms* and the real numbers are a *field.*

Warning. As mentioned above, a, b, and c may represent the same or different real numbers.

For example,

$$2(2 + 3) = 2 \cdot 2 + 2 \cdot 3.$$

Also, although multiplication is distributive over addition, it is not the case that addition is distributive over multiplication. That is, in general,

$$a + bc \neq (a + b)(a + c).$$

For example,

$$2 + (2 \cdot 3) \neq (2 + 2)(2 + 3).$$

The reason why it is worth pointing out that addition and multiplication on real numbers obey these axioms is that these operations do not satisfy all these axioms on all sets of numbers. For example, in the set of all integers there are not always multiplicative inverses: the inverse of 2 is $\frac{1}{2}$, which is not an integer. We also observe that subtraction is really addition of additive inverses and division is really multiplication of multiplicative inverses.

Next we summarize the rules of exponents:

1. For a real number x and positive integer n we write

$$\underbrace{xx\ldots x}_{n \text{ times}}$$

as x^n.

Example

$$2 \cdot 2 \cdot 2 = 2^3.$$

2. $x^0 = 1$ for any real number x.
3. For a negative integer n and a nonzero real number x, $x^n = 1/x^{-n}$. If n is a negative integer, $n \neq -1$, $0^n = 0$.

Example

$$3^{-2} = \frac{1}{3^2} = \frac{1}{3 \cdot 3}.$$

4. If y is a real number such that $y^n = x$ for some real number x, we write $y = x^{1/n}$.

Examples

$$2^5 = 32 \quad \text{so} \quad 2 = 32^{1/5}$$

$$(-2)^3 = -8 \quad \text{so} \quad -2 = (-8)^{1/3}$$

$$2^2 = 4 \quad \text{so} \quad 2 = 4^{1/2}.$$

Note that $(-2)^2 = 4$ also, but we usually reserve the notation $x^{1/n}$ for the positive root.

5. If p/q is a rational number, x a real number,

$$x^{p/q} = (x^{1/q})^p = (x^p)^{1/q}.$$

Example

$$8^{5/3} = (8^{1/3})^5 = 2^5 = 32.$$

6. For rational numbers r and s, x a real number,

$$x^r x^s = x^{r+s}, \qquad (xy)^r = x^r y^r,$$

$$\frac{x^r}{x^s} = x^{r-s}, \qquad (x \neq 0),$$

$$(x^r)^s = x^{rs}$$

(In these operations $x^{1/n}$ for $x < 0$ and n an even integer is not defined.)

Examples

$$(3x^2)^3 = 27x^6,$$

$$\sqrt{5} \cdot \sqrt{2} = \sqrt{10},$$

$$\frac{2y^3}{6y^5} = \tfrac{1}{3}\, y^{-2},$$

$$(6x^3)(2x^{-2})^4 = 3.2^5 x^3 x^{-8} = 96x^{-5}.$$

Warnings

1. Do not confuse exponents and coefficients. To multiply we add exponents and multiply coefficients.

Example

$$(3x^2)(2x^{-3}) = 6x^{-1}.$$

2. It is not the case that $(a + b)^n = a^n + b^n$, for all a and b, nor is it the case that $\sqrt{a + b} = \sqrt{a} + \sqrt{b}$, for all a and b.

Examples

$$(2 + 3)^2 = 5^2 = 25 \neq 2^2 + 3^2 = 13.$$

$$\sqrt{4 + 9} = \sqrt{13} \neq 2 + 3.$$

We have been using the *order relation* $<$ (less than) which is defined on the real numbers. It is defined by

$$a < b \text{ if and only if } b - a \text{ is positive.}$$

The order relation is subject to certain axioms:

1. *Trichotomy.* For any two real numbers a and b, precisely one of the following holds:
$$a < b, \qquad a = b, \qquad b < a.$$
2. *Transitivity*
$$\text{if} \quad a < b \quad \text{and} \quad b < c \quad \text{then} \quad a < c.$$

Moreover, the operations of addition and multiplication are connected with the order relation as follows:

01. $a > 0$, $b > 0$, imply that $ab > 0$.
02. $a < b$, c any real number, imply that $a + c < b + c$.

We are accustomed to dealing with equations, but the use of the order relation leads to inequalities. For example,

$$x - 2 < 3.$$

An identity such as

$$2x \equiv \frac{4x}{2}$$

is true for all values of x, but an equation holds only for certain values of the variable. The set of values for which an equation holds is called the *solution set* of the equation. For example,

$$2x - 1 = 3,$$

is true only when $x = 2$. The solution set of the equation is $\{2\}$. Similarly, an inequality involving a variable holds only for certain values of that variable. Again the values for which the inequality holds constitute the *solution set*.

The inequality

$$x - 2 < 3$$

is valid only when $x < 5$. Thus the solution set is

$$\{x \mid x < 5\}.$$

In solving an equation we may add the same quantity to both sides, or multiply both sides by the same nonzero constant, without altering the solution set. That is, these operations produce *equivalent equations*, equations which have the same solution sets. Can we do the same things to an inequality without changing its solution set?

We may add any quantity to both sides of an inequality since, by (02),

$$a < b,\ c \text{ any real number imply that } a + c < b + c.$$

On the other hand, if $a < b$ and $c > 0$ we have $ac < bc$, but if $a < b$, $c < 0$, then $ac > bc$ (see Problem 9). That is, multiplication by a negative quantity reverses the direction (called the *sense*) of the inequality.

Examples

1. Find the solution set of

$$3x - 2 < 4x + 1.$$

Solution. First we add 2 to both sides of the inequality and then we subtract $4x$ from both sides to obtain

$$-x < 3.$$

Then we multiply by -1 to obtain

$$x > -3.$$

The solution set is $\{x \mid x > -3\}$.

2. Find the solution set of

$$5x - 1 \leq 2x + 3.$$

Solution. The $\leq$ can be handled in the same way as $<$. However, one cautionary word: $\leq$ means $<$ *or* $=$. Thus if $x = y$, then $x \leq y$ is satisfied. On the other hand, if $x < y$, then $x \leq y$ is still satisfied. Proceeding with the solution, we subtract $2x$ and add 1 to each side of the inequality to obtain

$$3x \leq 4.$$

Then we divide by 3 (multiply by $\frac{1}{3}$), giving

$$x \leq \tfrac{4}{3}.$$

The solution set is

$$\{x \mid x \leq \tfrac{4}{3}\}$$

PROBLEMS

1. What are the additive and multiplicative inverses (if any) of the following real numbers:

a) 3.	b) -4.	c) $\frac{1}{2}$.	d) $\dfrac{-3}{4}$.
e) π.	f) $-\sqrt{2}$.	g) 0.	h) $\frac{5120}{6954}$.
i) 2.	j) $\frac{1}{9}$.	k) -3.	l) $\dfrac{-4}{5}$.
m) c/d.	n) $-4b$.		

2. Use the laws of exponents to write in simplest form:

a) $\left(\frac{2}{3}y^2\right)^5$.   b) $\frac{5x^2}{15x}$.   c) $(4a^3)^{3/2}$.

d) $\left(\frac{3xy}{4x^3y^2}\right)^3$.   e) $\left(\frac{169}{13}a^5\right)^2$.   f) $\left(\frac{1}{a^5}\right)^{-2}$.

g) $\left(\frac{7xy}{14x^3y^2}\right)^{1/5}$.   h) $(3x^3)(6x^2)^{-2}(2x^{-1})^5$.

3. *Prove:* If a and b are elements of a field and $ab = 0$, then either $a = 0$ or $b = 0$.

4. The relation "earlier than" satisfies the transitivity and trichotomy laws. Can you think of other examples of relations satisfying these rules?

5. Some ordinary expressions have extraordinary mathematical meanings. For example, "some" means "at least one," as in "there is some element in $\{1, 2, 3\} \cap \{2, 4\}$." "Precisely one" is a way of saying "at least one" *and* "at most one." Thus the intersection just mentioned contains precisely one element, whereas $\{1, 2, 3\} \cap \{2, 3, 4\}$ contains at least one element and $\{1, 2, 3\} \cap \{4, 5\}$ contains at most one element. Tell whether each of the following sets has at least one, at most one, or precisely one element.
   a) all integers which are irrational numbers.
   b) all integers greater than 2.
   c) all integers between 0 and $\sqrt{2}$.
   d) all integers which are rational numbers.

6. "Less than" and "greater than" are also subject to misinterpretation. For example, the set of all real numbers no greater than 2 includes 2: $\{x \mid x \leq 2\}$. Write the following in set notation.
   a) The set of all real numbers at least as great as -1.
   b) The set of all real numbers between 2 and 4.
   c) The set of all real numbers at most as great as 1.
   d) The set of all real numbers not as great as 1.

7. Solve for x:

a) $2x - 1 < 3$.   b) $-2x \leq 4$.   c) $6x < 5x + 1$.

d) $\frac{x}{2} < 3x - 4$.   e) $2x \geq 5 - 3x$.   f) $3x + 2 < 5$.

g) $-6x \leq 5$.   h) $5 - \frac{x}{2} > 3$.   i) $3x \geq 5 - 2x$.

8. A club is to make an excursion for which they have \$150 budgeted. If there are 20 members, what may the cost per person be so that they may stay within their budget?

9. Using the field axioms, 01, 02 and the definition of $<$, show that:

a) $a < b,\ c > 0 \Rightarrow ac < bc$.   b) $a < b,\ c < 0 \Rightarrow ac > bc$.

10. If an average meal costs at least 40 cents, what must a family of four budget for food for a week? (Three meals per day per person.)

1.4 ABSOLUTE VALUE

A notion which always causes trouble for the student of elementary mathematics is that of absolute value. The difficulties arise because he wants to rely on his intuition instead of on a definition and its consequences. Let us begin with the definition for any real number a. We define the *absolute value* of a by

$$|a| = \begin{cases} a & \text{if} \quad a \geq 0 \\ -a & \text{if} \quad a < 0. \end{cases}$$

Note that for any a, $|a| \geq 0$ since in the first case $|a| = a \geq 0$, and in the second $|a| = -a$, which is greater than zero, since $a < 0$ implies $-a > 0$.

So long as we have a number, the definition causes no trouble.

Anyone can find that

$$|5| = 5, \qquad |-2| = 2, \qquad |-\tfrac{1}{2}| = \tfrac{1}{2}.$$

However, it is the notion that $|a|$ can be $-a$ that introduces the difficulty. The first reaction may be to regard $-a$ as a negative number because of the minus sign and therefore to reject the possibility that $|a| = -a$, since one has been led to believe that absolute value is always positive. One need only keep firmly in mind that if a is itself negative, $-a$ is in fact positive.

Let us look at some properties of absolute value. For any real numbers x and y,

1. $|x| = |-x| = \sqrt{x^2}$. Here we use the usual convention that $\sqrt{\ }$ means the positive square root. Note that the square of any real number is positive and thus that there is no real number which is the square root of a negative number. For example, $\sqrt{-4}$ is not a real number. We use $\sqrt{4}$ to denote 2 even though $(-2)^2$ also is 4. If we want to indicate both square roots we write $\pm\sqrt{4} = \pm 2$.
2. $|xy| = |x||y|$.
3. $|x + y| \leq |x| + |y|$ (the triangle inequality).
4. If $|x| < a$, then $-a < x < a$.

We illustrate the use of some of these properties.

1. Find the solution set of

$$|x - 3| < 6.$$

Solution. By property 4, we write this as

$$-6 < x - 3 < 6.$$

Now we really have two inequalities but we may deal with them simultaneously by adding 3 to each "side":

$$-3 < x < 9.$$

Putting this in interval notation yields

$$(-3, 9)$$

as the solution set.

2. Find the solution set for

$$|2x + 2| \le 2|x| + 2.$$

Solution. By property 2, $2|x| = |2x|$ and by property 3,

$$|2x + 2| \le |2x| + |2|.$$

Thus for all x,

$$|2x + 2| \le 2|x| + |2| = 2|x| + 2.$$

The solution set is the set of all real numbers.

PROBLEMS

1. Write in interval notation:
   a) $\{x \mid |x| < 6\}$. b) $\{x \mid |x| \le 1\}$. c) $\{x \mid |x| < -1\}$.
   d) $\{x \mid |x - 2| < 1\}$. e) $\{x \mid |2x - 1| < 5\}$.
2. Solve for x:
   a) $|x - 2| \le 3$. b) $|2x + 1| < 2$.
   c) $|x - 4| + |x + 1| \le |2x - 3|$. d) $|4x - 5| > 0$.
   e) $|x + 1| = 5$. f) $|2x + 4| \ge 2$.
   g) $|5 - x| \ge -1$. h) $|7 - 2x| \ge 1$.
3. Carry out the indicated operations if possible:
   a) $2|y|$. b) $|x^2|\,|x - 4y|$. c) $|x| + |y|$. d) $\dfrac{|2x|}{|3x^2|}$.
4. Absolute value is used to define distance. The *distance* between two real numbers a and b is defined to be $|a - b|$. The *diameter* (or *length*) of the interval (a, b) is the distance between a and b. This is also the diameter of $(a, b]$, $[a, b)$, and $[a, b]$.
   a) Find the distance between 3 and 5.
   b) Find the distance between -1 and 2.
   c) Find the distance between 3 and -7.
   d) Find the distance between $\sqrt{2}$ and π.
   e) Denoting the distance between a and b by $d(a, b)$, show that
      i) $d(a, b) = 0$ if and only if $a = b$.
      ii) $d(a, b) = d(b, a)$.
      iii) $d(a, c) \le d(a, b) + d(b, c)$.
   f) Find the diameter of $(-\sqrt{2}, \sqrt{2})$.
   g) Find the diameter of $(-2, 5]$.
   h) Find the distance between -1 and 3.
   i) Find the distance between 0 and 4.
   j) Find the distance between 0 and -3.
   k) Find the distance between -1 and -5.

5. Write the solution set of each of the following in interval notation:

a) $|x + 2| < 3.$ b) $|2x - 1| \leq 4.$ c) $\left|\frac{x + 5}{2}\right| < \frac{1}{2}.$ d) $|3 - 2x| \leq 1.$

6. When does equality hold in the triangle inequality?

1.5 GRAPHING

In the preceding section we found solution sets to inequalities in terms of intervals. It is easy to graph such solution sets on the real line. For example, the graph of the inequality $|x - 3| < 6$ is given by Fig. 1.3 since the solution set is $(-3, 9)$. Suppose we are now asked to graph the solution set of

$$x + y = 2.$$

What we want is a set of pairs of numbers, one value for x and one value for y, whose sum is 2. The pairs 1 and 1, 0 and 2, -3 and 5 are in the solution set. However, to graph such pairs clearly we need something more than the real line.

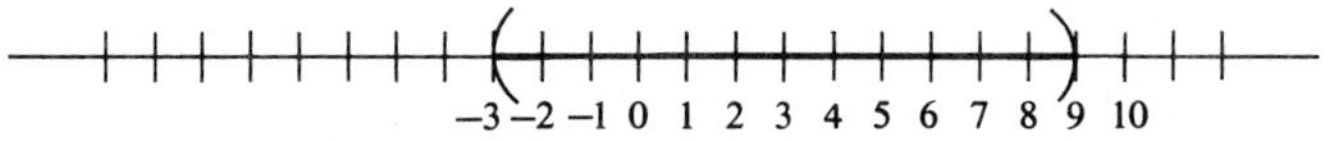

Fig. 1.3

If we draw two "real lines" perpendicular to each other and define the *coordinates* of any point in the plane by projecting the point onto the lines (called *axes*) in turn (Fig. 1.4), we have a one-to-one correspondence between pairs of real numbers and points in the plane. We adopt the convention of giving the coordinates of the points in the order

(horizontal, vertical),

where the horizontal coordinate is called the *abscissa* and the vertical the *ordinate*. Usually the horizontal axis is labeled the x-axis and the vertical axis the y-axis, but there is nothing sacred about this convention. Since the points (2, 3) and (3, 2)

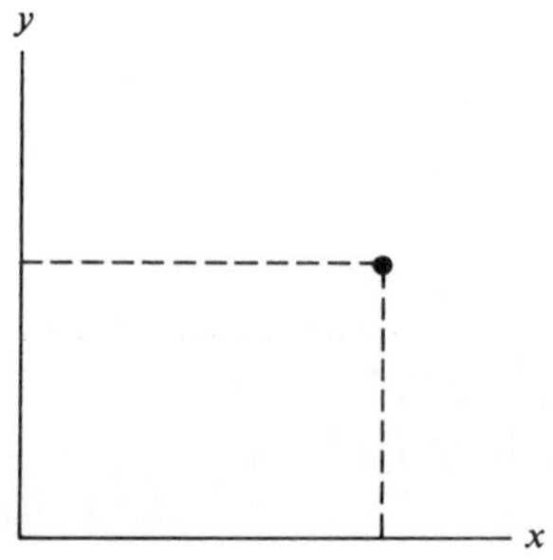

Fig. 1.4

are clearly not the same point, we really have a correspondence between points in the plane and *ordered* pairs of real numbers (Fig. 1.5). That is,

$$(x_1, y_1) = (x_2, y_2) \qquad \text{precisely when} \qquad x_1 = x_2 \qquad \text{and} \qquad y_1 = y_2.$$

The coordinate system we have established for the plane is called the *rectangular* or *Cartesian* coordinate system and is by no means the only possible one. However, it is the simplest for our purposes.

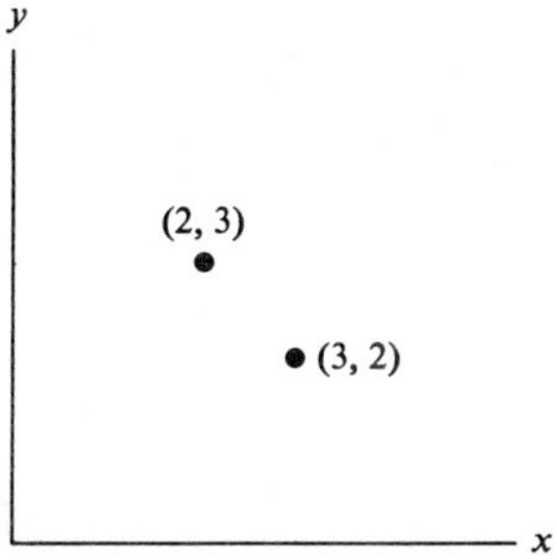

Fig. 1.5

Now to graph the solution set of $x + y = 2$. We want to represent the set of all pairs of real numbers (x, y) subject to the condition $x + y = 2$. It happens that the graph of any equation in which the only exponent of x and y is one is a straight line (Fig. 1.6). We have

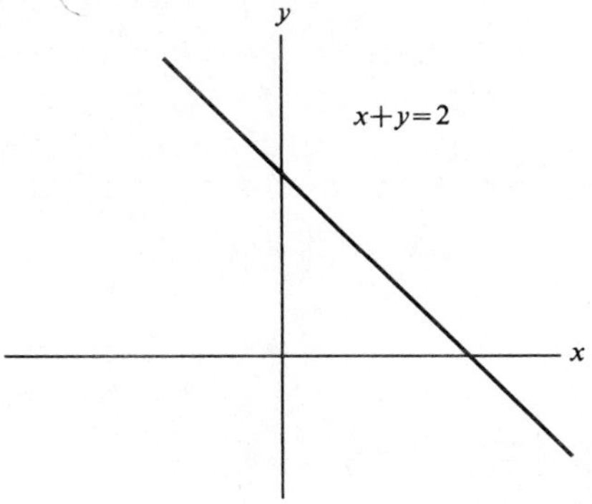

Fig. 1.6

If we know that the condition we are to graph can be represented by a straight line, it suffices to find two pairs of points and join them to form the entire graph. Hence, let $x = 0$; then $y = 2 - x = 2$. Let $x = 1$; then $y = 2 - 1 = 1$. Join (0, 2) and (1, 1) to get the graph of the solution set of $x + y = 2$.

More generally, if we have an equation involving x and y, we assign values to one of the variables and compute the corresponding values of the other. Then we plot the pairs of values as points in the plane with reference to the coordinate axes. Of course, except in simple cases such as straight lines, one does not always know the shape of the line joining the pairs of points. In Chapter 6, we develop some techniques for facilitating the sketching of curves.

Examples

1. Graph the solution set of

$$y = x^2 + 2x - 1.$$

Solution. It is the case that the graph of an equation in which the x and y terms have higher degree exponents is a smooth curve, but not a straight line. Therefore, we plot the following pairs of points and join them by a smooth curve (Fig. 1.7).

$x = 0$	$y = -1$	$(0, -1)$
$x = 1$	$y = 2$	$(1, 2)$
$x = -1$	$y = -2$	$(-1, -2)$
$x = -2$	$y = -1$	$(-2, -1)$
$x = -3$	$y = 2$	$(-3, 2)$.

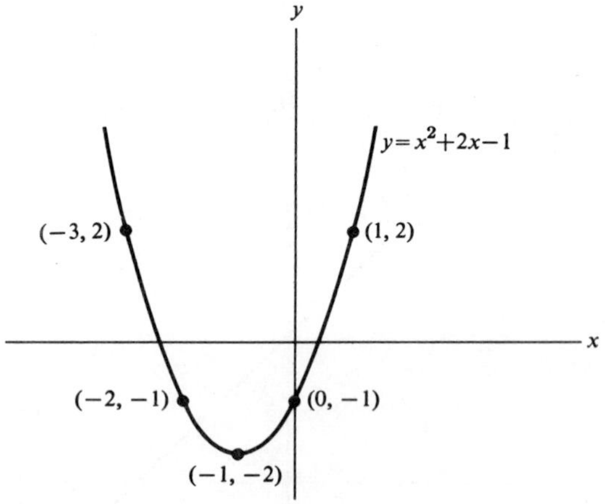

Fig. 1.7

2. On the real line, the graph of the inequality $|x - 3| < 2$ is the interval $(1, 5)$ (Fig. 1.8). In the plane we have

$$\{(x, y) \mid |x - 3| < 2\}$$

represented by the shaded area (see Fig. 1.9), with the broken lines indicating that those points for which $x = 1$ or 5 are not included.

3. Graph the set of points

$$\{(x, y) \mid x \geq 0, y \leq 0\}.$$

The solution is given by Fig. 1.10.

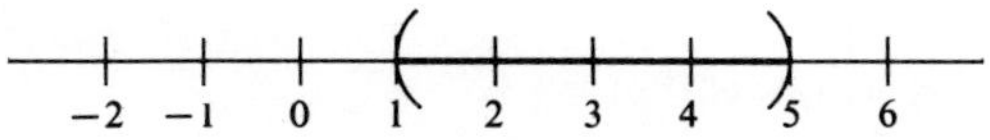

Fig. 1.8

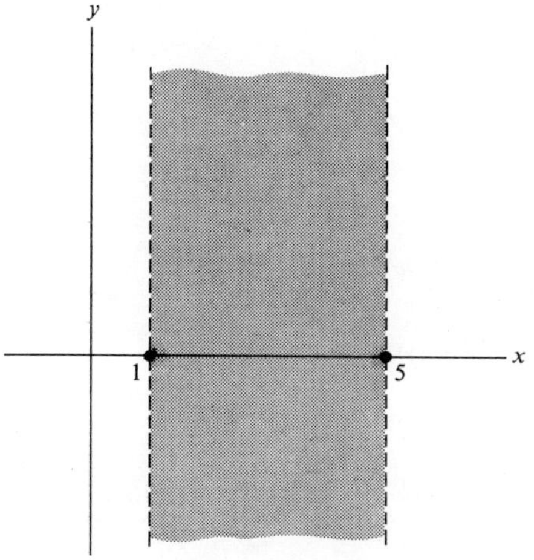

Fig. 1.9

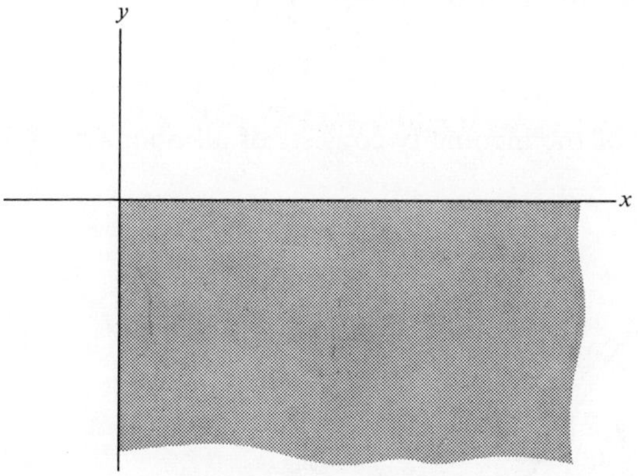

Fig. 1.10

4. Graph the solution set of the inequality

$$y < x - 1.$$

Solution. We graph the equality $y = x - 1$, a straight line. The solution set of the inequality consists of those points of the plane lying below this line (Fig. 1.11).

5. Graph the solution set of the inequality

$$3x + 2y \geq 2.$$

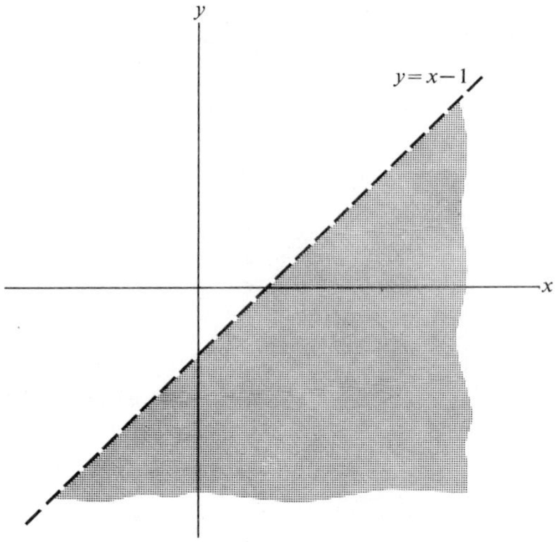

Fig. 1.11

Solution. First we solve for y

$$y \geq \frac{-3}{2} x + 1.$$

Then we graph the solution set of the equation

$$y = \frac{-3}{2} x + 1$$

and the solution set of the inequality consists of all points on and above this line (Fig. 1.12).

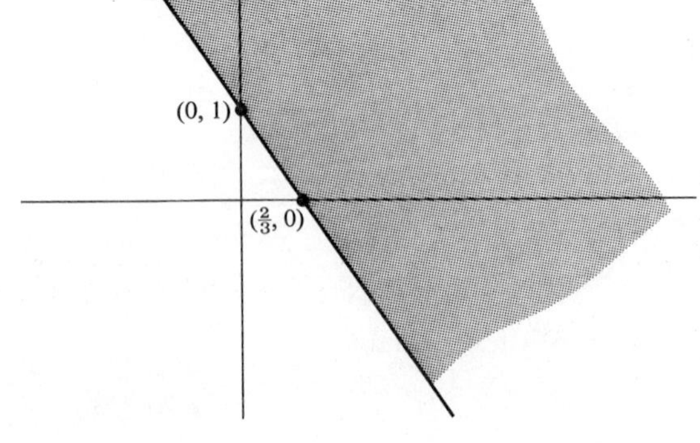

Fig. 1.12

6. Graph the solution set of

$$|x - y| < 2.$$

Solution. We have

$$-2 < x - y < 2.$$

The left inequality yields

$$y < x + 2$$

and the right

$$y > x - 2.$$

Thus the solution set (Fig. 1.13) consists of the points between the straight lines

$$y = x - 2 \quad \text{and} \quad y = x + 2.$$

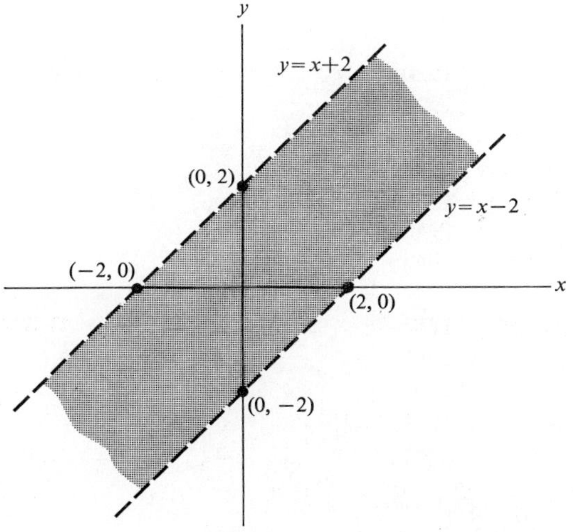

Fig. 1.13

PROBLEMS

1. Find the value of y:
   a) if $3y = 2x - 1$ and $x = 0$.
   b) if $3x + 4y = 6$ and $x = 1$.
   c) if $y^2 = x^2 - 4$ and $x = 4$.
   d) if $\sqrt{y} - 2 = x$ and $x = 4$.
2. Graph the intervals of Problem 2 of Section 1.2 on the real line.
3. Graph the solution set of:
   a) $y = 3x^2 - x + 1$.
   b) $y = 2x - 1$.
   c) $3x + 4y \leq 2$.
   d) $|x - y| > 1$.
   e) $|x - 2y| \leq 4$.
   f) $x + y = 2$.
   g) $3x - y = 3$.
   h) $y = 5$.
   i) $2x + 2y + 3 = 0$.
   j) $x^2 - x + 3 = 0$.
   k) $x + y \leq 2$.
   l) $3x - y \geq 1$.
   m) $2y + x \leq 5$.
   n) $x \geq 0$.
   o) $|2x + y| < 4$.

4. Graph the following sets:
   a) $\{(x, y) \mid x > 2, y = 0\}$.
   b) $\{(x, y) \mid y = 3x + 1\}$.
   c) $\{(x, y) \mid x^2 = y\}$.
5. The concentration of H_2SO_4 in a certain solution is given by

$$c = 0.04 + 0.01t,$$

   where t is the time in seconds. Graph the time as the abscissa and the concentration as the ordinate. What values may be assigned to t to get a realistic value for c?
6. Sometimes when data is collected at intervals rather than continuously, the best picture of what is happening is given by plotting various points and connecting them by line segments rather than by a smooth curve. Graph the following information in this way.

Year	Percentage of registered voters who voted in a certain precinct
1920	80
1924	74
1928	70
1932	85
1936	68
1940	71
1944	60
1948	65
1952	60
1956	54
1960	58
1964	52
1968	59

7. The cost of an article is given by

$$c = 7x + 49y,$$

   where x represents labor per hour and y transportation per unit. If the cost must be kept below 100, graph the set of values possible for x and y.
8. If the cost of an article is given by $c = 3x + 5y$, graph the set of values possible for x and y if the cost is less than 45.
9. If the force acting on an object is inversely proportional to the square of the distance of the object from the force, write an equation representing this relation and graph the solution set.

1.6 PROOF AND INDUCTION

One of the major questions confronting the neophyte in mathematics is: What constitutes a proof? We shall not attempt to answer this query in any great detail or with any great rigor, but a few remarks are in order. Basically a proof proceeds from what is known to be true—either as an axiom, as a previously proved theorem, or as an hypothesis—to what it is we want to prove, following the rules of logic. Each step must have a reason, a logical justification. At no stage of a proof may one assume the conclusion to be true; but one must never lose sight of what is to be proved.

Having made these general remarks, let us discuss a few specifics. The type of statement which is probably most common in mathematics (and which certainly gives the most trouble!) is the implication. Referring to implications, Bertrand Russell once remarked that mathematics is the science in which we never know what we are talking about nor whether what we say is true. In a statement of the form "if P, then Q," for example, "If it rains, I shall go to town," we are concerned only with what happens if it does rain—not whether it rains nor what happens if it doesn't. Therefore, if we want to prove that the implication "If P, then Q," written $P \Rightarrow Q$, is true, we assume that the antecedent P is true and show that Q must be also.

An easy example is the following. *Prove:* If n is a positive integer less than 5, then n^2 is less than 20.

We only care what happens if n is 1, 2, 3, or 4. We assume that $n = 1, 2, 3, 4$ in turn and verify the statement directly. Note that we have not proved that the square of an integer n is always less than 20, nor that 1, 2, 3, 4 are the only integers, but rather that *if* n is a positive integer less than 5, *then* its square is less than 20.

If we want to prove that $P \Rightarrow Q$ it suffices to prove that not $Q \Rightarrow$ not P. For $P \Rightarrow Q$ tells us that if P is true, Q cannot be false. Therefore whenever Q is false, P is also (not $Q \Rightarrow$ not P).

Example. *Prove:* If n is an integer and n^2 is even, then n is even.

Solution. First we must understand that "n even" means that $n = 2k$, for some integer k and that "n odd" means $n = 2j + 1$, for some integer j. Moreover, every integer is either even or odd, but not both.

Proof. Suppose n is odd.

Then $n = 2k + 1$ and $n^2 = 4k^2 + 4k + 1 = 2j + 1$, where j is the integer $2k^2 + 2k$.

Thus n^2 is odd.

We have proved not $Q \Rightarrow$ not P, where Q is "n is even" and P is "n^2 is even." Hence we may conclude $P \Rightarrow Q$, which is what we were to prove.

Another "roundabout" method of proof is *proof by contradiction.* A contradiction is a statement which cannot be true.

Example. "P and not P" is a contradiction.

Now suppose we want to prove a statement S. We assume that S is *not* true. We then show that this leads to a contradiction and from that we conclude that S is true.

Example. Prove that $\sqrt{2}$ is irrational.

Proof. Suppose that $\sqrt{2}$ is rational, then

$$\sqrt{2} = \frac{p}{q}, \qquad p, q \text{ integers}, \quad q \neq 0.$$

Suppose this expresses $\sqrt{2}$ in lowest terms; that is, p and q have no common factors.

Now we have

$$2 = \frac{p^2}{q^2},$$

$$2q^2 = p^2.$$

This means that p^2 is even. The proof in the previous example shows that p must be even, say $p = 2k$. Therefore

$$2q^2 = 4k^2$$

$$q^2 = 2k^2$$

Thus q^2, and also q, is even. But then p and q have 2 as a common factor. This, together with the hypothesis that p and q have no common factors, yields a contradiction. Therefore $\sqrt{2}$ cannot be rational.

One more type of statement needs some consideration. $P \Leftrightarrow Q$ is used to mean $P \Rightarrow Q$ *and* $Q \Rightarrow P$ and is read "P if and only if Q".

Example. Prove that n is even if and only if n^2 is even.

Proof. We have shown that n^2 even $\Rightarrow n$ even, so we need only show that n even $\Rightarrow$ n^2 even. But n even $\Rightarrow n = 2k$, for some integer k. Therefore $n^2 = 4k^2 = 2(2k^2)$ which is even.

We close this chapter with a discussion of a method of proof which almost invariably appears to the uninitiated to be a species of black magic. Mathematical induction is based on a peculiar property of the positive integers: namely, if any subset S of the positive integers is so defined that the integer m is contained in it, and whenever an integer n is in S, then the integer $n + 1$ is also in S, then all integers greater than or equal to m must be in S.

This property is of great assistance in proving statements about all positive integers. For example, suppose that we want to prove that for any positive integer n, the sum of all positive integers less than or equal to n is $\frac{1}{2}n(n+1)$. It is easy to see that if

$n = 1$
$$\tfrac{1}{2}n(n+1) = \tfrac{1}{2}(2) = 1,$$
$n = 2$
$$\tfrac{1}{2}n(n+1) = 1(3) = 1 + 2,$$
$n = 3$
$$\tfrac{1}{2}n(n+1) = \tfrac{3}{2}(4) = 6 = 1 + 2 + 3.$$

It is also easy to see that since there are infinitely many positive integers it is not possible to verify the statement for all of them. Thus we adopt the following procedure.

Principle of Mathematical Induction. If P_n is a statement about a positive integer n, then in order to prove that P_n is true for all positive integers n, it suffices to show that

1) P_1 is true,
2) For all n, $P_n \Rightarrow P_{n+1}$.

Intuitively it is clear that this procedure should work since if P_1 is true and if $P_1 \Rightarrow P_2$ is true, then P_2 is true, and if P_2 is true and if $P_2 \Rightarrow P_3$ is true, then P_3 is true, etc. This can never lead to a proof for all n, however, since it would taken an infinite number of steps. If we accept that the property of positive integers mentioned above holds, then the validity of the principle of mathematical induction follows without any appeal to an infinite process. If we prove (1), we have 1 in the set S of positive integers for which P_n is true, and by (2) if n is in S, then $n + 1$ is also; thus all integers greater than or equal to one are in the set S for which P_n is true. But this means that P_n is true for all positive integers.

Let us return to the example above. To simplify things we introduce *summation notation*. Suppose $f(i)$ is some expression involving an integer i. It could be that $f(i)$ is just i. Then we write

$$\sum_{i=1}^{n} f(i) = f(1) + f(2) + \cdots + f(n-1) + f(n)$$

to indicate the sum of the expressions $f(i)$ for each i from 1 to n.

Examples

1.
$$\sum_{i=1}^{n} i = 1 + 2 + 3 + \cdots + n$$

and for $n = 5$

$$\sum_{i=1}^{5} i = 1 + 2 + 3 + 4 + 5.$$

2.
$$\sum_{i=1}^{n} i^2 = 1 + 4 + 9 + \cdots + n^2$$

and for $n = 4$

$$\sum_{i=1}^{4} i^2 = 1 + 4 + 9 + 16.$$

3.
$$\sum_{i=1}^{n} \frac{1}{2}(i + 3) = 2 + \frac{5}{2} + 3 + \cdots + \frac{n+3}{2}$$

and if $n = 5$

$$\sum_{i=1}^{5} \frac{1}{2}(i + 3) = 2 + \frac{5}{2} + 3 + \frac{7}{2} + 4.$$

The i as used in these examples is called an *index*. Other letters may be used as indices; moreover, the values of the index need not start at 1 nor end at n. We could have, for example

$$\sum_{k=0}^{n} (2k - 1) = -1 + 1 + 3 + \cdots + 2n - 1,$$

$$\sum_{n=3}^{m+1} n^2 = 9 + 16 + 25 + \cdots + (m + 1)^2.$$

To restate the problem which began the discussion: Prove that

$$\sum_{i=1}^{n} i = \tfrac{1}{2}n(n + 1).$$

Solution. We proceed in two steps. P_n is

$$\sum_{i=1}^{n} i = \tfrac{1}{2}n(n + 1).$$

1. We have verified that P_1 is true.

2. We must prove that $P_n \Rightarrow P_{n+1}$ is true. We need only check the case when P_n is true. Suppose that P_n is true, i.e., suppose that

$$\sum_{i=1}^{n} i = \tfrac{1}{2}n(n + 1).$$

It is always wise to examine what it is we want to prove, namely:

$$\sum_{i=1}^{n+1} i = \tfrac{1}{2}(n + 1)((n + 1) + 1).$$

We note that

$$\sum_{i=1}^{n+1} i = \sum_{i=1}^{n} i + (n + 1),$$

and we therefore substitute in this equation what we have assumed $\sum_{i=1}^{n} i$ to be equal to, namely, $\frac{1}{2}n(n + 1)$

$$\begin{aligned}\sum_{i=1}^{n+1} i &= \tfrac{1}{2}n(n + 1) + n + 1 \\ &= \frac{n^2 + n + 2n + 2}{2} \\ &= \tfrac{1}{2}(n + 1)(n + 2) \\ &= \tfrac{1}{2}(n + 1)\big((n + 1) + 1\big)\end{aligned}$$

as desired.

Having proved that

$$\sum_{i=1}^{n} i = \tfrac{1}{2}n(n + 1)$$

for *all* positive integers n, it is possible to state, for example, that the sum of the first one hundred positive integers is $\frac{1}{2}(100)(101) = 5050$.

Since the proof above is the canonical one used to introduce mathematical induction, the student is led on occasion to believe that proof by mathematical induction consists of adding $n + 1$ to both sides of an equation. This is not the case. However, there is a technique which frequently does work in carrying out the second step in a proof by mathematical induction, namely, addition of the $(n + 1)$st term. Let us look at another example:

Prove that

$$\sum_{i=1}^{n} i^2 = \frac{n}{6}(n + 1)(2n + 1).$$

Solution.

1. P_1 is true since $1 = \frac{1}{6}(2)(3) = 1$.
2. Suppose that P_n is true, i.e., that

$$\sum_{i=1}^{n} i^2 = \frac{n}{6}(n + 1)(2n + 1).$$

We want to show that

$$\sum_{i=1}^{n+1} i^2 = \frac{n + 1}{6}\big((n + 1) + 1\big)\big(2(n + 1) + 1\big).$$

Let us write

$$\sum_{i=1}^{n+1} i^2$$

as

$$\sum_{i=1}^{n} i^2 + (n + 1)^2$$

and substitute what we have assumed $\sum_{i=1}^{n} i^2$ to be:

$$\sum_{i=1}^{n+1} i^2 = \frac{n(n+1)(2n+1)}{6} + (n+1)^2. \tag{1.1}$$

This has the same effect as taking the equation we assumed to be true, namely,

$$\sum_{i=1}^{n} i^2 = \frac{n}{6}(n+1)(2n+1),$$

and adding to it not $n + 1$ but rather the $(n + 1)$st term of the sum—in this case $(n + 1)^2$.

Let us proceed to simplify the right-hand side of (1.1):

$$\begin{aligned}\frac{n}{6}(n+1)(2n+1) + (n+1)^2 &= \frac{2n^3 + 3n^2 + n + 6n^2 + 12n + 6}{6}\\ &= \frac{2n^3 + 9n^2 + 13n + 6}{6}\\ &= \frac{(n+1)(n+2)(2n+3)}{6}\\ &= \frac{n+1}{6}((n+1)+1)(2(n+1)+1).\end{aligned}$$

Thus

$$\sum_{i=1}^{n+1} i^2 = \frac{n+1}{6}((n+1)+1)(2(n+1)+1)$$

as desired.

PROBLEMS

1. It is possible to define logical connectives by means of truth tables. The procedure is to assign truth values (true or false) to the statements being connected and give the truth value of the composite statement. For example, $P \Rightarrow Q$ is defined by

P	Q	$P \Rightarrow Q$
T	T	T
T	F	F
F	T	T
F	F	T

Note that all possible combinations of truth and falsity for P and Q must be listed. The truth tables for the other connectives are:

P	Q	P and Q
T	T	T
T	F	F
F	T	F
F	F	F

P	Q	P or Q
T	T	T
T	F	T
F	T	T
F	F	F

P	not P
T	F
F	T

Using these basic tables we can determine the truth value of more complicated statements. Defining equivalent statements to be those having the same truth table, show that $P \Rightarrow Q$ and not $Q \Rightarrow$ not P are equivalent.

2. Are each of the following pairs of statements equivalent?
   a) $P \Rightarrow Q$, (not Q) or P.
   b) $P \Rightarrow Q$, $Q \Rightarrow P$.
   c) $P \Rightarrow (Q \Rightarrow P)$, $Q \Rightarrow Q$.
   d) not (P or Q), not P and not Q.
3. Make a truth table for

   (not Q) and ($P \Rightarrow R$).

4. If $P \Rightarrow Q$, we know that P cannot be true and Q false. Hence the truth of Q is *necessary* for the truth of P. On the other hand, if we know $P \Rightarrow Q$ is true, to prove that Q is true it is *sufficient* to show that P is true. Translate the following statements into symbolism.
   a) If it does not rain, I shall not stay home.
   b) Either I shall stay at home and work or I shall go to town.
   c) For our team to win, it is necessary that Hobson play.
   d) It suffices to show that n^2 is even in order to show that n is even.
   e) It is neither necessary nor sufficient that you play the sonata in order for the evening to be a success.
5. Two sets S and T are said to be *equal* if they have exactly the same elements. *Prove:* $S = T$ if and only if $S \subset T$ and $T \subset S$.
6. Write the following in summation notation:
   a) $1 + 2 + 4 + 8 + \cdots + 2^n$.
   b) $1 + 3 + 5 + 7 + \cdots + 15$.
   c) $-1 + 2 + 7 + 14 + 23$.
   d) $1 + 4 + 9 + \cdots + n^2$.
   e) $6 + 8 + 10 + \cdots + (2n + 2)$.
   f) $-2 - 4 + 0 + 16 + \cdots + (n^3 - 3n^2)$.
7. Write out the first five terms of each of the following sums:

   a) $\sum_{i=1}^{15} (2i^2 + 4)$.
   b) $\sum_{j=5}^{n} (j - 1)$.
   c) $\sum_{p=0}^{14} (p^3 - 5)$.
   d) $\sum_{k=1}^{n} (k^2 - 5k)$.
   e) $\sum_{n=1}^{17} (n - 5)$.

8. Prove, using mathematical induction.

a) $\sum_{i=1}^{n} (3i - 1) = \frac{1}{2}(3n^2 - n)$. b) $\sum_{i=1}^{n} (2i^2 + 4) = \frac{1}{3}(2n^3 + 5n^2 + 4n + 2)$.

9. Try to discover a formula for $\sum_{i=1}^{n} (2i - 1)$ and verify it by mathematical induction.

10. Discover a formula for

$$\sum_{i=1}^{n} (3i + 2)$$

and verify it by mathematical induction.

11. Prove that if n^2 is odd, n is odd.

12. Prove that if x is the multiplicative inverse of a, then a is the multiplicative inverse of x.

13. Prove that if x and y are multiplicative inverses of a, then $x = y$.

14. Prove that the additive inverse of $a + b$ is $(-a) + (-b)$.

2 / THE SOLUTION OF ALGEBRAIC EQUATIONS

This chapter is intended as a concise summary of material on the solution of algebraic equations needed for the calculus of later chapters. Much of the material will probably be familiar to you and hence can be skimmed or skipped.

If

$$x^2 + 2x - 3 = 0, \tag{2.1}$$

a *solution* of this equation is a number which, when substituted for x, makes the equation a true statement. A solution of (2.1) is said to be a *root* of $x^2 + 2x - 3$. For example, $1 + 2 - 3 = 0$, so 1 is a solution of (2.1) and a root of $x^2 + 2x - 3$. As it stands, (2.1) is neither a true nor a false statement; solving the equation simply means finding for what values of x it is a true statement. In general, the solution set of an equation may have no elements, a finite number of elements, or an infinite number of elements.

2.1 LINEAR EQUATIONS

A *linear equation* in a single variable is an equation of the form

$$Ax + B = 0 \ (A, B \text{ real numbers}, A \neq 0).$$

The solution to this equation is given by

$$x = \frac{-B}{A}.$$

Example. Find the solution of $2x + 5 = 0$.

Solution

$$x = \frac{-5}{2}.$$

Linear equations in two variables are equations of the form

$$Ax + By + C = 0 \quad (A, B, C \text{ real numbers, } A \text{ and } B \text{ not both } 0) \tag{2.2}$$

We shall make use of the following theorem, which we state without proof.

Theorem 1. The graph of the solution set of a linear equation in two variables is a straight line and conversely any straight line represents the solution set of a linear equation.

If $B \neq 0$, we can rewrite (2.2) as

$$y = mx + b, \tag{2.3}$$

where $m = -A/B$ and $b = -C/B$. Suppose (x_1, y_1) and (x_2, y_2) are two points on the straight line whose equation is given by (2.3). Then

$$y_1 = mx_1 + b$$

and

$$y_2 = mx_2 + b.$$

Subtracting yields

$$y_1 - y_2 = m(x_1 - x_2),$$

so that

$$m = \frac{y_1 - y_2}{x_1 - x_2}.$$

Thus m is the difference in the y-values divided by the difference in x-values, or as we usually say: m is the change in y with respect to the change in x, or the *rise* in the line divided by the *run*. In particular, if x increases by 1, y changes by m. We call m the *slope* of the line

$$y = mx + b$$

(Fig. 2.1). If in (2.2), $B = 0$, the slope is not defined and

$$Ax + C = 0$$

represents a straight line parallel to the y-axis.

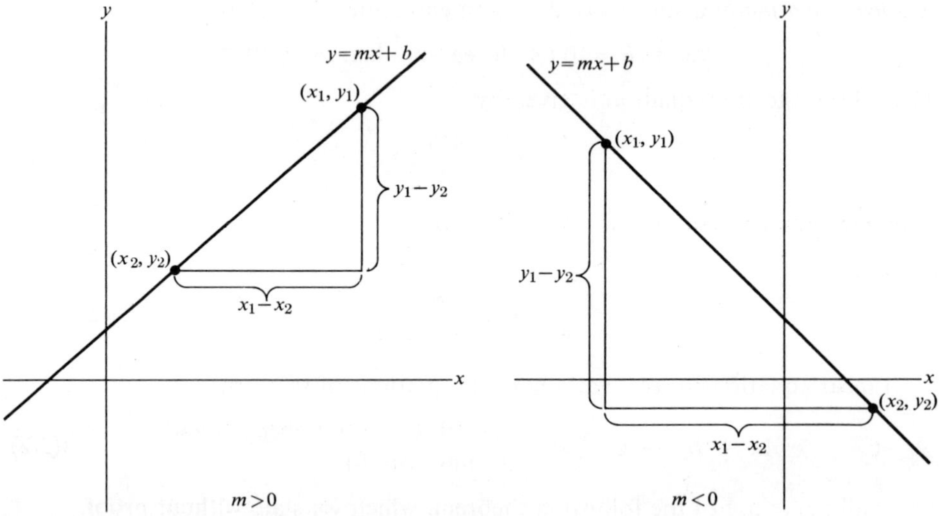

Fig. 2.1

From (2.3) we see that when $x = 0$, $y = b$; that is, b is the value of y where the straight line crosses the y-axis. We call b the *y-intercept*. Equation (2.3) is called the *slope-intercept* form of the equation of a straight line.

To put an equation which is in the form of (2.2) into the slope-intercept form, we put the y term on one side and the other terms on the other and divide by the coefficient of y.

Examples

1. $3x + 2y + 1 = 0$ in slope-intercept form is

$$y = -\tfrac{3}{2}x - \tfrac{1}{2},$$

the slope being $-\frac{3}{2}$ and the y-intercept $-\frac{1}{2}$. To graph the equation easily (Fig. 2.2) we can plot the y-intercept and then increase x by 1 and decrease y by $\frac{3}{2}$ to locate a second point.

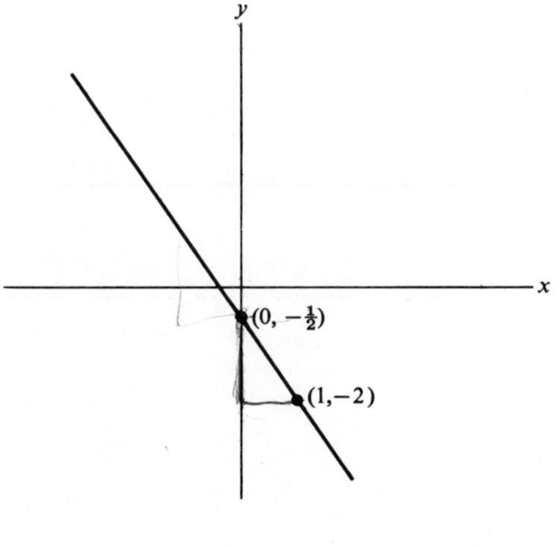

Fig. 2.2

2. $3x = 4$. This is the equation of a vertical line (Fig. 2.3)

$$x = \tfrac{4}{3}.$$

3. $y - 2 = 0$. This is the equation of a horizontal line; in slope-intercept form it is

$$y = 0x + 2,$$

that is, the slope is 0 and the y-intercept is 2 (see Fig. 2.4).

4. Find the equation of the straight line with slope 2 and y-intercept -1 (see Fig. 2.5).

Solution. $y = 2x - 1$.

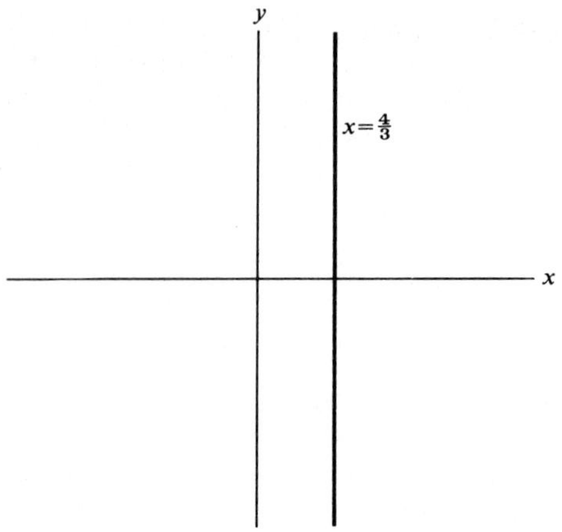

Fig. 2.3

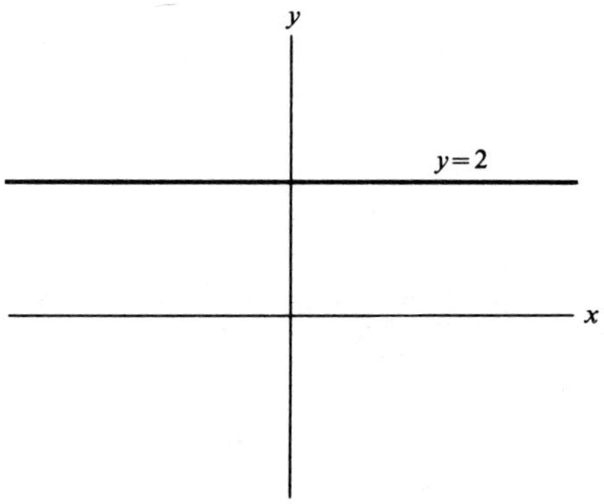

Fig. 2.4

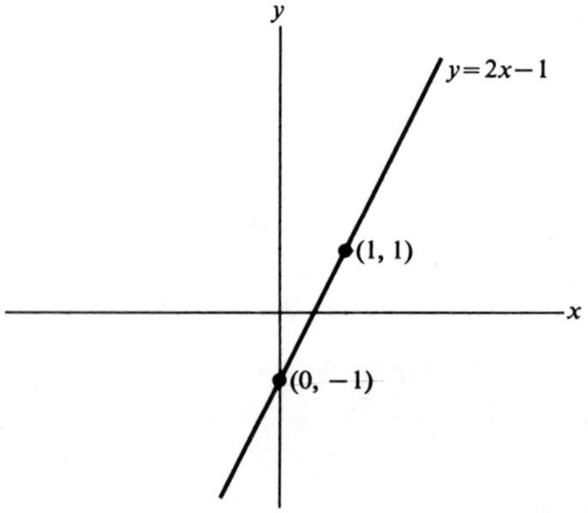

Fig. 2.5

Another standard form for the equation of a straight line is the *point-slope* form:

$$y - y_1 = m(x - x_1),$$

where (x_1, y_1) is any point on the line and m is again the slope.

Examples

1. $y - 4 = 3x - 2$ is the equation of a straight line of slope 3 passing through the point $(\frac{2}{3}, 4)$, Fig. 2.6.

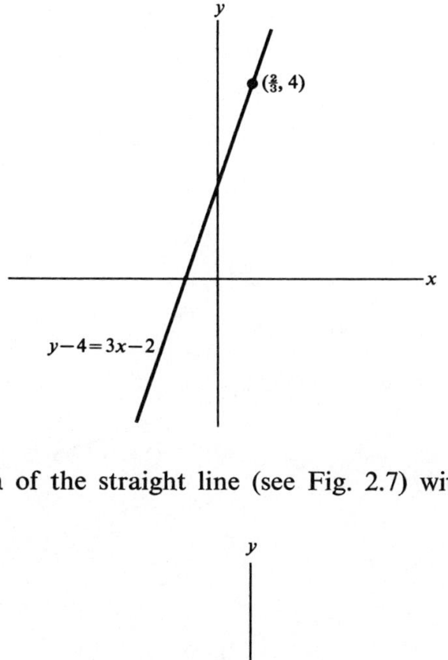

Fig. 2.6

2. Find the equation of the straight line (see Fig. 2.7) with slope $-\frac{1}{2}$ passing through $(2, -3)$.

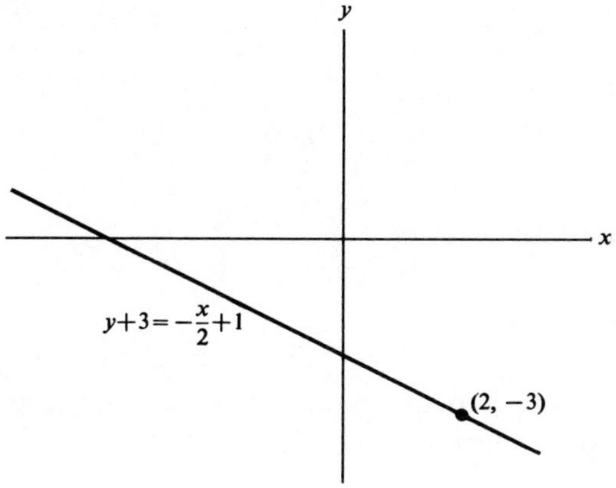

Fig. 2.7

Solution

$$y - (-3) = -\tfrac{1}{2}(x - 2),$$

$$y + 3 = \frac{-x}{2} + 1.$$

Suppose we want the equation of the straight line passing through (x_1, y_1) and (x_2, y_2). Then the slope is given by

$$\frac{y_2 - y_1}{x_2 - x_1} = m$$

so that in point-slope form we have

$$y - y_1 = \frac{y_2 - y_1}{x_2 - x_1}(x - x_1)$$

or alternatively

$$y - y_2 = \frac{y_2 - y_1}{x_2 - x_1}(x - x_2).$$

Example. Find the equation of the straight line (see Fig. 2.8) passing through $(2, -1)$ and $(-1, 0)$.

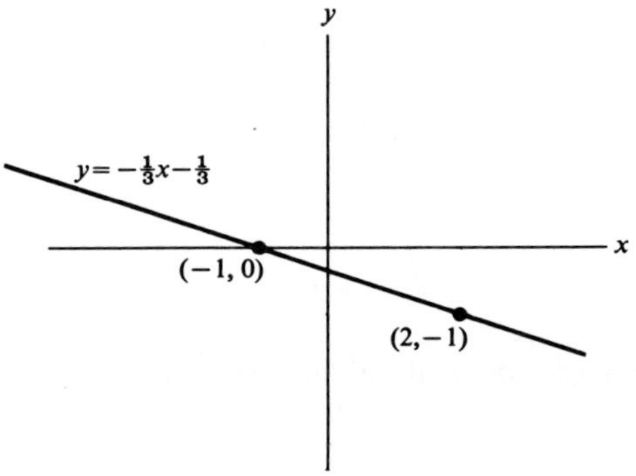

Fig. 2.8

Solution. The slope is given by

$$\frac{-1 - 0}{2 - (-1)} = \frac{-1}{3}.$$

Therefore the equation is

$$y + 0 = -\tfrac{1}{3}(x + 1)$$

or simply

$$y = -\tfrac{1}{3}x - \tfrac{1}{3}.$$

We remark, without proof, that parallel lines have the same slope and perpendicular lines have slopes which are the negative reciprocals of each other.

Examples

1. Find the equation of the line passing through the origin parallel to $3x - 2y = 3$.

Solution. First we find that the slope of the given line in Fig. 2.9 is $\frac{3}{2}$ by writing its equation as

$$y = \tfrac{3}{2}x - \tfrac{3}{2}.$$

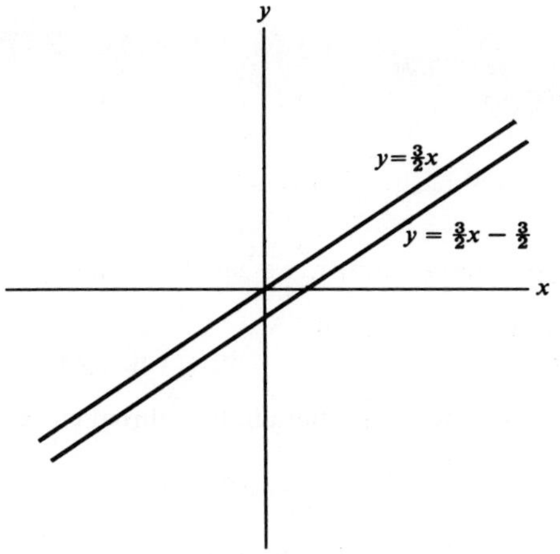

Fig. 2.9

Then the equation of the desired line is

$$y = \tfrac{3}{2}x.$$

since its slope is $\frac{3}{2}$ and its y-intercept 0.

2. Find the equation of the line through $(-1, 1)$ perpendicular to $2y + 4x = 6$.

Solution. The slope-intercept form of the given line is

$$y = -2x + 3,$$

so that its slope is -2. The slope of the line in Fig. 2.10 we want is the negative reciprocal of -2, namely $\frac{1}{2}$. Therefore, using the point-slope form, the equation we want is

$$y - 1 = \tfrac{1}{2}(x + 1)$$

or more simply

$$y = \tfrac{1}{2}x + \tfrac{3}{2}.$$

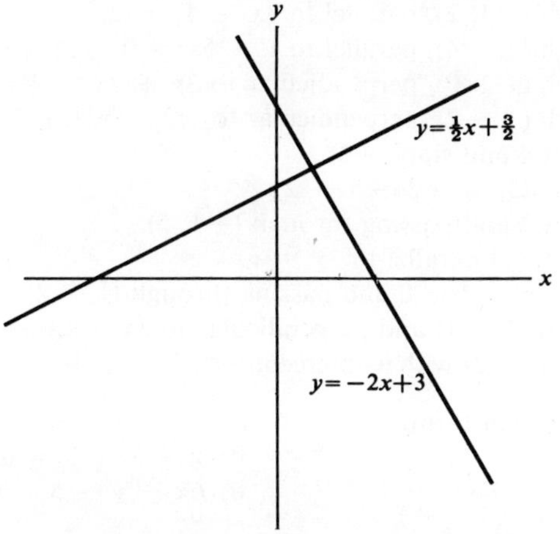

Fig. 2.10

PROBLEMS

1. Find the solution set of:
   a) $3x - 1 = 0.$ b) $4x + 2 = 0.$ c) $x/2 = 1.$
   d) $5x = -2.$ e) $\frac{3x}{2} + \frac{1}{2} = 0.$

2. If we add 4 to three times a number, we get 13. What is the number?

3. If $2y - 3x = 2$, what is the value of
   a) y when x is 2? b) x when y is -3?

4. The formula for conversion of Centigrade to Fahrenheit temperature is

$$F = \tfrac{9}{5}C + 32.$$

   If the temperature is 18°C, what is the temperature in degrees Fahrenheit? If it is 200°F, find the temperature in degrees Centigrade.

5. Graph the solution set of
   a) $3x - 2y = 0.$ b) $y + 4x = 2.$
   c) $3y - 2x + 1 = 0.$ d) $y/2 - 2x/3 + 4 = 0.$
   e) $2 - 2x = 5.$

6. Find the slope and y-intercept of the straight lines whose equations are given in Problem 5.

7. Find the equation of the straight line:
   a) with slope $\frac{1}{2}$, y-intercept 0. b) with slope -1, y-intercept 2.
   c) with slope -2, y-intercept -1. d) with slope $\frac{2}{3}$, y-intercept -2.
   e) with slope 2, passing through (3, 4).
   f) with slope $-1/2$, passing through $(-1, 2)$.
   g) with slope -3, passing through (0, 4).
   h) passing through (2, 1) and $(-3, 4)$.
   i) passing through (0, 1) and $(-2, 0)$.
   j) passing through $(-1, 2)$, parallel to $3x - 4y = 2$.
   k) passing through $(2, -4)$, parallel to $y - 5x = 0$.
   l) passing through $(-1, 0)$, perpendicular to $3y + 2x - 4 = 0$.
   m) passing through $(3, -2)$, perpendicular to $y/2 - x/3 + 1 = 0$.
   n) with y-intercept 4 and slope -1.
   o) passing through (3, 4) and $(-1, -2)$.
   p) with y-intercept 2 and passing through $(-1, 5)$.
   q) with y-intercept -3 parallel to $3x + y = 1$.
   r) parallel to $y - x + 2 = 0$ and passing through $(1, -2)$.
   s) passing through $(2, -1)$ and perpendicular to $3x + 2y - 1 = 0$.
   t) passing through (1, 2) with x-intercept -5.

8. Write in slope-intercept form:
   a) $3x + 2y = 1.$ b) $y - x + 3 = 0.$
   c) $2x = 5y + 1.$ d) $6x - y - 5 = 0.$

2.2 COMPLEX NUMBERS

The nice feature of linear equations with real coefficients is that they always have a real number as a solution. However, quadratic equations, that is, equations involving x^2, are not so pleasant. For example, there is no real number which satisfies the equation

$$x^2 + 1 = 0,$$

since the square of every real number is nonnegative. Hence we construct a system called the *complex numbers.*

We introduce a new entity, which we denote by i, whose square is -1:

$$i^2 = -1 \qquad (-i)^2 = -1$$

$$i = \sqrt{-1} \qquad -i = -\sqrt{-1}.$$

Using this symbol, we construct a new system

$$C = \{a + bi \mid a, b \quad \text{real numbers}\},$$

with

$$a + bi = c + di \quad \text{if and only if} \quad a = c \quad \text{and} \quad b = d.$$

We define the operation of addition by

$$(a + bi) + (c + di) = (a + c) + (b + d)i,$$

and multiplication by

$$(a + bi)(c + di) = (ac - bd) + (ad + bc)i.$$

Under these operations C satisfies the field axioms listed in Chapter 1, with the additive identity

$$0 + 0i,$$

$$(0 + 0i) + (a + bi) = a + bi,$$

and multiplicative identity

$$1 + 0i,$$

$$(1 + 0i)(a + bi) = a + bi.$$

The additive inverse of $a + bi$ is $-a + (-b)i$:

$$(a + bi) + (-a + (-b)i) = 0 + 0i.$$

The multiplicative inverse is found by observing that if

$$(a + bi)(x + yi) = 1 + 0i \quad (a \text{ and } b \text{ not both } 0),$$

then

$$ax - by = 1,$$

and

$$bx + ay = 0,$$

so that

$$x = \frac{a}{a^2 + b^2}, \; y = \frac{-b}{a^2 + b^2}.$$

Therefore

$$\frac{a}{a^2 + b^2} + \frac{-b}{a^2 + b^2}\, i$$

is the multiplicative inverse of $a + bi$ (not both a and b zero).

If a and b are both zero, $a + bi$ has no multiplicative inverse. We can consider the real numbers as a subset of the complex numbers by identifying the real number a with $a + 0i$.

By adding a new symbol i we have produced a solution to the equation

$$x^2 + 1 = 0.$$

In fact we have two: $\pm i$. The question arises as to whether this one new symbol will suffice or whether in order to solve, say

$$x^2 + 4 = 0,$$

we need add $j = \sqrt{-4}$, etc. It turns out that i is sufficient. For example, the solution to $x^2 + 4 = 0$ is given by $x = \pm 2i$.

Although there is no elementary proof of the fact, all *polynomial equations*, i.e., equations of the form

$$a_n x^n + a_{n-1} x^{n-1} + \cdots + a_1 x + a_0 = 0 \qquad (a_i \text{ real numbers}), \tag{2.4}$$

have solutions of the form $a + bi$, a, b real numbers, $i = \sqrt{-1}$. (This is known as the *fundamental theorem of algebra*.)

The *degree* of a term of a polynomial is its exponent (a_i is the coefficient of $a_i x^i$ in (2.4) and i is the exponent). An equation of the form (2.4) is of *degree n* if $a_n \neq 0$; that is, the degree of a polynomial equation is the degree of the term (with nonzero coefficient) which is of highest degree. If we count repeated roots* according to multiplicity, a polynomial of degree n has exactly n roots.

If we complicate things by considering polynomial equations with complex numbers as coefficients, for example,

$$(3 + 2i)x^2 + ix - 3 + 5i = 0,$$

it is still the case that there are precisely n roots, all of them complex numbers, where n is the degree of the equation.

---

* For example, $x^2 + 2x + 1$ has -1 as a repeated root of multiplicity 2.

PROBLEMS

1. Find the sum of:
   a) $3 + 5i$ and $-2 + 3i$.
   b) $-i$ and $2 + 4i$.
   c) 3 and $1 + i$.
   d) $6x + 5iy$ and $4x - 3iy$.
2. Find the product of:
   a) -2 and $-2 + 3i$.
   b) $2 + 3i$ and $4 - 2i$.
   c) $3i$ and $-1 - i$.
   d) $3x - 2iy$ and $-2x + 2iy$.
3. Find the sum and product of:
   a) $3 - 5i$ and $4 + 2i$.
   b) $7i$ and 6.
   c) $\sqrt{2}$ and $-3i$.
   d) $4 - 3i$ and $6i$.
   e) $3 + 2i$ and -5.
   f) $-2 + i$ and $3 - 2i$.
4. Find the multiplicative inverse of:
   a) $3 - 2i$.
   b) $3i$.
   c) $2 + 2i$.
   d) $2x - 6iy$.
   e) $-2 + 3i$.
   f) $-5i$.
   g) $2 - 4i$.
5. Find the solution set of:
   a) $4x^2 - 1 = 0$.
   b) $x^2 + 9 = 0$.
6. What is the degree of each of the following equations?
   a) $3x^3 - 2x^2 + x - 4 = 0$.
   b) $-3x^2 + 4x^4 - x = -2$.
   c) $3x^3 - 0x^5 + x + 4 = 0$.

2.3 SOLUTION OF QUADRATIC AND HIGHER DEGREE EQUATIONS

A *quadratic equation* in one variable is an equation of the form

$$ax^2 + bx + c = 0, \qquad a, b, c \text{ real numbers}, \qquad a \neq 0. \tag{2.5}$$

The general solution to the quadratic equation (2.5) is given by

$$x = \frac{-b \pm \sqrt{b^2 - 4ac}}{2a},$$

which is called the *quadratic formula*.

If it happens that $b^2 - 4ac > 0$, then the solutions are real and unequal. If $b^2 - 4ac = 0$, there is a real double solution; if $b^2 - 4ac < 0$, there are two complex solutions. The graph of $y = ax^2 + bx + c$ is a *parabola*.

Examples

1. Solve $2x^2 - 4x + 6 = 0$.

Solution

$$a = 2, \qquad b = -4, \qquad c = 6$$

so that

$$x = \frac{4 \pm \sqrt{16 - 48}}{4} = 1 \pm \sqrt{2}\, i.$$

2. Solve $3x^2 + 4x - 3 = 0$.

Solution

$$x = \frac{-4 \pm \sqrt{16 + 36}}{6} = \frac{-2 \pm \sqrt{13}}{3}$$

3. Solve $x^2 - 4x + 4 = 0$.

Solution

$$x = \frac{4 \pm \sqrt{16 - 16}}{2} = 2.$$

In a particular case it may be easier to solve a quadratic equation by *factoring*, that is, by writing the second-degree polynomial as the product of linear factors. However, this method is really a trial-and-error process and, therefore, if a solution is not immediately apparent, one is well advised to apply the quadratic formula, since it always works. The basis of a solution by factoring is that for polynomials with real or complex numbers as coefficients, the product of two polynomials is zero, if and only if at least one of them is zero; moreover, a is a solution to the polynomial equation

$$a_n x^n + a_{n-1} x^{n-1} + \cdots + a_1 x + a_0 = 0,$$

that is,

$$a_n a^n + a_{n-1} a^{n-1} + \cdots + a_1 a + a_0 = 0,$$

if and only if $(x - a)$ is a factor of the polynomial.

Examples

1. Solve $5x^2 - x - 6 = 0$.

Solution

$$5x^2 - x - 6 = (5x - 6)(x + 1) = 0,$$

$$5x - 6 = 0 \quad \text{or} \quad x + 1 = 0,$$

$$x = \tfrac{6}{5} \quad \text{or} \quad x = -1.$$

2. Solve $x^2 - 5x + 4 = 0$.

Solution

$$x^2 - 5x + 4 = (x - 4)(x - 1) = 0,$$

$$x - 4 = 0 \quad \text{or} \quad x - 1 = 0,$$

$$x = 4 \quad \text{or} \quad x = 1.$$

There are general formulas, analogous to the quadratic formula, for solutions to third- and fourth-degree polynomial equations in one variable (but none for fifth degree or higher). Because of their complexity they are not really useful. Thus, factoring and approximation are the methods commonly used for higher-degree polynomial equations. There are ways to improve on the hit-and-miss method of factoring. First of all we note that factoring is just the reverse of multiplying and we write the polynomial as a product. Hence a factor is a divisor of the polynomial. For example:

$$6x^3 + 7x^2 - x - 2 = (2x - 1)(x + 1)(3x + 2)$$

$$\begin{array}{r|l}
 & 3x^2 + 5x + 2 \qquad = (x + 1)(3x + 2) \\ \hline
2x - 1 & 6x^3 + 7x^2 - x - 2 \\
 & 6x^3 - 3x^2 \\ \cline{2-2}
 & \quad 10x^2 - x \\
 & \quad 10x^2 - 5x \\ \cline{2-2}
 & \qquad 4x - 2 \\
 & \qquad 4x - 2 \\ \cline{2-2}
\end{array}$$

and similarly for the other linear factors. If the original polynomial has integer coefficients, the goal of factoring is to find, if they exist, roots which are rational numbers.

If in the polynomial

$$a_nx^n + a_{n-1}x^{n-1} + \cdots + a_1x + a_0,$$

$a_n = 1$, the polynomial is said to be *monic*. If

$$x^n + a_{n-1}x^{n-1} + \cdots + a_1x + a_0 = (x - r_1)\cdots(x - r_n),$$

is a factorization of a monic polynomial with integer coefficients and r_i is an integer, $i = 1, \ldots, n$, then $r_1 \ldots r_n = a_0$, so that an integer root of the polynomial must be a divisor of a_0.

Example. In $x^3 - 4x + 4$, the only possible integer roots are the factors of 4, namely ± 1, ± 2, and ± 4.

If the polynomial we wish to factor is not monic, the problem is slightly more complicated, but the principle is the same. We state without proof:

Theorem 2. If a/b, a, b integers, $b \neq 0$, is a solution of

$$a_nx^n + a_{n-1}x^{n-1} + \cdots + a_1x + a_0 = 0, \qquad a_i \text{ an integer}, \quad i = 1, \ldots, n, \tag{2.6}$$

then a is a divisor of a_0 and b is a divisor of a_n.

Example. ± 1, $\pm \frac{1}{2}$, $\pm \frac{1}{3}$, $\pm \frac{1}{6}$, $\pm \frac{2}{3}$ are the only possible rational roots of $6x^3 + 2x^2 - x + 2$.

To verify whether $x + r$ is a factor of (2.6), we perform the division process

$$x + r \overline{\big)\, a_nx^n + \cdots + a_1x + a_0,}$$

and see whether there is a remainder; if not, $x + r$ is a factor of (2.6) and $-r$ is a root.

However, this division process can be very tedious, and so we use the method of *synthetic division.* In long division of polynomials, the powers of x play no significant role, except as place holders; moreover, the repetition of terms in the subtraction process is not necessary.

Example

$$
\begin{array}{r|rrrrrr}
 & 3x^4 & -\ 3x^3 & +\ 8x^2 & -\ 16x & +\ 30 & \\
\hline
x + 2 & 3x^5 & +\ 3x^4 & +\ 2x^3 & & -\ 2x & -\ 4 \\
 & 3x^5 & +\ 6x^4 & & & & \\
\cline{2-3}
 & & -\ 3x^4 & +\ 2x^3 & & & \\
 & & -\ 3x^4 & -\ 6x^3 & & & \\
\cline{3-4}
 & & & 8x^3 & +\ 0x^2 & & \\
 & & & 8x^3 & +\ 16x^2 & & \\
\cline{4-5}
 & & & & -\ 16x^2 & -\ 2x & \\
 & & & & -\ 16x^2 & -\ 32x & \\
\cline{5-6}
 & & & & & 30x & -\ 4 \\
 & & & & & 30x & +\ 60 \\
\cline{6-7}
 & & & & & & -\ 64 \text{ remainder}
\end{array}
$$

could be written

$$
\begin{array}{r|rrrrrr}
 & 3 & -3 & 8 & -16 & 30 & \\
\hline
2 & 3 & 3 & 2 & 0 & -2 & -4 \\
 & & 6 & & & & \\
\cline{3-3}
 & & -3 & & & & \\
 & & & -6 & & & \\
\cline{4-4}
 & & & 8 & & & \\
 & & & & 16 & & \\
\cline{5-5}
 & & & & -16 & & \\
 & & & & & -32 & \\
\cline{6-6}
 & & & & & 30 & \\
 & & & & & & 60 \\
\cline{7-7}
 & & & & & & -64
\end{array}.
$$

To save space we write this as

$$
\begin{array}{r|rrrrr|l}
 & 3 & -3 & 8 & -16 & 30 & -64 \text{ remainder} \\
\hline
2 & 3 & 3 & 2 & 0 & -2 & -4 \\
 & & 6 & -6 & 16 & -32 & 60
\end{array},
$$

where we have also eliminated the terms in the division process which appear in the quotient. Since we have really obtained the numbers in the quotient (after the first one) as a result of subtraction, it is more convenient to write:

$$\begin{array}{r|rrrrrr}
2 & 3 & 3 & 2 & 0 & -2 & -4 \\
 & & 6 & -6 & 16 & -32 & 60 \\
\hline
 & 3 & -3 & 8 & -16 & 30 & \boxed{-64}
\end{array}$$

Note that -64 is the value of $3x^5 + 3x^4 + 2x^3 - 2x - 4$ for x equals -2.

Let us summarize the method of synthetic division. To divide

$$a_nx^n + a_{n-1}x^{n-1} + \cdots + a_1x + a_0 \qquad \text{by} \qquad x + a,$$

we proceed as follows:

1. Write the coefficients, putting in a zero for missing terms.
2. Write a to the left in a box.
3. Bring down the a_n.

$$\begin{array}{r|lllll}
a & a_n & a_{n-1} & \cdots & a_1 & a_0 \\
 & & & & & \\
\hline
 & a_n & & & &
\end{array}$$

4. Multiply a by a_n and subtract from a_{n-1}.

$$\begin{array}{r|lllll}
a & a_n & a_{n-1} & \cdots & a_1 & a_0 \\
 & & aa_n & & & \\
\hline
 & a_n & a_{n-1} - aa_n & & &
\end{array}$$

5. Multiply $a_{n-1} - aa_n$ by a and subtract from a_{n-2}, continuing until a remainder is obtained. If the remainder is zero, $x + a$ is a factor; in any case the remainder is the value of the polynomial for $x = -a$.

$$\begin{array}{r|lllll}
a & a_n & a_{n-1} & \cdots & a_1 & a_0 \\
 & & aa_n & & & \\
\hline
 & a_n & a_{n-1} - aa_n & & & \boxed{\text{remainder}}
\end{array}$$

6. Write the quotient with a_n as the coefficient of x^{n-1}, $a_{n-1} - aa_n$ as the coefficient of x^{n-2}, etc.

Example

1. Divide $x^6 + 2x^4 + x^2 - 2$ by $x - 4$.

Solution

$$\begin{array}{r|rrrrrrr}
4 & 1 & 0 & 2 & 0 & 1 & 0 & -2 \\
 & & 4 & -16 & 72 & -288 & 1156 & -4624 \\
\hline
 & 1 & -4 & 18 & -72 & 289 & -1156 & 4622
\end{array}$$

The quotient is

$$x^5 - 4x^4 + 18x^3 - 72x^2 + 289x - 1156,$$

remainder 4622.

Warning. There is an alternative method of synthetic division (in fact, there are several) in which one divides by $-a$ and adds at each stage instead of subtracting. Do not interchange parts of each method.

2. Factor $6x^3 - 5x^2 + x + 12$.

Solution. Possible rational roots are ± 1, ± 2, ± 3, ± 4, ± 6, and ± 12, and ratios of these. Synthetic division by $x + 1$ yields:

$$\begin{array}{r|rrrr} 1 & 6 & -5 & 1 & 12 \\ & & 6 & -11 & 12 \\ \hline & 6 & -11 & 12 & \fbox{0} \end{array}$$

Hence $x + 1$ is a factor. We can continue by synthetic division, factor $6x^2 - 11x + 12$ by inspection, or use the quadratic formula.

3. Find the value of $6x^5 - 5x^3 + 2x^2 + 3$ if $x = 3$.

Solution

$$\begin{array}{r|rrrrrr} -3 & 6 & 0 & -5 & 2 & 0 & 3 \\ & & -18 & -54 & -147 & -447 & -1341 \\ \hline & 6 & 18 & 49 & 149 & 447 & \fbox{1344} \end{array}$$

The value is 1344.

The sad fact is that most polynomials with integer coefficients do not have rational roots. We have various methods for approximating irrational roots. The basic observation is that if one graphs a polynomial (Fig. 2.11) by substituting values of x and letting y be the corresponding values of the polynomial, and notes that for some value, say x_1, the value of the polynomial is positive and for another value, say x_2, it is negative, then between x_1 and x_2 there is a value of x which is a root.

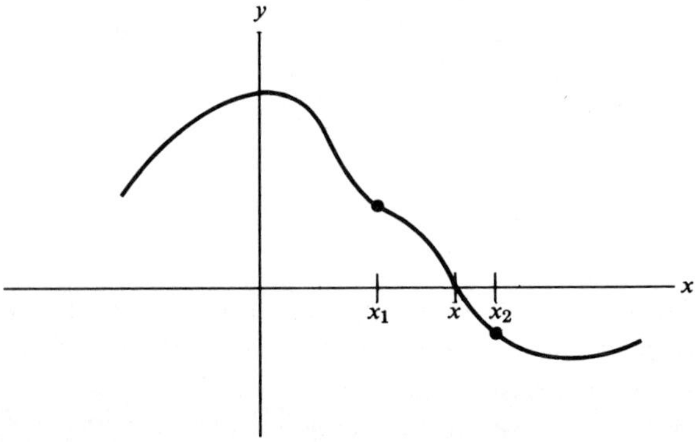

Fig. 2.11

Example. Approximate the roots of $x^3 + 2x^2 - x - 1$.

Solution. The only possible rational roots are ± 1, but a quick check shows that neither is a root. However, we have

x	y
-3	-7
-2	1
-1	1
0	-1
1	1

Hence the polynomial has roots between 0 and 1, between -1 and 0, and between -3 and -2 (Fig. 2.12). To get a better approximation we could substitute intermediate values. In the case of the root between 0 and 1 we have:

x	y
$\frac{1}{2}$	$-\frac{7}{8}$
$\frac{3}{4}$	$-\frac{13}{64}$
$\frac{9}{10}$	$\frac{449}{1000}$

Therefore, the root is between $\frac{3}{4}$ and $\frac{9}{10}$.

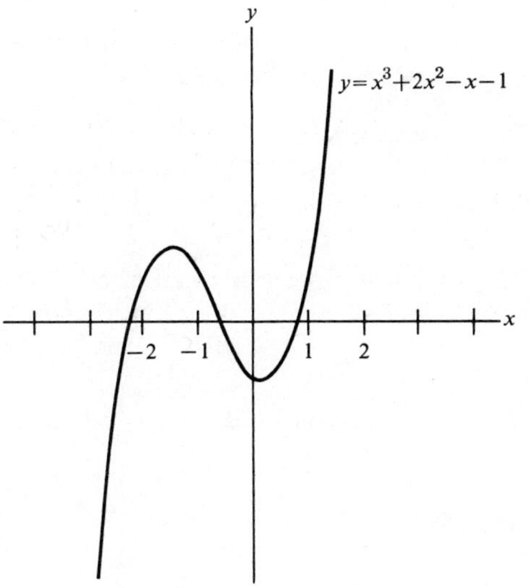

Fig. 2.12

Various methods for facilitating approximation of roots can be adapted for computer use. A principle to keep in mind when seeking roots is that complex roots of polynomials with real coefficients occur in pairs; if $a + bi$ is a root, then $a - bi$ is a root also. Therefore a polynomial of degree 5 has 1, 3, or 5 real roots.

PROBLEMS

1. Solve for x:
   a) $3x^2 - 4x + 2 = 0.$
   b) $x^2/2 - 5x - 4 = 0.$
   c) $2x^2 + 2x = 3.$
   d) $cx^2 + dx - 5e = 0.$
   e) $x^2 - 5x - 14 = 0.$
   f) $2x^2 - 5x + 1 = 0.$
   g) $2x^2 + 5x + 3 = 0.$
   h) $x^2 - 14x + 1 = 0.$
   i) $3x^2 - 5x = 0.$
   j) $x^2 + 2x + 5 = 0.$
2. Solve by factoring:
   a) $x^2 - x - 12 = 0.$
   b) $3x^2 + x - 2 = 0.$
   c) $12x^2 - 7x + 1 = 0.$
3. If
   $$y = 3x^2 + 4x + 1,$$
   find
   a) y when $x = 0.$
   b) y when $x = 1.$
   c) y when $x = -1.$
   d) x when $y = 0.$
   e) x when $y = -1.$
4. A plane takes one hour to fly 500 mi with the wind and $1\frac{1}{2}$ hrs to fly back. What is the air speed of the plane?
5. The outside dimensions of a picture frame are 4×6 ft. If the area of the frame is equal to the area of the picture it encloses, find the width of the frame.
6. Use synthetic division to find the roots of:
   a) $2x^3 - x^2 - 1.$
   b) $12x^4 + 16x^3 + x^2 - 4x - 1.$
   c) $2x^4 + 3x^3 + 2x^2 - 1.$
   d) $x^3 + 4x^2 + x - 6.$
   e) $x^4 + x^3 - 3x^2 - 5x - 2.$
   f) $2x^3 + 11x^2 + 2x - 15.$
7. Use synthetic division to find the value of the given polynomials:
   a) $3x^4 + 22x^3 - 19x^2 - 22x + 16, x = -8.$
   b) $x^5 - 4x^3 + 5x^2 - 5, x = 1.$
   c) $x^3 - x^2 + x - 1, x = -10.$
   d) $3x^4 - 2x^3 + 5x^2 - 4, x = -6.$
   e) $x^5 - 3x^4 + 2x^3 - 5x + 1, x = 3.$
   f) $x^4 + 5x^3 - 4x^2 + 5x + 17, x = 5.$
8. Approximate a real root (correct to one decimal place) of:
   a) $x^3 + x - 1.$
   b) $x^3 - 13x^2 + 7x - 1.$
   c) $x^5 + x^4 + x^3 - 2x^2 - 2x - 2.$
   d) $x^3 - 5x^2 + 2x - 1.$
   e) $x^4 - 2x^3 - x + 1.$
9. A ball is thrown upward. Its distance s above ground at time t is given by
   $$s = -16t^2 + 40t + 6.$$
   Find when it hits the ground.

2.4 ALGEBRAIC EQUATIONS

The quotient of polynomials

$$\frac{a_nx^n + \cdots + a_1x + a_0}{b_mx^m + \cdots + b_1x + b_0}$$

is a *rational expression.* An equation involving such expressions is called a *rational equation.* An equation involving square roots, cube roots or higher-order roots is called a *radical equation.* Any equation involving polynomials, rational expressions and radicals only is an *algebraic equation.*

Examples. The following are algebraic equations:

1. $$\frac{x-2}{x^2-1} = 2.$$

2. $$\sqrt{x^2+2x} - 5 = 0.$$

3. $$2x^3 + 3x - \frac{2}{x-1} = 0.$$

The procedure for solving these equations is quite simple—convert them to polynomial equations.

1. Solve

$$\frac{1}{x} + \frac{x+1}{x-2} = 3.$$

Solution. Multiplying both sides by $x(x-2)$ yields

$$x - 2 + x^2 + x = 3x^2 - 6x,$$

$$2x^2 - 8x + 2 = 0,$$

$$x = \frac{8 \pm \sqrt{64-16}}{4} = 2 \pm \sqrt{3}.$$

Warning. Squaring both sides of an equation may introduce an extraneous solution. Thus it is necessary to check the solutions of the polynomial equation to see whether they are in fact solutions of the original equation.

2. Solve

$$\sqrt{x+2} + 3x = 8.$$

Solution. First we put all terms not involving the radical on one side of the equation:

$$\sqrt{x+2} = -3x + 8.$$

Squaring yields

$$x + 2 = 9x^2 - 48x + 64,$$

$$9x^2 - 49x + 62 = 0,$$

$$(x-2)(9x-31) = 0,$$

$$x = 2, \qquad x = \tfrac{31}{9}.$$

Substituting these values in turn in the original equation shows that 2 is a solution, but $\frac{31}{9}$ is not.

PROBLEMS

1. Simplify

   a) $\dfrac{x^2 - y^2}{x + y}$. b) $\dfrac{x + (1/y)}{x - (1/y)}$. c) $\dfrac{3x - 9y}{x^2 - 4xy + 3y^2}$.

2. Perform the indicated operations and simplify:

   a) $\dfrac{x^2 - y^2}{x^3 - y^3}\Big/\dfrac{x + y}{x}$. b) $\dfrac{z}{x + y} - \left(\dfrac{z}{x} + \dfrac{z}{y}\right)$.

   c) $x + y + \dfrac{3x^2}{x - y}$.

3. Solve:

   a) $\sqrt{x + 7} = \sqrt{3x + 1}$.

   b) $\sqrt{4x^2 + 6x + 6} - 2\sqrt{x^2 + x - 1} = 2$.

   c) $\sqrt{x + 1} - \sqrt{x} = 2$.

4. Find two real numbers whose difference is one and which are such that the sum of their reciprocals is also one.

5. Solve for x:

   a) $\dfrac{3x^2 - 6x + 1}{x^2 - 2} = 4$. b) $\dfrac{x + 1}{x - 2} - \dfrac{x - 2}{x + 5} = 1$.

   c) $\dfrac{5}{x - 2} + \dfrac{3}{x - 1} = -5$. d) $\dfrac{ax + bx}{6} - \dfrac{5x}{cx} = 0$.

   e) $\sqrt{x - 1} = \sqrt{3x + 2}$. f) $\sqrt{2x + 1} + \sqrt{x} = 1$.

   g) $\sqrt{3x^2 - x} = \sqrt{5x + 1}$. h) $\dfrac{3x^2 + 1}{x - 5} = 1$.

   i) $\dfrac{x + 1}{7x - 5} = x$. j) $x - 5 = 1/x$.

   k) $\dfrac{1}{x^2 - 1} - \dfrac{2x}{x + 1} = 5$.

6. Given that two numbers differ by 11 and their positive square roots differ by 1, find the numbers.

2.5 SYSTEMS OF EQUATIONS

In Section 1 we saw that a linear equation in a single variable has a unique solution, but the solution set of a linear equation in two variables consists of all the points on a straight line. However, a system of two linear equations in two variables

$$\begin{aligned} ax + by &= e \\ cx + dy &= f \end{aligned} \tag{2.7}$$

may have a unique solution, that is, a pair of values for x and y which satisfy both equations. If such a solution exists, its geometric representation is as the intersection of the straight lines, shown in Fig. 2.13, represented by the two equations.

Example. Solve

$$x + y = 1,$$
$$2x - y = 2.$$

Solution. Graphing the two lines, we find that their intersection is at (1, 0). Therefore, the values $x = 1$ and $y = 0$ satisfy both of the equations.

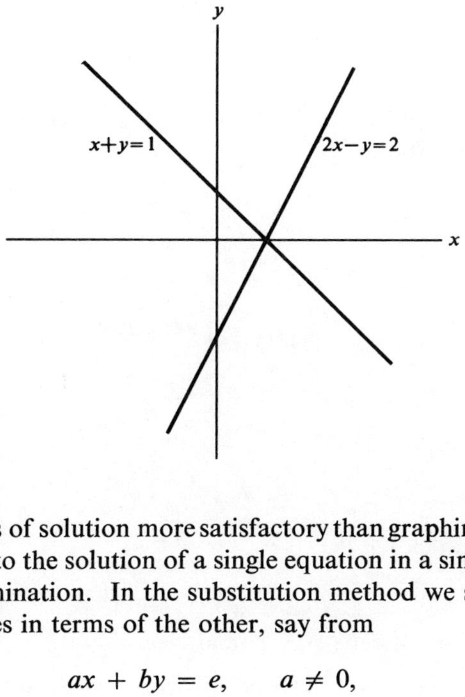

Fig. 2.13

There are methods of solution more satisfactory than graphing. Several involve reducing the problem to the solution of a single equation in a single variable, either by substitution or elimination. In the substitution method we solve one equation for one of the variables in terms of the other, say from

$$ax + by = e, \qquad a \neq 0,$$

we obtain

$$x = \frac{e - by}{a}.$$

(If $a = 0$, we simply substitute $y = e/b$ in the second equation of (2.7)). Then we substitute in the other equation.

$$c\left(\frac{e - by}{a}\right) + dy = f.$$

This yields

$$\frac{ad - bc}{a} y = f - \frac{ce}{a}$$

$$y = \left(\frac{af - ce}{a}\right)\left(\frac{a}{ad - bc}\right).$$

This method shows us the condition which is necessary and sufficient for the system (2.7) to have a unique solution, namely

$$ad - bc \neq 0$$

Example. Solve the system of linear equations

1. $$3x - 2y = 1,$$

2. $$-4x + y = 0.$$

We check that $ad - bc = 3 - 8 \neq 0$.

From Example 1,

$$x = \frac{1 + 2y}{3}.$$

Substituting yields

$$-4\left(\frac{1 + 2y}{3}\right) + y = 0$$

$$\frac{-5}{3} y = \frac{4}{3}$$

$$y = \frac{-4}{5}.$$

From this we obtain

$$x = \frac{1 - 8/5}{3} = \frac{-1}{5}.$$

The elimination of one variable involves multiplying one or both of the equations of the system by constants in order to have the same coefficient, up to sign, for one of the variables. Then the equations are added (or subtracted) to eliminate that variable.

Examples

1. Solve

$$-x + 2y = -4,$$

$$2x + 3y = 1.$$

Solution. Multiplying the first equation by 2 and adding it to the second gives

$$\begin{aligned} -2x + 4y &= -8 \\ 2x + 3y &= 1 \\ \hline 7y &= -7 \\ y &= -1 \end{aligned}$$

Substituting this value for y in either of the original equations yields $x = 2$.

2. Solve

$$3x - 2y = 1,$$

$$-6x + 4y = 3.$$

Solution. Any attempt to eliminate one variable by substitution or elimination eliminates them both, and results in a contradiction of the form $0 = c$, where c is a nonzero constant. This is due to the fact that the equations represent parallel lines and consequently there is no point of intersection (Fig. 2.14). That there is no unique solution could have been determined at the beginning by checking that $ad - bc = 12 - 12 = 0$.

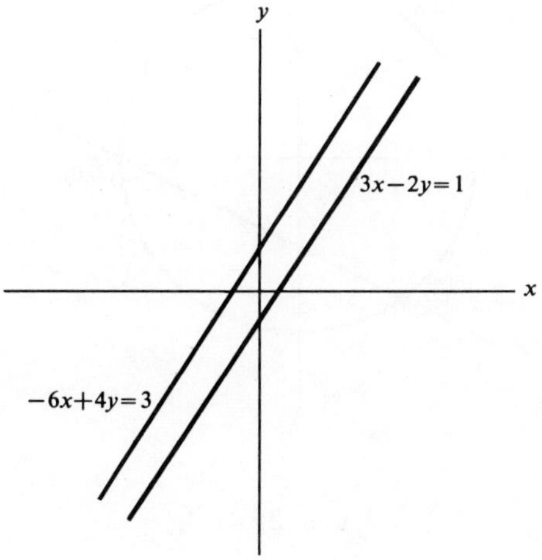

Fig. 2.14

3. Solve

$$2x - y = 1,$$

$$-2x + y = -1.$$

Solution. Again $ad - bc = 0$. However these equations represent the same line so the solution set consists of every point on that line (Fig. 2.15).

4. The method of substitution can also be used if one equation is linear and the other quadratic (Fig. 2.16).

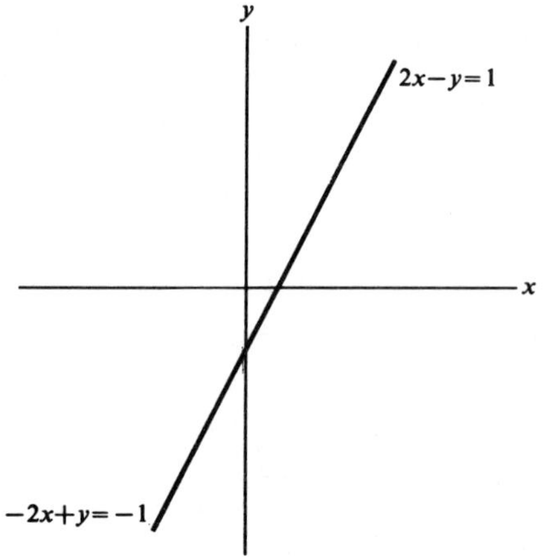

Fig. 2.15

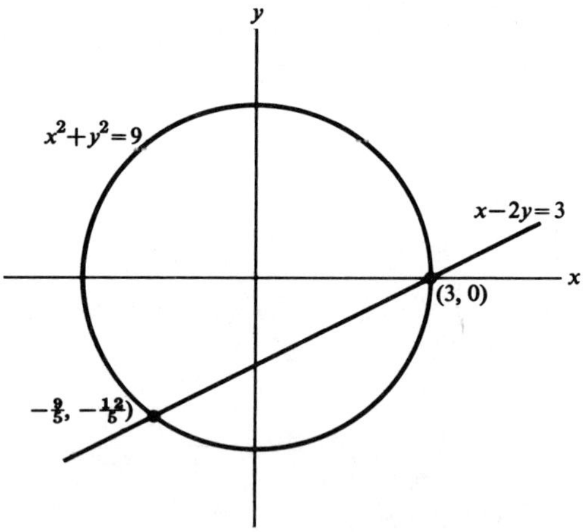

Fig. 2.16

Solve

$$x - 2y = 3,$$
$$x^2 + y^2 = 9.$$

Solution. From the first equation, $x = 3 + 2y$. Then

$$(3 + 2y)^2 + y^2 = 9,$$
$$5y^2 + 12y = 0,$$
$$y = 0, \qquad y = -\tfrac{12}{5}.$$

The solutions are $x = 3$, $y = 0$ and $x = -\frac{9}{5}$, $y = -\frac{12}{5}$, the points of intersection of the circle and the straight line.

The thorough consideration of systems of equations will be postponed until Chapter 10.

PROBLEMS

1. Solve the following systems of equations:

a) $x - 2y = 3,$
$2x + 3y = 5.$

b) $7x + 3y = 4,$
$x + y = 0.$

c) $x^2 + y^2 = 25,$
$x - 3y = -5.$

d) $2x - 4y = 5,$
$8y - 4x = -10.$

e) $2z - 4y = 4,$
$2x - z = 2,$
$3x - 2y - z = 5.$

f) $2x - 2y = 5,$
$x + 3y = 1.$

g) $5x + 3y = 1,$
$x - 2y = -1.$

h) $4x^2 + y^2 = 16,$
$x + 2y = 4.$

i) $2x + y = 1,$
$-2x - y = 0.$

2. The sum of the digits of a two-digit number is 14 and the product of the digits is 48. Find the number.
3. Find the intersection of the straight lines
$$3x - 4y = 2,$$
and
$$x + 2y = 0.$$
4. Find the intersection of $2x^2 + y^2 = 16$ and $y - 3x = 3$.
5. Suppose we want 100 ml of a solution which is to be 40 percent alcohol. We have available a 90 percent solution. How much water and how much 90 percent solution should be used to obtain the desired solution?
6. The area of a rectangular plot of land is 15,000 ft^2 and its perimeter is 500 ft. What are its dimensions?
7. The sum of the digits of a 2-digit number is 11 and the product of the digits is 30. Find the number.

2.6 APPLICATIONS

The major difficulty encountered in applying the mathematical principles and techniques one has learned is that of translating English into mathematics. On its most basic level it is what might be termed the "word problem dilemma." There is, unfortunately, no easy set of rules to lay down which, if followed, would automatically resolve the difficulty. There are a few guidelines which may help.

1) Read the problem thoroughly.
2) Write down what information is given.
3) Identify what it is you are to find.

4) Try to combine the information into an equation(s) whose solution will yield that which you have identified as being desired. Of course, there may be more than one way to do this, or you may not have sufficient information to do it at all.

Interest

1. An auto insurance policy can be paid in one annual payment of \$195 or in semiannual payments of \$103 each. What is the annual rate of interest paid by those electing the second payment plan?

Solution. First we must determine on what amount interest is to be paid. Since the first payment is made at once, the customer is in effect "borrowing" for a period of six months the difference between this payment and the annual premium. The "interest" is the difference between the annual premium and the sum of the two payments.

$$\begin{aligned} \text{Principal} &= P = 92 = \text{amount borrowed} \\ \text{Interest} &= I = 11 \\ \text{Time in years} &= t = \tfrac{1}{2}. \end{aligned}$$

Simple interest is computed according to the formula

$$I = Prt,$$

where r is the rate per year—the quantity we are trying to find. We have

$$\begin{aligned} 11 &= 92(\tfrac{1}{2})r, \\ r &= 23.9 \text{ percent.} \end{aligned}$$

2. A savings bond bought for \$17.50 is worth \$25 at the end of nine years. How does this amount compare with what \$17.50 invested at $5\frac{1}{2}$ percent annual interest compounded quarterly would be worth at the end of nine years?

Solution. We want to find the principal at the end of nine years for a \$17.50 investment at the rate described. We have 36 compounding periods. The formula for principal invested at compound interest is given by

$$P = P_0(1 + tr)^n,$$

where P_0 is the initial principal, r is the rate per year, t the number of years per compounding period, and n the number of compounding periods.

We have

$$\begin{aligned} P &= 17.50(1 + \tfrac{1}{4}(0.055))^{36} \\ &= 17.50(1.01375)^{36} \\ &= 17.50(1.6305) \\ &= 28.53. \end{aligned}$$

(see appendix B for computation by use of logarithms)

Thus the savings bond is worth less.

3. A loan of \$2000 for a new car is issued at 5 percent "discount" interest. This means that 5 percent is added to the principal and this amount is paid back in a certain number of monthly installments, in this case 12. Compute the equivalent rate of simple interest.

Solution. The total amount to be paid back is \$2100, in 12 installments of \$175 each.

\$2000 is borrowed for the first month.
2000 − 175 is borrowed for the second month.
⋮
2000 − 11(175) is borrowed for the twelfth month.

As before

$$I = Prt$$

$$100 = \sum_{n=0}^{11} (2000 - 175n)r(\tfrac{1}{12})$$

$$= \frac{r}{12}(2000 + 1825 + 1650 + 1475 + 1300 + 1125$$

$$+ 950 + 775 + 600 + 425 + 250 + 75)$$

$$= 1037.5r$$

$$r = 9.65 \text{ percent.}$$

Mixtures

1. A car's radiator contains eight gallons of a solution containing 15 percent antifreeze. What percentage of antifreeze must the two gallons of liquid which are added contain in order to bring the percentage of antifreeze in the radiator solution to 20 percent?

Solution. The radiator currently contains

$$8 \times 0.15 = 1.2 \text{ gallons of antifreeze.}$$

We need $10 \times 0.20 = 2$ gallons. Thus we must add 0.8 gallons of antifreeze.
Let x = percent antifreeze in the solution to be added. Then

$$2x = 0.8,$$

$$x = 0.4,$$

or 40 percent.

2. Two existing solutions are 80 percent and 50 percent alcohol, respectively. How much of each solution is required to produce 1000 ml of a 60 percent solution?

Solution. Let

$$x = \text{amount of 80 percent solution,}$$
$$y = \text{amount of 50 percent solution.}$$

Then

$$x + y = 1000 \text{ ml (solution equation),}$$
$$0.8x + 0.5y = 0.6(1000) \text{ (alcohol equation).}$$

Solving the first equation for x, we have

$$x = 1000 - y.$$

We substitute this in the other equation:

$$\begin{aligned} 0.8(1000 - y) + 0.5y &= 600 \\ -0.3y &= -200 \\ y &= 667 \text{ ml} \\ x &= 333 \text{ ml.} \end{aligned}$$

3. Two blends of coffee are worth \$1.20 and \$0.95 per pound. How much of the less expensive mix should be added to 2 lb of the \$1.20 coffee to obtain a mix worth \$1.05 per pound?

Solution. Let x = amount to be added. Then $2 + x$ = amount of new mixture. Thus

$$\begin{aligned} 0.95x + 2(1.20) &= (2 + x)1.05 \\ 0.1x &= 0.30 \\ x &= 3 \text{ lb.} \end{aligned}$$

Quadratics

1. The productivity P of workers in a factory can be expressed as a quadratic polynomial in which the variable is the temperature T in degrees Fahrenheit. The following information is available:

T	P
62	1000
70	1200
74	1100.

Find the productivity at 68°. At what temperature is productivity the highest?

Solution. The values given for T and P must satisfy the polynomial equation

$$P = aT^2 + bT + c,$$

where a, b, c are real numbers.

This leads to three equations:

1) $$1000 = 3844a + 62b + c,$$

2) $$1200 = 4900a + 70b + c,$$

3) $$1100 = 5476a + 74b + c.$$

By extension of the method for two linear equations in two variables, we use elimination to solve this system. Subtracting in turn (2) and (3) from (1):

4) $$-200 = -1056a - 8b,$$

5) $$-100 = -1632a - 12b.$$

Continuing, we multiply (4) by 3, and (5) by 2, and subtract:

$$\begin{array}{rcl} -600 &=& -3168a - 24b \\ -200 &=& -3264a - 24b \\ \hline -400 &=& 96a \end{array}$$

$$a = \frac{-25}{6}$$

$$b = 575$$

$$c = \frac{-111800}{6}.$$

The desired polynomial equation is

$$P = \frac{-25}{6}T^2 + 575T - \frac{111800}{6}. \tag{2.8}$$

To find the productivity at 68° we substitute $T = 68$ to get $P = 1200$.

To find the largest possible value for P we observe that, for large values of T, the first term in the equation is negative and the second is positive. Thus we want to maximize the effect of the second and minimize that of the first. For any quadratic polynomial

$$ax^2 + bx + c,$$

it turns out (for a proof, see Chapter 6) that this occurs when $x = -b/2a$. Thus we have the largest productivity when

$$T = \frac{-b}{2a} = 575(\tfrac{6}{50}) = 69°.$$

The productivity at this temperature is 1204.2.

2. It has been determined that the assets of a company are given by the following polynomial equation:

$$A = -3t^2 + 15t + 10{,}000,$$

where t is time in years. Will the company go bankrupt any time in the next five years?

Solution. We want to know whether there exists a real number $t < 5$ such that

$$-3t^2 + 15t + 10{,}000 = 0.$$

Using the quadratic formula, we find that the solutions to this equation are

$$t = \frac{-15 \pm \sqrt{225 + 120{,}000}}{6}.$$

We can discard the negative root, leaving us with

$$t = \frac{-15 + 346.7}{6} > 5.$$

So the answer is no.

3. We close this chapter with a classic type of problem. Find two positive integers such that their sum is 20 and the square of one divided by the other exceeds twice the divisor by 2.

Solution. Let x and y be the integers. Then

$$x + y = 20$$

$$\frac{x^2}{y} = 2y + 2.$$

Solving the first equation for x in terms of y yields

$$x = 20 - y.$$

Substituting this in the second, we have

$$\begin{aligned}(20 - y)^2 &= 2y^2 + 2y,\\ 0 &= y^2 + 42y - 400,\\ 0 &= (y + 50)(y - 8),\end{aligned}$$

$y = -50$ is not possible, so we have $y = 8$ and $x = 12$.

PROBLEMS

1. A demand law is given by:

$$p = 35 - \tfrac{1}{2}x,$$

where p is the price and x the quantity. Find p if $x = 13$. Find x if $p = 25$.

2. A life insurance policy has an annual premium of \$500 or quarterly premiums of \$135. What is the annual rate of interest paid by those paying quarterly premiums?

3. In how many years must a hundred-dollar savings bond (bought for \$75) mature in order to yield an annual simple interest rate of $5\frac{1}{4}$ percent?
4. How much interest will \$1000 earn at the end of two years if it is invested at 4 percent interest compounded quarterly?
5. A person borrows \$1000 for two years at 6 percent per year discount interest. He repays the loan in 24 equal monthly installments. What is the rate of simple interest he is paying?
6. Two blends of flour sell for 3 and 5 cents per lb, respectively. How much of each should be mixed to produce 10 lb of a mixture which will sell for 3.8 cents per lb?
7. A \$100,000 estate is to be divided among the wife, son, and daughter of the deceased. Each is to receive \$10,000 initially, and the rest is to be divided so that the wife receives three times as much as either of the children. How much does each receive?
8. The cost to a dealer of a TV set was \$75. He marked it up to a selling price and later offered it for sale at 25 percent off the selling price. If he still made a 10 percent profit on his original cost, what was the selling price?
9. The first \$7000 of taxable income is taxed at a rate of 12 percent, the next \$2000 at 15 percent, and the next \$2000 at 18 percent. If a person makes \$14,000 a year and has deductions amounting to \$3700, what percentage of his income is paid as income tax?
10. How many gallons of milk containing 4 percent butterfat and of cream containing 20 percent butterfat should be combined to give 5 gallons of milk containing 6 percent butterfat?
11. Market equilibrium corresponds to the point at which the demand curve and the supply curve intersect. The corresponding price p per unit quantity and quantity x represent the equilibrium price and quantity. If a demand law is given by

$$4x + 9p = 48,$$

and a supply law by

$$p = x/9 + 2,$$

find the equilibrium price and quantity.

12. Revenue is given by

$$R = px,$$

where p is price and x is quantity. If a demand law is given by

$$p = 50 + x/2,$$

find the quantity if the revenue is 600.

13. A demand law is given by

$$p = 50 - \frac{x^2}{1000},$$

where p is the price and x the quantity. Find p if $x = 50$. Find x if $p = 25$.

14. How much interest is earned at the end of 3 years by \$1000 invested at 5 percent interest compounded semiannually?

15. If \$500 is borrowed for 6 months at 12 percent discount interest, what is the rate of simple interest?
16. Two varieties of tea sell for \$1.25 and \$1.60 per lb, respectively. How much of each should be used to make 10 lb of a blend to sell for \$1.50?
17. The dealer cost of a refrigerator is \$100. After marking it up and then offering it at one-third off the selling price, he makes a 15 percent profit on his original cost. What was the selling price?
18. If a demand law is given by $3x + 10p = 72$ and a supply law by $p = x/10 + 3$, find the equilibrium price and quantity.
19. How much of 40 percent and 90 percent alcohol solutions should be mixed to given 1000 ml of a 75 percent solution?

3 / FINITE PROBABILITY

In this chapter we are concerned with what is customarily termed "finite mathematics." In reality we shall be solving problems by counting—a primitive method. However, we use techniques of analysis which make our work somewhat easier. The important thing to remember, since there are many variations on the situations we present, is to attack a problem in a systematic manner.

3.1 PERMUTATIONS AND COMBINATIONS

Suppose that there are five students and five chairs. How many different ways can they be seated?

In the first chair we can seat any one of the five, leaving any one of the remaining four for the second seat. For each of the choices for the second seat there are three possibilities for the third. This gives us $5 \cdot 4 \cdot 3 = 60$ possible arrangements so far. Now in each of these arrangements either of the remaining two students can sit in the fourth seat, but the occupant of the fifth seat is determined by those who go before. Therefore, the total number of different arrangements is $5 \cdot 4 \cdot 3 \cdot 2 \cdot 1 = 120$.

We make use of factorial notation:

$$n! = n(n-1)(n-2)\cdots 2 \cdot 1.$$

Examples

$$5! = 5 \cdot 4 \cdot 3 \cdot 2 \cdot 1 = 120$$

$$\frac{6!}{3!} = \frac{6 \cdot 5 \cdot 4 \cdot 3 \cdot 2 \cdot 1}{3 \cdot 2 \cdot 1} = 6 \cdot 5 \cdot 4 = 120$$

$$\frac{2!}{5!} = \frac{2 \cdot 1}{5 \cdot 4 \cdot 3 \cdot 2 \cdot 1} = \frac{1}{5 \cdot 4 \cdot 3} = \frac{1}{60}$$

$$n(n-1)! = n(n-1)(n-2)\cdots 2 \cdot 1 = n!$$

$$\frac{n!}{(n-1)!} = \frac{n(n-1)(n-2)\cdots 2 \cdot 1}{(n-1)(n-2)\cdots 2 \cdot 1} = n$$

$$\frac{(k+3)!(k-1)!}{(k!)^2} = \frac{(k+3)(k+2)(k+1)k(k-1)\cdots 2\cdot 1(k-1)\cdots 2\cdot 1}{k(k-1)(k-2)\cdots 2\cdot 1\cdot k(k-1)(k-2)\cdots 2\cdot 1}$$
$$= \frac{(k+3)(k+2)(k+1)}{k}.$$

A diagram such as Fig. 3.1, which is used to analyze all the possibilities in a given situation, is called a *tree diagram.*

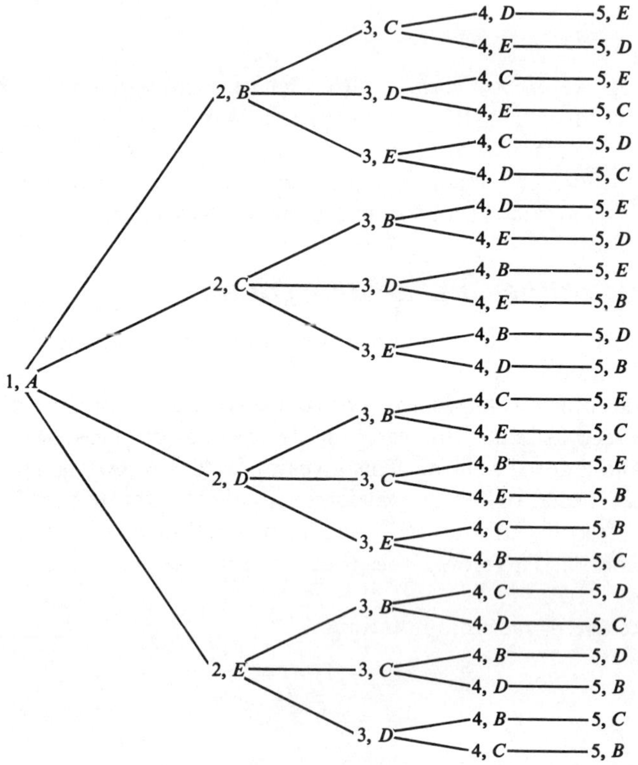

Fig. 3.1

Note that in the example of the chairs the *order* in the arrangements is important. If we do not care in what order they sit, all of the arrangements are the same. We call arrangements in which order is important *permutations.* We prove the following theorem.

Theorem 1. The number of permutations of n elements is $n!$.

Proof: By induction. For $n = 1$, the result is obvious. Suppose the theorem holds for n; that is, suppose that n elements can be arranged in $n!$ ways. Now suppose we have $n + 1$ elements to be arranged. After each one of them is selected for the first spot there are n spots and n elements remaining. By the induction hypothesis there are $n!$ ways to do this. But there are $n + 1$ ways to pick the first element so the total number of permutations is $(n + 1)n! = (n + 1)!$. This concludes the proof.

Now suppose there are five students, but only three seats. Then we have $5 \cdot 4 \cdot 3$ possible ways to seat them. In general there are

$$n \cdot (n - 1) \cdots (n - r + 1),$$

or $n!/(n - r)!$ ways to arrange n elements in r places. This is the number of permutations of n elements taken r at a time.

Examples

1. In a poll of the voters of his district a Congressman asks his constituents to select from six issues on a ballot the three which are most important to them and to rank them in order. In how many different ways may the ballot be marked?

Solution. A voter may list any one of six issues in the first place, any of the remaining five in second, and any one of four in the third position. Thus

$$6 \cdot 5 \cdot 4 = 120$$

different rankings are possible.

2. How many three letter prefixes for license plates are possible?

Solution. In this case we select three element subsets of a 26-element set in which order is important. However, while permutations allow each element of the set to appear only once, here we are permitted to use any of the 26 elements more than once, e.g., XXX is an allowable arrangement. It is clear that for the first place we have twenty-six choices; for each of these we have twenty-six choices for the second place, and for each of these $26 \cdot 26$ choices there are twenty-six possibilities for the third place. Thus the total is

$$26^3 = 17{,}576.$$

Theorem 2. The number of arrangements of n elements into r-element ordered subsets (with repetition allowed) is n^r.

Proof. For the first element of each subset, n choices are possible. For the second place there are also n choices, making a total of n^2 possibilities for the first two spots. Continuing this process, we see that there are n^r arrangements for the r places.

Example. How many different ways can a 5-question true-false test be answered?

Solution. Here $m = 2$ and $r = 5$; so from Theorem 2 we have $2^5 = 32$ as the answer (Fig. 3.2).

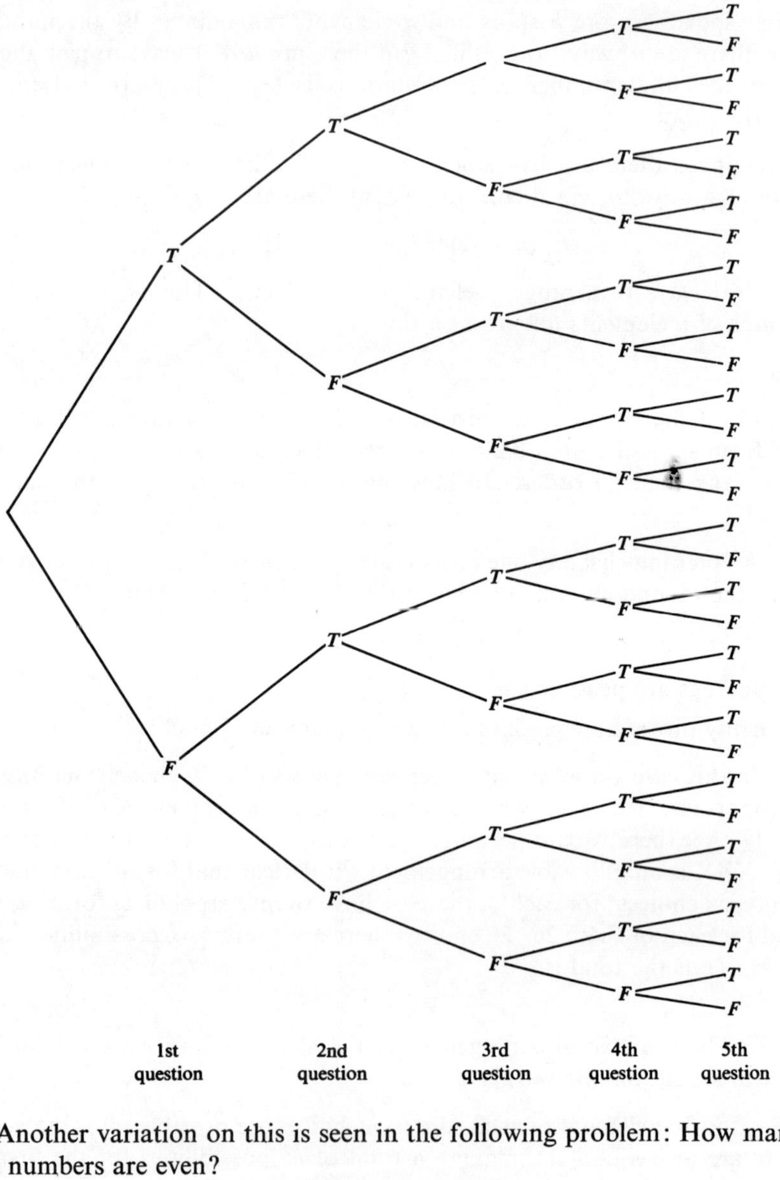

Fig. 3.2

Another variation on this is seen in the following problem: How many three-digit numbers are even?

Solution

Possible entries in the hundred's column 9
(e.g., 010 is a two-digit, not a three-digit number)
Possible entries in the ten's column 10
Possible entries in the one's column 5 (0, 2, 4, 6, 8)
Total $= 9 \cdot 10 \cdot 5 = 450$.

In fact, if there are r ways of doing a thing, each followed by s ways of doing something else, then there are rs ways of doing the two things. More generally, for any number n of operations which may be done in $r_1, \ldots, r_n$ ways respectively, the number of ways of performing the operations successively is the product $r_1 \ldots r_n$.

Examples

1. There are four serious contenders for the Republican nomination for president and two for the Democratic nomination. If the voters' preference is to be tested on each possible pair, on how many pairs must they be questioned?

Solution. Each Republican candidate can be paired with either of the Democrats. Hence there are

$$4 \cdot 2 = 8$$

possibilities.

2. There are four routes available from an airport to city hall, three routes from city hall to the auditorium, and five routes from the auditorium to the Grand Hotel. The chief of police, wanting to keep secret the route of a VIP from the airport to the Grand Hotel via the city hall and auditorium, makes up dummy caravans of five cars each to cover the other routes. How many cars must be assigned to this task?

Solution. The possible number of routes is

$$4 \cdot 3 \cdot 5 = 60.$$

Thus 59 dummy caravans and 295 cars are needed.

Sometimes a tree diagram is the best method of solution. Each branch represents a possibility, so we simply count up the branches.

Example. A victory in a playoff results for the team which first wins three games. In how many different ways can the playoff be resolved?

Solution. Suppose the teams are A and B. We construct a diagram (Fig. 3.3), shown on p. 68. The circles indicate the third win for one team in any sequence. Counting the circles gives the possible outcomes.

What would happen if instead of arranging the five students mentioned at the beginning of this section in various orders, we were only interested in whether they sat on blue or red chairs and that there were three blue and two red chairs?

There are $5 \cdot 4 \cdot 3$ ways of putting five students in three blue (or any other color) chairs. But we are not interested in the order in which they sit, so for our purposes each arrangement with the same three students in the blue chairs is the same. There are $3!$ ways for three students to sit in three chairs, so we must divide by $3!$:

$$\frac{5 \cdot 4 \cdot 3}{3!} = 10.$$

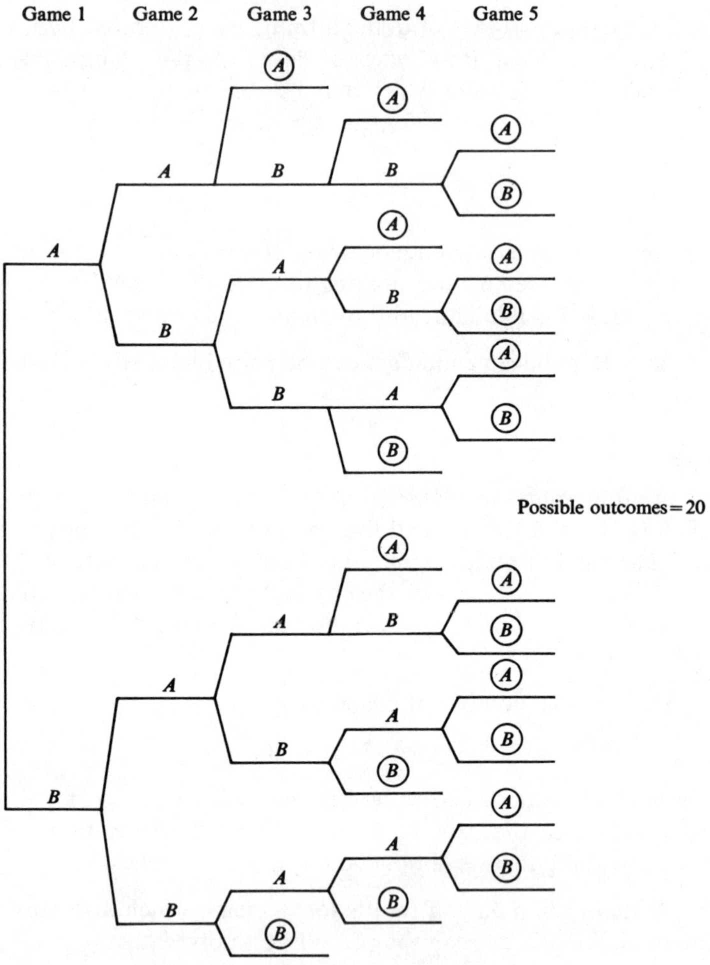

Fig. 3.3

Since there are only red chairs remaining, and the order in which the other two students sit is unimportant, we are finished.

In general, although we do not prove it, there are

$$\frac{n!}{r!\,(n - r)!}$$

ways of choosing n elements r at a time if order is not important. Arrangements in which order is not a consideration are called *combinations*. We use the symbol

$$\binom{n}{r} \quad \text{for} \quad \frac{n!}{r!\,(n - r)!}.$$

For example, suppose there are ten members of a class from whom a three-man committee is to be chosen. The number of ways this can be done is

$$\binom{10}{3} = \frac{10!}{3!\,7!} = \frac{10 \cdot 9 \cdot 8}{3 \cdot 2} = 120.$$

That is, there are a possible 120 different three-man committees in the sense that none of these have all three members the same.

The splitting up of the group of ten into three committee members and seven nonmembers is a partition of the original group. A *partition* of a set S is a collection of subsets of S such that every element of S is contained in exactly one element of the partition.

A partition may have more than the two elements of the committee example above. For example, suppose we had wanted to select a three-man committee and a two-man committee from our group of ten, with the provision that no one serve on both committees. Then we have a three-element partition

$$\{C_1, C_2, N\}$$

where C_1 is the three-man committee, C_2 is the two-man committee, and N consists of the noncommittee members. In how many different ways can we make such a partition?

We state without proof:

$$\begin{array}{l} \text{Number of ways of partitioning } n \\ \text{elements into subsets of} \\ n_1, \ldots, n_r \text{ elements} \end{array} = \frac{n!}{n_1! \cdots n_r!}$$

$$n_1 + \cdots + n_r = n$$

Thus in the above case there are

$$\frac{10!}{3!\,2!\,5!} = \frac{10 \cdot 9 \cdot 8 \cdot 7 \cdot 6}{3 \cdot 2 \cdot 2} = 2520$$

different ways of selecting the committees.

Examples

1. In how many ways can a three-man committee be selected from a ten-member group if the first person selected is to be chairman and the second is to be secretary?

Solution. We may view this as a permutation problem: How many three-element subset permutations are there of a set of ten elements?

$$\frac{10!}{7!} = 720.$$

Or we may choose to regard it as a combination problem where we consider three one-element subsets and one seven-element subset:

$$\frac{10!}{1!\,1!\,1!\,7!} = 720.$$

2. How many ways can three heads and two tails be obtained in five tosses of a coin?

Here the problem is to partition five elements into a three-element set and a two-element set. We have

$$\binom{5}{3} = \frac{5!}{3!\,2!} = 10.$$

This is an illustration of a case in which at first glance it might appear that permutations rather than combinations are involved. However, the order is really fixed (first toss, second toss, etc.) and we are combining the set of tosses into a three-element set (called "heads") and a two-element set (called "tails") (see Fig. 3.4).

3. Nine secretaries are to be assigned to three-man offices. How many different ways can this be done?

Solution

$$\frac{9!}{3!\,3!\,3!} = 1680.$$

Suppose two of the secretaries (say A and B) refuse to share the same office. Then how many ways can the assignments be made?

Solution. Suppose secretary A is assigned to one room and secretary B to another. Then the remaining secretaries may be assigned in

$$\frac{7!}{2!\,2!\,3!} = 210$$

different ways.

The remaining question is: In how many ways can two secretaries be assigned to separate rooms if there are three rooms? This may be restated as: In how many ways can three objects be selected two at a time? However, we need to count, for

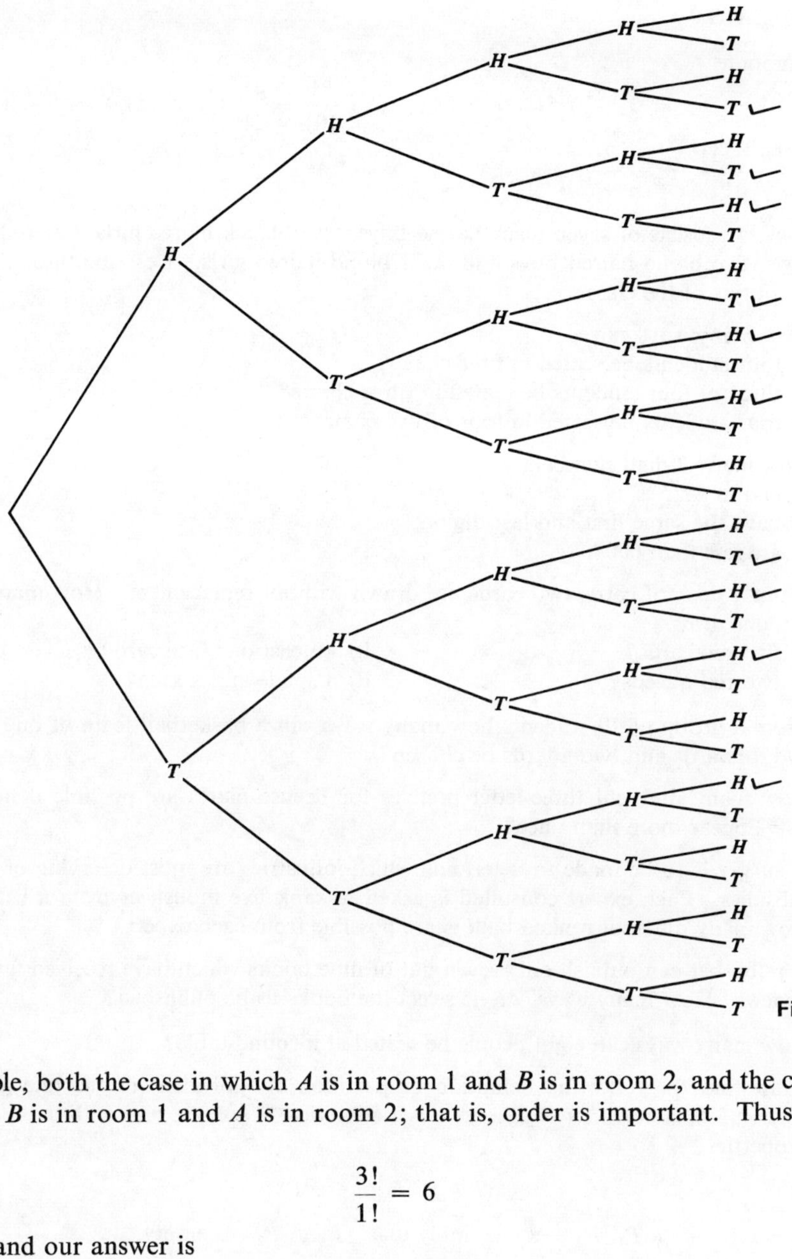

Fig. 3.4

example, both the case in which A is in room 1 and B is in room 2, and the case in which B is in room 1 and A is in room 2; that is, order is important. Thus there are

$$\frac{3!}{1!} = 6$$

ways and our answer is

$$210 \cdot 6 = 1260$$

ways of arranging the secretaries without A and B being assigned to the same office.

PROBLEMS

1. Compute

   a) $6!$. b) $5!/3!$. c) $\dfrac{8!}{10!} - \dfrac{4!}{2!}$. d) $\dfrac{(n+1)!}{n!}$.

   e) $\dfrac{(n-1)!}{n!} - \dfrac{2!}{(n+2)!}$.

2. A class consists of seven black-haired boys, seven black-haired girls, two red-haired boys, five blond-haired boys and eight blond-haired girls. Describe three different partitions of the class.

3. How many ways can
   a) four students be seated in four chairs?
   b) three of four students be seated in three chairs?
   c) four students be seated in four of five chairs?

4. How many 3-digit numbers
   a) end in 4?
   b) have the same first and last digits?
   c) are less than 600?

5. From a deck of cards two cards are drawn without replacement. How many ways can one draw
   a) two red cards?
   b) at least one face card?
   c) two red queens?
   d) a spade and a king?

6. From a group of 10 students how many ways can a basketball team of one center, two forwards and two guards be chosen?

7. How many different three-letter prefixes for license plates are possible if no letter may appear more than once?

8. A survey is to be made to determine which industries are most deserving of federal subsidies. Each expert consulted is asked to rank five industries from a list of 12. How many different ranked ballots are possible from each expert?

9. A publisher can publish only seven out of nine books which have received favorable reviews. How many ways can he select the books to be published?

10. How many ways can eight people be seated at a round table?

11. Show that the following definition of partition is equivalent to the one given in Section 3.1: A *partition* $\mathscr{P}$ of a set S is a collection of subsets of S with the following properties:

    a) $$S = \cup\{P \mid P \in \mathscr{P}\}$$

    b) $P_1, P_2 \in \mathscr{P}$ imply that $P_1 \cap P_2$ is empty.

12. A student must select a program of four courses including one of three English courses, one of two mathematics courses, one of four foreign language courses, and one of three social science courses. How many different programs are possible?

13. How many "shortest" routes are there from X to Y?

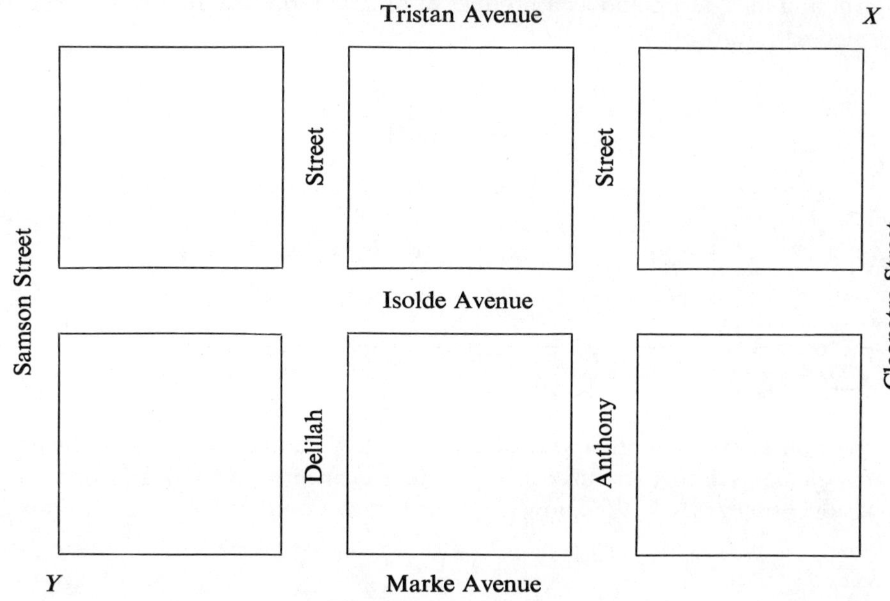

3.2 THE BINOMIAL THEOREM

Frequently we need to carry out operations of the form

$$(x + 3y)^4 \qquad (2a + 3b)^5.$$

We want a more efficient method for doing this than multiplying the many factors in each power and collecting like terms. Our technique is to apply the binomial theorem.

Let us look at the expansion of

$$(x + y)^n,$$

for some small values of n:

$$\begin{aligned}
(x + y)^0 &= 1,\\
(x + y)^1 &= x + y,\\
(x + y)^2 &= x^2 + 2xy + y^2,\\
(x + y)^3 &= x^3 + 3x^2y + 3xy^2 + y^3,\\
(x + y)^4 &= x^4 + 4x^3y + 6x^2y^2 + 4xy^3 + y^4.
\end{aligned}$$

The coefficients are obtained by adding like terms. Thus the question is: How many ways can the combination x^jy^{n-j} occur for a particular n and j? For example, in the expression for $(x + y)^4$, one gets an x^3y term as a product of an

x^3 term and a y term, as a product of an x^2y term and an x term, as a product of a yx^2 term and an x term, and as a product of an xyx term and an x term. This is illustrated as follows:

$$\begin{array}{r} x + y \\ x + y \\ \hline x^2 + xy + yx + y^2 \end{array}$$

$$\begin{array}{r} x^2 + xy + yx + y^2 \\ x + y \\ \hline x^3 + x^2y + xyx + xy^2 + yx^2 + yxy + y^2x + y^3 \end{array}$$

$$\begin{array}{r} x^3 + x^2y + xyx + xy^2 + yx^2 + yxy + y^2x + y^3 \\ x + y \\ \hline x^4 + x^3y + x^2yx + x^2y^2 + xyx^2 + xyxy + xy^2x + xy^3 + yx^3 + yx^2y + yxyx \\ + yxy^2 + y^2x^2 + y^2xy + y^3x + y^4 \end{array}$$

The question above is another way of asking. How many ways can one divide a set of n elements into two subsets, one with j elements (the x's) and one with $n - j$ elements (the y's)? We know from the last section that the answer to this is

$$\frac{n!}{j!\,(n-j)!} = \binom{n}{j} = \binom{n}{n-j}.$$

In particular, for the case of x^3y we have

$$\binom{4}{3} = \frac{4!}{3!\,1!} = 4.$$

We are now ready to state *The Binomial Theorem*:

$$(x + y)^n = \sum_{i=0}^{n} \binom{n}{n-i} x^{n-i}y^i.$$

The proof, which we omit, is by induction.

The numbers $\binom{n}{n-i}$ are called *binomial coefficients*. They are easily remembered by observing the following array of coefficients of the expansions of $(x + y)^n$:

1
1 1
1 2 1
1 3 3 1
1 4 6 4 1
1 5 10 10 5 1
1 6 15 20 15 6 1
...

called *Pascal's triangle*. The binomial coefficient

$$\binom{n}{r} = \binom{n}{n-r}$$

is the $(r + 1)$st entry in the $(n + 1)$st row. If it is not equal to 1 it is obtained by adding the two entries above on either side.

Examples

1. Find $(x + 3y)^4$.

Solution. From the binomial theorem we have

$$\begin{aligned}(x + 3y)^4 &= x^4 + 4x^3(3y) + 6x^2(3y)^2 + 4x(3y)^3 + (3y)^4\\ &= x^4 + 12x^3y + 54x^2y^2 + 108xy^3 + 81y^4.\end{aligned}$$

2. Find $(2a - 3b)^5$.

Solution

$$\begin{aligned}(2a - 3b)^5 &= (2a)^5 + 5(2a)^4(-3b) + 10(2a)^3(-3b)^2 + 10(2a)^2(-3b)^3\\ &\quad + 5(2a)(-3b)^4 + (3b)^5\\ &= 32a^5 - 240a^4b + 720a^3b^2 - 1080a^2b^3 + 810ab^4 - 273b^5.\end{aligned}$$

3. Find the coefficient of x^4y^6 in $(x + 2y)^{10}$.

Solution. We want

$$\binom{10}{4} \cdot 2^6. \qquad \binom{10}{4} = \frac{10!}{4!\,6!} = \frac{10 \cdot 9 \cdot 8 \cdot 7}{4 \cdot 3 \cdot 2} = 210.$$

Thus the desired coefficient is $64 \cdot 210 = 13440$.

PROBLEMS

1. Find $(x + 3y)^4$.
2. Find $(3z - 2w)^6$.
3. Find the coefficient of
   a) x^4y^5 in $(x + y)^9$.
   b) x^3y^3 in $(3x - y)^6$.
   c) x^5y^3 in $(2x + 4y)^8$.
   d) x^3y^{12} in $(ax + by)^{15}$.
4. Find $(x/2 - 2y)^5$.
5. Find $(3a - 4c)^3$.
6. Find the coefficient of
   a) x^3y^5 in $(2x + y)^8$.
   b) x^2y^5 in $(x - 2y)^7$.
   c) a^2b^3 in $(3a + 2b)^5$.
   d) x^6y^8 in $(x + y)^{14}$.

3.3 PROBABILITY

Probability is a much misunderstood and much maligned subject. The U.S. Weather Bureau announces "The probability of rain is 20 percent." The gambler confidently assures all in earshot that the laws of chance guarantee that his losing streak will end. Our task in this and the following section is to develop at least enough probability theory to enable us to analyze such statements from a mathematical point of view.

The study of probability has a two-fold origin—the desire to systematize gambling and the attempt to compensate for the inaccuracy of astronomical measurements.

Let us consider some act we want to perform one or more times, for example, tossing a coin. Each performance is called a *trial* and the collection of trials is said to be an *experiment*. Thus if the coin is tossed three times, each toss is a trial and the set of three tosses is an experiment.

The result of a trial is called an *outcome* and the set of all possible outcomes of the trials of an experiment is a *sample space* for the experiment. For example, if a coin is tossed once, the sample space of this experiment, which has only one trial, is {heads, tails} (briefly denoted as $\{H, T\}$). If a coin is tossed three times the sample of the experiment is:

$$\{HHH, HHT, HTH, HTT, THH, THT, TTH, TTT\}.$$

Now let us consider subsets of a sample space defined by some specific condition; we call these *events*. For example, in the experiment consisting of three tosses of a coin, the event "at least two heads appear" is

$$\{HHH, HHT, HTH, THH\}.$$

We want to assign probabilities to events. First we assign a *probability measure* to a sample space as follows:

1. A positive number (*weight*) is assigned to each element of the sample space.
2. The sum of the weights must equal 1.

If X is a subset of a sample space, the *measure of* X, $m(X)$, is the sum of the weights of its elements. In particular, the *probability* of an event S, denoted by $p(S)$, is defined to be its measure as a subset of the sample space.

If we roll a die, we may write the sample space as $\{1, 2, 3, 4, 5, 6\}$. Because of symmetry, we assume that each face has the same probability of turning up and assign $\frac{1}{6}$ to each of the outcomes. Thus the probability of the event "an even number turns up" is

$$\tfrac{1}{6} + \tfrac{1}{6} + \tfrac{1}{6} = \tfrac{1}{2}.$$

The example above is an *equiprobable measure*; that is, each element of the sample space has the same weight assigned to it. We look at another equiprobable measure.

Example. Five spades including the queen are divided between East and West. What is the probability that East has the queen and exactly one other spade?

Solution. Let the event S be defined by "East has the queen and exactly one other spade." The sample space is the collection of ways five spades can be divided.

From the last section we know that $\binom{5}{j}$ is the number of ways that East can have j spades, so

$$\begin{aligned}\text{number of possibilities} &= \binom{5}{0} + \binom{5}{1} + \binom{5}{2} + \binom{5}{3} + \binom{5}{4} + \binom{5}{5}\\ &= 1 + 5 + 10 + 10 + 5 + 1\\ &= 32.\end{aligned}$$

That is, the sample space has 32 elements. Now the question is: Which of these possible outcomes is in S? East can have two spades including the queen by having the queen and any one of four other cards, so there are four elements in the event S. Assuming that each arrangement is equally likely (equiprobable measure), we assign $\frac{1}{32}$ as the weight of each outcome. Now the probability of S is

$$p(S) = \tfrac{1}{32} + \tfrac{1}{32} + \tfrac{1}{32} + \tfrac{1}{32} = \tfrac{1}{8}.$$

In general, if we have an equiprobable measure

$$p(S) = \frac{\text{number of elements in } S}{\text{number of elements in the sample space}}.$$

Warning. This is valid only for equiprobable measures.

We now discuss a probability measure which is not equiprobable.

Example. Suppose that candidates N, H, and W are running for an office. Suppose that N and H are equally likely to win, but that W is only $\frac{1}{3}$ as likely to win as N is. What is the probability that H does not win?

Solution. Our sample space is $\{N, H, W\}$. We assign weight p to N, weight q to H, and weight r to W. We have

$$\begin{aligned}p &= q,\\ r &= \tfrac{1}{3}p,\\ p + q + r &= 1.\end{aligned}$$

Solving these equations by substitution, we obtain

$$\begin{aligned}p + p + \tfrac{1}{3}p &= 1,\\ \tfrac{7}{3}p &= 1,\\ p &= \tfrac{3}{7} = q,\\ r &= \tfrac{1}{7}.\end{aligned}$$

The event defined by "H does not win" is $\{N, W\}$ so the probability of this statement is $\frac{3}{7} + \frac{1}{7} = \frac{4}{7}$.

We need several properties of probability measures before proceeding.

Theorem 3. For any subsets A and B of a sample space,

$$m(A \cup B) = m(A) + m(B) - m(A \cap B).$$

Proof. Suppose A and B are disjoint, that is, their intersection is empty. Then the weight of each element of $A \cup B$ is added exactly once to give that its measure is the sum of the weights of the elements of A and the elements of B.

If A and B are not disjoint, the weight of an element of their intersection is counted both in $m(A)$ and in $m(B)$. Thus to eliminate the duplication we must subtract the sum of the weights of the elements of $A \cap B$, namely $m(A \cap B)$.

To translate the theorem into a statement about probability we observe that an event defined by a conjunction is the intersection of the events defined by its components and that an event defined by a disjunction is the union of the events defined by its components. For example, in the die experiment, if S_1 is the event "an even number turns up" and S_2 is "a number ≤ 3 turns up,"

$$S_1 = \{2, 4, 6\},$$

$$S_2 = \{1, 2, 3\}.$$

Then $S_1 \cup S_2 = \{1, 2, 3, 4, 6\}$, which is also the event "an even number or a number less than or equal to three turns up." Also, $S_1 \cap S_2 = \{2\}$, which is the event "an even number less than or equal to three turns up."

We have

Theorem 3′

$$p(S_1 \quad \text{or} \quad S_2) = p(S_1) + p(S_2) - p(S_1 \quad \text{and} \quad S_2).$$

A special case of Theorem 3′ is

Corollary

$$p(\text{not } S) = 1 - p(S).$$

(not S is the event defined by the negative of the statement which defines S.)

This is easy to see if we remember that every outcome is either in the event S or in the event not S (but not in both) so that $p(\text{not } S \text{ or } S) = 1$.

We also note the following easily verified properties of probability measures:

1. $p(S) = 0$ if and only if S is empty.
2. $p(S) = 1$ if and only if S is the entire sample space.
3. $0 \leq p(S) \leq 1$.

Examples

1. A bridge deck consists of 52 cards—13 each of spades, hearts, diamonds, and clubs. Each suit has an ace, a king, a queen, a jack, and one each of 2 through 10. The king, queen, and jack are called *face cards*. If a card is drawn at random from a bridge deck, what is the probability of drawing a spade or a face card?

Solution. Let S_1 be "A spade is drawn" and S_2 be "A face card is drawn." Since we may assume that it is equally likely that any card be drawn, we have an equiprobable measure. Thus

$$\begin{aligned} p(S_1) &= \frac{\text{number of elements in the event}}{\text{number of elements in the sample space}} \\ &= \frac{\text{number of spades in deck}}{\text{number of cards in deck}} \\ &= \tfrac{13}{52} = \tfrac{1}{4}. \end{aligned}$$

$$\begin{aligned} p(S_2) &= \frac{\text{number of suits} \times \text{number of face cards per suit}}{52} \\ &= \frac{4 \cdot 3}{52} = \frac{3}{13}. \end{aligned}$$

$$\begin{aligned} p(S_1 \text{ and } S_2) &= \frac{\text{number of spade face cards}}{52} \\ &= \tfrac{3}{52}. \end{aligned}$$

$$\begin{aligned} p(S_1 \text{ or } S_2) &= p(S_1) + p(S_2) - p(S_1 \text{ and } S_2) \\ &= \tfrac{1}{4} + \tfrac{3}{13} - \tfrac{3}{52} \\ &= \tfrac{11}{26}. \end{aligned}$$

2. Four of 25 items are known to be defective. If three items are selected at random, what is the probability that:

   a) none is defective?

   b) at least one is defective?

   c) all are defective?

Solution. The sample space consists of all three-element subsets of the set of 25 items. There are

$$\frac{25!}{22!\,3!} = 2300$$

such subsets, each equally likely to be selected. If none of the items selected is faulty, they must be chosen from the 21 nondefective items. This can be done

$$\frac{21!}{18!\,3!} = 1330$$

ways. Thus the probability that none of the selected items is faulty is $\frac{1330}{2300}$.

The probability that at least one is defective is the probability of not (none is defective) so it is

$$1 - \frac{1330}{2300} = \frac{970}{2300}.$$

Finally, if all are defective they must be selected from the four defective items. This can be done in

$$\frac{4!}{3!\,1!} = 4$$

ways. Therefore the required probability is $\frac{4}{2300}$.

3. Four salesmen leave their briefcases in an outer office. If the secretary returns them at random, what is the probability that at least one man will get his own briefcase? That each man will get his own briefcase?

Solution. This involves an extension of the formula for the probability for one of two events $\big(p(S_1 \text{ or } S_2)\big)$ given above. The formula can be derived by considering a counting problem: How do we count the number of elements in the union of m sets?

Denoting the number of elements in a set S by $n(S)$, we have, in the case of two sets (Fig. 3.5),

$$n(S_1 \cup S_2) = n(S_1) + n(S_2) - n(S_1 \cap S_2),$$

and for three (Fig. 3.6),

$$\begin{aligned} n(S_1 \cup S_2 \cup S_3) = {} & n(S_1) + n(S_2) + n(S_3) - n(S_1 \cap S_2) - n(S_1 \cap S_3) \\ & - n(S_2 \cap S_3) + n(S_1 \cap S_2 \cap S_3), \end{aligned}$$

where the final term adds back in those elements which have been subtracted twice.

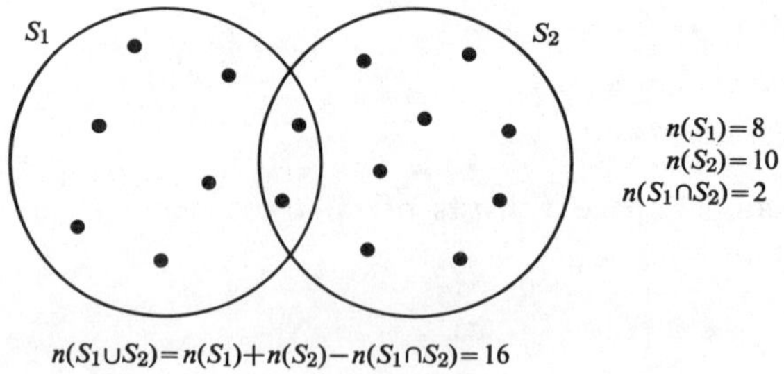

Fig. 3.5

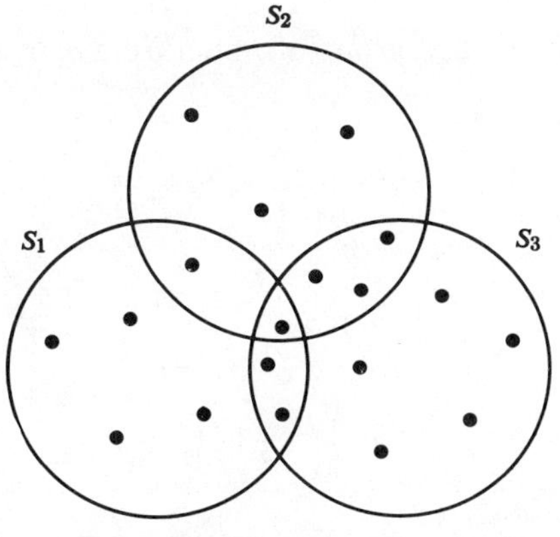

$n(S_1)=8$
$n(S_2)=8$
$n(S_3)=11$
$n(S_1 \cap S_2)=2$
$n(S_1 \cap S_3)=3$
$n(S_2 \cap S_3)=4$
$n(S_1 \cap S_2 \cap S_3)=1$

$$\begin{aligned} n(S_1 \cup S_2 \cup S_3) &= n(S_1)+n(S_2)+n(S_3)-n(S_1 \cap S_2)-n(S_1 \cap S_3) \\ &\quad -n(S_2 \cap S_3)+n(S_1 \cap S_2 \cap S_3) \\ &= 19 \end{aligned}$$

Fig. 3.6

The general formula, which we shall not prove, is

$$n(S_1 \cup \cdots \cup S_m) = \sum_{i=1}^{m} n(S_i) - \sum_{i \neq j} n(S_i \cap S_j) + \sum_{\substack{i \neq j \neq k \\ i \neq k}} n(S_i \cap S_j \cap S_k)$$
$$- \cdots \pm n(S_1 \cap \cdots \cap S_m).$$

From this we obtain the corresponding probability formula

$$p(S_1 \text{ or } S_2 \text{ or } \cdots \text{ or } S_m) = \sum_{i=1}^{m} p(S_i) - \sum_{i \neq j} p(S_i \text{ and } S_j)$$
$$+ \cdots \pm p(S_1 \text{ and } \cdots \text{ and } S_m). \qquad (*)$$

Now let us look at the briefcase problem for m men, with S_i being "The ith man gets his own briefcase back." We are interested in computing the probability that at least one man gets his own back so we want

$$p(S_1 \quad \text{or} \quad S_2 \quad \text{or} \ \cdots \ \text{or} \quad S_m).$$

How do we compute each term in the sum given by (*)? Each of these terms is just the probability that some subset of the men get their own briefcases back, with the jth term giving the probability that j men get their own back.

Since m briefcases can be distributed in $m!$ ways and if j men must get their own, the distribution can be done $(m - j)!$ ways, the probability that a particular set of j men get their own briefcases is

$$\frac{(m - j)!}{m!}.$$

But there are $\binom{m}{j}$ ways of choosing a group of j men from m men so the jth term in the sum (*) will be given by

$$\binom{m}{j} \cdot \frac{(m-j)!}{m!} = \frac{m!}{(m-j)!\,j!} \cdot \frac{(m-j)!}{m!} = \frac{1}{j!}.$$

Thus we have

$$p(S_1 \text{ or } S_2 \text{ or } \cdots \text{ or } S_m) = 1 - \frac{1}{2!} + \frac{1}{3!} - \cdots \pm \frac{1}{m!}$$

$$= \sum_{i=1}^{m} (-1)^{i+1} \frac{1}{i!}.$$

In our problem $m = 4$ so we have

$$p(S_1 \text{ or } S_2 \text{ or } S_3 \text{ or } S_4) = 1 - \tfrac{1}{2} + \tfrac{1}{6} - \tfrac{1}{24} = \tfrac{15}{24}.$$

It turns out that as the number of men and briefcases increases the probability that at least one man gets his own briefcase approaches

$$1 - \frac{1}{e} = 0.632121\ldots,$$

where e is an important irrational constant whose acquaintance we shall make in Chapter 8.

The question of the probability of each man receiving his own briefcase still remains. But this is easy since $p(S_1 \text{ and } S_2 \text{ and } S_3 \text{ and } S_4)$ is the last term in the sum above, namely $\frac{1}{24}$.

Note that while we have been giving probabilities as fractions, these can be converted to decimals or percentages; that is, probabilities of $\frac{1}{4}$, 0.25 and 25 percent are all the same.

PROBLEMS

1. Construct the sample space for two tosses of a die. Is this the same as for the toss of a pair of dice?
2. Chekov, Adams, and Mann are entered in an automobile race. Construct the set of possible outcomes.
3. A coin is tossed three times. Find the probabilities of the following events.
   a) More tails than heads occur. b) Exactly one head occurs.
   c) The same side turns up on every toss.
4. A card is drawn at random from a bridge deck. What is the probability
   a) that a red ace is drawn? b) an ace or a king is drawn?
   c) a face card is not drawn?
5. Two cards are drawn from a bridge deck without replacement. What is the probability that
   a) a spade and a king are drawn? b) both cards are of the same suit?
   c) neither card is a face card?

6. Six of fifteen pages of a German assignment are to be selected at random to be put on a translation exam. A student must translate two of the six. If a student studies only the first ten pages, what is the probability that at least two of the pages he has studied will appear?

7. If eleven people are seated at random at a circular table, what is the probability that a particular pair of people are seated next to each other?

8. A sample of three items is selected at random from 100 items. If there are five faulty items in the 100, what is the probability that
   a) no faulty items are in the sample? b) all items in the sample are faulty?
   c) at least one item in the sample is faulty?

9. From an urn containing three white, two blue, and four red balls, a ball is drawn and replaced and a ball is again drawn. What is the probability that
   a) exactly one red ball is drawn? b) both balls drawn are red?
   c) both balls drawn are blue? d) both blue balls are drawn?
   e) two balls of different colors are drawn?

10. What is the probability that of thirty people at least two have the same birthday? (Day and month, not year.)

11. Each of four divisions of a government agency is carrying out two research projects. Three projects are selected at random to be dropped for budgetary reasons. What is the probability that two of the discontinued projects are from the same division?

12. Four boys and three girls are lined up randomly for a photograph. What is the probability that boys and girls alternate in the picture?

13. Consumer products are classified in national income analyses as durable goods, nondurable goods, and services. Their proportions are 0.11, 0.51, and 0.38, respectively. What is the probability that a consumer good selected at random is a durable good? That two goods selected at random are durable goods?

14. A coin is tossed four times. Find the probability of each of the following events.
   a) Exactly two heads occur. b) More heads than tails occur.
   c) At least one head occurs.

15. Two cards are drawn at random from a bridge deck. What is the probability that
   a) two hearts are drawn?
   b) at least one card is a face card?
   c) the queen of spades and the ace of hearts are drawn?
   d) a queen followed by a heart is drawn?

16. A sample of 10 items is selected at random from 1000 items. If 0.5 percent of the 1000 are faulty, what is the probability that in the sample
   a) no items are faulty? b) no more than one item is faulty?
   c) more than half the items are faulty?

17. An urn contains seven black balls and five white balls. If two balls are drawn at random what is the probability that
   a) the first ball drawn is white? b) at least one ball drawn is white?
   c) exactly one ball drawn is white? d) no balls drawn are white?
   e) both balls drawn are white?

3.4 CONDITIONAL PROBABILITY

Sometimes we compute the probability of a statement and then obtain additional information. We want to know whether this alters the probability. Let us consider the following example.

We choose a family at random from the set of all families with exactly two children. What is the probability of the family chosen having two boys if we know that

1) at least one child is a boy
2) the first child is a boy

(assuming that the probabilities of having a boy or a girl are equal)?

Solution. If we had no additional information, the sample space would be

$$\{BB, BG, GB, GG\}$$

and the probability that there would be two boys would be $\frac{1}{4}$. Under assumption (1) the set is reduced to

$$\{BB, BG, GB\},$$

and the probability that there are two boys is $\frac{1}{3}$. Under assumption (2) the sample space consists of

$$\{BB, BG\},$$

and the probability that there are two boys is $\frac{1}{2}$.

The basic technique we employed in the above example was to observe that since the size of the sample space was reduced by the additional information, the probability was increased. We formalize the notion.

Suppose we have calculated that the probability of a statement S is $p(S)$ Then we receive additional information, say that event T occurs. This reduces the original sample space $\mathscr{L}$ to the set which is the event T and we need to assign a new measure m' to subsets of T. Now if X is a subset of T it certainly is a subset of $\mathscr{L}$ so that we know its original probability measure $m(X)$; while its new measure $m'(X)$ may be different, the relation of its measure to the measure of the other subsets of T remains unchanged. Since probability reflects measure, the relation of the probabilities also is unchanged. That is, if X and Y are subsets of T, we must have

$$\frac{m(X)}{m(Y)} = \frac{m'(X)}{m'(Y)}.$$

We also know that since T is our new sample space $m'(T) = 1$. Thus we have

$$\frac{m(X)}{m(T)} = \frac{m'(X)}{1},$$

or

$$m'(X) = \frac{m(X)}{m(T)}.$$

We can now determine the probability of S, given that T occurs. We write this as $p(S \mid T)$ and call it the *conditional probability* of S, given T. The set defined by the condition defining S has possibly been reduced, since only those elements in the event S *and* in T will be included. The new event is $S \cap T$ and we have

$$p(S \mid T) = m'(S \cap T) = \frac{m(S \cap T)}{m(T)} = \frac{p(S \text{ and } T)}{p(T)}.$$

The only difficulty we might encounter is that $m(T)$ might be zero. But then T is empty. Hence, in computing conditional probability we assume that the given event is not empty.

Example. What is the probability of getting at least 10 in two rolls of a die, given that a six occurs on the first roll?

Solution. Let S be "The sum is ≥ 10" and T be "Six occurs on the first roll."

$$\text{Number of possible outcomes} = 36,$$

$$\begin{aligned} S &= \{(4, 6), (5, 5), (5, 6), (6, 4), (6, 5), (6, 6)\} \\ T &= \{(6, 1), (6, 2), (6, 3), (6, 4), (6, 5), (6, 6)\} \\ S \cap T &= \{(6, 4), (6, 5), (6, 6)\}. \end{aligned}$$

To each possible outcome we assign a weight of $\frac{1}{36}$ so that

$$\begin{aligned} p(S \text{ and } T) &= \tfrac{3}{36} = \tfrac{1}{12} \\ p(T) &= \tfrac{1}{6} \\ p(S \mid T) &= \frac{p(S \text{ and } T)}{p(T)} = \tfrac{1}{12} \cdot \tfrac{6}{1} = \tfrac{1}{2}. \end{aligned}$$

While if the original measure is equiprobable, the new measure assigned in computing conditional probability must be equiprobable also, conditional probability can also be treated in other cases.

Example. Suppose we have a loaded die such that the probability of a number turning up is proportional to that number.

1) What is the probability of getting an odd number?
2) What is the probability of getting an odd number if the number thrown is greater than 3?

Solution. The sample space is $\{1, 2, 3, 4, 5, 6\}$ and the weights satisfy the following:

$$w_1 + w_2 + w_3 + w_4 + w_5 + w_6 = 1 \ (w_i \text{ is the weight assigned to } i),$$

$$w_i = iw_1, \qquad i = 2, \ldots, 6.$$

Simplifying, we have

$$w_1 + 2w_1 + 3w_1 + 4w_1 + 5w_1 + 6w_1 = 21w_1 = 1,$$

and

$$w_i = i/21, \qquad i = 1, \ldots, 6.$$

Let S be "An odd number turns up." The truth set of S is $\{1, 3, 5\}$.

$$p(S) = \tfrac{1}{21} + \tfrac{3}{21} + \tfrac{5}{21} = \tfrac{9}{21} = \tfrac{3}{7}.$$

Let T be "A number greater than three turns up." The truth set of T is $\{4, 5, 6\}$ and

$$p(T) = \tfrac{4}{21} + \tfrac{5}{21} + \tfrac{6}{21} = \tfrac{5}{7}.$$

The truth set of S and T is $\{5\}$ and

$$p(S \text{ and } T) = \tfrac{5}{21}.$$

Thus the probability of S, given T, is

$$p(S \mid T) = \frac{p(S \text{ and } T)}{p(T)} = (\tfrac{5}{21})(\tfrac{7}{5}) = \tfrac{1}{3}.$$

We can now give a precise statement of what is meant by independence in the context of probability. The outcome of each toss of a coin is unaffected by the results of previous tosses. Thus it is intuitively clear that

"A head turns up on toss number 10,"

and

"A head turns up on toss number 11,"

should be independent events. On the other hand,

"A face card is drawn,"

and

"A queen is drawn,"

do not seem to be independent. If events are to be independent, the probability of one should not be affected by the occurrence of the other; precisely, we want S and T are *independent* if:

$$p(S \mid T) = p(S)$$

and

$$p(T \mid S) = p(T).$$

Since

$$p(S \mid T) = \frac{p(S \text{ and } T)}{p(T)} \quad \text{and} \quad p(T \mid S) = \frac{p(S \text{ and } T)}{p(S)},$$

we must have

$$\frac{p(S \text{ and } T)}{p(T)} = p(S),$$

or equivalently

$$\frac{p(S \text{ and } T)}{p(S)} = p(T).$$

Hence we need not require that $p(T \mid S) = p(T)$ since it follows from $p(S \mid T) = p(S)$.

We also observe that we have

$$p(S \quad \text{and} \quad T) = p(S)p(T), \tag{3.1}$$

if and only if S and T are independent. Condition (3.1) above is sometimes taken as the definition of independence.

Example. Let S be defined by "The same side occurs on three tosses of a coin" and T be defined by "At most one head occurs in three tosses of a coin." It may not be intuitively clear whether S and T are independent. Let us compute $p(S$ and $T)$.

Solution. The set of outcomes is

$$\{HHH, HHT, HTH, HTT, THH, THT, TTH, TTT\}.$$

The event S is $\{HHH, TTT\}$ and T is $\{HTT, THT, TTH, TTT\}$.

Assigning the equiprobable measure, we have

$$p(S) = \tfrac{1}{4} \qquad \text{and} \qquad p(T) = \tfrac{1}{2}.$$

The truth set of S and T is $\{TTT\}$ so that

$$p(S \quad \text{and} \quad T) = \tfrac{1}{8} = p(S)p(T).$$

Therefore S and T are independent.

This example is a special case of what we call a *stochastic process*—that is, a sequence of experiments where the outcome of each experiment depends on some chance element. In a stochastic process we shall always assume that there are a finite number of experiments and a finite number of possible outcomes for each experiment. The example above was an *independent trials process*—the outcome of each toss is independent of the previous tosses. However, stochastic processes need not have this property.

Let us look at a tree diagram (Fig. 3.7) for a stochastic process. Suppose that we have three experiments; the outcomes for the first are a and b, for the second c, e, and f, and for the third g and h. We let p_a denote the probability of outcome a, $p_{c|a}$ the probability of outcome c given that a occurs, $p_{g|c|a}$ the probability of outcome g given that c and a occur, etc. To meet the requirements for a probability measure on the set of outcomes of each experiment we must have:

$$p_a + p_b = 1,$$

$$p_{c|a} + p_{e|a} + p_{f|a} = 1, \qquad p_{c|b} + p_{e|b} + p_{f|b} = 1,$$

$$p_{g|c|a} + p_{h|c|a} = 1, \qquad p_{g|e|a} + p_{h|e|a} = 1,$$

$$p_{g|f|a} + p_{h|f|a} = 1, \qquad p_{g|c|b} + p_{h|c|b} = 1,$$

$$p_{g|e|b} + p_{h|e|b} = 1, \qquad p_{g|f|b} + p_{h|f|b} = 1.$$

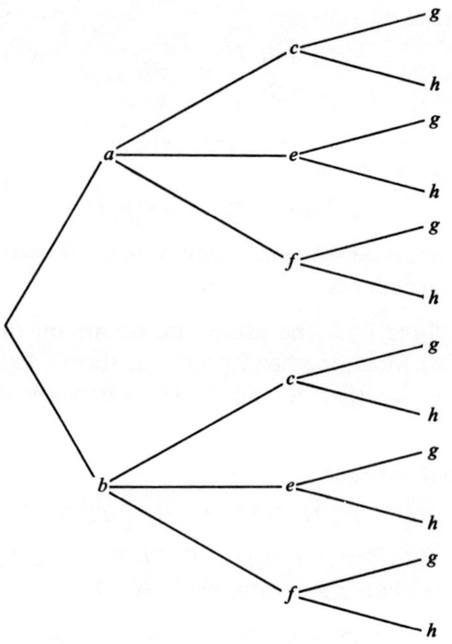

Fig. 3.7

Next we assign a weight to each path through the tree diagram by

$$p_{acg} = p_a p_{c\,|\,a} p_{g\,|\,c\,|\,a},$$

where p_{acg} is the path through a, c, and g, and similarly for each of the 12 paths. These assignments determine what we call a *tree measure*. We want to verify that it satisfies the requirements for a probability measure on the set of 12 outcomes for the sequence of experiments considered as a single experiment.

All the weights are positive since they are the products of positive numbers. We also have

$$\begin{aligned}
p_{acg} + p_{ach} + p_{aeg} + p_{aeh} + p_{afg} + p_{afh} + p_{bcg} + p_{bch} + p_{beg} + p_{beh} + p_{bfg} + p_{bfh} \\
= p_a p_{c|a} p_{g|c|a} + p_a p_{c|a} p_{h|c|h} + p_a p_{e|a} p_{g|e|a} + p_a p_{e|a} p_{h|e|a} \\
+ p_a p_{f|a} p_{g|f|a} + p_a p_{f|a} p_{h|f|a} + p_b p_{c|b} p_{g|c|b} + p_b p_{c|b} p_{h|c|b} \\
+ p_b p_{e|b} p_{g|e|b} + p_b p_{e|b} p_{h|e|b} + p_b p_{f|b} p_{g|f|b} + p_b p_{f|b} p_{h|f|b} \\
= p_a(p_{c|a}(p_{g|c|a} + p_{h|c|a}) + p_{e|a}(p_{g|e|a} + p_{h|e|a}) \\
+ p_{f|a}(p_{g|f|a} + p_{h|f|a})) + p_b(p_{c|b}(p_{g|c|b} + p_{h|c|b}) \\
+ p_{e|b}(p_{g|e|b} + p_{h|e|b}) + p_{f|b}(p_{g|f|b} + p_{h|f|b})) \\
= p_a(p_{c|a} + p_{e|a} + p_{f|a}) + p_b(p_{c|b} + p_{e|b} + p_{f|b}) \\
= p_a + p_b \\
= 1.
\end{aligned}$$

Example. The classic example of this sort of probability problem involves urns and black and white balls. Suppose there are two urns, the first with three black balls and two white balls, the second with one black ball and two white balls. If an urn is chosen at random and a ball is drawn, what is the probability of its being white?

Solution. We use an equiprobable measure to make the original assignments. Then we observe that two branches of the tree (see Fig. 3.8) lead to the desired result so we take the sum of the weights of each to get the probability of a white ball being drawn.

$$\begin{aligned}\text{Probability of a white ball being drawn} &= p_{1W} + p_{2W}\\ &= p_1 p_{W|1} + p_2 p_{W|2}\\ &= (\tfrac{1}{2})(\tfrac{2}{5}) + (\tfrac{1}{2})(\tfrac{2}{3})\\ &= \tfrac{8}{15}.\end{aligned}$$

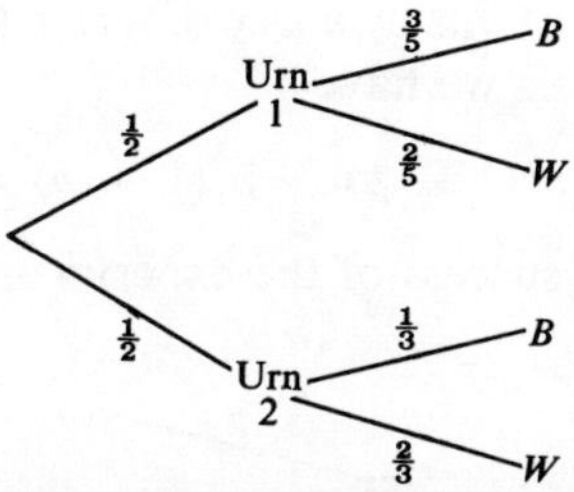

Fig. 3.8

One's intuition might lead one to believe that the probability would be $\frac{1}{2}$ since there are the same total number of white balls as black. Beware of intuition in probability problems. Note that this is not an independent trials process since the probability of getting a white ball at the second stage depends on the outcome of the first experiment—the choice of urns.

Example. A company employee is to be selected to represent his fellow workers. If among three of the five departments four-fifths of the workers are Democrats and in the remaining two departments two-thirds of the workers are Republicans, what is the probability that the man chosen is a Democrat if the department from which he is selected is chosen at random and he is then selected from this department at random (Fig. 3.9)?

Solution

$$\begin{aligned}\text{Probability of being a Democrat} &= (\tfrac{3}{5})(\tfrac{4}{5}) + (\tfrac{2}{5})(\tfrac{1}{3})\\ &= (\tfrac{12}{25}) + (\tfrac{2}{15})\\ &= \tfrac{46}{75}.\end{aligned}$$

Note that this probability is independent of the actual number of men in each department.

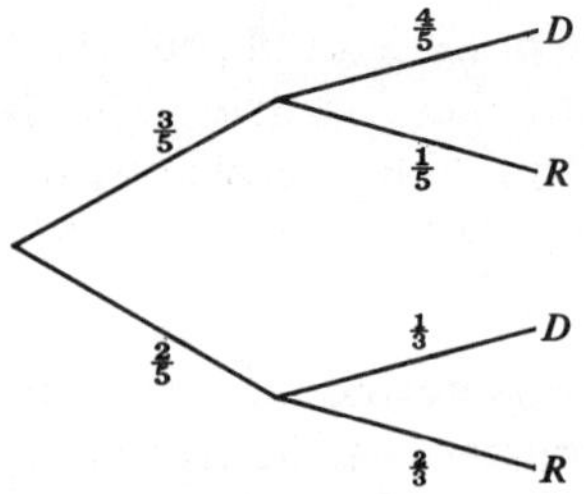

Fig. 3.9

In conclusion we present one more useful formula which is a special case of a stochastic process (another special case, Markov chains, is considered in Chapter 11). Suppose we have an independent trials process of the same experiment repeated a number of times.

Suppose, moreover, that there are only two possible outcomes, which we shall call "success" and "failure." We want to know the probability of exactly k successes in n trials. We can do this by a tree diagram (see Fig. 3.10). For example, for two successes in three trials we have

$$\text{Probability} = p^2(1 - p) + p^2(1 - p) = 2p^2(1 - p),$$

where p is the probability of success of the experiment.

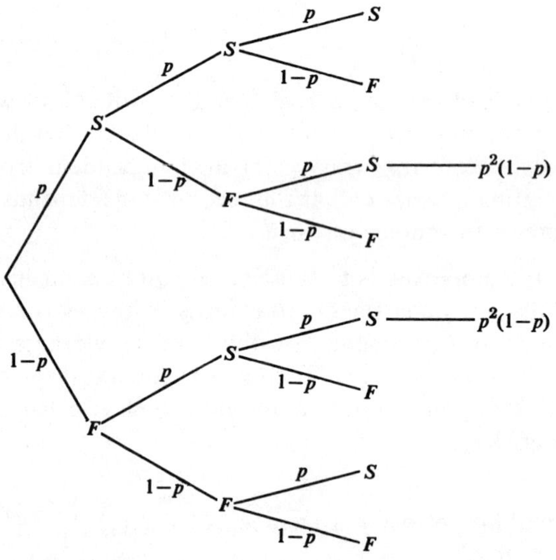

Fig. 3.10

Although we shall not prove it, in general there will be a probability of $p^k(1 - p)^{n-k}$ for each of the branches leading to k successes in n trials. Thus the question is: How many branches lead to this result? This is just a matter of determining ways of choosing n elements k at a time—the binomial coefficient $\binom{n}{k}$. Hence the probability of k successes in n trials is

$$\binom{n}{k} p^k(1 - p)^{n-k}.$$

Example. Suppose that the probability of a missile hitting a target is $\frac{1}{3}$. How many missiles should be directed at a target if we want the probability for at least one hit to be at least 0.8?

Solution. We want

$$\sum_{j=1}^{n} \binom{n}{j} (1/3)^j(2/3)^{n-j} \geq 0.8.$$

Suppose $n = 3$. Then

$$\begin{aligned}\sum_{j=1}^{3} \binom{3}{j} (1/3)^j(2/3)^{3-j} &= 3(1/3)^1(2/3)^2 + 3(1/3)^2(2/3)^1 + 1(1/3)^3\\ &= 4/9 + 2/9 + 1/27\\ &= 19/27 < 0.8.\end{aligned}$$

Suppose $n = 4$. Then

$$\begin{aligned}\sum_{j=1}^{4} \binom{4}{j} (1/3)^j(2/3)^{4-j} &= 4(1/3)(2/3)^3 + 6(1/3)^2(2/3)^2 + 4(1/3)^3(2/3) + (1/3)^4\\ &= 32/81 + 8/27 + 8/81 + 1/81\\ &= 65/81 = 0.802 > 0.8.\end{aligned}$$

Thus we must direct at least four missiles at the target.

Let us look at the probability of j heads appearing in ten tosses of a coin. We have

$$p(0) = \binom{10}{0} (1/2)^0(1/2)^{10} = 1/2^{10} \sim 0.001$$

$$p(1) = \binom{10}{1} (1/2)^1(1/2)^9 = 10/2^{10} \sim 0.01$$

$$p(2) = \binom{10}{2} (1/2)^2(1/2)^8 = 45/2^{10} \sim 0.045$$

$$p(3) = \binom{10}{3} (1/2)^3(1/2)^7 = 120/2^{10} \sim 0.12$$

$$p(4) = \binom{10}{4} (1/2)^4(1/2)^6 = 210/2^{10} \sim 0.20$$

$$p(5) = \binom{10}{5} (1/2)^5(1/2)^5 = 252/2^{10} \sim 0.25$$

$$
\begin{aligned}
p(6) &= p(4) \sim 0.20 \\
p(7) &= p(3) \sim 0.12 \\
p(8) &= p(2) \sim 0.045 \\
p(9) &= p(1) \sim 0.01 \\
p(10) &= p(0) \sim 0.001.
\end{aligned}
$$

Thus we see that the most probable number of heads is $5 = \frac{1}{2} \cdot 10 = pn$, but that the probability of five heads occurring is only 0.25 (Fig. 3.11).

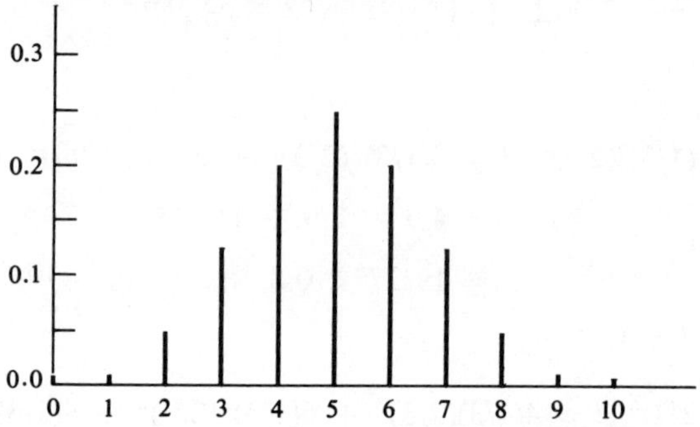

Fig. 3.11

In general, in n trials the numbers of successes which occur with the greatest probability are those near np. One important measure is the probability of the proportion j/n of successes differing from p by less than a given amount. That is, for some positive quantity ε we are interested in

$$p(|(j/n) - p| < \varepsilon).$$

The "law of large numbers" states that for any $\varepsilon > 0$ this probability tends to 1 as n increases without bound.

This is the "law of chance" (often misinterpreted) to which a persistent loser frequently appeals. Suppose an (unbiased) coin has been tossed 400 times with "heads" appearing 300 times. The gambler increases his bet on "tails" on the grounds that the most probable number of tails in 401 tosses is near 200 and so a tail really should appear on the next toss. How does this follow from the law of large numbers? It doesn't. The law of large numbers simply tells us that if we are going to toss a coin n times and we want the probability to be at least 0.99 that the fraction of tails will be within 0.001 of 0.5 we can choose a sufficiently large n to assure this. It does not tell us how large n must be, nor that for any n we are certain to have exactly half tails, nor that the probability of tails on any one toss is increased. In fact, coin tossing is an independent trials process and the probability of tails on toss number 401 is the same as on toss 1.

The quantity $\sqrt{np(1 - p)}$ is called a *standard deviation*. It turns out that

$$p(|j - pn| < \sqrt{np(1 - p)}),$$

is always approximately 0.68. Therefore, the variation of 1 standard deviation in the number of successes from pn is quite probable (over 0.3). However,

$$p(|j - pn| < 2\sqrt{np(1 - p)}) \sim 0.95$$

$$p(|j - pn| < 3\sqrt{np(1 - p)}) \sim 0.99^+,$$

so that a variation of more than three standard deviations is very unlikely.

In the case of our coin-tosser we note that

$$\sqrt{np(1 - p)} = \sqrt{400(\tfrac{1}{2})(\tfrac{1}{2})} = 10.$$

Therefore a deviation of 100 from the most probable number of tails is so unlikely as to lead one to suspect that the coin is not in fact unbiased; however, we emphasize that such a large deviation is *possible* with an unbiased coin.

Let us examine some additional illustrations of the concepts of this chapter.

Examples

1. There are two kinds of genes which determine the color of one's eyes: X and Y. Each person has two color-determining genes; if both are X, he has blue eyes; otherwise his eyes are brown. Assume that $\frac{1}{4}$ of the people have two X genes, $\frac{1}{4}$ have two Y genes and the rest have one of each. If a person has brown eyes, what is the probability that he has two Y genes? A child gets one gene from his mother and one from his father, each selected randomly from the parent's two genes. What is the probability that the child of a blue-eyed mother and a brown-eyed father will have blue eyes?

Solution. To answer the first question, let S be "He has two Y genes" and T be "He has brown eyes." Then the probability we want is

$$p(S \mid T) = \frac{p(S \text{ and } T)}{p(T)} = \frac{1/4}{3/4} = \frac{1}{3}.$$

In the second part we have a stochastic process which is not an independent trials process. Given that the father has brown eyes there are two possibilities: two Y genes or one Y gene and one X gene. The gene selected from the mother must be an X, so that need not enter our considerations.

We draw a tree diagram (Fig. 3.12) based on the fact that it is given that the father has brown eyes. From the above calculation the probability that he has two Y's is $\frac{1}{3}$ and that he has one X and one Y is $\frac{2}{3}$. The random selection of the child's gene gives a probability of $\frac{1}{2}$ for each branch at the next stage. Let p_X be the probability of an X gene for the child, $p_{(X,Y)}$ the probability of one X and one Y for the father.

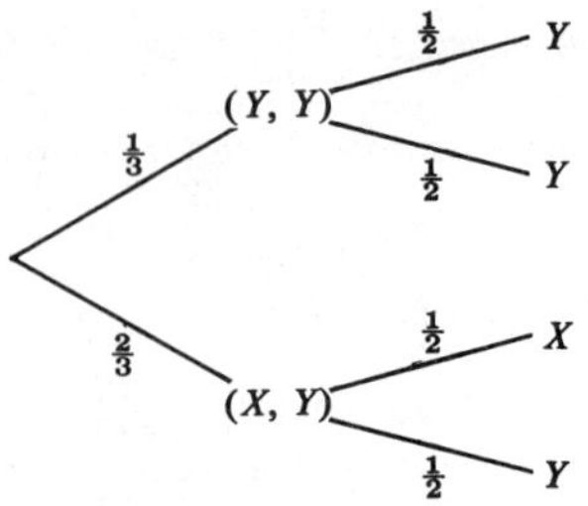

Fig. 3.12

One path through the tree produces an X to match the mother's X and produce blue eyes:

$$\begin{aligned} p_{(X,Y)X} &= p_{(X,Y)}p_{X\,|\,(X,Y)} \\ &= (\tfrac{2}{3})(\tfrac{1}{2}) \\ &= \tfrac{1}{3}. \end{aligned}$$

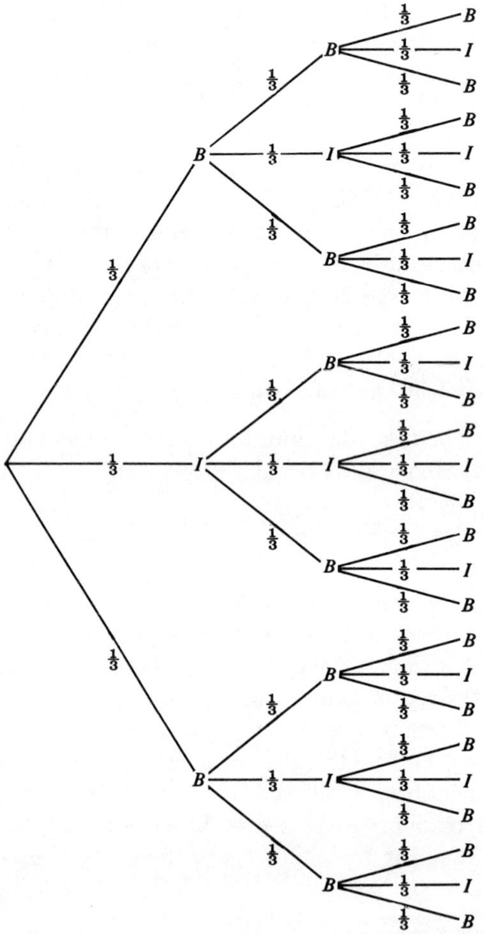

Fig. 3.13

2. A student claims to be able to distinguish instant coffee from brewed coffee. He is given a series of three tests in each of which he is given two cups of brewed coffee and one of instant and asked to pick out the instant. If he gets two or more correct, his claim is considered to be established. By means of a tree diagram find the probability that he will be able to establish his claim by guessing on every trial.

Solution. The tree diagram in Fig. 3.13 has an equiprobable measure at each stage. Seven paths lead to at least two correct guesses. The probability for each path is $(\frac{1}{3})(\frac{1}{3})(\frac{1}{3})$; therefore the probability that he can establish his claim by guessing is $\frac{7}{27}$.

We still must analyze the remaining example at the beginning of Section 3. A 20 percent chance of rain says that a probability of $\frac{1}{5}$ has been assigned to the statement "It will rain." Exactly how the meteorologists make this assignment is not our concern; in fact it is usually experimentally determined, that is, in the past it has rained once out of five times when the conditions were the same as when the prediction is being made.

SUMMARY

Partitions of n elements into sets of $n_1, \ldots, n_r$ elements

$$\frac{n!}{n_1! \cdots n_r!}$$

Combinations of n elements taken r at a time

$$\binom{n}{r} = \frac{n!}{(n-r)!\,r!}$$

Permutations of n elements taken r at a time

$$\frac{n!}{(n-r)!}$$

Arrangements of n elements into a set of r elements

$$n^r$$

(repetitions allowed).

Probability

$$p(S_1 \text{ or } S_2) = p(S_1) + p(S_2) - p(S_1 \text{ and } S_2)$$

$$p(S_1 \mid S_2) = \frac{p(S_1 \text{ and } S_2)}{p(S_2)}$$

Probability of k successes in n independent trials

$$\binom{n}{k} p^k(1 - p)^{n-k}$$

where p is the probability of success in a single trial.

....

PROBLEMS

1. A machine produces bolts, 0.1 percent of which are defective. What is the probability that two bolts in a row are defective?
2. The probability that an individual is over 30 and makes more than \$15,000 is 0.18 and the probability that an individual makes over \$15,000 is 0.21; what is the probability that an individual is over 30 if his income is over \$15,000?
3. A manufacturer produces 1000 items, of which 40 are defective. If two items are selected at random and tested, what is the probability that the first is satisfactory and the second defective?
4. If two cards are drawn at random from a bridge deck, what is the probability that an ace is followed by a king
   a) if the first card is replaced before the second is drawn?
   b) if the first card is not replaced before the second is drawn?
5. If 20 of 10,000 items are defective, what is the probability that of a 4-item sample
   a) all are defective?
   b) none are defective?
   c) exactly one is defective?
   d) at least one is defective?
6. In an independent trials process the probability for success on each individual trial is $\frac{1}{4}$. Calculate the probability of exactly five successes in nine trials.
7. Urn A contains two red balls, three white balls and four blue balls. Urn B contains three white balls and two red balls. An urn is chosen at random and from it one ball is drawn at random. Find the probability of each of the statements:
   a) urn A is chosen.
   b) urn B is chosen and a red ball is drawn.
   c) urn B is chosen and a blue ball is drawn.
   d) urn A is chosen and the ball is not blue.
   e) the ball drawn is not red.
   f) the ball drawn is white, given that urn B is chosen.
   g) urn B is chosen, given that the ball drawn is white.
8. A student takes a six-question true-false exam. What is the probability that he will get all answers correct if
   a) he guesses?
   b) he guesses, knowing that three questions are true and three questions are false?
   c) he guesses, knowing that there will not be three questions in a row with the same answer?

9. Two cards are drawn in succession (without replacement) from a bridge deck. What is the probability that
   a) at least one card is a jack?
   b) both cards are jacks?
   c) both cards are jacks, given that the first card is a jack?
   d) both cards are jacks, given that one card is a jack?
   e) both cards are jacks, given that the first card is the jack of diamonds?
10. The fraction of registered Republicans, Democrats, and Independents in two cities is
    1st city: 0.25 Republicans, 0.45 Democrats, 0.30 Independents.
    2nd city: 0.45 Republicans, 0.45 Democrats, 0.10 Independents.
    A city is chosen at random and two voters are selected successively and at random from this city. Find the probability that
    a) two Republicans are chosen?
    b) the second voter chosen is a Democrat?
    c) neither voter is an Independent?
11. A die is thrown twice. What is the probability that
    a) the sum of the numbers is 7?
    b) the sum of the numbers is 7, given that the first number is less than 4?
    c) the sum of the numbers is 7, given that the first number is a 3?
12. Twenty-five of 2000 items are defective. If 10 items are selected at random, what is the probability that
    a) all are defective?
    b) all are defective, given that at least one is defective?
    c) none are defective?
    d) none are defective, given that at least one is defective?
13. A bus company's accident probability is 0.01. What is the probability of no accidents if
    a) one bus is sent out? b) 10 buses are sent out?
    c) 100 buses are sent out?
    [*Hint:* Independent trials process.]
14. Urn A contains two red and two white balls; urn B contains three red and one white ball. An urn is selected at random and a ball is drawn from it. What is the probability that
    a) a white ball is drawn?
    b) a white ball is drawn, given that urn A is chosen?
    c) urn B is selected, given that the ball is white?
15. What is the probability of 40 heads in 100 tosses of a coin?

4 / FUNCTIONS AND CONTINUITY

In this chapter we introduce the concept which is the demarcation between algebra and analysis. Elementary algebra of the sort studied in high school and up to this point in this book is essentially a finite subject, and while recent developments in abstract algebra have been of tremendous importance, the basic material has been known for over a thousand years. On the other hand, analysis—to which calculus is an introduction—concerns itself with infinite processes, and offers variations on the concept of limit. Moreover, calculus is comparatively recent—it has been in existence only about three hundred years—while the precise formulation of the fundamental notion of limit is even more recent.

Since man's real-world experiences are always finite in nature, the process of taking a limit introduces certain difficulties. However, we rely on the mathematicians' usual tools—precise definitions and the use of logic.

The intuitive idea of function is much older than calculus—indeed the ancient Greek mathematicians understood the concept involved—and is basic both to analysis and to algebra, at least in their sophisticated manifestations. The dependence of one quantity on another is a straightforward notion, but the precise nature of the dependence—in particular whether one quantity in fact determines another—can present complications.

4.1 FUNCTIONS

The rates for parking at a certain lot are 60 cents for the first hour or fraction thereof and 40 cents for each additional hour or fraction thereof, up to a 3-dollar maximum in a 24-hr period. Thus the length of time parked determines the amount which must be paid. We say that the amount owed is a *function* of the time parked. For example, parking for $3\frac{1}{2}$ hours incurs a fee of 60 cents + 3(40 cents) = \$1.80.

Suppose a piece of equipment loses $\frac{1}{20}$ of its value each year of its operation; this process is called *depreciation*. If its original value is V_0, then its value V at the end of t years is

$$V = (1 - \tfrac{1}{20}t)V_0;$$

its value is a function of its years of operation because for any value of t,

$0 \le t \le 20$, a unique value of V is determined. This kind of depreciation is called *straight line* depreciation since the graph of V is a straight line.

Similarly, in an equation such as

$$y = 3x^2 + 2x - 1,$$

we can assign a value to x and find a corresponding value for y; if for each number assigned to x we get a unique value for y, we say that y is a *function* of x. We write

$$y = f(x).$$

Thus in this example, $f(1) = 3(1) + 2(1) - 1 = 4, f(0) = -1$, etc.

The table

x	y
0	3
1	4
2	-1
6	2
-1	6

also defines y as a function of x. We sometimes call x an *independent variable* and y a *dependent variable*, since its value is determined by the value of x. The use of x and y is not significant. We could just as well use any other letters as variables as the important thing is the *rule* which assigns a value to y for each value of x. For example, if

$$y = x^2 + 2x - 4 \quad \text{and} \quad s = t^2 + 2t - 4,$$

then we write $y = f(x)$ and $s = f(t)$. If we do not need to emphasize the variable, we may simply refer to the function f. We may also denote a function by some letter other than f. (Although the same letter should be used for the same function throughout a particular discussion.) For example, if

$$y = \frac{1}{x^2 + 2},$$

we can write $y = g(x)$ and say that y is a function of x.

If $f(x)$ is a polynomial in x, it is called a *polynomial function*. Similarly, if $f(x)$ is the quotient of polynomials, it is called a *rational function*. Not all functions are defined by algebraic expressions. For example, recall that the table given above defines a function.

If we have the equation

$$y^2 = -x^2 - 4x + 1, \tag{4.1}$$

the value of y still depends on that of x, but each value of x does not uniquely determine a value of y; for one value of x there may be more than one value of y.

For example, if $x = -4$, $y = \pm 1$. In this case we call the dependence of y upon x a *relation*; Eq. (4.1) does not define a function.

In the graph of the function

$$y = 3x^2 + x + 1,$$

we observe that any line perpendicular to the x-axis cuts the graph in no more than one place (see Fig. 4.1). This is to be expected since for each value of x there is only one value of y. In fact, any graph of a function has this property (Fig. 4.2). Sometimes this is an easy way to determine whether we have a function.

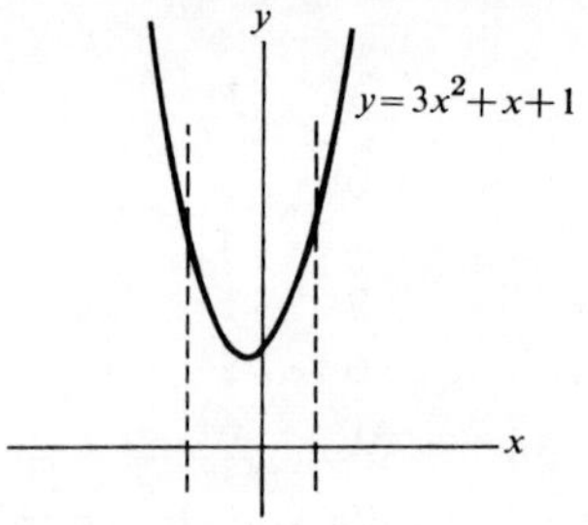

Fig. 4.1

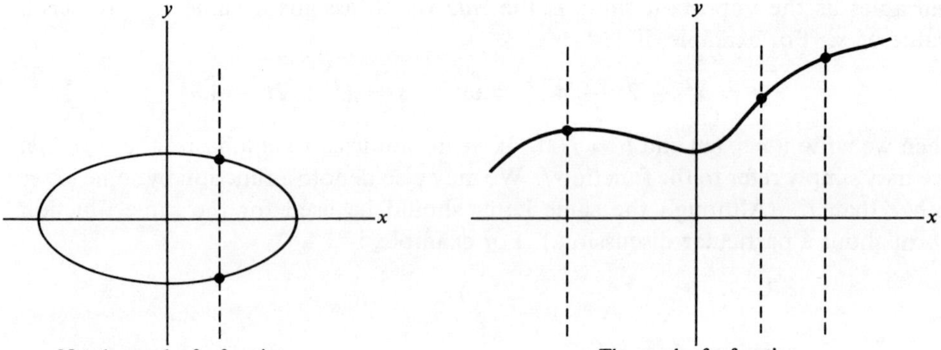

Fig. 4.2

Suppose

$$y = \frac{1}{x}.$$

Then certainly every value of $x \neq 0$ determines precisely one value of y. However, $x = 0$ does not determine any value of y. We say that the function $f(x) = 1/x$ is *undefined* for $x = 0$. Similarly, since we are dealing only with real numbers, if

$$y = \sqrt{x},$$

then we exclude from consideration all negative values of x. If $y = f(x)$, we call the set of numbers which may be assigned to x the *domain* of the function f. For example, if $y = \sqrt{x}$, the domain is the nonnegative real numbers.

In the depreciation problem, $V = f(t) = (1 - \frac{1}{20}t)V_0$, the domain of the function is $[0, 20]$, the set of allowable values for t. In the example

$$y = \sqrt{x},$$

we also note that the set of possible values of y is limited since by convention $\sqrt{\ }$ denotes the positive square root. We call the set of numbers which occur as values of y the *image* of the function. For $y = \sqrt{x}$, the image is the set of nonnegative real numbers. If the set X is the domain of the function $y = f(x)$, we denote the image by $f(X)$. Elements of $f(X)$ are written as $f(x)$ for some $x \in X$; that is,

$$f(X) = \{y \mid y = f(x) \quad \text{for some} \quad x \in X\}.$$

Thus, if

$$f(x) = \sqrt{x},$$

$$f(X) = \text{the set of nonnegative real numbers.}$$

If

$$f(x) = x^2 + 2,$$

$$f(X) = \text{the set of all real numbers} \geq 2.$$

Another related notion is that of a *range* of a function. A range is not uniquely determined and can be any set which contains the image. Since we are considering only real numbers, the image of any function must be a subset of the real numbers. Therefore, the set R of all real numbers qualifies as a range of any of our functions. Functions which have subsets of the real numbers as both their domain and image are called *real-valued functions of a real variable*. If $f(x)$ is a real-valued function of a real variable whose domain is D, we write

$$f\colon \mathrm{D} \to R \qquad \text{or} \qquad f\colon R \to R.$$

Observation of the graph of a function is useful in determining the domain and image. If we project the graph onto the x-axis we obtain the domain. In Fig. 4.3 this gives us all the real numbers except 0. Projection onto the y-axis gives us the image (Fig. 4.4). If

$$y = x^2 + 1,$$

the image is the set of all real numbers greater than or equal to 1.

We have been thinking of a function as a machine into which we put a real number, say a, and from which we get a real number, $f(a)$:

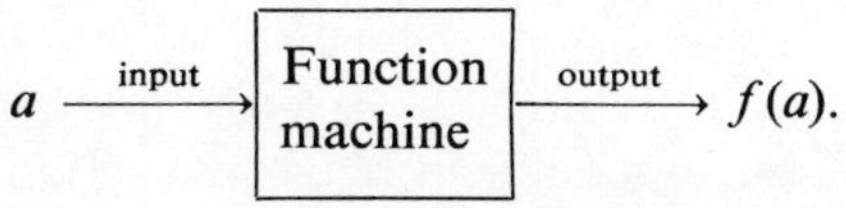

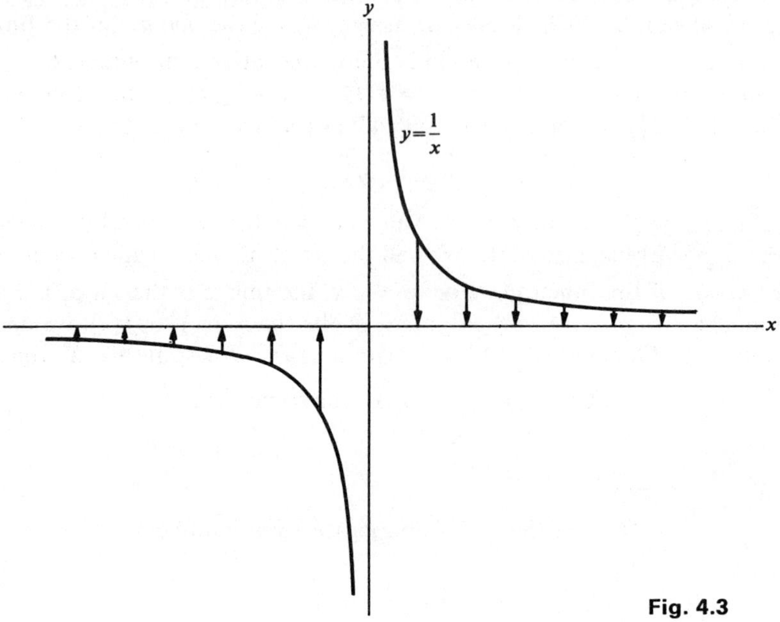

Fig. 4.3

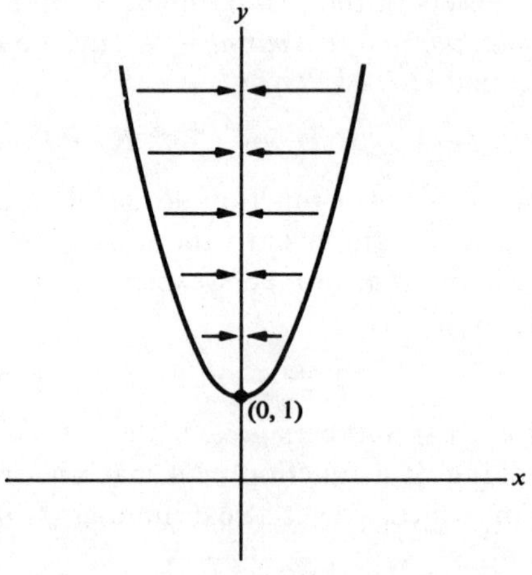

Fig. 4.4

Using the terminology of domain and image we can think of a function in another way: as a *mapping* which takes a set of numbers onto another set of numbers (see figure).

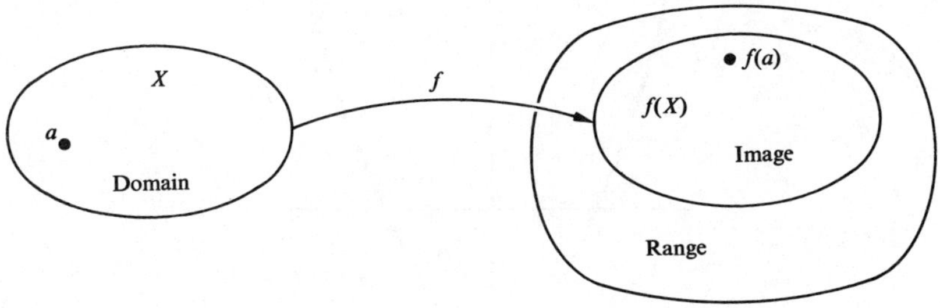

Examples

1. If

$$y = f(x) = x^3 + x^2 - 1,$$

then each value for x gives a unique value for y so $f(x)$ is a function (Fig. 4.5). Also, as x takes on all real values, so does y. Therefore,

$$\text{image } f = \text{domain } f = \text{real numbers.}$$

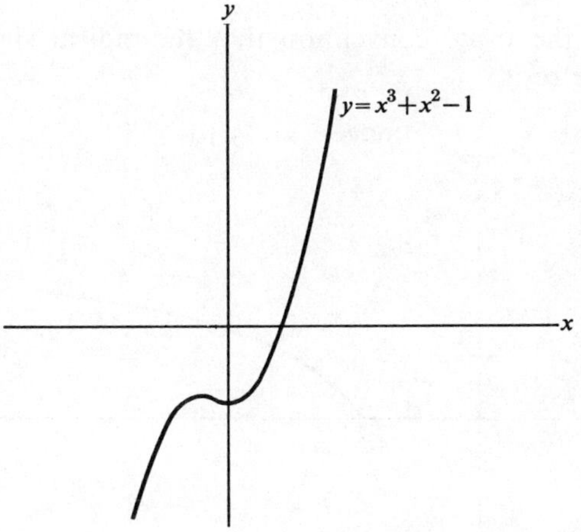

Fig. 4.5

2. If

$$y = f(x) = x^2 - 2,$$

each value of x determines a unique value of y. However, $x^2 \geq 0$ for all real x so

$$y \geq -2.$$

Thus image $f = \{y \mid y \geq -2\}$. Domain f = the real numbers.

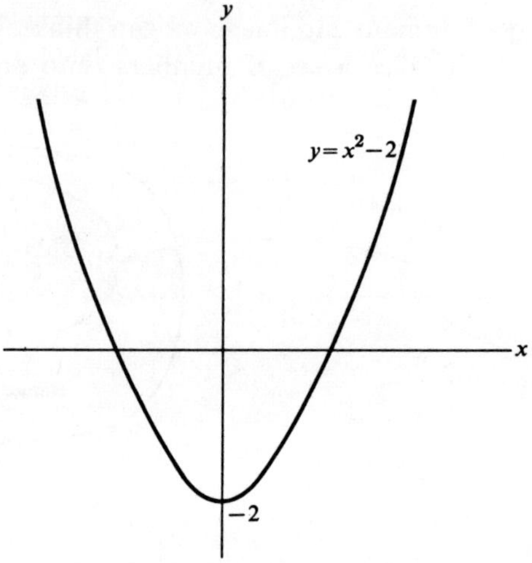

Fig. 4.6

3. If

$$y = f(x) = \sqrt{x - 2},$$

then f is a function of x and

$$\text{domain } f = \{x \mid x \geq 2\}$$

and, adopting the usual convention that the radical sign indicates the nonnegative square root,

$$\text{image } f = \{y \mid y \geq 0\} \qquad \text{(as seen in Fig. 4.7).}$$

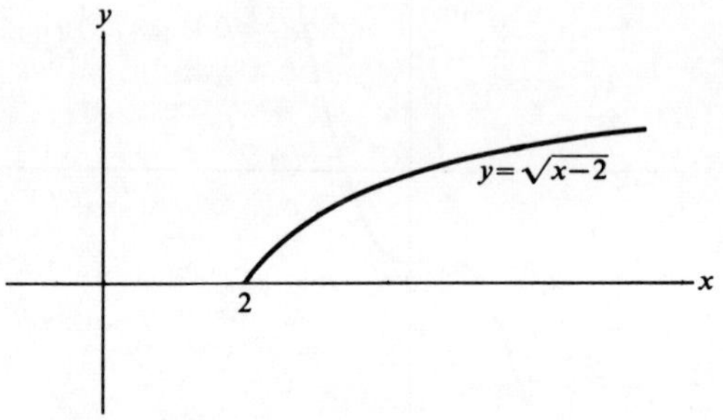

Fig. 4.7

4. If

$$y = \sqrt{4 - x^2},$$

the quantity under the radical sign must be nonnegative. Thus we must have $x^2 \leq 4$ or $-2 \leq x \leq 2$. Therefore, the domain of the function is $[-2, 2]$. On the other hand, y must also be nonnegative, but can never be greater than $\sqrt{4} = 2$ so that the image is $[0, 2]$ (Fig. 4.8).

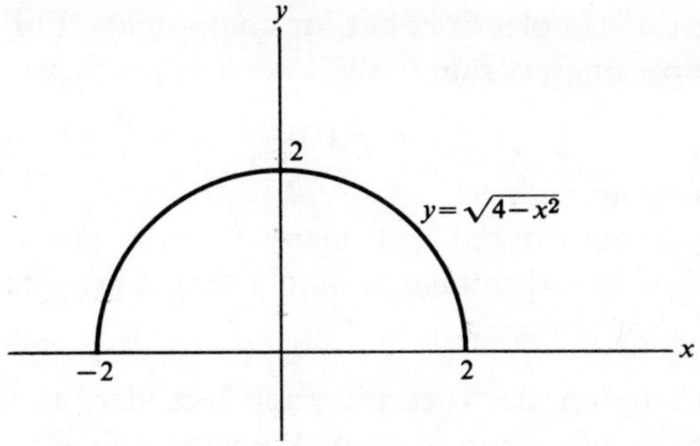

Fig. 4.8

5. If

$$y \geq x^2 + 1,$$

then y is not a function of x because, for example, if $x = 1$, then y could have any value ≥ 2. That is, a value of x does not determine a unique value of y. (Fig. 4.9).

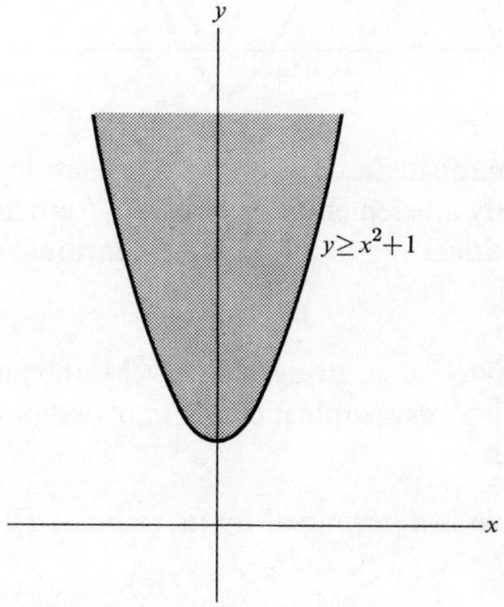

Fig. 4.9

6. It is important to be able to translate problems into functional notation. For example, if we are told that the interest is determined by the amount borrowed, we may write

$$I = f(N),$$

where I denotes interest and N the amount borrowed. The precise form which the function takes is then determined by additional information as to the rate and type of interest: simple, discount, or compound. For example, if we are told that the simple interest rate is eight percent per year we write

$$I = 0.08N$$

for the interest for one year.

In Chapter 2 we worked out many interest problems without using functional notation; its advantage is simply that it provides us with a general way of thinking about interest.

7. The graph drawn by an electrocardiograph recorder, as shown in Fig. 4.10, represents voltage as a function of time. We write $v = f(t)$. At each time t we may read off the unique value of the voltage at that time.

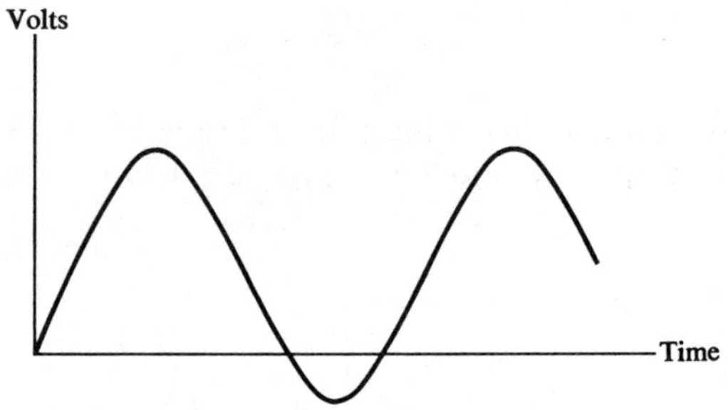

Fig. 4.10

Giving a mathematical description to a situation as we have done in the examples above is really an elementary exercise in "model building." The equation $v = f(t)$ is a mathematical model of an electrocardiograph. The equation

$$y = ax + b,$$

is a mathematical model of a straight line. Mathematical models can be very complicated, relying on very sophisticated knowledge of both mathematics and the area of application.

8. The area of a circle is determined by its radius. Thus we may write

$$A = f(r),$$

where A represents area and r radius. We know that

$$f(r) = \pi r^2.$$

Hence given a value for the radius, we determine uniquely the corresponding area.

Warning. This problem is an example of the determination of the domain by external conditions. Since a nonpositive radius has no meaning, we restrict the domain to the positive reals.

9. The retail price RP of merchandise depends on its wholesale price WP. We have, at least in businesses which expect to stay in operation,

$$RP \geq WP.$$

However, this relation does not determine the retail price, so we cannot accurately say that the retail price is a function of the wholesale price. The relation is shown in Fig. 4.11 where $x = WP$, $y = RP$, and the shaded area indicates possible values of RP.

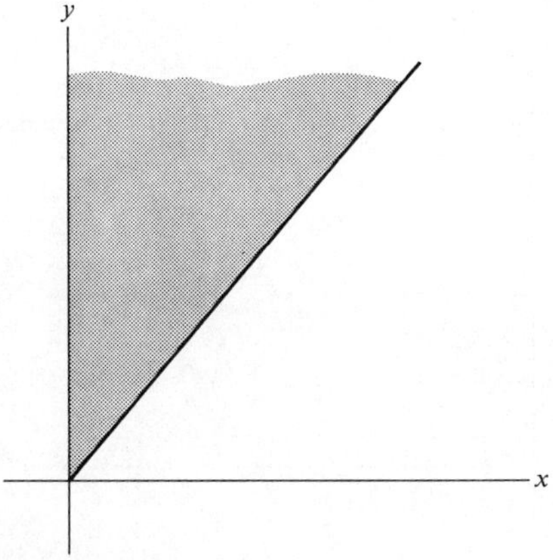

Fig. 4.11

10. Let $y = f(x)$ be defined by

x	y
0	3
-1	2
1	-1
2	3

Then domain $f = \{0, -1, 1, 2\}$ and image $f = \{3, 2, -1\}$. Note that

x	y
0	3
-1	2
0	-1
1	5

does not define y as a function of x since one value of x (namely 0) yields two different values for y.

11. Let us take an example of the mechanical embodiment of the idea of a function. In a potentiometer an electrical conductor is attached to a rotating shaft. The conductor makes contact with a strip of varying electrical resistance; the amount

of resistance encountered depends on the position of the shaft. Thus the resistance of the potentiometer is a function of the position. This principle is used in volume control knobs on TV sets; changes in resistance produce changes in loudness.

12. The postal rates for regular mail are 8 cents for each ounce or fraction thereof. The rate is a function of the weight; its graph is illustrated in Fig. 4.12. The domain is the positive real numbers; the image is the set of positive integer multiples of 8.

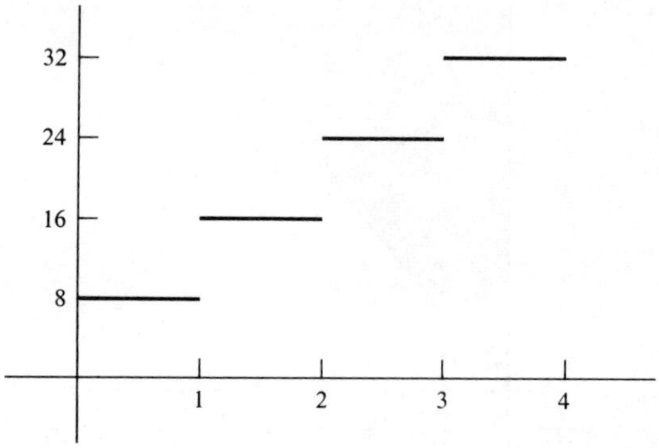

Fig. 4.12

Suppose $f(x)$ and $g(x)$ are real-valued functions of a real variable. Then we define a new function, the *composite function*, by

$$(f \circ g)(x) = f(g(x)).$$

That is, we substitute $g(x)$ for x in the expression for $f(x)$. For example, if

$$f(x) = x^2 - 4x + 1,$$
$$g(x) = 2x + 3,$$

then

$$\begin{aligned}(f \circ g)(x) &= f(g(x)) \\ &= (2x + 3)^2 - 4(2x + 3) + 1 \\ &= 4x^2 + 12x + 9 - 8x - 12 + 1 \\ &= 4x^2 + 4x - 2.\end{aligned}$$

Warning. It is not the case that

$$f \circ g = g \circ f.$$

For example, above we have

$$(f \circ g)(x) = 4x^2 + 4x - 2,$$

but

$$\begin{aligned}(g \circ f)(x) &= g(f(x)) \\ &= 2(x^2 - 4x + 1) + 3 \\ &= 2x^2 - 8x + 5.\end{aligned}$$

In addition, the reader is warned that $(f \circ g)(x)$ is defined if and only if $g(x)$ is in the domain of f. Thus if

$$f(x) = \frac{1}{x - 2}$$

and

$$g(x) = 2x^2 + 1,$$

2 is not in the domain of f so if $g(x) = 2$, that is, $x = \pm 1/\sqrt{2}$, $(f \circ g)(x)$ is not defined.

Examples

1. If

$$s(t) = \sqrt{3t^2 - 2t}$$

and

$$r(t) = 3t - 4,$$

then

$$\begin{aligned}(s \circ r)(t) &= s(r(t)) \\ &= \sqrt{3(3t - 4)^2 - 2(3t - 4)} \\ &= \sqrt{27t^2 - 72t + 48 - 6t + 8} \\ &= \sqrt{27t^2 - 78t + 56}.\end{aligned}$$

Moreover,

$$(r \circ s)(t) = r(s(t)) = 3\sqrt{2t^2 - 2t} - 4.$$

2. If

$$f(x) = \frac{2x^2 - 4x - 1}{2x}$$

and

$$g(x) = \frac{1}{x},$$

then

$$(f \circ g)(x) = f(g(x)) = \frac{(2/x^2) - 4/x - 1}{2/x} = \frac{x(2 - 4x - x^2)}{2x^2}$$

$$= \frac{2 - 4x - x^2}{2x}$$

and

$$(g \circ f)(x) = \frac{1}{(2x^2 - 4x - 1)/2x} = \frac{2x}{2x^2 - 4x - 1}.$$

PROBLEMS

1. In each of the following determine whether y is a function of x.

a) $y = 3x^3 - 2x + 1$.

b) $y = 5x$.

c)

x	y
1	3
2	5
3	4
4	-5

d)

x	y
0	0
1	2
-1	5
0	4
2	-1

e) $y^2 + x^2 = 4$.

f) $y = |x|$.

g) $y = [x]$, where $[x]$ is the greatest integer $\leq$ the real number x.

h) $y = \begin{cases} x^2 + 2 & \text{when} \quad x > 0 \\ 2x - 5 & \text{when} \quad x \leq 0. \end{cases}$

i) $y = \sqrt{x + 2}$.

j) $y = 1/x^2$.

k) $4x^2 - y^2 = 1$.

l) $y = x^2 - 4$.

m) $y = 5x - 9$.

n) $y^2 = 4x + 1$.

o) $y = \dfrac{1}{x - 2}$.

2. Classify each of the following functions as a polynomial function, a rational function, or neither.

a) $f(x) = x^2 - 5x + 2$.

b) $g(x) = (5x - 4)/(2x - 1)$.

c) $s(t) = \sqrt{t^2 - 2}$.

d)

x	$f(x)$
6	10
5	14
4	2
3	3
-2	6
1	0

e) $f(x) = (x^2 + 2)/(5x + 1)$.

f) $f(x) = x^3 - 1$.

g) $f(x) = 2^x$.

3. Determine the domain and image of each of the *functions* in Problem 1.

4. The cost of domestic airmail is 11 cents per ounce or fraction of an ounce. Is the cost c a function of the weight w? If so, graph the function.

5. Does the following table define a function? If so, what is a function of what?

Deaths from heart disease per 100,000	
65	New York
42	District of Columbia
84	Kansas
79	Illinois
69	California

6. In the following system of equations, C is the money value of aggregate consumption, I is the money value of aggregate investment and a, b, u, and v are constants.

$$Y = C + I,$$
$$C = a + bY,$$
$$I = u + vY.$$

Solve this system to get Y, C, and I as functions of the constants only.

7. If $f(x)$ and $g(x)$ are as given, form the composite functions $f \circ g$ and $g \circ f$ and discuss their domains,

a) $f(x) = x^2 - 1$, $g(x) = 3x + 2$.
b) $f(x) = \sqrt{x + 2}$, $g(x) = x^4$.
c) $f(x) = (2x^2 + 1)/(x - 5)$, $g(x) = 2x$.
d) $f(x) = \sqrt{x^2 - 4}$, $g(x) = 5x - 4$.
e) $f(x) = 2x + 1$, $g(x) = \sqrt{x + 1}$.
f) $f(x) = 5 - x^2$, $g(x) = 1/2x$.
g) $f(x) = \sqrt{2x - 1}$, $g(x) = 3x^2 + 1$.
h) $f(x) = \dfrac{1}{2x - 3}$, $g(x) = \sqrt{5x}$.
i) $f(x) = |x - 1|$, $g(x) = x^2 + 2$.

8. Automobiles depreciate most rapidly in their first year. Suppose a \$5000 model is worth \$3500 at the end of a year and loses value at the rate of \$500 a year for six more years and for two more years after that at the rate of \$100 a year. Is the value a function of the time the auto is owned? If so, graph this function. Is this straight line depreciation?

4.2 THE CONCEPT OF LIMIT

Consider the behavior of the function

$$f(x) = x^2$$

as $|x|$ becomes very small. Since whenever

$$0 \leq |x| < 1, \qquad \text{we have} \qquad 0 \leq x^2 \leq |x|,$$

$f(x)$ also becomes very small (Fig. 4.13).

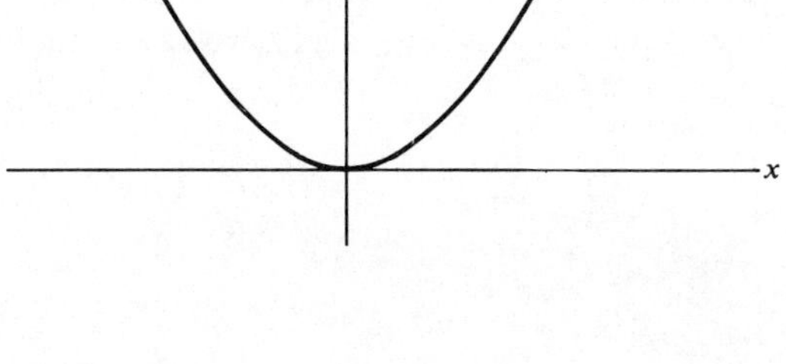

Fig. 4.13

For example, if we want $f(x) = x^2 < \frac{1}{4}$, we need only pick values of x such that $|x| < \frac{1}{2}$. We express this relation as

$$f(x) \to 0 \quad \text{as} \quad x \to 0.$$

Similarly, for values of x near 1, the values of

$$f(x) = 2x^2 + x + 4,$$

are near 7. Moreover, to get $f(x)$ nearer 7 we need only get x nearer to 1. We need a precise formulation of this "nearness." We recall from Chapter 1 that a neighborhood of a real number is any open interval (a, b) containing that number and that the diameter of (a, b) is $b - a$.

Then, in the second example, what we want to say is that if we pick a neighborhood of 7 of arbitrarily small diameter, it is possible to find a whole neighborhood of 1 which is carried by the mapping f into the neighborhood of 7 (Fig. 4.14).

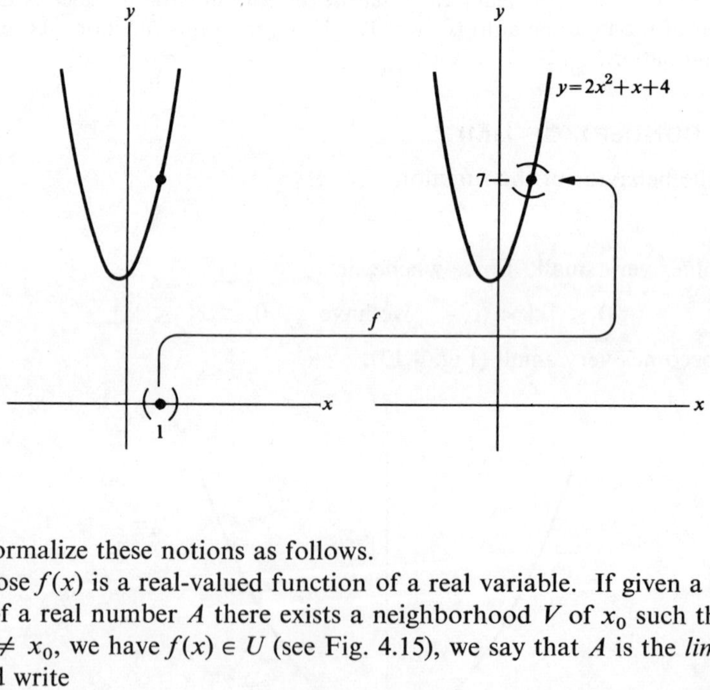

Fig. 4.14

We formalize these notions as follows.

Suppose $f(x)$ is a real-valued function of a real variable. If given a neighborhood U of a real number A there exists a neighborhood V of x_0 such that for all $x \in V$, $x \neq x_0$, we have $f(x) \in U$ (see Fig. 4.15), we say that A is the *limit of $f(x)$ at x_0*, and write

$$\lim_{x \to x_0} f(x) = A.$$

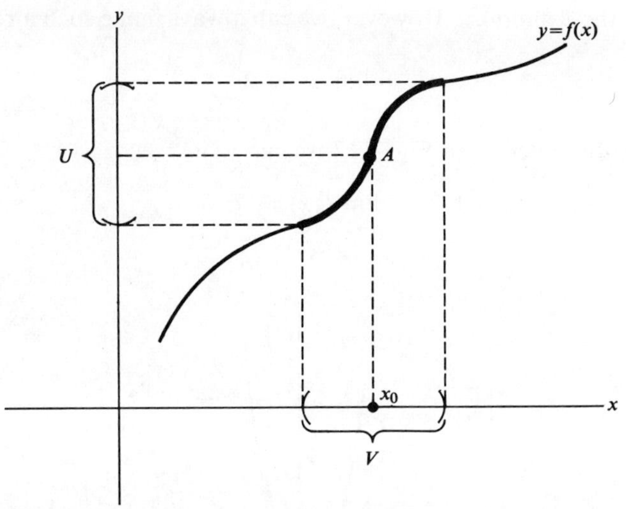

Fig. 4.15

In the first example above we note that if we have a small neighborhood about 0 on the y-axis (Fig. 4.16) then a small neighborhood can be drawn about 0 on the x-axis so that all values of $f(x)$ for the x in this neighborhood lie in the neighborhood on the y-axis. Decreasing the diameter of the original neighborhood requires that the diameter of the neighborhood on the x-axis also be decreased to meet the

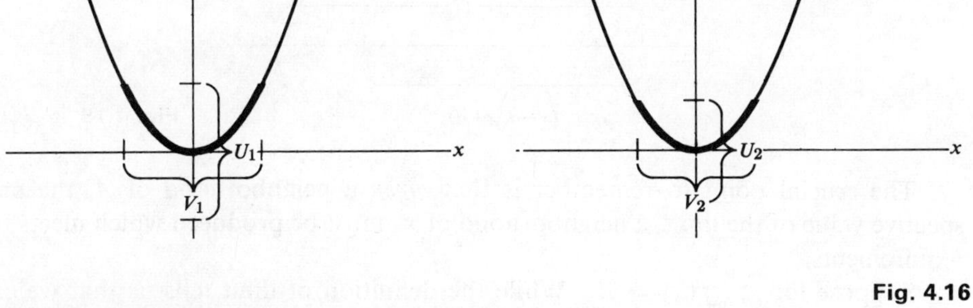

Fig. 4.16

conditions of the definition. However, we can always make such a reduction so we write

$$\lim_{x \to 0} x^2 = 0.$$

Similarly, for the function in Fig. 4.17 we can write

$$\lim_{x \to 1} f(x) = 7.$$

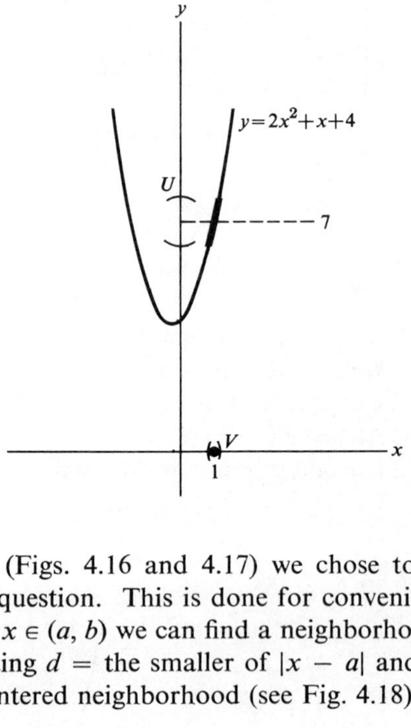

Fig. 4.17

Note: In the graphs (Figs. 4.16 and 4.17) we chose to use neighborhoods centered at the points in question. This is done for convenience, but there is no loss of generality, since if $x \in (a, b)$ we can find a neighborhood centered at x and contained in (a, b) by letting $d =$ the smaller of $|x - a|$ and $|x - b|$ and taking $(x - d, x + d)$ as our centered neighborhood (see Fig. 4.18).

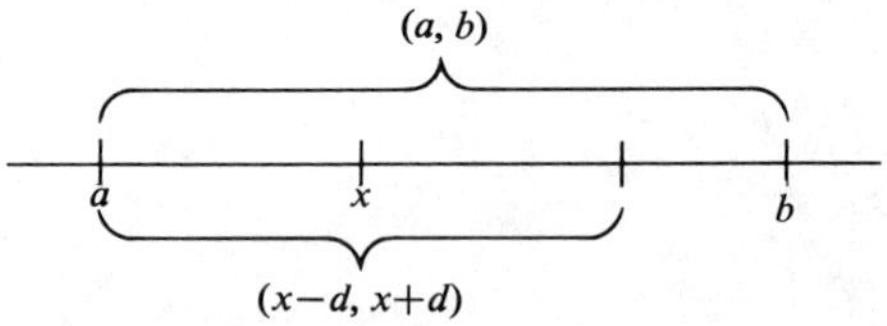

Fig. 4.18

The crucial point to remember is that *given* a neighborhood of A, the prospective value of the limit, a neighborhood of x_0 must be produced which meets the requirements.

Suppose $\lim_{x \to x_0} f(x) = A$. While the definition of limit tells us that we can get values of $f(x)$ arbitrarily close to A by picking values of x sufficiently close to

x_0, it does not tell us how close "sufficiently close" is; that is, having been given a neighborhood U of A, we are guaranteed that a neighborhood V of x_0 exists such that $x \in V$ implies that $f(x) \in U$, but we are not told which neighborhoods qualify as V. That must be worked out in a particular problem if the information is required. In the case of $f(x) = x^2$ it was easy—if the diameter of the given neighborhood is ε we pick as V a neighborhood of diameter $\sqrt{\varepsilon}$ or less. It isn't always so easy, but on the other hand we rarely need such information.

Warning. Limits do not always exist. Consider the following function:

$$f(x) = \begin{cases} x + 2, & x \geq 0, \\ -x, & x < 0. \end{cases}$$

No matter how small a neighborhood of 0 we pick there will be values of x in it which give values of $f(x) < 0$ and $f(x) > 2$. Thus

$$\lim_{x\to 0} f(x) \neq 0 \qquad \text{and} \qquad \lim_{x\to 0} f(x) \neq 2,$$

for if we pick a neighborhood of either 0 or 2 which excludes the other, values of x close to 0 give us values of $f(x)$ outside the neighborhood. Similarly, if we conjecture that any other real number A is such that

$$\lim_{x\to 0} f(x) = A,$$

we need only choose a neighborhood of A so small as to exclude at least one of 0 and 2 to see that our conjecture is false (Fig. 4.19).

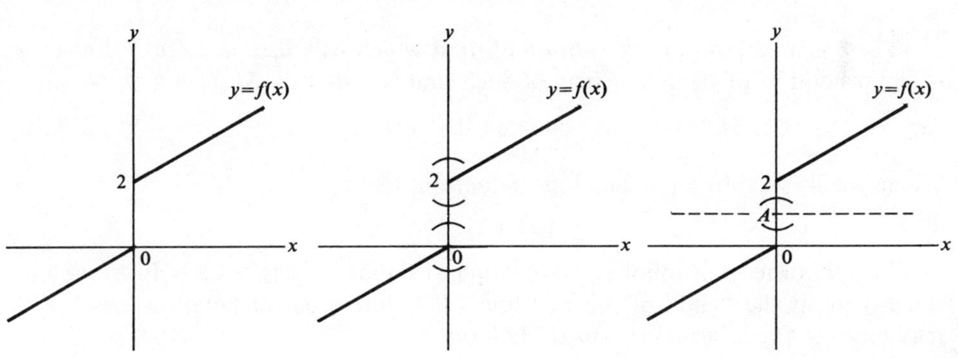

Fig. 4.19

We note that in the definition of limit we excluded $x = x_0$ from the condition. This is so that functions such as the following can be dealt with in a reasonable manner.

Let

$$f(x) = x^2 + 3x + 2, \qquad x \neq 1,$$
$$f(1) = 0.$$

Then

$$\lim_{x \to 1} f(x) = 6$$

since no matter how small a neighborhood of 6 we pick we can choose a neighborhood of 1 such that for all x in this neighborhood, excluding 1 itself, $f(x)$ is in the original neighborhood of 6 (Fig. 4.20).

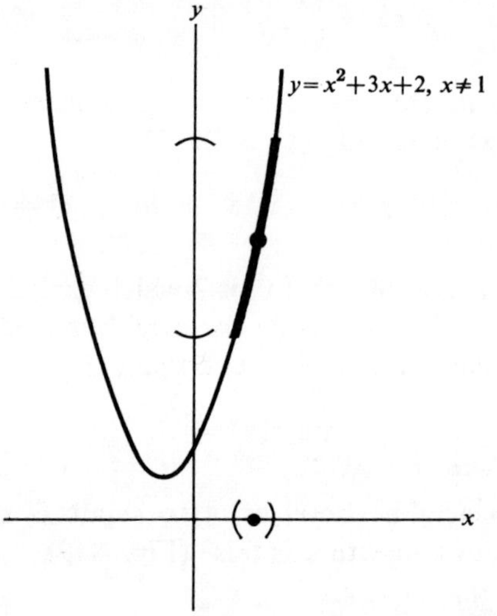

Fig. 4.20

There is a variation in the notion of limit which will also be useful. If given a neighborhood U of A, there exists M such that for all $x > M$, $f(x) \in U$, we say

$$\lim_{x \to +\infty} f(x) = A.$$

We are really adapting our previous definition, using

$$\{x \mid x > M\},$$

as a "neighborhood of infinity." We emphasize that $+\infty$ is not a point stuck out somewhere at the "end" of the real line. It is just a useful notation; $x \to +\infty$ really means "as x becomes arbitrarily large."

Analogously,

$$\lim_{x \to -\infty} f(x) = A$$

if given a neighborhood U of A, there is an M such that for all $x < M$, $f(x) \in U$.

In the following examples the results we state without proof seem clear from the graphs of the functions (see Figs. 4.21 through 4.26).

Examples

1. $\lim_{x\to+\infty} \frac{1}{x} = 0 = \lim_{x\to-\infty} \frac{1}{x}$.

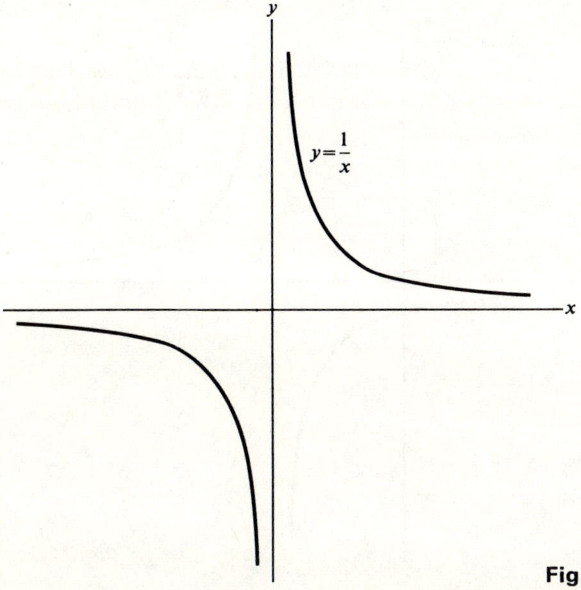

Fig. 4.21

2. $\lim_{x\to+\infty} 2x$ fails to exist; that is, as x becomes arbitrarily large so does $2x$.

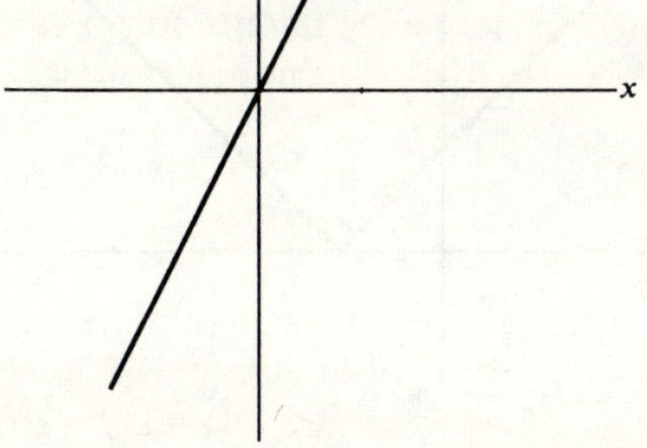

Fig. 4.22

3. $\lim_{x \to 2} 1/(x - 2)$ fails to exist since as $x \to 2$, $1/(x - 2)$ becomes arbitrarily large in absolute value.

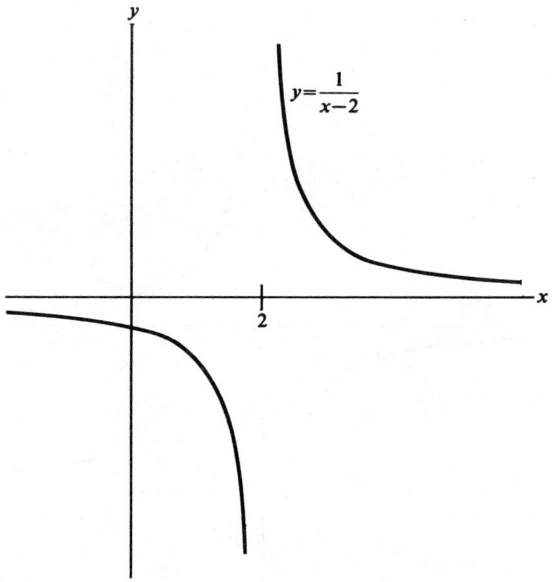

Fig. 4.23

4. $\lim_{x \to -2} \sqrt{x^2 - 2x + 1} = 3.$

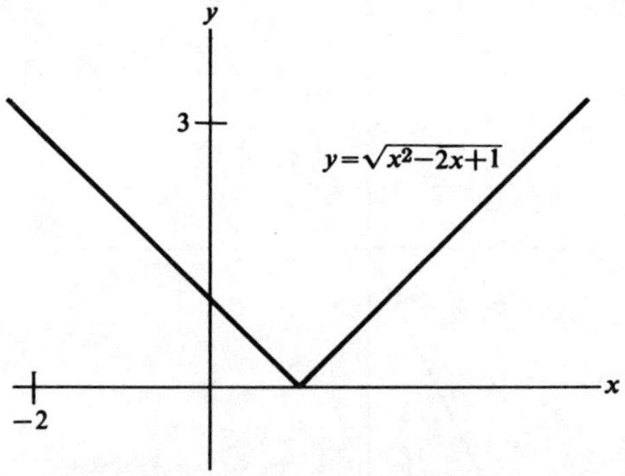

Fig. 4.24

5. Let $f(x)$ be defined by $f(x) = \begin{cases} 3x^2 + 2x + 1, & x \geq 0, \\ -4, & x < 0. \end{cases}$

$\text{Lim}_{x \to 0} f(x)$ fails to exist.

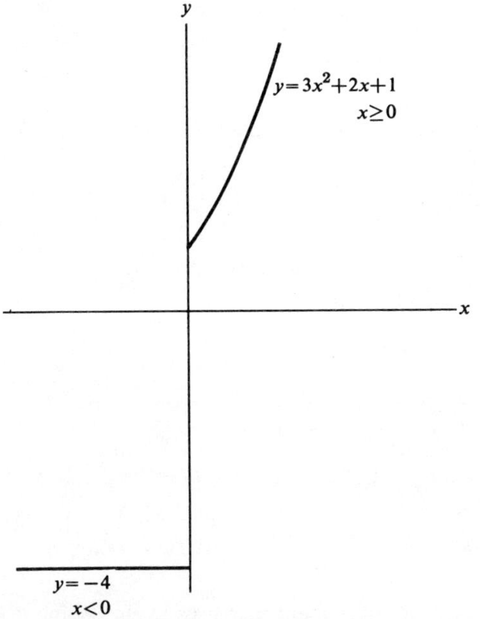

Fig. 4.25

6. If $f(x) = 3x^2 + 2x + 1, \quad x \neq 0,$

$f(0) = 0,$

then $\lim_{x \to 0} f(x) = 1$ since for all x, $x \neq 0$, in a sufficiently small neighborhood of 0, $f(x)$ is in an arbitrarily small neighborhood of 1.

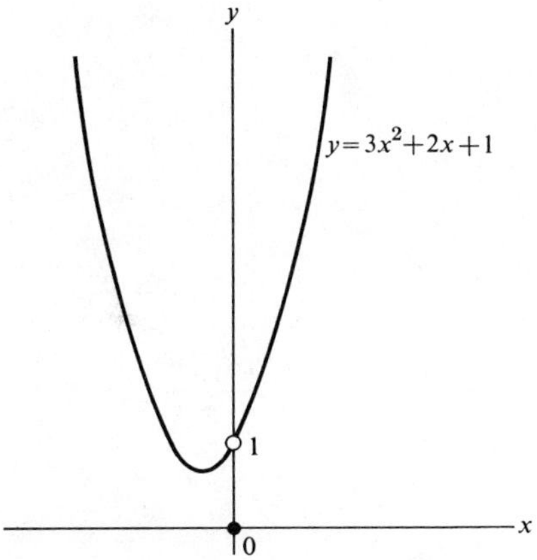

Fig. 4.26

In fact, in all "reasonable" functions: polynomial functions, rational functions, algebraic functions; the limit as $x \to x_0$ is obtained simply by substituting the value of the function at x_0. In Section 4, we shall give a name to this class of well-behaved functions.

PROBLEMS

1. For each of the following functions, determine $\lim_{x\to 0} f(x)$ if it exists and tell whether it is equal to $f(0)$.

   a) $f(x) = 2x^2 - 5x + 1.$

   b) $f(x) = \dfrac{2x + 1}{x}.$

   c) $f(x) = \text{sign } x = \begin{cases} 1 & \text{if } x \geq 0 \\ -1 & \text{if } x < 0 \end{cases}.$

   d) $f(x) = \begin{cases} 2x + 1 & \text{if } x \geq 0 \\ x & \text{if } x < 0 \end{cases}.$

   e) $f(x) = \sqrt{5x^2 + 4x + 5}.$

   f) $f(x) = \sqrt{x^2 - 4x - 2}.$

   g) $f(x) = \begin{cases} x^2 + 5x + 1 & x \neq 0 \\ 5 & x = 0 \end{cases}.$

2. In each of the following, what can be said about $\lim_{x\to +\infty} f(x)$?

   a) $f(x) = 2x^3 - 4x + 1.$

   b) $f(x) = 1/(x - 1).$

   c) $f(x) = \sqrt{x^2 + 2}.$

   d) $f(x) = 2/\sqrt{x}.$

   e) $f(x) = 2 - 5/x^2.$

3. The percentage return on investment increases as the selling price of a certain item is increased, but the rate of increase slackens off, never reaching more than 12 percent regardless of the selling price. Express this information in limit notation and sketch the graph of the percentage return as a function of the selling price.

4. Find, if it exists, each of the following limits.

   a) $\lim_{x\to 5} x^2 - 5x + 1.$

   b) $\lim_{x\to -2} \dfrac{2x^2}{5x}$

   c) $\lim_{x\to -\infty} \dfrac{1}{x - 1}.$

   d) $\lim_{x\to \sqrt{2}} 2x^2 - 2x + 3.$

   e) $\lim_{x\to -2} \sqrt{5x^2 + 2x - 1}.$

   f) $\lim_{x\to 1} [x]$ (greatest integer function).

   g) $\lim_{x\to a} x^2 + 4x - 5.$

   h) $\lim_{x\to 2} (x^2 - 5x + 3).$

   i) $\lim_{x\to 0} \dfrac{1}{x}.$

   j) $\lim_{x\to -1} \dfrac{2x + 1}{4x + 1}.$

   k) $\lim_{x\to 2} \sqrt{x^2 + 4x}.$

   l) $\lim_{x\to 2} \dfrac{3x + 1}{x - 2}.$

   m) $\lim_{x\to +\infty} (2x + 1).$

   n) $\lim_{x\to +\infty} \dfrac{1}{3x}.$

5. If $f(x) = 3x^2 + 4x - 1$, find $\lim\limits_{h\to 0} \dfrac{f(3 + h) - f(3)}{h}$.

6. In each of the following cases tell whether $\lim\limits_{x\to 1} f(x) = f(1)$.

a) $f(x) = \dfrac{3x + 1}{x}$.

b) $f(x) = x^2 + 2x + 1,\ x \neq 1$
$f(1) = 2$.

c) $f(x) = \sqrt{x + 1}$.

d) $f(x) = 2x - 1,\ x \neq 1$
$f(1) = 0$.

4.3 PROPERTIES OF LIMITS

Suppose

$$\lim_{x\to x_0} f(x) = A \qquad \text{and} \qquad \lim_{x\to x_0} g(x) = B.$$

What can be said about

$$\lim_{x\to x_0} (f(x) + g(x))?$$

We observe the graphs in Fig. 4.27.

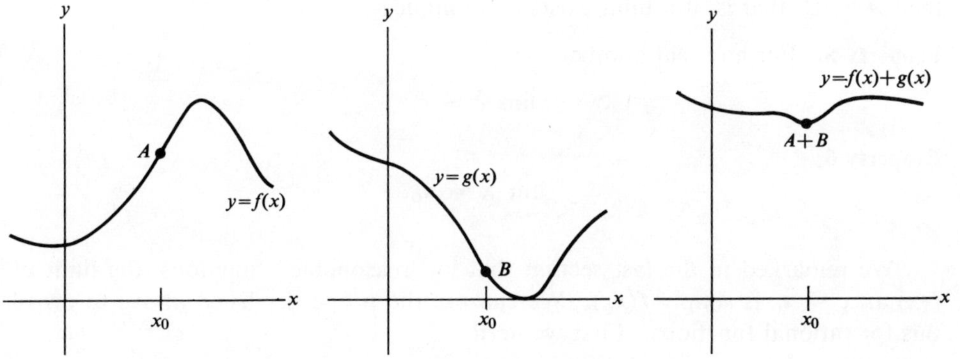

Fig. 4.27

It seems from the graph of the combined functions that

$$\lim_{x\to x_0} (f(x) + g(x)) = A + B.$$

We state the following properties without proof (for proofs see Appendix A):

Property 1. If

$$\lim_{x\to x_0} f(x) = A \qquad \text{and} \qquad \lim_{x\to x_0} g(x) = B$$

then

$$\lim_{x\to x_0} (f(x) + g(x)) = A + B.$$

Property 2. If

$$\lim_{x\to x_0} f(x) = A \qquad \text{and} \qquad \lim_{x\to x_0} g(x) = B,$$

then

$$\lim_{x\to x_0} f(x)g(x) = AB.$$

In particular, if c is a real number, then

$$\lim_{x\to x_0} cf(x) = c \lim_{x\to x_0} f(x).$$

Property 3. If

$$\lim_{x\to x_0} f(x) = A \qquad \text{and} \qquad \lim_{x\to x_0} g(x) = B \neq 0,$$

then

$$\lim_{x\to x_0} \frac{f(x)}{g(x)} = A/B.$$

Property 4. If

$$\lim_{x\to x_0} f(x) = A \qquad \text{and} \qquad \lim_{x\to x_0} f(x) = B,$$

then $A = B$; that is, if a limit exists it is unique.

Property 5. For any real number c

$$\lim_{x\to x_0} c = c.$$

Property 6

$$\lim_{x\to x_0} x = x_0.$$

We remarked in the last section that in "reasonable" functions, the limit of $f(x)$ as $x \to x_0$ is simply $f(x_0)$. We can use the properties listed above to prove this for rational functions. First we need

Theorem 1

$$\lim_{x\to x_0} x^n = x_0^n.$$

Proof. By Property 6 this statement is true for $n = 1$. Suppose that it is true for n, i.e., suppose that

$$\lim_{x\to x_0} x^n = x_0^n.$$

By Property 2,

$$\lim_{x\to x_0} x^{n+1} = \left(\lim_{x\to x_0} x^n\right)\left(\lim_{x\to x_0} x\right)$$

and by the induction assumption and the case $n = 1$ the right-hand side is equal to $x_0^n x_0 = x_0^{n+1}$, as desired.

Theorem 2. If

$$p(x) = a_n x^n + \cdots + a_1 x + a_0,$$

then

$$\lim_{x \to x_0} p(x) = a_n x_0^n + \cdots + a_1 x_0 + a_0.$$

Proof

$$\begin{aligned}
\lim_{x \to x_0} p(x) &= \lim_{x \to x_0} (a_n x^n + \cdots + a_1 x + a_0) \\
&= \lim_{x \to x_0} a_n x^n + \cdots + \lim_{x \to x_0} a_1 x + \lim_{x \to x_0} a_0 \qquad \text{(Property 1)} \\
&= \lim_{x \to x_0} a_n x^n + \cdots + \lim_{x \to x_0} a_1 x + a_0 \qquad \text{(Property 5)} \\
&= a_n \lim_{x \to x_0} x^n + \cdots + a_1 \lim_{x \to x_0} x + a_0 \qquad \text{(Property 2)} \\
&= a_n x_0^n + \cdots + a_1 x_0 + a_0 \qquad \text{(Theorem 1).}
\end{aligned}$$

We use Theorem 2 and Property 3 to find the limit of a rational function.

Theorem 3. If

$$\frac{p(x)}{q(x)} = \frac{a_n x^n + \cdots + a_1 x + a_0}{b_m x^m + \cdots + b_1 x + b_0},$$

then

$$\lim_{x \to x_0} \frac{p(x)}{q(x)} = \frac{p(x_0)}{q(x_0)} \qquad \text{if } q(x_0) \neq 0.$$

Examples

1.

$$\begin{aligned}
&\lim_{x \to 2} (3x^2 + 5x - 5)(2x^3 - 2x^2 - 1) \\
&\quad = \Big(\lim_{x \to 2} (3x^2 + 5x - 5)\Big)\Big(\lim_{x \to 2} (2x^3 - 2x^2 - 1)\Big) \\
&\quad = \Big(\lim_{x \to 2} 3x^2 + \lim_{x \to 2} 5x + \lim_{x \to 2} (-5)\Big)\Big(\lim_{x \to 2} 2x^3 + \lim_{x \to 2} (-2x^2) + \lim_{x \to 2} (-1)\Big) \\
&\quad = (12 + 10 - 5)(16 - 8 - 1) \\
&\quad = 17 \cdot 7 \\
&\quad = 119.
\end{aligned}$$

2.

$$\lim_{x \to -1} \frac{5x^2 - 2x + 1}{x^3 - 5x} = \frac{\lim_{x \to -1} (5x^2 - 2x + 1)}{\lim_{x \to -1} (x^3 - 5x)} = \frac{8}{4} = 2.$$

PROBLEMS

1. Find the following limits:

a) $\lim_{x\to 3} -2x$.
b) $\lim_{x\to 0} -2x$.
c) $\lim_{t\to 5} \dfrac{t^2+3t-1}{t+1}$.

d) $\lim_{y\to 4} (y^2-4y+1)$.
e) $\lim_{x\to 7} \dfrac{7x^2-7x}{2x+1}$.
f) $\lim_{x\to b} (3x^3-4x+1)$.

g) $\lim_{x\to -1} \dfrac{3x^3-4x+1}{2x^2-4}$.
h) $\lim_{t\to 0} \dfrac{t^4-5t^3-5}{5t^5+4}$.
i) $\lim_{x\to x_0} \dfrac{4x^3+4x^2-7}{x^2-x+1}$.

2. If $\lim_{x\to 2} f(x) = -6$ and $\lim_{x\to 2} g(x) = 1$, find $\lim_{x\to 2}(f(x)+g(x))$, $\lim_{x\to 2} f(x)g(x)$, $\lim_{x\to 2} f(x)/g(x)$.
3. Give examples of functions $f(x)$ and $g(x)$ such that $\lim_{x\to 0} f(x)$ and $\lim_{x\to 0} g(x)$ do not exist, but $\lim_{x\to 0}(f(x)+g(x))$ does exist.
4. Give examples of functions $f(x)$ and $g(x)$ such that at least one of $\lim_{x\to 0} f(x)$ and $\lim_{x\to 0} g(x)$ does not exist, but $\lim_{x\to 0} f(x)g(x)$ does exist.

4.4 CONTINUOUS FUNCTIONS

Recall the example of Section 2:

$$f(x) = 3x^2+2x+1, \qquad x \neq 0,$$
$$f(0) = 0.$$

We observe that although $\lim_{x\to 0} f(x)$ exists it is not equal to $f(0)$. This is indicated by the "gap" in the graph (see Fig. 4.26). Similarly, we see from the graph of

$$y = \frac{1}{x}$$

(Fig. 4.21) that there is a "gap" at $x = 0$. We formulate this notion precisely and give a name to the property which these functions lack.

A function $f(x)$ is *continuous* at x_0 if

$$\lim_{x\to x_0} f(x) = f(x_0),$$

that is, if for any neighborhood U of $f(x_0)$ there is a neighborhood V of x_0 such that for all $x \in V$, $f(x) \in U$. Note that this definition implies

1) $\lim_{x\to x_0} f(x)$ exists;
2) $f(x_0)$ exists, that is, x_0 is in the domain of f;
3) they are equal.

Intuitively, the notion of continuity is that of unbrokenness of the graph of the function.

Examples

1. $$f(x) = 3x^2 + 2x + 1,$$

is continuous for all real x (see Fig. 4.28).

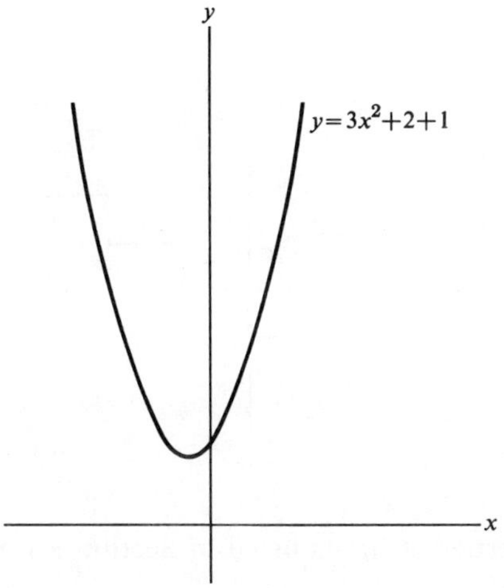

Fig. 4.28

2. $$f(x) = \frac{1}{x^2}$$

is continuous whenever $x \neq 0$ (Fig. 4.29).

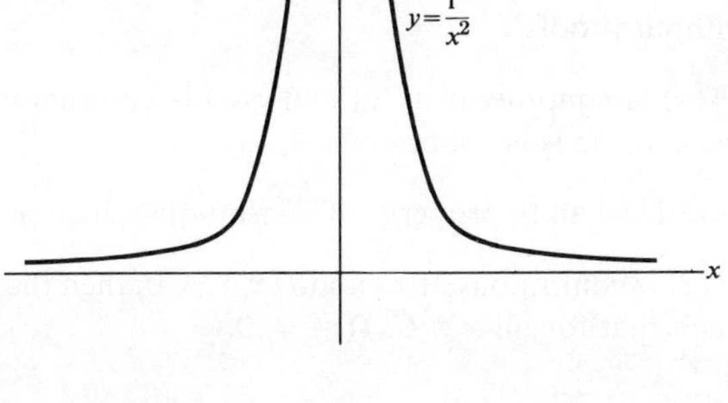

Fig. 4.29

3. $f(x) = [x]$ = integer part of x, that is, the largest integer less than or equal to x, is continuous except at the integers (Fig. 4.30).

If a function is continuous for all real values of x, we say that it is *continuous*. If $f(x)$ is continuous at every $x \in (a, b)$, $f(x)$ is continuous *on* (a, b).

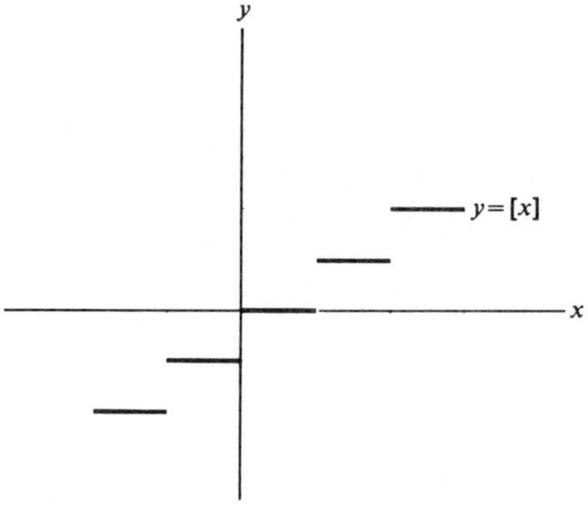

Fig. 4.30

From the properties of limits listed in Section 3 we obtain at once:

Theorem 3. If $f(x)$ and $g(x)$ are continuous at x_0, then

1) $f(x) + g(x)$ is continuous at x_0.
2) $f(x)g(x)$ is continuous at x_0; in particular, if c is a real number, $cf(x)$ is continuous at x_0.
3) $f(x)/g(x)$ is continuous at x_0 if $g(x_0) \neq 0$.

In particular, if $p(x)$ and $q(x)$ are polynomial functions, $q(x_0) \neq 0$,we have

$$\lim_{x \to x_0} \frac{p(x)}{q(x)} = \frac{p(x_0)}{q(x_0)},$$

so that the rational function $p(x)/q(x)$ is continuous at x_0 if $q(x_0) \neq 0$. Thus we have a large class of continuous functions on hand: constants, polynomial functions, rational functions.

We state without proof:

Theorem 4. If $f(x)$ is continuous at x_0 and $g(x)$ is continuous at $f(x_0)$, then the composite function $g(f(x))$ is continuous at x_0.

We prove another handy property of continuous functions.

Theorem 5. If $f(x)$ is continuous at x_0 and $f(x_0) \neq 0$, then there exists a neighborhood V of x_0 such that for all $x \in U$, $f(x) \neq 0$.

Proof. Suppose $f(x_0) = \varepsilon > 0$. (The proof is similar if $f(x_0) < 0$.) Since $f(x)$ is continuous at x_0,

$$\lim_{x \to x_0} f(x) = f(x_0).$$

But $(0, 2\varepsilon)$ is a neighborhood of $f(x_0)$, so from the definition of limit, there is a neighborhood V of x_0 such that for all $x \in V$, $f(x) \in (0, 2\varepsilon)$. But this means that for all $x \in V$, $f(x) > 0$ and in particular $f(x) \neq 0$, as required (Fig. 4.31).

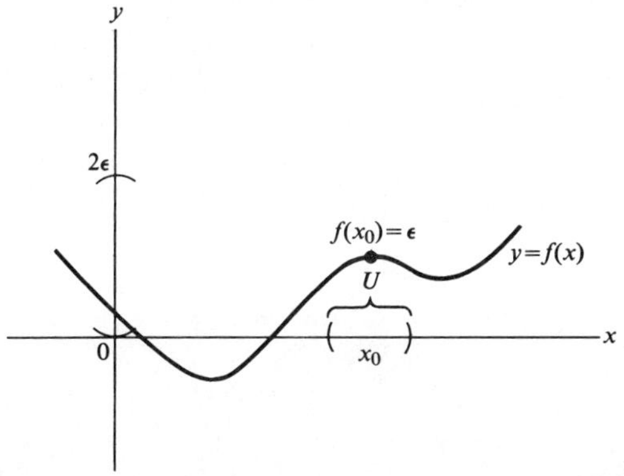

Fig. 4.31

We state without proof a very useful theorem.

Intermediate Value Theorem. If a and b are real numbers, $a < b$, such that there are real numbers x_1 and x_2 in the domain of a continuous function $f(x)$ with $f(x_1) = a$, $f(x_2) = b$, and if $c \in (a, b)$, then there is an x_0 between x_1 and x_2 with

$$f(x_0) = c.$$

Intuitively this theorem means that to get from a to b the graph must pass through all intermediate points (Fig. 4.32), i.e., the graph of a continuous function is unbroken. We made use of this theorem in Chapter 2 in locating roots of polynomials. If

$$f(x) = x^3 + x^2 - 4x + 1,$$

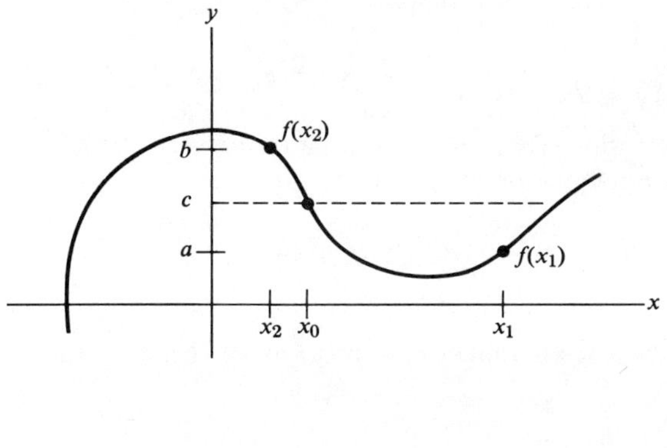

Fig. 4.32

then $f(0) = 1$ and $f(1) = -1$ so that by the intermediate value theorem, there is a $c \in (0, 1)$ such that

$$f(c) = 0.$$

Thus c is a root of the polynomial between 0 and 1.

PROBLEMS

1. Is each of the following functions continuous?

a) $f(x) = 5x^3 - 4x + 1$.

b) $f(x) = \begin{cases} 3x^2 & \text{if } x \geq 0 \\ -3x^2 & \text{if } x < 0 \end{cases}$.

c) $f(x) = 2x^3/(x - 1)$.

d) $f(x) = |x|$.

e) $f(x) = \dfrac{1}{x^2 + 1}$.

f) $f(x) = x/2$.

g) $f(x) = 2/x$.

h) $f(x) = \dfrac{x + 2}{x^2 + 1}$.

i) $f(x) = \dfrac{|x|}{x + 1}$.

j) $f(x) = 3x^3 - 5x + 1$.

k) $f(x) = |x - 1|$.

2. For what values of x does each of the following functions fail to be continuous?

a) $f(x) = 6x^2 + 5x - 4$.

b) $f(x) = [x]$ (greatest integer function).

c) $f(x) = \dfrac{3x^2 - 5x}{x + 1}$.

d) $f(x) = \begin{cases} x^2 - 1 & \text{if } x \geq 0 \\ -x^2 + 1 & \text{if } x < 0 \end{cases}$.

3. Is the function defined by

$$f(x) = \begin{cases} \dfrac{x^2 - 1}{x - 1} & x \neq 1 \\ 2 & x = 1 \end{cases}.$$

continuous at $x = 1$?

4. If $y = f(x)$ is positive for $x = 1$ and negative for $x = 2$ and is continuous, why is it the case that the equation $f(x) = 0$ has at least one solution between $x = 1$ and $x = 2$?

5. If

$$f(x) = \frac{x^2 - 2x - 3}{x + 1}, \qquad x \neq 1,$$

is it possible to assign a value to $f(-1)$ to make $f(x)$ continuous at $x = -1$?

6. What does the intermediate value theorem tell us about the possible number of real roots of a cubic polynomial?
7. Is it possible to define a function which is continuous and which agrees with $f(x)$ everywhere that $f(x)$ is continuous, if $f(x)$ is

a) $1/x^2$.

b) $2x - 1$, $x > 0$,
x^2, $x \leq 0$.

c) $2x - 1$, $x \neq 0$,
0, $x = 0$.

d) $\dfrac{|x|}{x}$.

e) $\dfrac{2x}{x}$.

5 / THE DERIVATIVE

The two central notions of calculus are the derivative and the integral. In fact, it was the fashion for many years to think of all of calculus as being divided into two parts—differential and integral—each dealing with one of these. However, it turns out that basic to both these notions is the single concept of limit, which we introduced in the last chapter. Moreover, the fundamental theorem of calculus, which we discuss in Chapter 7, connects the processes of differentiation and integration.

Calculus was invented in the late seventeenth century independently by Newton in England and Leibniz in Germany. Newton used the idea of velocity and Leibniz that of a tangent to develop the derivative. Calculus was put on a firm logical foundation many years later, mainly by Cauchy and Weierstrass.

In this chapter we use an intuitive geometric approach to introduce the notion of derivative and then use limits for the formal definition. We derive the properties of the derivative and the rules of differentiation for some simple functions.

5.1 RATES OF CHANGE, SLOPE

In Chapter 2 we discussed the slope of a straight line. If

$$y = mx + b,$$

then the slope is

$$m = \frac{y_2 - y_1}{x_2 - x_1},$$

where (x_1, y_1) and (x_2, y_2) are any two points on the line (Fig. 5.1). This ratio is constant and not dependent upon the selection of the pair of points. We can refer to the slope as the "rate of change of y with respect to x" since it tells us how much y changes as x changes by a specified amount.

Now let us consider an equation where y is not a linear function of x, say

$$y = f(x) = 3x^2 + 2.$$

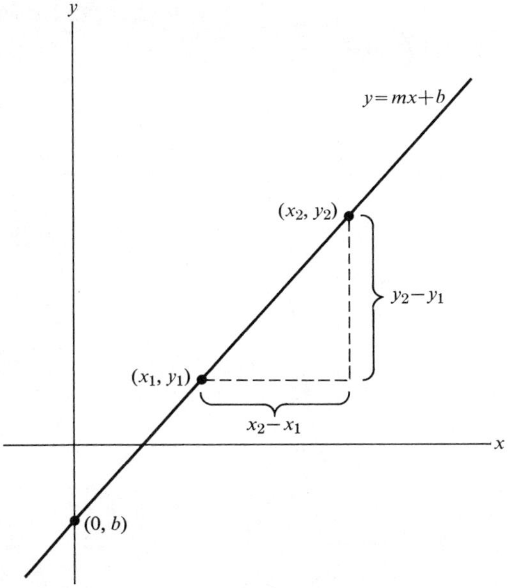

Fig. 5.1

As x increases from 0 to 1, y increases from 2 to 5, but as x increases from 1 to 2, y increases from 5 to 14. Thus in one case

$$\frac{y_2 - y_1}{x_2 - x_1} = \frac{f(x_2) - f(x_1)}{x_2 - x_1} = 3$$

and in the other

$$\frac{y_2 - y_1}{x_2 - x_1} = \frac{f(x_2) - f(x_1)}{x_2 - x_1} = 9.$$

Hence the "rate of change of y with respect to x," defined in this way, is not constant for all pairs of points on the graph of this quadratic equation (Fig. 5.2).

Suppose we then decide to fix one of the points on the graph of the function, say

$$(x_1, f(x)) = (1, 5),$$

and consider the rate of change at (1, 5) (see Fig. 5.3). However, this still results in ambiguities since, for example,

$$\frac{14 - 5}{2 - 1} = 9, \quad \frac{29 - 5}{3 - 1} = 12, \quad \frac{2 - 5}{0 - 1} = 3, \quad \frac{5 - 5}{-1 - 1} = 0, \quad \frac{14 - 5}{-2 - 1} = -3.$$

We observe that each of these ratios is the slope of the straight line joining two points on the curve, called the *secant* line. Why do we get different rates of change at a fixed point x_1? The problem is that we are computing an average rate of change of the function over the interval $[x_1, x_2]$ rather than the rate

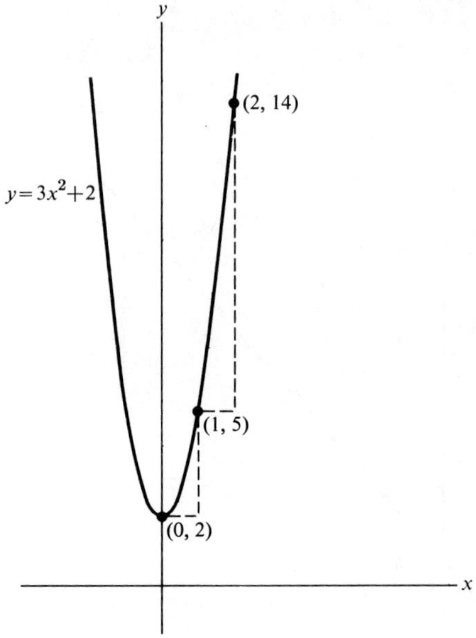

Fig. 5.2

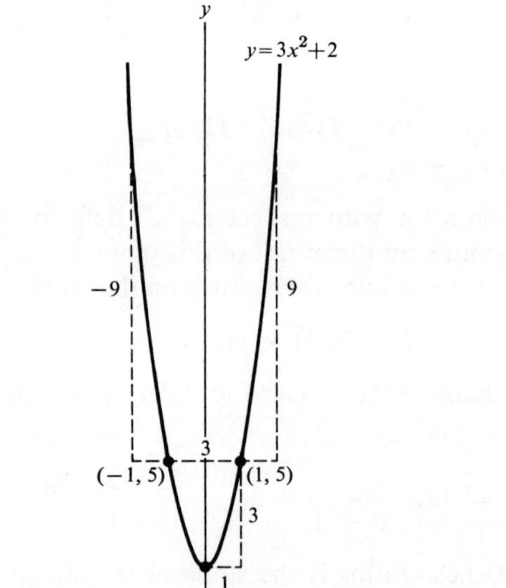

Fig. 5.3

of change *at* x_1. Thus, if we say that we have traveled 100 miles in two hours and that therefore our velocity (rate of change of distance with respect to time) was 50 mph, we mean that our *average* velocity over the two-hour period was 50 mph. At any moment we may have been traveling 30 mph, 60 mph, or at some other speed.

Consequently, we need a better notion of "rate of change of y with respect to x"—one which gives us an unambiguous answer at a particular point, and more important, since we have learned that this is what distinguishes mathematics from arithmetic, one which works at an arbitrary point (x, y), for appropriate functions $y = f(x)$. It seems reasonable to use the limiting value of the slopes of the secant lines through a fixed point $(x_0, f(x_0))$ and a varying point $(x, f(x))$. For "nice" functions, this limiting value is the slope of the line touching the curve at $(x_0, f(x_0))$, i.e., the tangent line.

To return to the graph of our function, suppose we imagine holding the point $(x_0, f(x_0))$ fixed and moving $(x, f(x))$ towards it from the left or right (Fig. 5.4). Then the line joining the points $(x_0, f(x_0))$ and $(x, f(x))$ approaches the line tangent to the curve at $(x_0, f(x_0))$, providing that the curve is reasonably well-behaved. In other words, if all goes well, the ratio

$$\frac{y - y_0}{x - x_0} = \frac{f(x) - f(x_0)}{x - x_0}$$

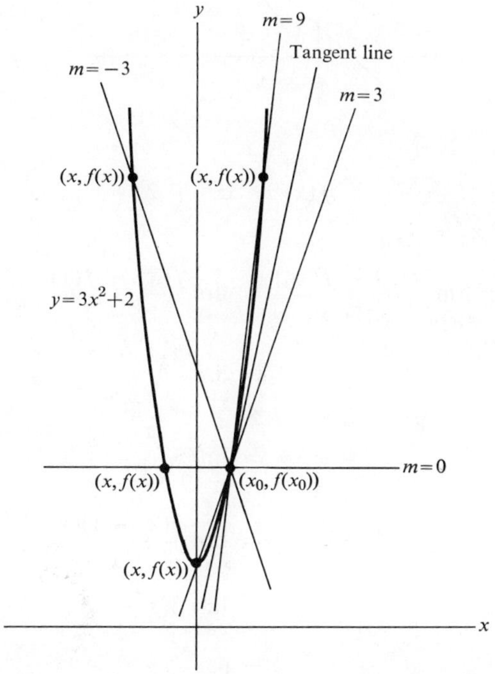

Fig. 5.4

should approach the slope of the tangent line (Fig. 5.5); thus we are interested in the quotient

$$\frac{f(x) - f(x_0)}{x - x_0}$$

as x approaches x_0, that is,

$$\lim_{x\to x_0} \frac{f(x) - f(x_0)}{x - x_0} = \text{slope of tangent line to } y$$

$$= f(x) \text{ at } (x_0, f(x_0)).$$

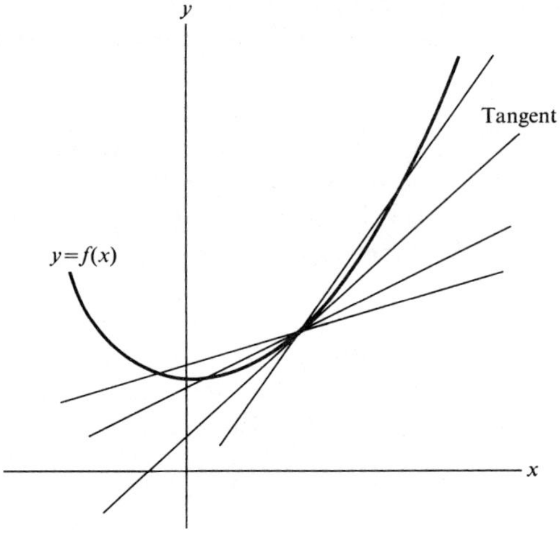

Fig. 5.5

Returning to

$$f(x) = 3x^2 + 2,$$

we observe that if $x_0 = 1$,

$$\begin{aligned}\lim_{x\to x_0} \frac{f(x) - f(x_0)}{x - x_0} &= \lim_{x\to 1} \frac{f(x) - f(1)}{x - 1}\\ &= \lim_{x\to 1} \frac{3x^2 + 2 - 5}{x - 1}\\ &= \lim_{x\to 1} \frac{3x^2 - 3}{x - 1}\\ &= \lim_{x\to 1} \frac{3(x - 1)(x + 1)}{x - 1}\\ &= \lim_{x\to 1} 3(x + 1)\\ &= 6.\end{aligned}$$

More generally for this function

$$\lim_{x\to x_0} \frac{f(x) - f(x_0)}{x - x_0} = \lim_{x\to x_0} \frac{3x^2 + 2 - (3x_0^2 + 2)}{x - x_0}$$

$$= \lim_{x\to x_0} \frac{3(x + x_0)(x - x_0)}{x - x_0}$$

$$= 6x_0.$$

Warning. For some functions $f(x)$ and some points x_0,

$$\lim_{x\to x_0} \frac{f(x) - f(x_0)}{x - x_0}$$

may fail to exist. Consider the graph of

$$f(x) = |x|,$$

at the point $(x_0, f(x_0)) = (0, 0)$ (Fig. 5.6). The line joining any point on the curve with positive x-coordinate to the origin is the straight line $f(x) = x$, of slope 1, whereas any point with negative x-coordinate is joined to the origin by a straight line of slope -1. Hence

$$\frac{f(x) - f(x_0)}{x - x_0} = \frac{|x| - 0}{x - 0} = 1 \quad \text{for} \quad x > 0$$

and

$$\frac{f(x) - f(x_0)}{x - x_0} = \frac{|x| - 0}{x - 0} = -1 \quad \text{for} \quad x < 0$$

and so

$$\lim_{x\to 0} \frac{f(x) - f(0)}{x - 0}$$

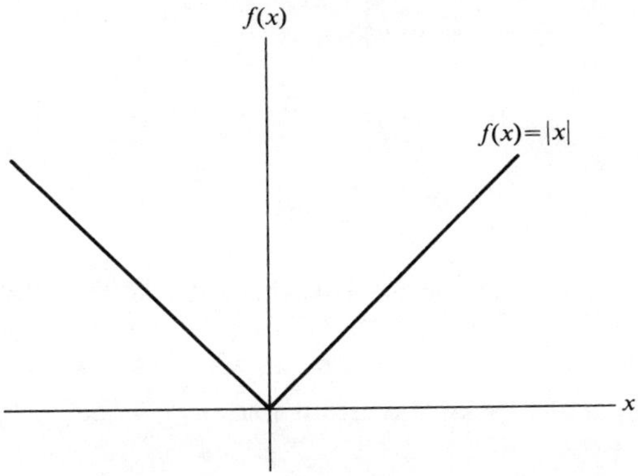

Fig. 5.6

fails to exist. Thus we added above the proviso "for appropriate functions $y = f(x)$."

Warning. It is very important that the limit of the ratio

$$\frac{f(x) - f(x_0)}{x - x_0}$$

not be confused with the limit of the function $f(x)$ itself. For example, we have

$$\lim_{x \to 0} f(x) = \lim_{x \to 0} |x| = 0,$$

but

$$\lim_{x \to 0} \frac{f(x) - f(0)}{x - 0}$$

fails to exist.

If we think of the ratio

$$\frac{f(x) - f(0)}{x - 0}$$

as a new function and sketch its graph (Fig. 5.7), we see that its limit as $x \to 0$ does not exist.

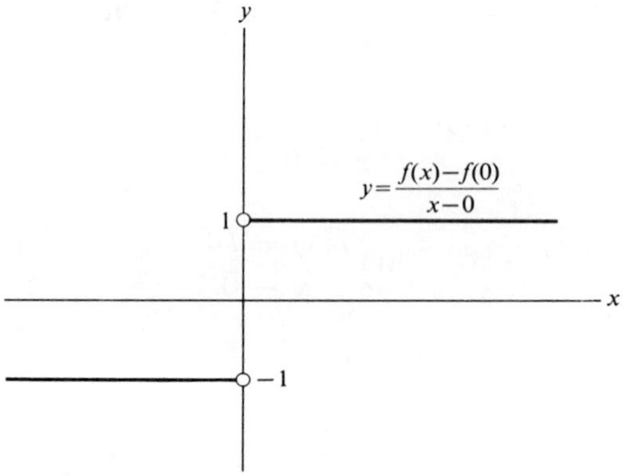

Fig. 5.7

PROBLEMS

1. Sketch the graph of $y = 2x^2 - 3x + 1$. Find the slope of the secant line through
   a) (0, 1) and (1, 0).
   b) (0, 1) and (−1, 6).
   c) (0, 1) and (2, 3).
   d) (0, 1) and (3, 10).
2. Find the limit of the slope of the secant line of $y = 2x^2 - 3x + 1$ through (0, 1) and (x, y) as $x \to 0$.

3. If $F(x) = 5x^2 - 2x + 1$, find

a) $\lim_{x \to 0} \frac{F(x) - F(0)}{x}$.

b) $\lim_{x \to -1} \frac{F(x) - F(-1)}{x + 1}$.

c) $\lim_{x \to 0} F(x)$.

d) $\lim_{x \to -1} F(x)$.

e) $\lim_{x \to x_0} \frac{F(x) - F(x_0)}{x - x_0}$.

f) $\lim_{x \to x_0} F(x)$.

4. Find the limit of the slope of the secant line of $y = 5x^2 - 4x - 1$ through (1, 0) and (x, y), as $x \to 1$.

5. If $F(x) = 3x^2 + 5x - 1$, find

a) $\lim_{x \to 0} F(x)$.

b) $\lim_{x \to 0} \frac{F(x) - F(0)}{x}$.

c) $\lim_{x \to 1} \frac{F(x) - F(1)}{x - 1}$.

d) $\lim_{x \to 1} F(x)$.

e) $\lim_{x \to x_0} F(x)$.

f) $\lim_{x \to x_0} \frac{F(x) - F(x_0)}{x - x_0}$.

5.2 DEFINITION OF DERIVATIVE

Now let us state the problem in more general terms. If

$$y = f(x)$$

is defined on the interval $[a, b]$ and if at the point $x_0 \in (a, b)$

$$\lim_{x \to x_0} \frac{y - y_0}{x - x_0} = \lim_{x \to x_0} \frac{f(x) - f(x_0)}{x - x_0}$$

exists, then this limit is called the *derivative of* $f(x)$ *with respect to* x *at* x_0. In this case we write

$$\lim_{x \to x_0} \frac{f(x) - f(x_0)}{x - x_0} = f'(x_0)$$

and say that the function $f(x)$ is *differentiable at* x_0.

We note that differentiability is, like continuity, a pointwise property. If $y = f(x)$ is differentiable for all real values of x, $f(x)$ is said to be *differentiable.* If $f(x)$ is differentiable at each $x \in (a, b)$, $f(x)$ is said to be differentiable *on* (a, b).

At those points in the domain of $f(x)$ at which it exists, the derivative defines a new function of x; that is, $f'(x)$ is a function defined for all x at which $f(x)$ is differentiable.

We also still use the notation for derivatives which was used by Leibniz in the belief that the derivative represented the quotient of "infinitesimal" quantities

dy and dx which were treated for computational purposes in the same way as finite quantities:

$$f'(x) = \frac{dy}{dx} = \frac{df}{dx}.$$

If we write $f(x) - f(x_0) = y - y_0$ as Δy (the increment in y) and $x - x_0$ as Δx (the increment in x), then we have

$$\frac{\Delta y}{\Delta x} \to \frac{dy}{dx} \quad \text{as} \quad \Delta x \to 0.$$

Although the concept of limit has clarified the notion of derivative and has eliminated the ambiguities involved in dealing with infinitesimals (called "ghosts of departed quantities" by Bishop Berkeley), the notation is still useful and Leibniz's manipulations can be shown to have a logical basis.

In the notation just given, the derivative is

$$\lim_{\Delta x \to 0} \frac{f(x + \Delta x) - f(x)}{\Delta x}.$$

Sometimes we replace Δx by a single letter representing a variable, say h, and write the derivative as

$$\lim_{h \to 0} \frac{f(x + h) - f(x)}{h}.$$

Note that the first notation which we introduced emphasizes the point x_0 at which the derivative is being computed, whereas these forms are independent of the particular point at which the derivative is being considered.

Let us compute some derivatives.

Examples

1. If $f(x) = 6x^2$, find $f'(x_0)$.

Solution

$$\begin{aligned} \lim_{x \to x_0} \frac{f(x) - f(x_0)}{x - x_0} &= \lim_{x \to x_0} \frac{6x^2 - 6x_0^2}{x - x_0} \\ &= \lim_{x \to x_0} \frac{6(x - x_0)(x + x_0)}{x - x_0} \\ &= \lim_{x \to x_0} 6(x + x_0) \\ &= 12x_0. \end{aligned}$$

2. Let $f(x) = 2x^2 - 3x + 1$. Find $f'(x)$ and $f'(2)$.

Solution. We have, for any x_0,

$$\lim_{x \to x_0} \frac{f(x) - f(x_0)}{x - x_0} = \lim_{x \to x_0} \frac{2x^2 - 3x + 1 - (2x_0^2 - 3x_0 + 1)}{x - x_0}$$

$$= \lim_{x \to x_0} \frac{2(x^2 - x_0^2) - 3(x - x_0)}{x - x_0}$$

$$= \lim_{x \to x_0} \left(\frac{2(x + x_0)(x - x_0)}{x - x_0} - \frac{3(x - x_0)}{x - x_0} \right)$$

$$= \lim_{x \to x_0} 2(x + x_0) + \lim_{x \to x_0} -3$$

$$= 4x_0 - 3.$$

Thus $f'(x) = 4x - 3$ and $f'(2) = 5$.

3. Let $f(x) = \sqrt{x}$. Find $f'(x)$.

Solution. We have

$$\lim_{x \to x_0} \frac{f(x) - f(x_0)}{x - x_0} = \lim_{x \to x_0} \frac{\sqrt{x} - \sqrt{x_0}}{x - x_0}.$$

At this point we look back at Examples 1 and 2 and observe that the crucial step was factoring the quantity $(x - x_0)$ out of the numerator and denominator. Here we do not have such a factor in the numerator, but we may remedy this by "rationalizing the numerator":

$$\lim_{x \to x_0} \frac{\sqrt{x} - \sqrt{x_0}}{x - x_0} \cdot \frac{\sqrt{x} + \sqrt{x_0}}{\sqrt{x} + \sqrt{x_0}} = \lim_{x \to x_0} \frac{x - x_0}{x - x_0} \cdot \frac{1}{\sqrt{x} + \sqrt{x_0}}$$

$$= \lim_{x \to x_0} \frac{1}{\sqrt{x} + \sqrt{x_0}}.$$

We use without proof the fact that if $f(x)$ is continuous, then $\sqrt{f(x)}$ is also, i.e.,

$$\lim_{x \to x_0} \sqrt{f(x)} = \sqrt{f(x_0)}.$$

In particular,

$$\lim_{x \to x_0} \frac{1}{\sqrt{x} + \sqrt{x_0}} = \frac{1}{2\sqrt{x_0}}.$$

Thus

$$f'(x) = \frac{1}{2\sqrt{x}}, \qquad \text{if} \quad x > 0.$$

The domain of $f(x)$ is the nonnegative real numbers, but the domain of $f'(x)$ is the positive real numbers. In other words, $\sqrt{x}$ is differentiable everywhere in its domain except at $x = 0$.

4. Let $f(x) = 1/2x$. Find $f'(x)$.

Solution. We have

$$\begin{aligned}\lim_{x\to x_0}\frac{f(x)-f(x_0)}{x-x_0} &= \lim_{x\to x_0}\frac{1/2x - 1/2x_0}{x-x_0}\\ &= \lim_{x\to x_0}\frac{x_0 - x}{2xx_0(x-x_0)}\\ &= \lim_{x\to x_0}\frac{-(x-x_0)}{2xx_0(x-x_0)}\\ &= \frac{-1}{2x_0^2}.\end{aligned}$$

Hence $f(x)$ is differentiable wherever it is defined, namely for all real numbers except 0, with $f'(x) = -1/2x^2$.

5. Let $f(x) = (2x - 1)/x$. Find $f'(x)$.

Solution. Let us use different notation for a change.

$$\begin{aligned}f'(x) &= \lim_{h\to 0}\frac{(2(x+h)-1)/(x+h) - (2x-1)/x}{h}\\ &= \lim_{h\to 0}\frac{2x(x+h) - x - 2x(x+h) + (x+h)}{x(x+h)h}\\ &= \lim_{h\to 0}\frac{1}{x(x+h)}\\ &= \frac{1}{x^2}.\end{aligned}$$

6. Find the equation of the line tangent to

$$f(x) = 3x^2 - 2x + 1$$

at (2, 9).

Solution. First we find the slope of the required tangent, i.e., $f'(2)$.

$$\begin{aligned}f'(2) &= \lim_{x\to 2}\frac{3x^2 - 2x + 1 - 9}{x-2} = \lim_{x\to 2}\frac{(3x+4)(x-2)}{x-2}\\ &= \lim_{x\to 2}(3x+4) = 10.\end{aligned}$$

Now using the point-slope form of the equation of a straight line, we get that the equation of the tangent line is

$$y - 9 = 10(x - 2).$$

Warning. It is clear from the function $y = |x|$ discussed above, that a function may be differentiable for some values of x and not for others, since only at the origin does this function fail to have a derivative. It is also important to realize that x and y are simply place-holding variables and that we could also speak of

$$u = f(v),$$

and calculate $f'(v) = du/dv$. If $y = f(x)$, we could also solve for x in terms of y (if possible) and find dx/dy. Other notations used for the derivative of $y = f(x)$ are $D_x y$ and $D_x f$.

The intuitive notion of continuity is that of an unbroken graph. It is clear from the example $f(x) = |x|$ that not all continuous functions are differentiable; we observe that the graph of $y = |x|$, while unbroken, has a sharp point at the origin. In fact, differentiability intuitively means smoothness. Functions whose graphs are of the form shown in Fig. 5.8a are not differentiable (the slopes of the tangent lines do not have a limiting value at certain points), while graphs which are smooth (Fig. 5.8b) represent differentiable functions.

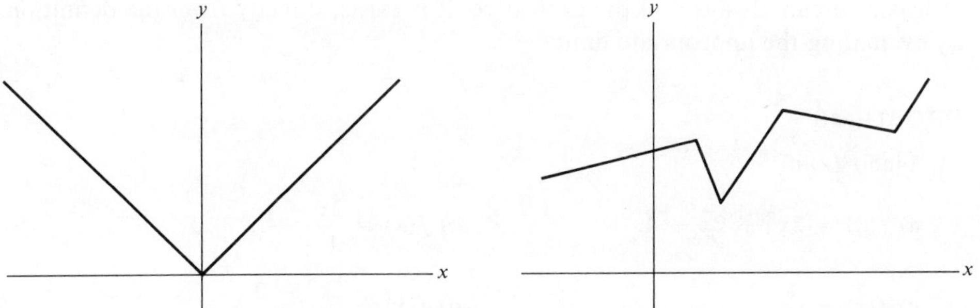

Fig. 5.8a

(a)

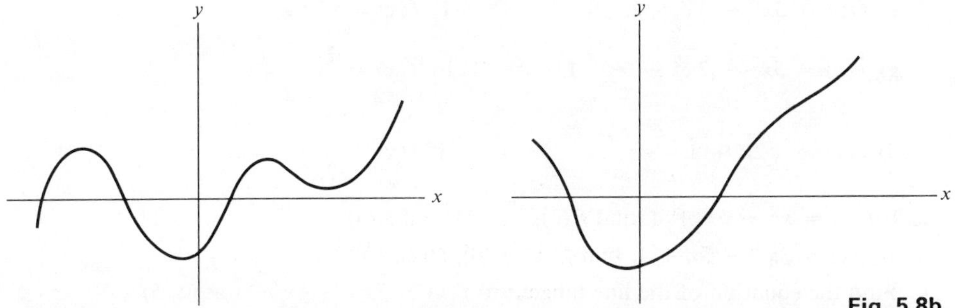

Fig. 5.8b

(b)

We have the following relation between continuity and differentiability.

Theorem 1. If $f(x)$ is differentiable at x_0, then $f(x)$ is continuous at x_0.

Proof. We must show that

$$\lim_{x \to x_0} f(x) = f(x_0).$$

We multiply $f(x)$ by 1 in a convenient form and add and subtract an appropriate quantity:

$$\lim_{x \to x_0} f(x) = \lim_{x \to x_0} \left[\left(\frac{f(x)}{x - x_0} - \frac{f(x_0)}{x - x_0}\right)(x - x_0) + f(x_0)\right].$$

Applying the rules for limits from the last chapter this is equal to

$$\lim_{x \to x_0} \frac{f(x) - f(x_0)}{x - x_0} \lim_{x \to x_0} (x - x_0) + \lim_{x \to x_0} f(x_0) = f'(x_0) \cdot 0 + f(x_0) = f(x_0).$$

As we have mentioned, the converse is not true. However, for many years after the invention of calculus the notion of limit was so hazy that mathematicians, lacking an adequate definition of either, generally believed that continuity and differentiability were synonymous.

In the next section we develop some rules which enable us to find the derivatives of certain functions very easily, but it is important to remember that, theoretically at least, we can always find the derivative, if it exists, directly from the definition by evaluating the appropriate limit.

PROBLEMS

1. Find $f'(x)$ if

a) $f(x) = 2x^2 + 2x - 4.$

b) $f(x) = \dfrac{2x - 1}{1 - x}.$

c) $f(x) = \dfrac{1}{x^2}.$

d) $f(x) = \dfrac{2x^2 - 3}{2x + 1}.$

e) $f(x) = 3x^3 - 3x + 2.$

f) $f(x) = \sqrt{x - 1}.$

g) $f(x) = 5x^3 - 2x^2 + x - 1.$

h) $f(x) = \dfrac{2x - 1}{\sqrt{x}}.$

i) $f(x) = \sqrt{2x + 1}.$

j) $f(x) = \dfrac{1}{2x^2 - 1}.$

2. If $f(x) = x^2 - 5x + 4$, find $f'(0)$, $f'(-1)$, and $f'(a)$.
3. If $f(x) = 3x^3 - 5x + 1$, find $f'(x)$, $f'(0)$, $f'(2)$, $f'(-2)$.
4. Find the equation of the line tangent to $f(x) = 3x^2 - 4x + 1$ at $(2, 5)$.
5. Find the equation of the line tangent to $f(x) = 2x^2 - 5x + 2$ at $(1, -1)$.

6. Find the equation of the line tangent to $f(x) = \frac{1}{2}x$ at $x = 4$.
7. Translate the following statements into equations involving derivatives:
   a) The level of difficulty of learning at time t decreases at a rate proportional to the product of the level of difficulty at time t and the total amount of practice.
   b) If $R(x)$ is the revenue function, the marginal revenue $R'(x)$ varies with the square of the production level x.
   c) The increase of a population is proportional to the size of the population.

5.3 PROPERTIES OF DERIVATIVES, DERIVATIVES OF POLYNOMIALS

It is clear that if

$$y = f(x) = mx + b$$

is a straight line not parallel to the y-axis, then y is everywhere differentiable as a function of x and

$$\frac{dy}{dx} = m,$$

the slope of the line. We can also compute dy/dx directly:

$$\lim_{x\to x_0}\frac{y - y_0}{x - x_0} = \lim_{x\to x_0}\frac{mx + b - (mx_0 + b)}{x - x_0} = \lim_{x\to x_0}\frac{m(x - x_0)}{x - x_0} = m.$$

We want to develop a systematic method for determining which other functions are differentiable and for computing their derivatives. Of course, $y = c$ is a special case of a straight line so any constant function is differentiable, with derivative 0; this is to be expected, since there is no change in y. Since polynomials are made up of sums and products of functions of the type $y = c$ and $y = x$, if we can show that sums and products of differentiable functions are differentiable we shall have that all polynomial functions are differentiable.

Theorem 2. If $f(x)$ and $g(x)$ are differentiable at $x = x_0$, then $f(x) + g(x)$ is differentiable at x_0 and

$$\boxed{\frac{d(f + g)}{dx} = \frac{df}{dx} + \frac{dg}{dx}.}$$

Proof

$$\lim_{x\to x_0}\frac{(f(x) + g(x)) - (f(x_0) + g(x_0))}{x - x_0} = \lim_{x\to x_0}\frac{f(x) + g(x) - f(x_0) - g(x_0)}{x - x_0}$$

$$= \lim_{x\to x_0}\left[\frac{f(x) - f(x_0)}{x - x_0} + \frac{g(x) - g(x_0)}{x - x_0}\right]$$

$$= \lim_{x\to x_0}\frac{f(x) - f(x_0)}{x - x_0} + \lim_{x\to x_0}\frac{g(x) - g(x_0)}{x - x_0}.$$

If by hypotheses these two limits exist, their sum does also, and $f(x) + g(x)$ is differentiable at x_0, with its derivative the sum of the derivatives of $f(x)$ and $g(x)$ at x_0.

The problem of the product is more difficult. We have

Theorem 3. If $f(x)$ and $g(x)$ are differentiable at $x = x_0$, then $f(x)g(x)$ is differentiable at $x = x_0$ and

$$\frac{d(fg)}{dx} = f\frac{dg}{dx} + g\frac{df}{dx}.$$

Proof.

We are interested in

$$\lim_{x \to x_0} \frac{f(x)g(x) - f(x_0)g(x_0)}{x - x_0}.$$

We add and subtract an appropriate term, namely $f(x)g(x_0)$, to enable us to factor the numerator. Thus

$$\lim_{x \to x_0} \frac{f(x)g(x) - f(x_0)g(x_0)}{x - x_0} = \lim_{x \to x_0} \frac{f(x)g(x) - f(x)g(x_0) + f(x)g(x_0) - f(x_0)g(x_0)}{x - x_0}$$

$$= \lim_{x \to x_0} \frac{f(x)[g(x) - g(x_0)]}{x - x_0} + \lim_{x \to x_0} \frac{g(x_0)[f(x) - f(x_0)]}{x - x_0}$$

$$= f(x_0)\frac{dg}{dx} + g(x_0)\frac{df}{dx},$$

since $\lim_{x \to x_0} f(x) = f(x_0)$, because $f(x)$ is differentiable and hence continuous at x_0.

Corollary. If c is a constant, then

$$\frac{d(cf)}{dx} = c\frac{df}{dx}.$$

Proof. This follows from the theorem and the fact that the derivative of a constant is zero.

Warning. The tendency is to want to carry the multiplicative property of limits

$$\lim_{x \to x_0} f(x)g(x) = \lim_{x \to x_0} f(x) \lim_{x \to x_0} g(x) \qquad \text{(if they exist)}$$

over to differentiation, but the novice is wise to convince himself early of the foolhardiness of such a course and to implant Theorem 3 firmly in his mind.

Example. Let $f(x) = 3x + 2$ and $g(x) = x - 5$, so that

$$\frac{df}{dx} = 3 \qquad \text{and} \qquad \frac{dg}{dx} = 1.$$

Then

$$\frac{d(fg)}{dx} = (3x + 2) \cdot 1 + (x - 5) \cdot 3 = 3x + 2 + 3x - 15 = 6x - 13.$$

Note that we cannot check this result without going back to the definition of derivative, since we do not have a general rule for differentiating higher powers of x. We remedy this by using Theorem 3 to derive such a formula.

Theorem 4. If $f(x) = x^n$, n a positive integer, then

$$f'(x) = nx^{n-1}.$$

Proof. By induction.

For $n = 1$, we have $f(x) = x$ so that $f'(x) = 1 = 1 \cdot x^0$.

Now we assume that $dx^n/dx = nx^{n-1}$. We want to show that $dx^{n+1}/dx = (n + 1)x^n$. By Theorem 3,

$$\frac{d(x \cdot x^n)}{dx} = xnx^{n-1} + x^n \cdot 1 = (n + 1)x^n,$$

giving us the desired result.

Putting these results together we get that polynomial functions defined on the real line are everywhere differentiable. Moreover, if

$$f(x) = a_0 + a_1x + a_2x^2 + \cdots + a_nx^n, \qquad a_i \text{ real numbers}, \qquad i = 1, \ldots, n,$$

then

$$f'(x) = a_1 + 2a_2x + \cdots + na_nx^{n-1}.$$

If we want the value of the derivative function at a particular x_0 we simply make the substitution $x = x_0$ to obtain

$$f'(x_0) = a_1 + 2a_2x_0 + \cdots + na_nx_0^{n-1}.$$

Examples

1. If $f(x) = 2x^2 + 5x - 1$, then

$$f'(x) = 4x + 5.$$

$f'(2) = 4 \cdot 2 + 5 = 13$ and $f'(-1) = 1$, and $f'(-\frac{5}{4}) = 0$.

2. If $f(x) = 5x^3 - 4x^2 + 2$, then

$$f'(x) = 15x^2 - 8x.$$

3. If $f(x) = -x^5 + 4x^4 - 3x^2 - x$, then

$$f'(x) = -5x^4 + 16x^3 - 6x - 1.$$

4. The position of a blood particle in an artery at time t is given by

$$s = 3t^2 + 2t + 3.$$

Its velocity is the derivative of the position function. Find its velocity at time $t = 10$.

Solution

$$v(t) = \frac{ds}{dt} = 6t + 2.$$

When $t = 10$, $v = 62$.

5. Find the vertex of the parabola $3x^2 - 2x + 1$.

Solution. The line tangent to a parabola at its vertex is horizontal. Since the slope of the tangent is the derivative of the function, the derivative at the vertex must be 0. Therefore we find,

$$f'(x) = 6x - 2,$$

set this equal to zero and solve for x:

$$6x - 2 = 0.$$

$$x = \tfrac{1}{3}.$$

We now know that at $x = \frac{1}{3}$ the tangent has zero slope. But only at the vertex does a parabola have a horizontal tangent so $(\frac{1}{3}, \frac{2}{3})$ must be the vertex.

PROBLEMS

1. Find dy/dx if
   a) $y = 6x^5 - 5x^4 + 2x - 5$.
   b) $y = 14x^3 - 7x - 17$.
   c) $y = -2x^6 - 4x^4 - x^2 + 5x$.
   d) $y = 7x^4 - 5x^3 - 1$.
   e) $y = -x^5 + 4x^3 + x^2 - 5$.
   f) $y = ax^3 + bx^2 + cx + d$.
   g) $y = 18x^4 + 5x^2 - 4x + 2$.
   h) $y = (19x^5 - 5x^2 - 4x)(3x^2 - 2x + 1)$.
2. Use the product formula to find the derivative of
   a) $f(x) = (6x^3 - 2x^2 - x + 5)(7x^4 - 6x^3 + 5x^2 + 2)$.
   b) $g(x) = (7x^2 - 5x + 16)(1/(2x - 1))$.
3. Find a formula for the derivative of $f(x)g(x)h(x)$ in terms of the derivatives of $f(x)$, $g(x)$, and $h(x)$.
4. If the position of a particle at time t is given by

$$s = t^3 - 4t^2 + t - 2,$$

find its velocity when $t = 10$.

5. At what point(s) does $f(x) = 2x^2 - 5x + 1$ have a tangent parallel to the x-axis?
6. A particle is projected vertically upward with a speed of 80 ft/sec. At the end of t seconds its height is given by

$$s = -16t^2 + 80t.$$

   a) How high does it rise?
   b) When does it hit the ground?
7. Assume that the profit function is given by

$$P(x) = px - C(x),$$

   where p is the constant sales price, x the number of units sold, and $C(x)$ is the production cost, defined by

$$C(x) = F + V(x/k)^{3/2},$$

   where F is fixed cost, V variable cost. Find the rate at which the profit changes. When does it start to decline?
8. For each of the following find the point where the tangent is horizontal.
   a) $y = 2x^2 - 5x + 3$.
   b) $y = -x^2 + 6x - 1$.
   c) $y = 3x^2 + 4x - 5$.
   d) $y = 2x^2 + 10x - 9$.
9. Find the equation of the line tangent to $5x^5 - 4x^2 + 3x - 4$ at $x = 1$.

5.4 TANGENTS

We recall that the derivative of $y = f(x)$ at a point x_0 represents the limit of the slopes of the lines joining other points on the curve to (x_0, y_0). Using our intuitive notion of a tangent line, we saw that the "limit" of the lines themselves is the line tangent to the curve at that point and in fact the derivative is the slope of that line. Indeed, the notion of tangency may be defined in terms of derivatives. That is, if $y = f(x)$ is differentiable at x_0, a line passing through $(x_0, f(x))$ with slope $f'(x_0)$ is *tangent to the curve* $y = f(x)$ *at* $(x_0, f(x_0))$. (See Fig. 5.9.)

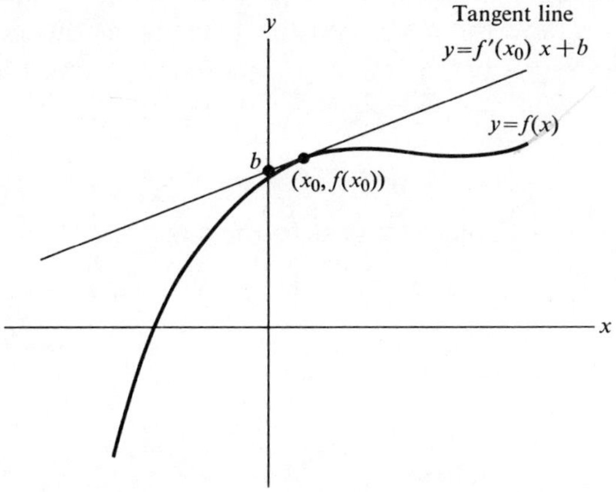

Fig. 5.9

This is more satisfactory than the intuitive notion of a tangent as a line "touching" a curve. It can be difficult, for example, to determine a tangent line by sight. Also, the line tangent at some point may cross the graph of the function at another point (Fig. 5.10).

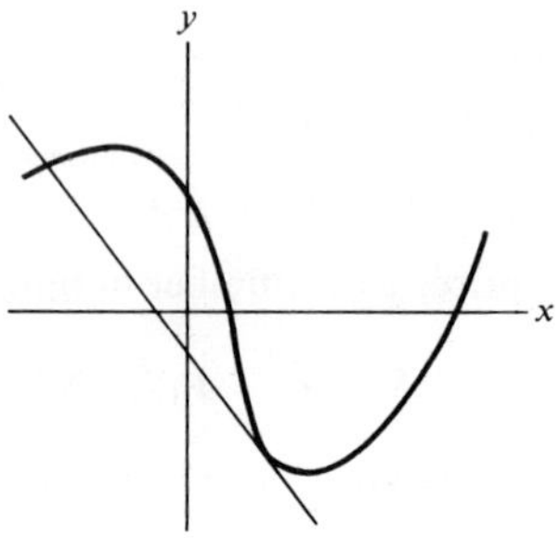

Tangent crosses

Fig. 5.10

Note that since in the definition above we have specified the slope and a point on the tangent line, the line is completely and uniquely determined, so that our definition makes sense.

Examples

1. Find the equation of the line tangent to the curve

$$y = 2x^2 + x - 1$$

at the point $x = 1$ (Fig. 5.11).

Solution. We have

$$\frac{dy}{dx} = 4x + 1$$

so at $x = 1$ the derivative is 5 and the slope of the tangent line is thus 5. When $x = 1$, $y = 2$, so we have the slope and the point (1, 2) on the line. Hence the equation of the line is

$$y - 2 = 5(x - 1),$$

or

$$y = 5x - 3.$$

2. Find the equation of the line tangent to the curve

$$y = x^3 - 4x^2 + x + 1,$$

at the point $x = -2$ (Fig. 5.12).

Solution. We have

$$\frac{dy}{dx} = 3x^2 - 8x + 1$$

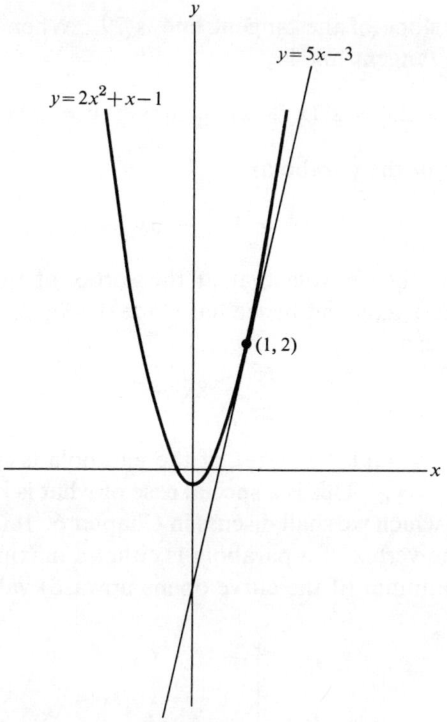

Fig. 5.11

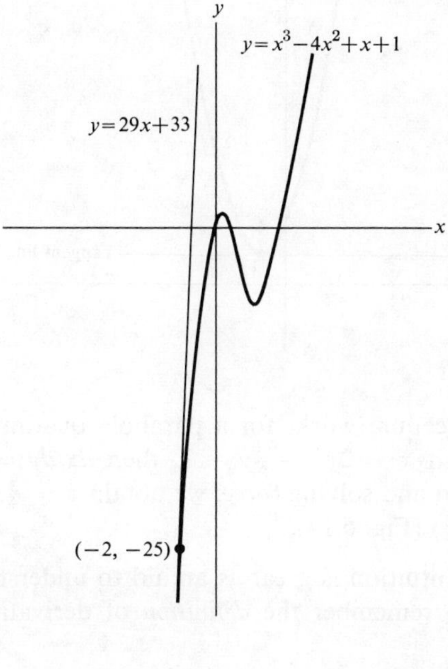

Fig. 5.12

so at $x = -2$, the slope of the tangent line is 29. When $x = -2$, $y = -25$, so the equation of the tangent line is

$$y + 25 = 29(x + 2) \qquad \text{or} \qquad y = 29x + 33.$$

3. Find the vertex of the parabola

$$y = 4x^2 - 3x + 1.$$

Solution. As previously we note that at the vertex of the parabola the tangent line is parallel to the x-axis and hence has slope 0 (Fig. 5.13). Thus we find dy/dx and set it equal to zero:

$$\frac{dy}{dx} = 8x - 3.$$

If $8x - 3 = 0$, $x = \frac{3}{8}$, and the vertex of the parabola is $(\frac{3}{8}, \frac{7}{16})$. The equation of the tangent line is $y = \frac{7}{16}$. This is a special case of what is known as the maximum-minimum problem, which we shall discuss in Chapter 6. It is clear that if a parabola is quadratic in x, the vertex of a parabola is either a maximum (if the curve opens downward) or a minimum (if the curve opens upward) value for y.

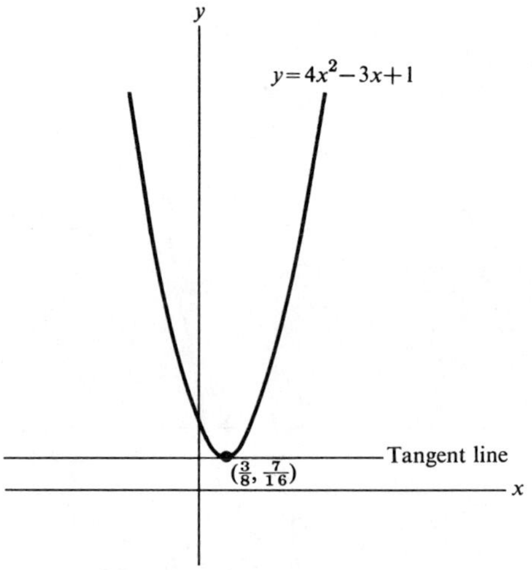

Fig. 5.13

An analogous procedure works for a parabola quadratic in y if we compute dx/dy. For example, if $x = 2y^2 - 3y + 2$, then $dx/dy = 4y - 3$. Setting the derivative equal to zero and solving for y, we obtain $y = \frac{3}{4}$. When $y = \frac{3}{4}$, $x = \frac{7}{8}$, so the vertex is at $(\frac{7}{8}, \frac{3}{4})$ (Fig. 5.14).

Warning. Geometric intuition is great as an aid to understanding differentiation, but it is important to remember the *definition* of derivative rather than always

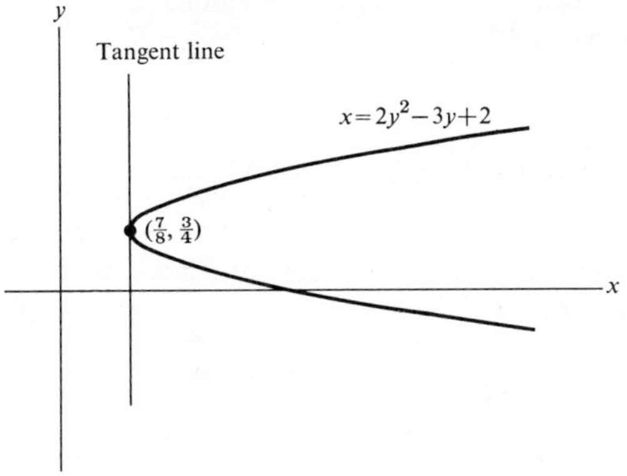

Fig. 5.14

thinking of derivatives in terms of tangents. Then when an exotic function appears—one whose graph defies visualization—we shall not be at a loss, but can go back to the definition.

PROBLEMS

1. Find the equation of the line tangent to
   a) $y = 3x^3 - 4x + 1$ at $x = 3$.
   b) $y = 1/(2x^2 + 1)$ at $x = 0$.
   c) $y = 4x^3 - 2x^2 - 1/x$ at $x = 2$.
   d) $y = -5x^2 + 6x - 1$ at $x = b$.
   e) $y = 5x^3 - 5x^2 + 1$ at $x = -1$.
   f) $y = 2x^2 - 2$ at $x = 0$.
   g) $y = 7x^2 + x + 1$ at $x = 2$.
   h) $y = 2x - 1$ at $x = 3$.
2. Find the vertex of each of the following parabolas:
   a) $y = x^2 - 5x + 1$.
   b) $y = -x^2 + 4x + 1$.
   c) $y = 5x^2 - 5x + 2$.
   d) $y = -3x^2 + 4x$.
   e) $y = -5x^2 + 2x - 1$.
   f) $y = 3x^2 - 6x$.
   g) $y = 15x^2 + 4$.
   h) $y = -x^2 + x - 1$.
3. If a ball is thrown from an 80 ft cliff and its motion is given by
$$s = -16t^2 + 44t + 80,$$
how high does it rise and when does it hit the ground?
4. The effectiveness of a fungicide is a function of its concentration in the atmosphere. Letting c be the concentration in parts per thousand, we have its effectiveness given by
$$E(c) = 10{,}000 - 5c - 49c^2.$$
At what concentration is the fungicide most effective?
5. Find the image of the following functions:
   a) $f(x) = -3x^2 + x + 3$.
   b) $f(x) = 5x^2 + 4x - 1$.
6. Find the maximum value of the following functions:
   a) $f(x) = -3x^2 + 2x$.
   b) $f(x) = -x^2 + x - 5$.

7. Find the minimum value of the following functions:
   a) $f(x) = 2x^2 + 5x + 4$.
   b) $f(x) = 7x^2 - 5x$.
   c) $f(x) = x^2/2 - 4$.

5.5 QUOTIENTS, NONINTEGER EXPONENTS

We now know how to find the derivative of any polynomial function; the next degree of sophistication is to consider rational functions, that is, the quotients of polynomials. We recall that if $f(x) \neq 0$ for all x in some neighborhood of x_0 and if $\lim_{x \to x_0} f(x)$ exists, then

$$\lim_{x \to x_0} \frac{1}{f(x)} = \frac{1}{\lim_{x \to x_0} f(x)}.$$

Theorem 5. If $f(x)$ is differentiable at x_0 and $f(x) \neq 0$ for all x in some neighborhood of x_0, then $1/f$ is differentiable at x_0 and

$$\boxed{\left(\frac{1}{f}\right)' = \frac{-f'}{f^2}.}$$

Proof. We have

$$\begin{aligned}
\lim_{x \to x_0} \frac{1/f(x) - 1/f(x_0)}{x - x_0} &= \lim_{x \to x_0} \frac{f(x_0) - f(x)}{f(x)f(x_0)(x - x_0)} \\
&= \lim_{x \to x_0} \frac{1}{f(x)f(x_0)} \lim_{x \to x_0} \frac{f(x_0) - f(x)}{x - x_0} \\
&= \frac{1}{f^2(x_0)} \left[- \lim_{x \to x_0} \frac{f(x) - f(x_0)}{x - x_0} \right] \\
&= \frac{-f'(x_0)}{f^2(x_0)}.
\end{aligned}$$

Example. Let

$$f(x) = \frac{1}{x^3 - x^2 + 1}.$$

Find $f'(2)$.

We note that the denominator of $f(x)$ is not zero at $x = 2$. Since it is a continuous function, by Theorem 5 of Chapter 4 we know that it is not zero in some neighborhood of 2. Hence the theorem applies and

$$f'(x) = \frac{-f'(x)}{f^2(x)} = \frac{-(3x^2 - 2x)}{(x^3 - x^2 + 1)^2}.$$

Thus

$$f'(2) = \frac{-(3 \cdot 4 - 2 \cdot 2)}{(8 - 4 + 1)^2} = \frac{-8}{25}.$$

The reason we require that $f(x)$ be nonzero in a neighborhood of x_0 is that in order to define a derivative of a function at x_0 we need that the function be defined in a neighborhood of x_0; it is necessary that $f(x) \neq 0$ for $1/f(x)$ to be defined.

Corollary 1. If $f(x)$ and $g(x)$ are differentiable at x_0 and $g(x) \neq 0$ for all x in some neighborhood of x_0, then $f(x)/g(x)$ is differentiable at x_0 and

$$\boxed{\left(\frac{f}{g}\right)' = \frac{gf' - fg'}{g^2}.}$$

Proof. We write f/g as $f \cdot (1/g)$ and apply Theorems 3 and 5 to get the result:

$$\left(\frac{f}{g}\right)' = \left(f \cdot \frac{1}{g}\right)' = f\frac{-g'}{g^2} + \frac{1}{g} \cdot f' = \frac{-fg' + gf'}{g^2}.$$

Example. Find the derivative of

$$\frac{x^2 - 2x + 3}{2x^2 - 1}.$$

Solution. Letting

$$f(x) = x^2 - 2x + 3 \quad \text{and} \quad g(x) = 2x^2 - 1,$$

we have

$$f'(x) = 2x - 2 \quad \text{and} \quad g'(x) = 4x,$$

so that

$$\begin{aligned}\left(\frac{f}{g}\right)' &= \frac{(2x^2 - 1)(2x - 2) - (x^2 - 2x + 3)(4x)}{(2x^2 - 1)^2} \\ &= \frac{4x^3 - 4x^2 - 2x + 2 - 4x^3 + 8x^2 - 12x}{4x^4 - 4x^2 + 1} \\ &= \frac{4x^2 - 14x + 2}{4x^4 - 4x^2 + 1}.\end{aligned}$$

This gives us the derivative of the original function except at the zeros of $2x^2 - 1$, namely $\pm\sqrt{2}/2$.

Theorem 5 also enables us to extend Theorem 4 to all integer exponents, positive or negative.

Corollary 2. If n is a negative integer, then

$$\frac{dx^n}{dx} = nx^{n-1}.$$

Proof. We write x^n as $1/x^{-n}$. Then $-n$ is a positive integer and

$$\frac{dx^{-n}}{dx} = (-n)x^{-n-1}$$

by Theorem 4. We combine this with Theorem 5 to get

$$\frac{dx^n}{dx} = \frac{d(1/x^{-n})}{dx} = \frac{-(-n)x^{-n-1}}{(x^{-n})^2} = \frac{nx^{-n-1}}{x^{-2n}} = nx^{n-1},$$

using the rules of exponents for raising to a power and dividing.

Example. Find $f'(-1)$ if

$$f(x) = \frac{1}{x^3} + 2x^2 - \frac{1}{4x^2}.$$

Solution. We can write $f(x)$ as $x^{-3} + 2x^2 - \frac{1}{4}x^{-2}$, so that

$$\begin{aligned} f'(x) &= -3x^{-3-1} + 2 \cdot 2x - \frac{(-2)}{4}x^{-2-1} \\ &= -3x^{-4} + 4x + \tfrac{1}{2}x^{-3} \\ &= \frac{-3}{x^4} + 4x + \frac{1}{2x^3}. \end{aligned}$$

Thus

$$f'(-1) = \frac{-3}{(-1)^4} + 4(-1) + \frac{1}{2(-1)^3} = -3 - 4 - \tfrac{1}{2} = -7\tfrac{1}{2}.$$

What about the derivative of $x^{p/q}$, where p and q are integers? In other words, can the formula for the derivative of powers of x be extended to rational exponents? Going even further, let us consider an expression of the form x^π. It is difficult to imagine what an irrational exponent should denote; we postpone that problem until Chapter 8, asking only: *if* we know what $f(x) = x^\pi$ means, what should $f'(x)$ be?

We state without proof:

Theorem 6. If $f(x) = x^a$, a a real number, and $f(x)$ is defined and differentiable, then

$$\frac{df}{dx} = \frac{dx^a}{dx} = ax^{a-1}.$$

Examples

1. If $f(x) = \sqrt{x}$, find $f'(2)$.

Solution. We have

$$f'(x) = \frac{1}{2\sqrt{x}} \qquad \text{so that} \qquad f'(2) = \frac{1}{2\sqrt{2}}.$$

2. Find the derivative of

$$\frac{x^2 - 3x + 4}{x^{5/2} + 1}.$$

Solution. Let

$$f(x) = x^2 - 3x + 4 \qquad \text{and} \qquad g(x) = x^{5/2} + 1.$$

Then

$$f'(x) = 2x - 3 \qquad \text{and} \qquad g'(x) = \tfrac{5}{2}x^{3/2},$$

so that

$$\left(\frac{f}{g}\right)' = \frac{(x^{5/2} + 1)(2x - 3) - (x^2 - 3x + 4)\frac{5}{2}x^{3/2}}{(x^{5/2} + 1)^2}$$

$$= \frac{-x^{7/2} + 9x^{5/2} - 20x^{3/2} + x - 6}{2(x^{5/2} + 1)^2},$$

for x nonnegative.

3. If

$$y = x^{\sqrt{2}} \qquad \text{find} \qquad \frac{dy}{dx}.$$

$$\frac{dy}{dx} = \sqrt{2}x^{\sqrt{2}-1},$$

whenever $x^{\sqrt{2}}$ is defined.

PROBLEMS

1. Find $f'(x)$ if

a) $f(x) = \dfrac{16x^3 - 5x^2 + 2x - 1}{3x^2 - 14}$. b) $f(x) = \dfrac{5x^{14}}{3x^{16} - 2x^2 + 1}$.

c) $f(x) = \dfrac{(3x^3 - 1)(2x - 5)}{(6x^2 - 1)(17x^4 - 5x + 5)}$.

d) $f(x) = (7x^3 - 5x + 4)(6x^3 - 17x^2 + 2x - 1)^2$.

e) $f(x) = 2x^{1/2}$. f) $f(x) = 7x^{4/3}5x^{2/3} + 2$.

g) $f(x) = (2x^{1/3} + 1)(6x^{-3/4} - 4x^{-1/2} + 5)$.

h) $f(x) = \dfrac{8x^a}{x - 1}$. i) $f(x) = \dfrac{8\sqrt{x}}{14x^{3/2}}$.

j) $f(x) = \dfrac{1}{2x - 3}$.

k) $f(x) = \dfrac{3x^3 - 5x^2 + 1}{4x^2 + 15x - 2}$.

l) $f(x) = \dfrac{\sqrt{x}}{14x^2 - 2x + 1}$.

m) $f(x) = \dfrac{(3x^3 - 1)(2x + 3)}{6x^2 - 17x + 1}$.

n) $f(x) = 16x^{2/3} - x^{-1/2} + 5$.

o) $f(x) = \dfrac{1}{x^{3/2} - 5}$.

2. If x is the proportion of votes won by a political party in a national election, political scientists claim (based on past results) that the proportion of seats won by that party in the U.S. House of Representatives is given by

$$f(x) = \frac{x^3}{3x^2 - 3x + 1}.$$

What are the domain and image of $f(x)$? Find $f'(x)$. Is $f(x)$ an increasing function of x? (That is, does $x_1 > x_2$ imply that $f(x_1) > f(x_2)$ for all $x_1, x_2 \in$ domain of f?)

5.6 HIGHER DERIVATIVES

Since the derivative of a function $f(x)$ with respect to x is itself a function of x, we can consider whether or not it is differentiable. If so, we may compute its derivative with respect to x, that is, df'/dx at x_0 is

$$\lim_{x \to x_0} \frac{f'(x) - f'(x_0)}{x - x_0},$$

if this limit exists. We call df'/dx the *second derivative* of f and write f'' or d^2f/dx^2. If the second derivative of $f(x)$ exists at x_0, we say that $f(x)$ is *twice differentiable* at x_0.

The second derivative measures the rate of change of the rate of change. The most obvious instance of the occurrence of second derivatives is acceleration. Acceleration is the rate of change of velocity. If the position of a particle at time t is given by

$$s = t^2 + 2t - 6,$$

then its velocity is defined to be the rate of change of position with respect to time and is given by:

$$v = \frac{ds}{dt} = 2t + 2.$$

Its acceleration is defined to be

$$a = \frac{dv}{dt} = \frac{d^2s}{dt^2} = 2.$$

Of course the process may be repeated to find the third and successive derivatives:

$$f'''(x) = f^{(3)}(x), f^{(4)}(x), \ldots, f^{(n)}(x), \ldots.$$

If a function $f(x)$ is n-times differentiable and its nth derivative is a continuous function, then $f(x)$ is said to be a C^n-*function*. Thus a function which is differentiable and whose derivative is continuous is a C^1-function. We also use C^0 to denote a continuous function. If a function is infinitely differentiable, that is, the nth derivative exists and is continuous for arbitrarily large n, it is said to be a C^∞-*function.*

Warning. Do not think that when the derivative of a function is zero, the derivative does not exist—zero exists. If the nth derivative of a function is 0, the function is n-times differentiable, and in fact is C^∞, since all successive derivatives are derivatives of a constant and hence exist (and equal 0). In particular, polynomials are C^∞.

Examples

1. If $f(x) = 3x^3 - 7x^2 + 6x$, find $f^{(n)}(x)$ for all positive integers n.

Solution

$$\begin{aligned} f^{(1)}(x) &= f'(x) = 9x^2 - 14x + 6 \\ f^{(2)}(x) &= 18x - 14 \\ f^{(3)}(x) &= 18 \\ f^{(n)}(x) &= 0, \qquad n \geq 4. \end{aligned}$$

2. If

$$y = \frac{x + 2}{x^{1/2}} \quad \text{find} \quad \frac{d^2y}{dx^2}.$$

Solution

$$\begin{aligned} \frac{dy}{dx} &= \frac{x^{1/2} - (x + 2)\frac{1}{2}x^{-1/2}}{x} \\ &= \frac{x^{1/2} - \frac{1}{2}x^{1/2} - x^{-1/2}}{x} \\ &= \frac{1}{2x^{1/2}} - \frac{1}{x^{3/2}} = \frac{1}{2}x^{-1/2} - x^{-3/2} \\ \frac{d^2y}{dx^2} &= \frac{1}{2}\left(\frac{-1}{2}\right)x^{-3/2} - \frac{-3}{2}x^{-5/2} \\ &= \frac{-1}{4}x^{-3/2} + \frac{3}{2}x^{-5/2} \\ &= \frac{-1}{4x^{3/2}} + \frac{3}{2x^{5/2}}. \end{aligned}$$

3. If the position of a particle at time t is given by

$$s = \frac{1}{t} + t^2,$$

find its acceleration when $t = 1$.

Solution. We have

$$v = \frac{ds}{dt} = \frac{-1}{t^2} + 2t,$$

and

$$a = \frac{dv}{dt} = \frac{2}{t^3} + 2.$$

When $t = 1$, $a = 4$.

PROBLEMS

1. If the position of a particle is given by

$$s = 3t^3 - 14t + 1,$$

find the velocity and acceleration.

2. If the velocity of a particle is given by

$$v = \frac{ds}{dt} = 2t^2 - 5t + 1,$$

find an expression for s. Is this the only possible function $s(t)$ whose derivative is $2t^2 - 5t + 1$?

3. If in Problem 2 we know that when $t = 0$, $s = 10$, is it possible to determine s uniquely?

4. If the acceleration of a particle is given by

$$a = \frac{dv}{dt} = \frac{d^2s}{dt^2} = 8t + 7,$$

find expressions for v and for s.

5. Find $f'(x)$, $f''(x)$, and $f'''(x)$.

a) $f(x) = 3x^2 - 4$.

b) $f(x) = \dfrac{1}{x^3 + 1}$.

c) $f(x) = 2x^3 - 4x + 5$.

d) $f(x) = -5x^4 + 2x^2 - 4x$.

e) $f(x) = \sqrt{x} + 5x - 1/x$.

f) $f(x) = -2x^3 + 4x^2 - 1$.

g) $f(x) = \dfrac{1}{2x^2 + 1}$.

h) $f(x) = \sqrt{x}$.

6. If the position of a particle is given by

$$s = 4t^3 - 7t^2 + 5,$$

find the velocity and acceleration.

7. If $v = ds/dt = 3t^2 + 4t - 1$ and when $t = 0$, $s = 0$, find an expression for s.

5.7 THE CHAIN RULE

We know how to differentiate sums and products of functions. But there is another way we can combine functions, namely by composition. Recall that if f and g are functions we define the composite function by

$$(g \circ f)(x) = g(f(x)),$$

whenever $f(x) \in$ domain g. (We recall that the order of composition is important.) We want to determine the derivative of the composite of differentiable functions f and g in terms of the derivatives of f and g.

To do this we shall need one of the most powerful rules of differentiation, one that will enable us to differentiate complicated expressions by considering them as composite functions. For example, if we want to know the derivative of

$$f(x) = (x^2 - 3x + 2)^9,$$

it is clear that we can raise the expression inside the parentheses to the ninth power and differentiate the resulting polynomial by the rules of Section 3. However, this is very tedious and provides lots of opportunities for errors. Suppose we define a new function $u = g(x)$ by

$$u = g(x) = x^2 - 3x + 2.$$

Then $f(x) = u^9$ and

$$\frac{df}{du} = 9u^8.$$

Unfortunately, we want the derivative with respect to x, not with respect to u, so more work needs to be done. For the present we only note that if we let $h(x) = x^9$, then $h(g(x)) = f(x)$; that is, $f(x)$ may be regarded as a composite function.

If we are given

$$f(x) = \sqrt{3x^2 + 1},$$

it is clear that we cannot use our present formulas to find $f'(x)$. However, we can define a new variable u as a function of x and find df/du:

$$u = 3x^2 + 1 \qquad f(x) = u^{1/2},$$

so $df/du = \frac{1}{2}u^{-1/2}$. Again, we may write $f(x)$ as a composite function by letting $h(x) = x^{1/2}$ and $u = g(x)$ so that $f(x) = h(g(x))$. Thus the question we must answer is: How do we change the variable with respect to which we are differentiating, or, equivalently, how do we differentiate a composite function?

Let us consider the derivative of the composite of differentiable functions f and g. By definition, at x_0

$$\frac{d(g(f(x)))}{dx} = \lim_{x \to x_0} \frac{g(f(x)) - g(f(x_0))}{x - x_0}.$$

Now suppose that for all x in some neighborhood of x_0, $f(x) \neq f(x_0)$ if $x \neq x_0$. Then we multiply the ratio whose limit we are considering by

$$\frac{f(x) - f(x_0)}{f(x) - f(x_0)}$$

to obtain

$$\frac{d(g(f(x)))}{dx} = \lim_{x \to x_0} \frac{g(f(x)) - g(f(x_0))}{f(x) - f(x_0)} \cdot \frac{f(x) - f(x_0)}{x - x_0}.$$

Since the continuity of $f(x)$ gives us that as $x \to x_0$, $f(x) \to f(x_0)$, we have

$$\lim_{x \to x_0} \frac{g(f(x)) - g(f(x_0))}{f(x) - f(x_0)} = g'(f(x_0)).$$

(If $f(x)$ is differentiable at x_0, and $g(x)$ is differentiable at $f(x_0)$.)

This suggests the *chain rule*:

$$\boxed{\frac{d(g(f(x)))}{dx} = g'(f(x))f'(x),}$$

whenever the right-hand side is defined; or if we let $f(x) = u$ we have

$$\frac{dg}{dx} = \frac{dg}{du} \cdot \frac{du}{dx}.$$

Since we assumed in the derivation above that $f(x) \neq f(x_0)$, we have not proved the validity of the chain rule in its full generality. However, we shall use it anyway. For a proof the interested reader may consult, for example, Lang: *A First Course in Calculus.*

Note that in the chain rule the derivatives behave *as if du* is being "cancelled" in a multiplication problem. There is nothing wrong with remembering the rule this way as long as it is clear that the behavior is illusory.

Examples

1. We return to the problem at the beginning of this section. Let $f(x)$ be given by $(x^2 - 3x + 2)^9$. Find $f'(x)$.

Solution. We let $u = x^2 - 3x + 2$. Then

$$\frac{df}{du} = \frac{d(u^9)}{du} = 9u^8.$$

Also,

$$\frac{du}{dx} = 2x - 3.$$

Thus by the chain rule

$$\frac{df}{dx} = \frac{df}{du} \cdot \frac{du}{dx} = 9u^8(2x - 3) = 9(x^2 - 3x + 2)^8(2x - 3).$$

2. The other problem we introduced was to find $f'(x)$ if

$$f(x) = \sqrt{3x^2 + 1}$$

Solution. Let $u = 3x^2 + 1$. Then

$$\frac{df}{du} = \frac{d(u^{1/2})}{du} = \tfrac{1}{2}u^{-1/2}.$$

Moreover $du/dx = 6x$, so by applying the chain rule we have

$$\frac{df}{dx} = \frac{df}{du} \cdot \frac{du}{dx} = \tfrac{1}{2}u^{-1/2}(6x) = \tfrac{1}{2}(3x^2 + 1)^{-1/2}(6x) = \frac{3x}{\sqrt{3x^2 + 1}}.$$

3. If

$$f(x) = \frac{x + 2}{\sqrt{3x^2 + x}},$$

find $f'(x)$.

Solution. Here the problem is more complicated. We must use the chain rule and the rule for quotients. Let

$$g(x) = \sqrt{3x^2 + x}, \qquad u = 3x^2 + x, \qquad h(x) = x + 2.$$

Then

$$\frac{du}{dx} = 6x + 1, \qquad g(x) = u^{1/2},$$

$$\frac{dh}{dx} = 1, \qquad \frac{dg}{du} = \tfrac{1}{2}u^{-1/2}.$$

From this we have

$$\frac{dg}{dx} = \frac{dg}{du} \cdot \frac{du}{dx} = \tfrac{1}{2}u^{-1/2}(6x + 1) = \tfrac{1}{2}(3x^2 + x)^{-1/2}(6x + 1).$$

Now we can use the quotient formula:

$$\frac{df}{dx} = \frac{g(dh/dx) - h(dg/dx)}{g^2}$$

$$= \frac{(3x^2 + x)^{1/2} \cdot 1 - (x + 2)\frac{1}{2}(3x^2 + x)^{-1/2}(6x + 1)}{3x^2 + x}$$

$$= \frac{1}{(3x^2 + x)^{1/2}} - \frac{6x^2 + 13x + 2}{2(3x^2 + x)^{3/2}}$$

$$= \frac{6x^2 + 2x - 6x^2 - 13x - 2}{2(3x^2 + x)^{3/2}}$$

$$= \frac{-11x - 2}{2(3x^2 + x)^{3/2}}.$$

Warning. While the process seems tedious and unnecessarily long, the reader is advised to go through the process of introducing a new variable and making the preliminary computations before applying the chain rule—at least until he has worked enough problems for the inclusion of du/dx to be automatic. An even more common error than using the product of derivatives as the derivative of a product is to omit the quantity du/dx—for example, if

$$f(x) = \sqrt{5x + 2},$$

the tendency is to forget that

$$\frac{d(5x + 2)}{dx} = 5.$$

The correct result is

$$f'(x) = \frac{5}{2\sqrt{5x + 2}}.$$

We combine the rule of differentiating powers of x with the chain rule to get a formula which simplifies the application of the chain rule to algebraic functions.

Theorem 7. If $f(x)$ is differentiable and a is a real number, then

$$\frac{df^a}{dx} = af^{a-1}\frac{df}{dx},$$

whenever both sides are defined.

Proof. We let $u = f(x)$.

$$\frac{du^a}{du} = au^{a-1},$$

so by the chain rule

$$\frac{df^a}{dx} = \frac{du^a}{dx} = \frac{du^a}{du}\frac{du}{dx} = au^{a-1}\frac{du}{dx} = af^{a-1}\frac{df}{dx}.$$

Examples

1. Find the derivative of $(3x^2 + 5x - 9)^{2/3}$.

Solution. Let

$$f(x) = 3x^2 + 5x - 9,$$

so that

$$\frac{df}{dx} = 6x + 5.$$

The function whose derivative we want to find is $f^{2/3}$. We have

$$\frac{df^{2/3}}{dx} = \tfrac{2}{3}f^{-1/3}\frac{df}{dx}$$

$$= \tfrac{2}{3}(3x^2 + 5x - 9)^{-1/3}(6x + 5).$$

2. Find the second derivative of $(t^2 - 8t + 2)^3$.

Solution. We let

$$f(t) = t^2 - 8t + 2,$$

so that

$$\frac{df}{dt} = 2t - 8.$$

The first derivative of the original function is

$$3(t^2 - 8t + 2)^2(2t - 8).$$

Now we treat this derivative as a product and find its derivative to give us the second derivative of the original function.

Let $g(t) = 3(t^2 - 8t + 2)^2$, $h(t) = 2t - 8$, $f(t) = t^2 - 8t + 2$. We have from above that $df/dt = 2t - 8$. We use Theorem 7 to find

$$\frac{dg}{dt} = 6(t^2 - 8t + 2)(2t - 8).$$

Finally, the second derivative is given by

$$g\frac{dh}{dt} + h\frac{dg}{dt} = 3(t^2 - 8t + 2)^2(2) + (2t - 8)(6)(t^2 - 8t + 2)(2t - 8).$$

PROBLEMS

1. Find the derivative of each of the following functions:

a) $(3x^3 - 4x + 1)^3$.

b) $\sqrt{5x^2 - 9x}$.

c) $(7x^5 - 4x - 1)^{-2}$.

d) $\sqrt{3x^3 - 5x + 1}$.

e) $(7x^2 - 4x + 2)^{1/2}(-4x + 1)^{-1/3}$.

f) $\sqrt{\dfrac{8x - 1}{3x + 2}}$.

g) $\dfrac{(16x^2 - 5x + 1)^3}{(4x^3 - 5x^2)^{2/3}}$.

h) $(6x^5 - 9x^2 - 4x + 1)^{15}$.

i) $(ax^4 - bx^3 - cx + d)^7$.

j) $\sqrt{(6x - 5)(4x + 1)(17x^4 + 1)^{13}}$.

k) $\sqrt{6x^2 + \sqrt{2x}}$.

l) $(5x^4 - 3x^3 + x - 1)^{14}$.

m) $\dfrac{1}{(6x^3 - 5x + 5)^3}$.

n) $\sqrt{3x^2 - 9x + 2}$.

o) $\sqrt{\dfrac{3x - 1}{3x^3 - 4x^2}}$.

2. If the position of a particle at time t is given by

a) $s = \sqrt{t^2 - 2t + 3}$,

b) $s = (3t^3 - 4t - 1)^{2/3}$,

c) $s = \dfrac{1}{(2t^4 - 5t)^2}$,

d) $s = \sqrt{3t^2 - 4t - 5}$,

find its velocity and acceleration.

3. Find the equation of the line tangent to the given curve at the given point.

a) $f(x) = (x^2 - 5x + 1)^6$ at $x = 2$.

b) $f(x) = 1/(\sqrt{2x} + 5)$ at $x = 2$.

c) $f(x) = (2x^2 + 4x + 4)^{3/2}$ at $x = 0$.

d) $f(x) = (2x^3 + 4x + 1)^{-1/3}$ at $x = 0$.

e) $f(x) = \sqrt{2x^2 + 1}$ at $x = 0$.

f) $f(x) = \dfrac{2x}{(x^2 + 3x + 1)^2}$ at $x = 0$.

5.8 INVERSE FUNCTIONS AND IMPLICIT DIFFERENTIATION

We can use the chain rule to derive a formula for the derivative of an inverse function. We know what it means to speak of an additive inverse. $-a$ is the additive inverse of a because

$$a + (-a) = 0.$$

Similarly, $1/a$ is the multiplicative inverse of $a \neq 0$ because

$$a\frac{1}{a} = 1.$$

However, on functions we have another operation, that of composition. It would be nice for functions to have inverses with respect to this operation, and of course it would be even nicer if, given the derivative of a function, we could find the derivative of its inverse. If $f(x) = y$, it would seem natural to define $f^{-1}(y) = x$. However, this may not be a good definition in the sense that there may be more than one x such that $f(x) = y$, and therefore $f^{-1}(y)$ will not be uniquely determined. Recall that in the definition of function we require that for each value of x there be a uniquely determined value of $f(x)$. Nothing is said about two different values of x giving the same value for $f(x)$; that is, a value of $f(x)$ need not uniquely determine a value for x. Therefore, making the definition

$$f^{-1}(y) = x \qquad \text{if} \quad f(x) = y,$$

may not make f^{-1} a function.

For example, if

$$f(x) = x^2 + 1,$$

$x = 1$ and $x = -1$ both yield $f(x) = 2$. Thus $f^{-1}(2)$ is both 1 and -1, an unsatisfactory situation.

Suppose, however, that $f(x)$ is a function such that

$$f(x_1) = f(x_2) \qquad \text{implies} \quad x_1 = x_2.$$

In other words, the uniqueness is reciprocal—for each value of $f(x)$ there is precisely one value of x. Then we say f is a *one-to-one function.* If f is a one-to-one function, then we define its *inverse* by

$$f^{-1}(y) = x \qquad \text{where} \qquad f(x) = y.$$

Since f is one-to-one, f^{-1} is also a function. Moreover,

$$f(f^{-1}(y)) = f(x) = y \qquad \text{and} \qquad f^{-1}(f(x)) = f^{-1}(y) = x.$$

Thus $f^{-1} \circ f$ and $f \circ f^{-1}$ are both the identity function $I(x) = x$ and f^{-1} is an inverse for the operation of composition of functions.

The derivative of the identity function $I(x) = x$ is 1, so since $I(y) = y = (f \circ f^{-1})(y)$, we have

$$\begin{aligned} 1 &= (f \circ f^{-1})'(y) = f'(f^{-1}(y))(f^{-1})'(y) \\ &= f'(x)(f^{-1})'(y) \\ &= \frac{dy}{dx} \cdot \frac{dx}{dy}. \end{aligned}$$

Thus

$$\frac{dx}{dy} = \frac{1}{dy/dx},$$

provided dy/dx is defined and not zero. In other words, if the derivative of the inverse function exists it is the reciprocal of the derivative of the function.

Warning. The derivatives of inverse functions are reciprocals, where inverse means with respect to the operation of composition. $1/f$ is the inverse of f with respect to multiplication, but it is *not* the case that $(1/f)' = 1/f'$. In fact, $(1/f)' = -f'/f^2$.

Examples

1. Let $y = 3x + 2$. Then $dy/dx = 3$, and since $y = f(x)$ is one-to-one, we have by the above that $dx/dy = \frac{1}{3}$. We can also obtain this result by actually determining the inverse function, that is, by solving for x in terms of y:

$$x = \frac{y - 2}{3}, \qquad \frac{dx}{dy} = \frac{1}{3}.$$

However, for more complicated functions application of the reciprocal rule sometimes greatly simplifies calculation of the derivative.

2. Consider an economic situation where demand and price are the only variables. Let D units be the quantity demanded and P the price per unit. We make several assumptions:

1) D and P are always nonnegative.

2) P is a one-to-one function of D and D is a one-to-one function of P.

3) D is a decreasing function of P.

Condition (3) gives what is called a "normal" demand function, one which goes down as the price goes up. That is, if the item is free ($P = 0$) D will be very large. As P increases, D decreases until a point P_m is reached which represents the highest price anyone will pay. The domain of the demand function is $[0, P_m]$. If

$$D(P) = -3P + 12,$$

(Fig. 5.15) then we can solve this equation for P:

$$P(D) = \frac{-D + 12}{3}.$$

These functions are inverses of each other since

$$P(D(P)) = P(-3P + 12) = \frac{-(-3P + 12) + 12}{3} = P,$$

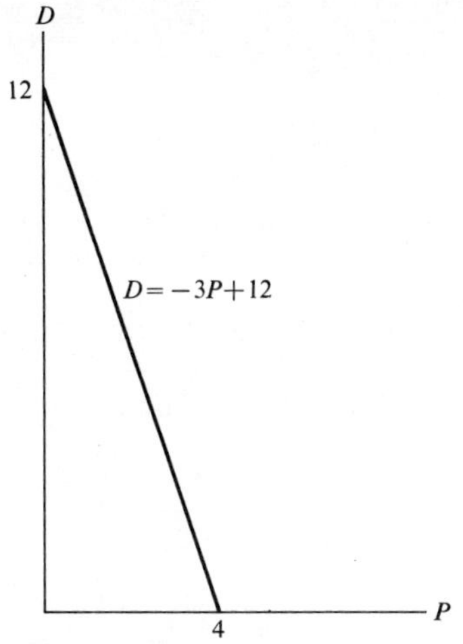

Fig. 5.15

and similarly $D(P(D)) = D$. We have

$$\frac{dD}{dP} = -3, \qquad \frac{dP}{dD} = \frac{-1}{3} = \frac{1}{dD/dP}.$$

Let us take another example subject to the same conditions. Suppose the demand is given by

$$D = -2P^2 - 4P + 50.$$

Solving for P, we obtain

$$P = \frac{4 \pm \sqrt{16 + 8(50 - D)}}{-4}.$$

As it stands, this is not a function, but Condition (1) tells us that P must be nonnegative, so we choose the minus sign in the $\pm$ to get P as a function of D:

$$P = -1 + \tfrac{1}{2}\sqrt{104 - 2D}.$$

Now

$$\frac{dD}{dP} = -4P - 4,$$

and

$$\frac{dP}{dD} = \tfrac{1}{2} \cdot \tfrac{1}{2}(104 - 2D)^{-1/2}(-2) = \frac{-1}{2\sqrt{104 - 2D}}.$$

At first glance, these do not appear to be reciprocals, but making the appropriate substitution yields

$$\frac{dD}{dP} = -4(-1 + \tfrac{1}{2}\sqrt{104 - 2D}) - 4 = -2\sqrt{104 - 2D}$$
$$= \frac{1}{dP/dD}.$$

We state the result we derived above as a theorem:

Theorem 8. If f and f^{-1} are differentiable, then

$$(f^{-1})' = \frac{1}{f'}.$$

The information about derivatives of inverses is more useful than it might appear since it is frequently possible to define an inverse by restricting the domain of a function. For example, if $f(x) = x^2$, then $f^{-1}(x) = \sqrt{x}$ is an inverse function for f when f is considered as a function on the nonnegative real numbers (here as usual $\sqrt{x}$ indicates the positive square root). In fact, if f is one-to-one on some neighborhood of x_0, we may define an inverse function on that neighborhood and consider its derivative.

We note that if f and g are inverse functions, that is, $f(g(x)) = x$ and $g(f(x)) = x$, then if we graph $y = f(x)$ and $y = g(x)$, the graphs are mirror images with respect to the line $y = x$ (Fig. 5.16).

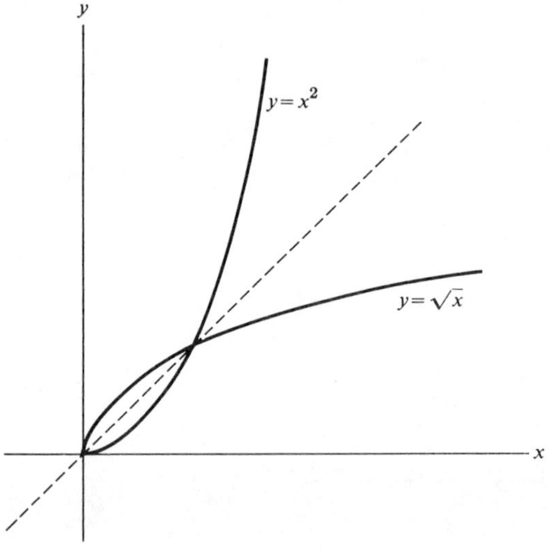

Fig. 5.16

Suppose we have a function given by

$$y^2 = x^2 + 1,$$

and we want to find dy/dx. It is clear that we can solve for y, obtaining

$$y = \pm\sqrt{x^2 + 1},$$

(that is, two different functions, one with the + and one with the −). Then

$$\frac{dy}{dx} = \pm\tfrac{1}{2}(x^2 + 1)^{-1/2}(2x).$$

It is, however, customary to take the positive square root in such a case.

On the other hand, using the chain rule, we simply consider y as *some* (differentiable) function of x, without solving the equation to determine precisely what function. Then if we let

$$u = y^2,$$

we have

$$\frac{du}{dx} = \frac{du}{dy}\frac{dy}{dx} = 2y\frac{dy}{dx}.$$

But

$$u = y^2 = x^2 + 1,$$

so that

$$\frac{du}{dx} = \frac{d(x^2 + 1)}{dx} = 2x.$$

Putting these two equations together we obtain

$$2y\frac{dy}{dx} = 2x,$$

or

$$\frac{dy}{dx} = \frac{x}{y},$$

which, when we substitute for y, is the same answer as above. However, if we are content with dy/dx as a function of both x and y rather than as a function of x alone, x/y is the answer easily obtained by using the chain rule and without ever solving for y.

This procedure is called *implicit differentiation* since y is given as an implicit rather than explicit function of x.

It is not really necessary to make the substitution of u above; we can simply differentiate both sides of the original equation with respect to x. Thus in general if

$$f(y) = g(x),$$

where f is a differentiable function of y, and g a differentiable function of x we have

$$\frac{df}{dx} = \frac{df}{dy}\frac{dy}{dx} = \frac{dg}{dx} \qquad \text{so that} \qquad \frac{dy}{dx} = \frac{dg}{dx}\Big/\frac{df}{dy}, \qquad \text{if} \quad \frac{df}{dy} \neq 0.$$

Examples

1. If

$$x^2 + 3xy + y^2 = 2,$$

find dy/dx.

Solution. We have

$$2x + 3x\frac{dy}{dx} + 3y + 2y\frac{dy}{dx} = 0,$$

and

$$\frac{dy}{dx} = \frac{-2x - 3y}{3x + 2y}.$$

Warning. There are two points to observe in the above example. They are (1) $3xy$ is a product and must be differentiated as such; and (2) we are considering y as an implicit function of x and applying the chain rule.

2. If

$$\frac{y^2}{x + 1} = x - 2 \qquad \text{find} \quad \frac{dy}{dx}.$$

Solution. Using implicit differentiation and the quotient rule we obtain

$$\frac{(x + 1)2y\,(dy/dx) - y^2}{(x + 1)^2} = 1 \qquad \text{or} \qquad \frac{dy}{dx} = \frac{(x + 1)^2 + y^2}{(x + 1)2y}.$$

3. Find where the tangents to the curve $4x^2 + 2xy + y^2 = 12$ are horizontal and where they are vertical.

Solution. We have

$$8x + 2x\frac{dy}{dx} + 2y + 2y\frac{dy}{dx} = 0$$

$$(x + y)\frac{dy}{dx} = -(4x + y)$$

$$\frac{dy}{dx} = -\frac{4x + y}{x + y}.$$

This gives us the slope of the tangent. When

$$\frac{dy}{dx} = 0, \qquad y = -4x.$$

Substituting in the original equation gives

$$4x^2 - 8x^2 + 16x^2 = 12 \qquad \text{so that} \qquad x^2 = 1.$$

Thus the tangent is horizontal at $(1, -4)$ and at $(-1, 4)$. When the denominator of dy/dx is zero, the tangent is vertical. This happens when $y = -x$. Substituting again,

$$4x^2 - 2x^2 + x^2 = 12 \qquad \text{so that} \qquad x^2 = 4.$$

Thus the tangent is vertical at $(2, -2)$ and at $(-2, 2)$.

Solving the equation for the intercepts gives

$$x = 0 \qquad y = \pm 2\sqrt{3}$$

$$y = 0 \qquad x = \pm\sqrt{3}.$$

We can use this information and that about the tangents to sketch the curve (see Fig. 5.17).

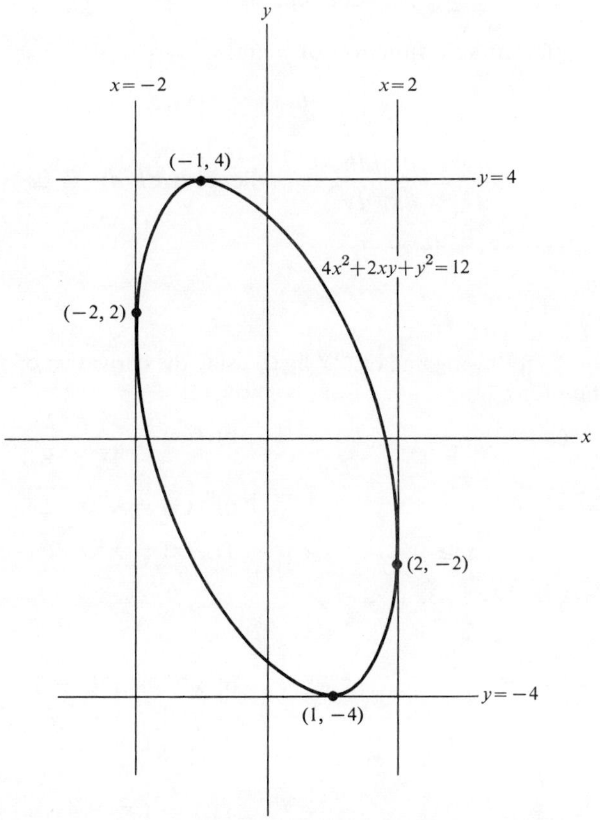

Fig. 5.17

SUMMARY OF DIFFERENTIATION RULES

Assume f, g, and u differentiable functions of x, c and a real numbers.

1. $$(f + g)' = f' + g'.$$

2. $$(fg)' = fg' + gf'.$$

3. $$(cf)' = cf'.$$

4. $$\left(\frac{f}{g}\right)' = \frac{gf' - fg'}{g^2}, \qquad \text{whenever } g \neq 0.$$

5. If h is a differentiable function of u,

$$\frac{dh}{dx} = \frac{dh}{du}\frac{du}{dx}.$$

6. $$\frac{du^a}{dx} = au^{a-1}\frac{du}{dx},$$ whenever both sides of the equation are defined.

7. If h is a differentiable function of y and

$$h(y) = f(x),$$

then

$$\frac{dy}{dx} = \frac{df}{dx}\bigg/\frac{dh}{dy}, \qquad \text{whenever } dh/dy \neq 0.$$

PROBLEMS

1. In each of the following find $(f^{-1})'$ by (i) using the derivative of f and (ii) by finding an equation for f^{-1}.

   a) $f(x) = 6x - 7.$

   b) $f(x) = \dfrac{3x + 2}{4x - 3}.$

   c) $f(x) = x^3 - 4.$

   d) $f(x) = \sqrt{2x - 3}.$

   e) $f(x) = x^2 + 1,\ x \geq 0.$

   f) $f(x) = x^2 + 2x - 3,\ x \geq -1.$

   g) $f(x) = x^2 + 2x - 3,\ x < -1.$

2. Find dy/dx:

   a) $x^2 + y^2 = 25.$

   b) $x^{1/2} + y^{1/2} = 1.$

   c) $y^2 = x^3 - 2.$

   d) $\dfrac{1}{y} + \dfrac{1}{x} = 1.$

   e) $y^2 = x^2 - x.$

   f) $y = (x^2 + 4x)^3.$

g) $x^2y^2 = x^2 + 2y^2$.

h) $y^2 = \sqrt{x} + 1/x^2$.

i) $y^2 + 2xy = x + y$.

j) $y = \dfrac{x^2 - 1}{x^2 + 1}$.

k) $y = x\sqrt{x^2 + 1}$.

l) $x^2 + xy + x^4 = 2$.

m) $4x^2 + y^2 = 16$.

n) $\dfrac{x^{1/2}}{2} - \dfrac{y^{1/2}}{2} = 1$.

o) $y^2 = x - y + 3xy$.

p) $\dfrac{1}{y} - 3x^2 + x - y = 1$.

3. A line is said to be *normal* to a curve at (x_0, y_0) if it is perpendicular to the line tangent to the curve at (x_0, y_0). Find the tangent and normal lines to the curves at the indicated points.

 a) $x^2 + 4xy + 2y^2 = 1$ at $(0, 1/\sqrt{2})$. b) $4x^2 - y^2 = 16$ at $(2, 0)$.

4. Find the inverse of $f(x)$ and the derivative of the inverse.

 a) $f(x) = \sqrt{3x + 1}$. b) $f(x) = 5x + 4$.

 c) $f(x) = x^2 - 2x - 8,\ x \geq 1$. d) $f(x) = 2x - 16$.

6 / APPLICATIONS OF THE DERIVATIVE

We have already seen one application of the derivative, namely finding a line tangent to a curve. In this chapter we look at others, most of which involve the determination of the behavior of a function by studying its derivative.

6.1 MAXIMA AND MINIMA

A source of many applications of differentiation is the problem of maximizing (or minimizing) some quantity. A typical instance would be: A certain number of square meters of sheet metal are available for the manufacture of tin cans. What should be the relation between the diameter and height of the can in order to enclose the greatest volume with a fixed amount of metal?

The definition of maximum coincides with our intuitive notion: If a point x_0 is such that $f(x_0) \geq f(x)$ for all x in the domain of f, we say that $f(x)$ has a *maximum* at x_0, or that $f(x_0)$ is the maximum value for the function. Similarly, if x_0 is such that $f(x_0) \leq f(x)$ for all $x \in$ domain f, we say that $f(x)$ has a *minimum* at x_0. We saw that the vertex (maximum or minimum y-coordinate) of a parabola could be determined by differentiating the equation of the curve, setting the derivative equal to zero and solving for x.

With certain refinements this technique of differentiating and setting the derivative equal to zero is what we use. The real difficulty in maximization problems usually arises in formulating the function to be differentiated. Partly this is the old "word-problem dilemma"—translating English to mathematics—discussed in Chapter 2. On the other hand, there are also situations in which the conditions of the problem cannot be expressed as a nice differentiable function. Some of the latter are treated in Chapter 10.

In the case of the parabola we are interested in points at which the derivative of a function is zero. We give a name to such points. If $f(x)$ is differentiable at x_0 with $f'(x_0) = 0$, then x_0 is a *critical point* of the function. Each parabola has one critical point; at that point the function achieves a maximum or minimum.

If $f(x_0) \geq f(x)$ for all x in some neighborhood of x_0, then $f(x)$ has a *local maximum* at x_0 and $f(x_0)$ is a local maximum for $f(x)$. *Local minimum* is defined analogously.

Examples

1. If $f(x) = 3x + 2$, then the function has neither a maximum, a minimum, local maxima, nor local minima (Fig. 6.1). If we restrict the domain of $f(x)$, and consider it to be defined only on $[-1, 1]$, then at -1 there is a minimum and at 1 a maximum. In fact, on any finite interval the left-hand endpoint is a minimum and the right-hand endpoint a maximum.

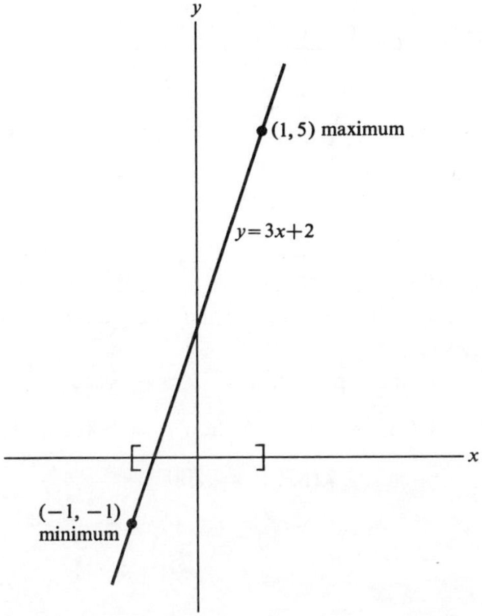

Fig. 6.1

2. If $f(x) = 3x^2 - 2$, then there is a minimum at 0 and no maximum (Fig. 6.2).
3. If $f(x) = x^3 - 9x + 1$, then

$$f'(x) = 3x^2 - 9.$$

When $f'(x) = 0$, $x = \pm\sqrt{3}$. By looking at the graph in Fig. 6.3 we observe that at neither of these points is there a maximum or minimum, but the value of $f(x)$ at $x = \sqrt{3}$ is less than its value at nearby values of x and, similarly, at $x = -\sqrt{3}$ the value of $f(x)$ is greater than at nearby values of x. Hence there is a local minimum at $\sqrt{3}$ and a local maximum at $-\sqrt{3}$. There is no maximum or minimum.

4. To run a plane costs \$300 per hour plus fuel. The cost of fuel per hour is proportional to the square of the speed and is \$200 per hour when the speed is 500 mph. The plane's maximum speed is 1000 mph. At what speed should the plane be flown to assure the minimum cost per mile?

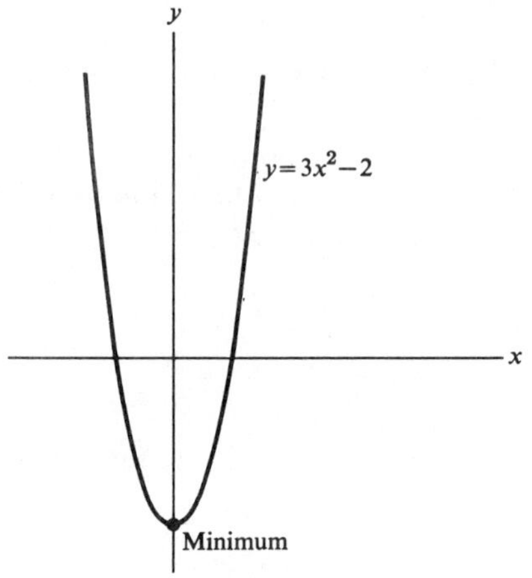

Fig. 6.2

Solution. The cost C of running the plane for a distance D is the cost per hour times the number of hours. The cost per mile is C/D; this is the quantity we want to minimize. We have

$$\text{cost per hour} = 300 + kv^2,$$

where v is the speed and k a constant. When $v = 500$, the fuel cost is 200 so

$$k(500)^2 = 200$$

$$k = \frac{1}{1250}.$$

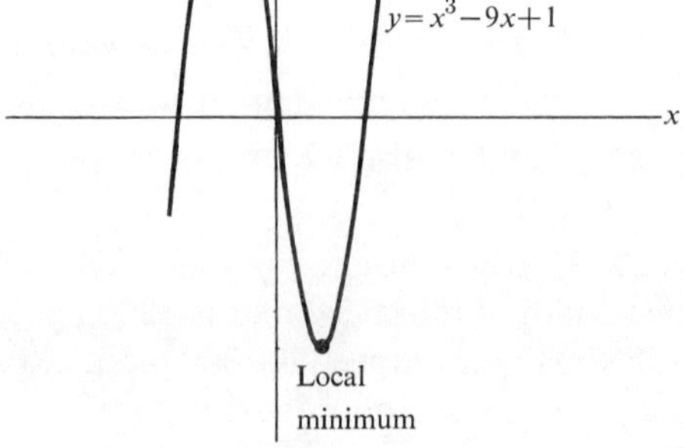

Fig. 6.3

Therefore,

$$\text{cost per hour} = 300 + \frac{v^2}{1250}.$$

Since $D = vt$, the number of hours is $t = D/v$, and the cost of running the plane D miles is

$$C = (\text{cost per hour}) \times (\text{number of hours})$$

$$C = \left(300 + \frac{v^2}{1250}\right)\frac{D}{v}.$$

We want to minimize

$$\frac{C}{D} = \left(300 + \frac{v^2}{1250}\right)\frac{1}{v}.$$

This is not a parabolic function, but if we proceed with the method we used for parabolas, we find

$$\left(\frac{C}{D}\right)' = \frac{-300}{v^2} + \frac{1}{1250}.$$

Setting this equal to zero yields

$$v = 612.$$

Although we know that the point where the derivative of a parabola is zero is actually a maximum or a minimum, we do not know that for other functions we can find a maximum or minimum in this fashion. We study this problem in the next section.

A question naturally arises: Is a maximum necessarily a local maximum? (same for minimum). The answer is a qualified Yes. Recall Example 1 above with the restricted domain. The maximum occurs at an endpoint; if one allows "one-sided" neighborhoods to qualify in the definition of local maximum, then the maximum is a local maximum (Fig. 6.4).

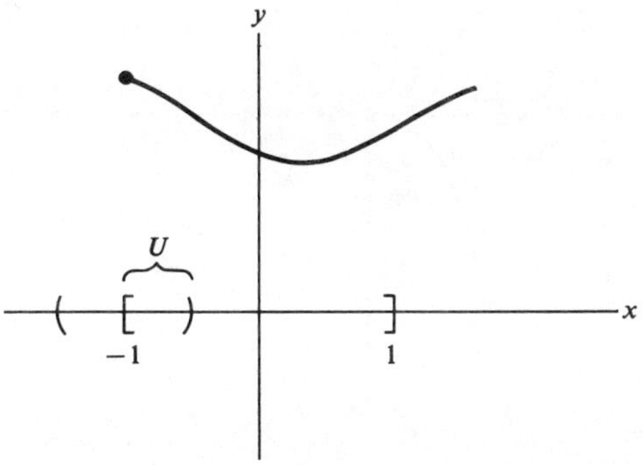

Fig. 6.4

Theorem 1. If $f(x)$ is defined on (a, b) and differentiable at $x_0 \in (a, b)$, and x_0 is a local maximum (or minimum) for $f(x)$, then $f'(x_0) = 0$.

Proof. We prove the maximum case. Consider

$$\frac{f(x) - f(x_0)}{x - x_0}$$

for values of x near x_0. If x_0 is a local maximum, there is a neighborhood U of x_0 such that for all $x \in U$, $f(x) - f(x_0)$ is always nonpositive. For values of x to the right of x_0, $x - x_0$ is positive so

$$\frac{f(x) - f(x_0)}{x - x_0} \leq 0.$$

On the other hand, if x is to the left of x_0, $x - x_0$ is negative, so that

$$\frac{f(x) - f(x_0)}{x - x_0} \geq 0.$$

But, since $f(x)$ is differentiable at x_0, we know that

$$\lim_{x \to x_0} \frac{f(x) - f(x_0)}{x - x_0}$$

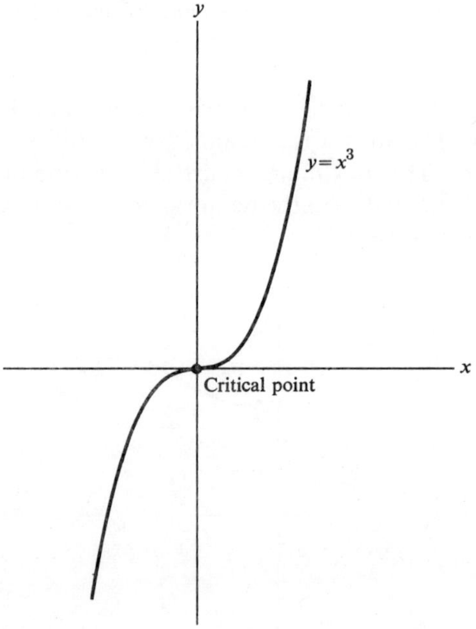

Fig. 6.5

exists. Hence it must be that

$$\lim_{x \to x_0} \frac{f(x) - f(x_0)}{x - x_0} = f'(x_0) = 0.$$

Warning. If $f(x)$ is defined on $[a, b]$ and has a local maximum or minimum at a or b, it is not necessarily the case that $f'(a)$ or $f'(b)$ is zero. In Example 1 above, the derivative has a value of 3 at the endpoints.

Warning. The converse of Theorem 1 is not true. Consider for example the function (Fig. 6.5)

$$f(x) = x^3.$$

Clearly $f'(0) = 0$, but 0 is neither a local maximum nor a local minimum.

PROBLEMS

1. Find the maximum and minimum values, if they exist, of the following functions:
   a) $f(x) = 1/x$. b) $f(x) = 3x^2 + 2x - 1$.
   c) $f(x) = 3x + 14$. d) $f(x) = x^3 - 5x + 2$.
   e) $f(x) = -7x^2 + 4x - 2$. f) $f(x) = \dfrac{x + 1}{2x - 2}$.
2. Find the maximum value (if it exists) of each of the functions of Problem 1 on the interval $[1, 2]$.
3. Find the local maxima and local minima (if any) of the following functions:
   a) $f(x) = -3x^2 + 2x - 4$. b) $f(x) = \dfrac{-1}{2x + 1}$.
   c) $f(x) = 3x^3 + 2x^2 - 5x + 1$. d) $f(x) = \sqrt{x + 1}$.
   e) $f(x) = 5x^2 + 1$. f) $f(x) = \dfrac{1}{\sqrt{2x - 3}}$.
4. Find the critical points of the following functions.
   a) $f(x) = 2x^2 + 3x - 4$. b) $f(x) = \sqrt{2x^2 + 1}$.
   c) $f(x) = 2x^3 - 2x^2 - 2x + 7$.
5. Find the maxima and minima of the given functions on the specified domains.
   a) $f(x) = 3x^2 + 2x - 1$ on $[0, 1]$. b) $f(x) = 3x^2 + 2x - 1$ on $[-1, 1]$.
   c) $f(x) = -3x + 2$ on $[0, 1]$. d) $f(x) = 2x^3 - 2x^2 + 5$ on $[-2, 2]$.

6.2 CRITICAL POINTS

In view of the warning in the last section, the question arises of how to determine which critical points are maxima and minima. One method would be to investigate values of a function at nearby points. This is not always satisfactory, however, because the points selected may not be sufficiently close to the critical point.

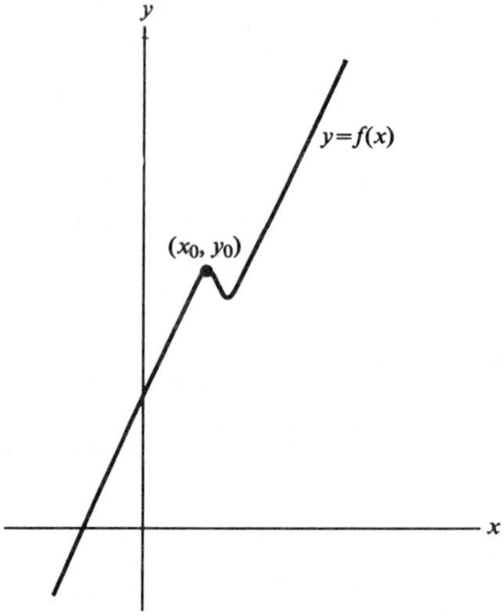

Fig. 6.6

For example, in Fig. 6.6, x_0 might appear to be neither a local maximum nor a local minimum for the function $f(x)$ since values of x very close to x_0 give values for $f(x)$ both larger and smaller than $f(x_0)$. Thus we resort to computing the second derivative in order to obtain more information. In Fig. 6.7 the slope of the tangent line is increasing as the graph of the function passes through the critical point $(x_0, f(x_0))$. This slope, which is the first derivative, has a positive rate of change given by the second derivative. At this critical point there is a local minimum. This is a general phenomenon. We state without proof the following:

Second Derivative Test. Suppose x_0 is a critical point for $f(x)$. Moreover, suppose $f'(x)$ is differentiable at x_0. Then if $f''(x_0) > 0$, $f(x)$ has a local minimum at x_0. If $f''(x_0) < 0$, $f(x)$ has a local maximum at x_0.

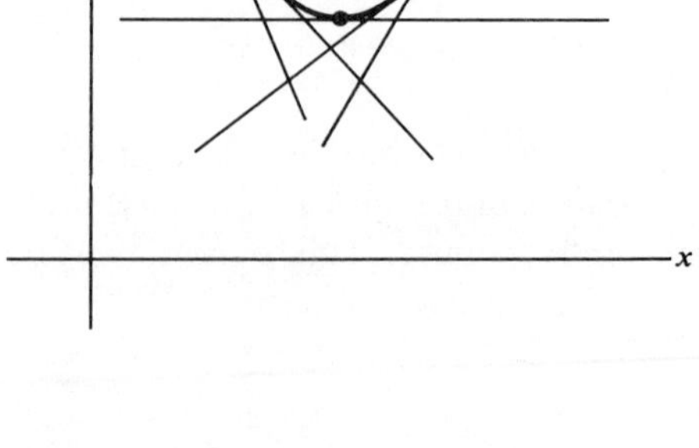

Fig. 6.7

Warning. If $f''(x_0)$ is not defined or if $f''(x_0) = 0$, this test fails to give any information. For example, if $f(x) = x^4$, $f'(0) = 0$, so zero is a critical point. But as $f''(0) = 0$ the test is of no help to us. However, the function has a minimum at 0. On the other hand, 0 is a critical point for $f(x) = x^3$, $f''(0) = 0$, but there is neither a local maximum nor a local minimum at 0 (Fig. 6.8).

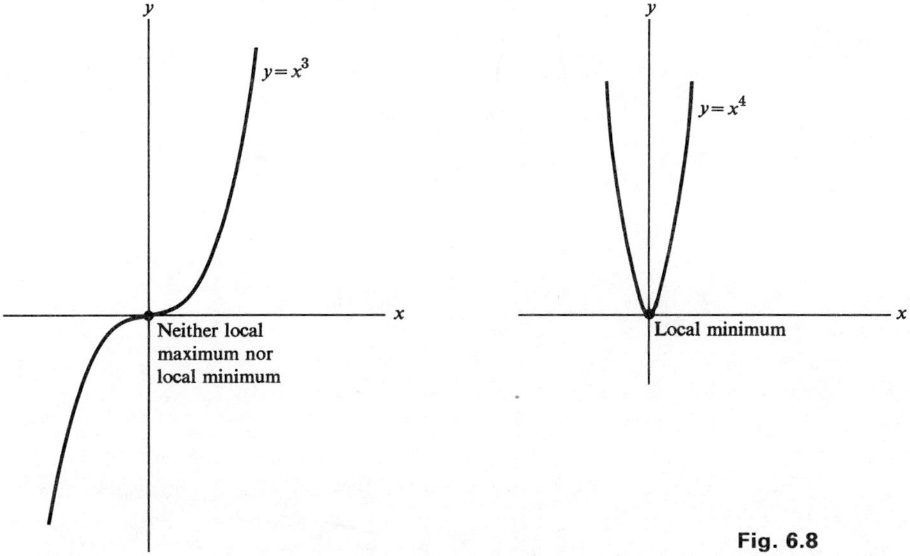

Fig. 6.8

Examples

1. Classify the critical points of $f(x) = x^3 - 4x^2 - 3x + 2$ (Fig. 6.9).

Solution. We have that

$$f'(x) = 3x^2 - 8x - 3.$$

Setting $3x^2 - 8x - 3$ equal to 0 yields the critical points $x = 3, -1/3$.

Now

$$f''(x) = 6x - 8,$$

therefore

$$f''(3) = 18 - 8 = 10 > 0,$$

so at $x = 3$ there is a local minimum, and

$$f''(-1/3) = -2 - 8 < 0,$$

so at $x = -1/3$ there is a local maximum.

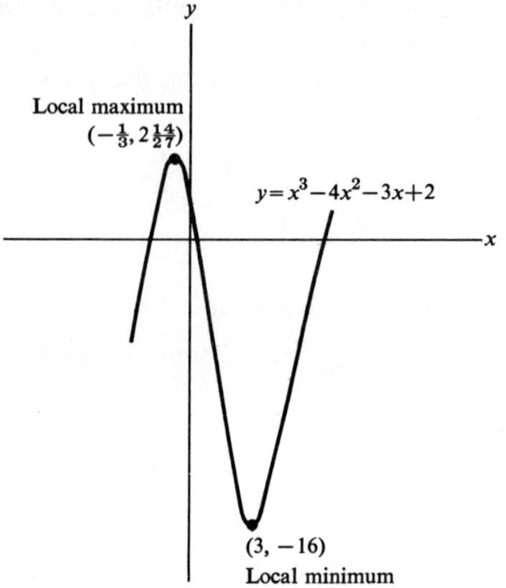

Fig. 6.9

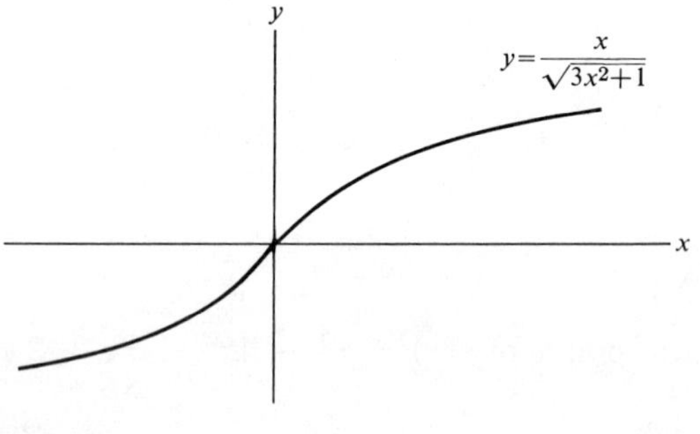

Fig. 6.10

2. Classify the critical points of

$$f(x) = \frac{x}{\sqrt{3x^2 + 1}}$$

(Fig. 6.10).

Solution. We obtain

$$f'(x) = \frac{\sqrt{3x^2 + 1} - x(3x^2 + 1)^{-1/2}(6x)/2}{3x^2 + 1}.$$

Setting $f'(x) = 0$, we have

$$\sqrt{3x^2 + 1} - \frac{3x^2}{\sqrt{3x^2 + 1}} = 0$$
$$3x^2 + 1 - 3x^2 = 0$$
$$1 = 0,$$

which is a contradiction. Thus there are no critical points.

3. Determine the critical points of

$$f(x) = x^4 + 32x - 5.$$

(See Fig. 6.11.)

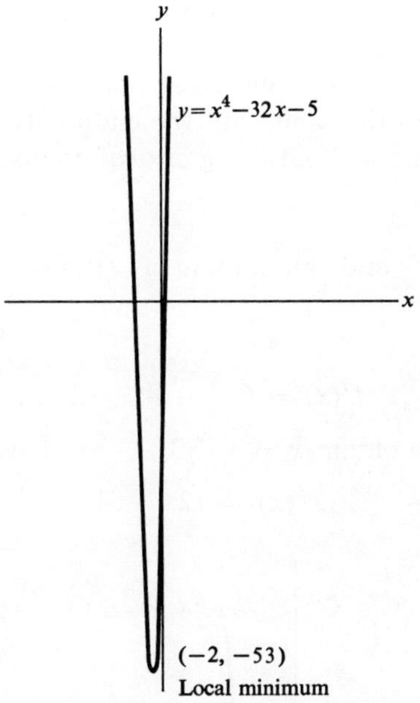

Fig. 6.11

Solution. We have

$$f'(x) = 4x^3 + 32,$$

so that when $f'(x) = 0$, $4(x^3 + 8) = 4(x + 2)(x^2 - 2x + 4) = 0$, and the critical point is $x = -2$, since the other roots are complex and hence not in the domain of $f(x)$. We find that

$$f''(x) = 12x^2,$$
$$f''(-2) = 48 > 0,$$

so there is a local minimum at $x = -2$.

The local minimum is $f(-2) = -53$. It is also a minimum, since for large values of $|x|$ the function is positive.

Suppose that we are interested in the maximum (or minimum) for a function $f(x)$ defined on the closed interval $[a, b]$. We have the following theorem, which we state without proof:

Theorem 2. If $f(x)$ is continuous on $[a, b]$, it has a maximum and a minimum there.

Warning. Functions defined on an interval which is not closed or on the whole real line need not have maximum or minimum there. For example, $f(x) = 1/x$ is defined on (0, 1), but it has no maximum or minimum there. Also, $f(x) = x^2$ considered as a function on the whole real line has a minimum at $x = 0$, but no maximum; while for $f(x) = -x^2$ the reverse is true.

To determine a maximum or a minimum for a function defined on $[a, b]$ we check its critical points to determine local maximum and minimum and then compare them with each other and with the endpoints, which, as we have seen, may be maxima or minima without being critical points.

Examples

1. Find the maximum and minimum of $f(x) = 2x^3 - 2x^2 - 2x + 5$ on $[-10, 10]$ (Fig. 6.12).

Solution. We find

$$f'(x) = 6x^2 - 4x - 2.$$

Letting $f'(x) = 0$, we obtain $x = -1/3, 1$. Moreover

$$f''(x) = 12x - 4$$

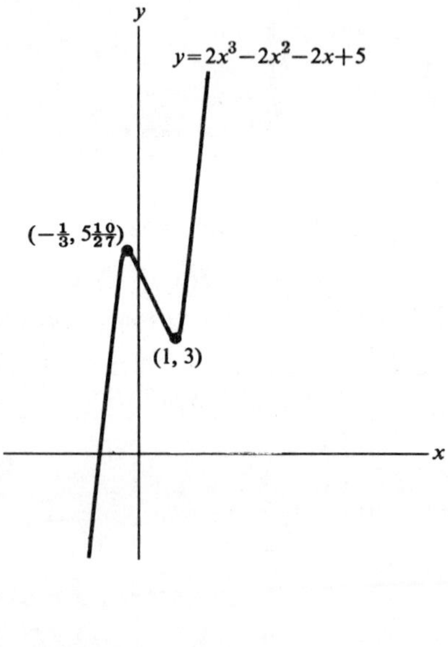

Fig. 6.12

so $f''(-1/3) = -8$, so there is a local maximum at $x = -1/3$. $f''(1) = 8$, so there is a local minimum at $x = 1$.

Now we compute the value of the function at the endpoints:

$$f(10) = 2(10)^3 - 2(10)^2 - 20 + 5 = 1785$$

$$f(-10) = 2(-10)^3 - 2(-10)^2 + 20 + 5 = -2175.$$

Finally we observe that $f(1) = 3$ and $f(-1/3) = 5\frac{10}{27}$ so the (absolute) maximum on $[-10, 10]$ occurs at 10 and the (absolute) minimum occurs at -10.

2. We return to the problem which introduced this section. (See Fig. 6.13.) Suppose we want to know what shape can will give us maximum volume for a fixed surface area. The cylindrical part of the can is a rectangle of dimension h by $2\pi r$, where h = height and r = radius. The top and bottom each have area πr^2. Thus the total steel required is

$$2\pi rh + 2\pi r^2.$$

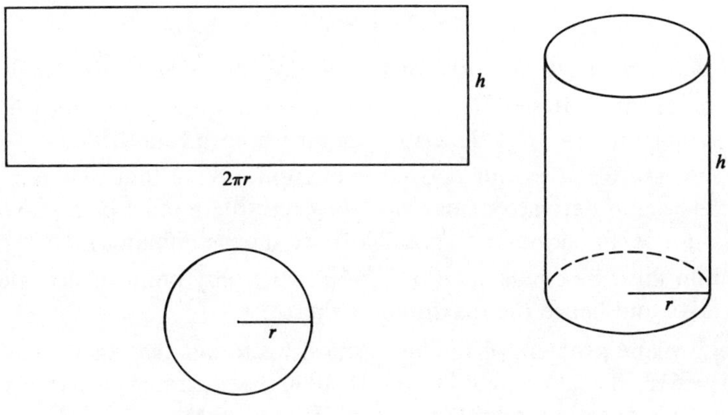

Fig. 6.13

This is constant so suppose we let

$$2\pi rh + 2\pi r^2 = k. \tag{1}$$

What are the allowable values of r? Clearly we must have $r \geq 0$ and $h \geq 0$. Also it is obvious that there is an upper limit on the value of r—it cannot be greater than $\sqrt{k/2\pi}$ (which occurs when $h = 0$). Hence if the volume is given as a function of r, it will be defined on a closed interval $[0, \sqrt{k/2\pi}]$ and assume a maximum there, by Theorem 2.

We know that the volume of a cylinder is

$$V = \pi r^2 h.$$

Since so far we can only handle functions of one variable, we solve equation (1) for h,

$$h = \frac{k - 2\pi r^2}{2\pi r},$$

and substitute this in the expression for V. Thus

$$V = \pi r^2 \cdot \frac{(k - 2\pi r^2)}{2\pi r} = \frac{rk - 2\pi r^3}{2}.$$

We now have a differentiable function representing the quantity which we want to maximize. We find

$$\frac{dV}{dr} = \frac{k - 6\pi r^2}{2}$$

and if

$$\frac{dV}{dr} = 0,$$

$$r = \pm\sqrt{k/6\pi}.$$

We can reject $-\sqrt{k/6\pi}$ as a value for the radius. Now either the maximum is at an endpoint of the interval $[0, \sqrt{k/2\pi}]$—clearly not the case, since if $r = 0$ or $h = 0$ the volume is 0—or at the critical point we have found.

Thus $r = \sqrt{k/6\pi}$ gives our desired maximum. Note that it is not necessary to check the second derivative since we know that there must be a maximum and that if it is not at an endpoint (a possibility we have eliminated) the derivative at the maximum must be equal to zero. $\sqrt{k/6\pi}$ is the only point in the interval with zero derivative and hence the maximum is there.

3. In the airplane problem of the last section we found that there was a critical point at $v = 612$. The domain is $[0, 1000]$ (1000 the maximum speed of the plane). Clearly $v = 0$ is not a reasonable solution. If $v = 1000$,

$$\frac{C}{D} = \left(300 + \frac{v^2}{1250}\right)\frac{1}{v} = 1.1 \text{ (approximately)},$$

whereas for $v = 612$

$$\frac{C}{D} = 0.99 \text{ (approximately)}.$$

Therefore $v = 612$ yields the desired minimum cost per mile.

4. An open rectangular box with square base is to be made from 12 square meters of material (see Fig. 6.14). What is the maximum possible volume of such a box?

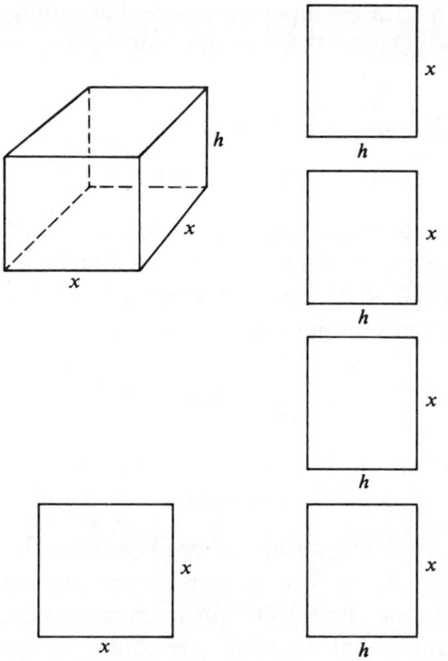

Fig. 6.14

Solution. Letting V be the volume we have

$$V = x^2h.$$

We must cut one square of area x^2 and four sides each of area xh. Thus $x^2 + 4xh = 12$. Observe that $0 \le x \le \sqrt{12}$. Solving for h we obtain

$$h = \frac{12 - x^2}{4x}.$$

$$V = x^2 \cdot \frac{12 - x^2}{4x} = 3x - \frac{x^3}{4}.$$

$$\frac{dV}{dx} = 3 - \tfrac{3}{4}x^2.$$

Letting $3 - \frac{3}{4}x^2 = 0$ gives $x = \pm 2$.

We can discard $x = -2$. If $x = 2$, $h = 1$, and so $x = 2$ yields a volume of 4 cubic meters. The endpoints of our domain of possible values for x, i.e., $x = 2\sqrt{3}$ and $x = 0$ give a volume of 0. Thus since we know the volume function has a maximum it must be at the critical point $x = 2$.

5. A group wants to charter a plane to Europe. An airline has told them that if they get 75 people to go they may charter a plane for \$200 per person. The fare

will decrease by 50 cents for each person above that number. That is, if 85 people go, each person will pay \$195. What number of passengers will result in the maximum revenue for the airline?

Solution. The revenue function is given by

$$R = R(n) = n(200 - 0.5n),$$

where n is the number of passengers. The domain of $R(n)$ is $[0, 400]$ and it is continuous so it attains its maximum there. For practical purposes it is really not defined everywhere in $[0, 400]$ since we cannot have fractional passengers, but that doesn't matter in our computations. Finding

$$\frac{dR}{dn} = 200 - n,$$

we get that $n = 200$ when $dR/dn = 0$. This gives the maximum as well as a local maximum since the revenue at the endpoints ($n = 0$ and $n = 400$) is 0.

6. A businessman sells 10,000 strings of beads a year. It costs \$10 per thousand to store the beds; each reorder costs \$2 plus \$8 per thousand. How many times a year, and in what size lots, should the businessman reorder in order to minimize his yearly costs? (Assume that the beads are sold at a constant rate.)

Solution. Let x = the lot size in thousands. Then $x/2$ is the average number of strings of beads on hand at a given time. The cost we are to minimize is the carrying cost plus the reorder cost times the number of reorders.

$$\text{Carrying cost} = 10\frac{x}{2}$$

$$\text{reorder cost} = \text{fixed} + \text{cost per thousand} = 2 + 8x$$

$$\text{number of reorders} = \frac{\text{number of sales}}{\text{lot size}} = \frac{10}{x}$$

$$\text{Cost} = \text{carrying cost} + (\text{reorder cost})(\text{number of reorders})$$

$$C(x) = 5x + (2 + 8x)\frac{10}{x}$$

$$= 5x + \frac{20}{x} + 80.$$

The derivative of this function is $5 - 20/x^2$. Setting this equal to zero yields $x = 2$; that is, the lot size is 2000.

The domain of the cost function is $[0, 10]$. $C(0)$ is not defined; $C(10) = 132$. But $C(2) = 100$, so 2000 is the lot size which minimizes the cost. Dividing it into 10,000 gives 5 as the number of reorders.

PROBLEMS

1. Classify the critical points of the following functions:

a) $f(x) = 6x^2 - 4x + 5$.

b) $f(x) = 2x^3 - 3x^2 + 2x - 3$.

c) $f(x) = \dfrac{x}{\sqrt{2x^2 + 4}}$.

d) $f(x) = x^4 + 2x^3 - 3x^2 - 4x + 5$.

e) $f(x) = \dfrac{2x - 5}{x}$.

f) $f(x) = 3x^2 - 4x + 4$.

g) $f(x) = 3x^3 + x^2 - 7x + 1$.

h) $f(x) = 2x/\sqrt{x^2 + 1}$.

i) $f(x) = \dfrac{2x + 4}{3x}$.

2. Find the maximum and minimum of

$$f(x) = 3x^3 - 2x^2 - 5x - 5,$$

on the interval $[-2, 2]$.

3. Find the maximum and minimum of $3x^3 - 2x^2 - 5x + 1$ on $[-1, 1]$.

4. Find the dimensions of the rectangle of maximum area which can be inscribed in a semicircle of radius 1.

5. Show that the rectangle of maximum area for a given perimeter is a square.

6. A page is to contain 24 square inches of printed material and a margin of $1\frac{1}{2}$ in. at the top and bottom and 1 in. on each side. Determine the dimensions which give the minimum total area for the page.

7. A piece of wire a meter long is to be cut into two pieces, one bent into a circle, one into a square. How should the cut be made in order that the pieces of wire enclose the maximum area?

8. Two farmers living on one side of a river agreed to build a bridge across the river. If the distances from the farms to the river are one and two miles respectively and if the perpendiculars from the farms to the stream are $1\frac{1}{2}$ miles apart, where should the bridge be located to minimize the sum of the straight line distances from the bridge to the farms?

9. A cabin is in the woods nine miles from the road at a point twelve miles along the road from the nearest town. If a man has a Land-Rover which averages 15 mph cross-country and 30 mph along the road, what route should be taken to minimize the time from town to the cabin?

10. The sum of three positive numbers is 30. The first plus twice the second plus three times the third add up to 60. Find the numbers which maximize the product of the three.

11. Find the dimensions of the cylindrical can (with open top) of maximum volume which can be made from 10 sq ft of sheet metal.

12. A revenue function is given by $R = 100 + 15n - n^2$. Find when the revenue is a maximum.

13. A businessman sells 5000 tables a year. It costs \$25 per thousand to store the tables; each reorder costs \$3 plus \$10 per thousand. How many times a year should the businessman reorder, and in what size lots, in order to minimize his yearly costs?
14. Find two numbers whose sum is 7 and whose product is as large as possible.
15. A cylindrical container is to have a volume of 1000 cc. Find its dimensions if its total area (top, bottom, and side) is to be a minimum.
16. A publisher offers a book at \$10 per copy for 100 copies or less. For each additional copy up to 200 copies total the price is reduced by a nickel for all copies over 100; e.g., if 110 copies are ordered, 100 will be at \$10 each and 10 at \$9.50 each. Find the number of copies which will produce the maximum revenue for the publisher.

6.3 THE MEAN VALUE THEOREM

We use the maxima-minima theory to prove

Theorem 3. (*Rolle's Theorem*). If f is differentiable on an interval containing a and b and $f(a) = f(b)$, then there exists $c \in (a, b)$ such that $f'(c) = 0$.

Proof. Since $f(x)$ is continuous, it has a maximum and a minimum on $[a, b]$. If they are both at the endpoints, then $f(a) = f(b)$ is both a maximum and a minimum and f is constant on $[a, b]$. But the derivative of a constant function is everywhere zero so we may take any point in (a, b) as the c such that $f'(c) = 0$.

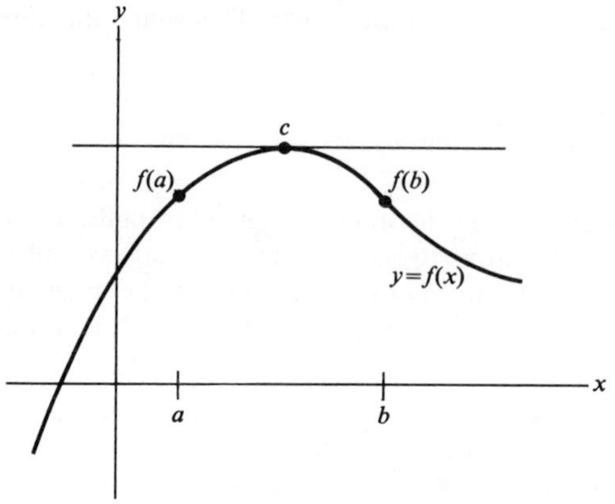

Fig. 6.15

If either the maximum or the minimum is not at an endpoint, it is at some point $c \in (a, b)$. But then by Theorem 1, $f'(c) = 0$. (See Fig. 6.15.)

This theorem is in turn used to prove the Mean Value Theorem.

Theorem 4. (*Mean Value Theorem*). If $f(x)$ is differentiable on an interval containing a and b, then there exists $c \in (a, b)$ such that

$$f(b) - f(a) = f'(c)(b - a).$$

Intuitively this theorem says that there must be some point $(c, f(c))$ on the graph of $f(x)$ (Fig. 6.16) such that the slope $f'(c)$ of the tangent line through $(c, f(c))$ is equal to the slope $(f(b) - f(a))/(b - a)$ of the line joining the endpoints $(a, f(a))$ and $(b, f(b))$. It appears that to prove the theorem we want to transform the line through $f(a)$ into a line parallel to the x-axis and to apply Rolle's theorem. That is, we simply define a new function in terms of the old, making certain that the new function satisfies the hypotheses of Rolle's theorem.

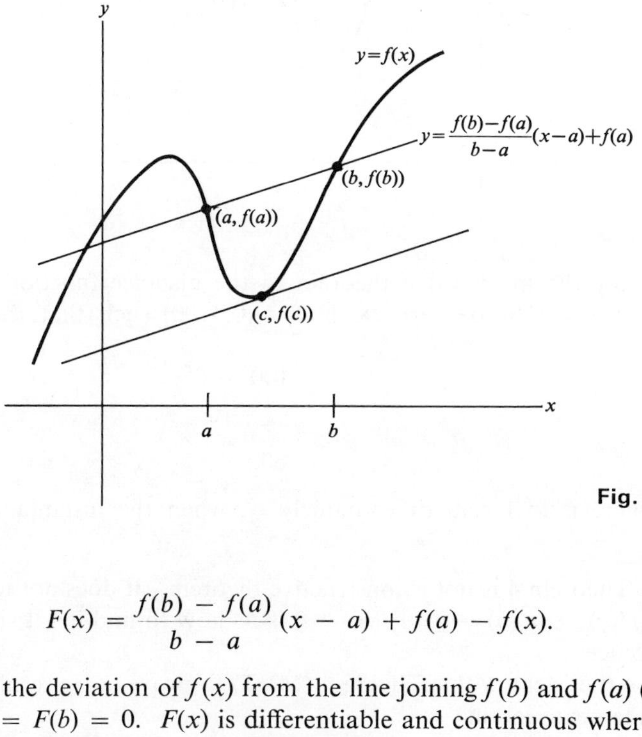

Fig. 6.16

Proof. Let

$$F(x) = \frac{f(b) - f(a)}{b - a}(x - a) + f(a) - f(x).$$

$F(x)$ measures the deviation of $f(x)$ from the line joining $f(b)$ and $f(a)$ (Fig. 6.17). We have $F(a) = F(b) = 0$. $F(x)$ is differentiable and continuous where required, since $f(x)$ is. So there exists $c \in (a, b)$ such that

$$F'(c) = 0.$$

But

$$F'(x) = \frac{f(b) - f(a)}{b - a} - f'(x) \quad \text{so that} \quad F'(c) = 0 = \frac{f(b) - f(a)}{b - a} - f'(c),$$

as was required to show.

Example. Suppose a car travels 100 miles in two hours, and that its distance is a differentiable function of time. Its average velocity is then 50 mph. Is its instantaneous velocity ever 50 mph?

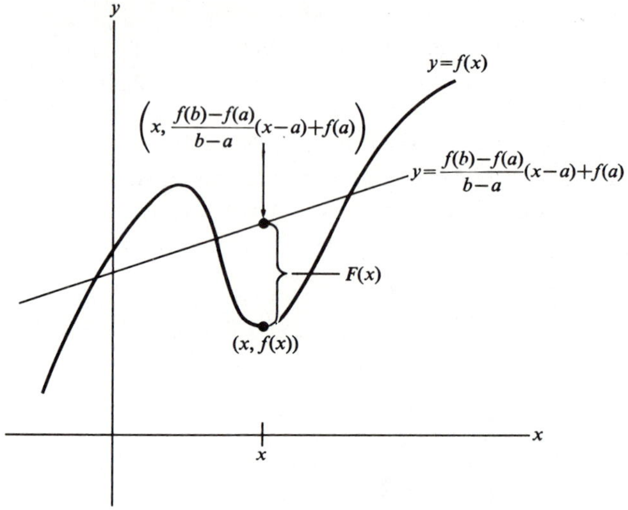

Fig. 6.17

We apply the mean value theorem to the distance function $s(t)$. We have $s(0) = 0$, $s(2) = 100$ so there exists $t_1 \in (0, 100)$ such that the instantaneous velocity

$$s'(t_1) = \frac{100 - 0}{2 - 0}$$
$$= 50.$$

That is, there is in fact a time, namely t_1, when the instantaneous velocity is 50 mph.

Warning. Theorem 4 is not a constructive theorem. It does not tell us what the c such that $f'(c) = \big(f(b) - f(a)\big)/(b - a)$ is or how to find it. It simply assures us of its existence.

We use the Mean Value Theorem (MVT) to prove some results which will be useful in the next section.

Theorem 5. If f is differentiable on an interval containing a and b, then

1) $f'(x) = 0$ for all $x \in (a, b)$ implies that $f(x)$ is constant on (a, b);
2) $f'(x) > 0$ for all $x \in (a, b)$ implies that $f(x)$ is *strictly increasing* on (a, b), i.e., $x_1 < x_2$ implies that $f(x_1) < f(x_2)$;
3) $f'(x) < 0$ for all $x \in (a, b)$ implies that $f(x)$ is *strictly decreasing* on (a, b), i.e., $x_1 < x_2$ implies that $f(x_1) > f(x_2)$.

Proof. We observe that by the MVT, for any $x_1, x_2 \in (a, b)$, $x_2 > x_1$, there is a $c \in (x_1, x_2)$ such that

$$f(x_2) - f(x_1) = (x_2 - x_1)f'(c). \tag{6.1}$$

For (1) we note that for any such c, $f'(c)$ must be 0 since $c \in (a, b)$; therefore, $f(x_2) - f(x_1) = 0$. But this is true for all pairs of points in (a, b), so $f(x)$ is constant on (a, b).

For (2) we have that in (6.1), $f'(c) > 0$ and $x_2 - x_1 > 0$, forcing $f(x_2) - f(x_1) > 0$. Thus for any pair of points $x_1, x_2 \in (a, b)$, $x_1 < x_2$ implies $f(x_1) < f(x_2)$; that is, $f(x)$ is strictly increasing on (a, b).

Part (3) is proved similarly.

We state without proof another result which can be proved by use of the MVT.

L'Hospital's Rule: If $f(x)$ and $g(x)$ are differentiable, $\lim_{x\to a} f(x) = \lim_{x\to a} g(x) = 0$ and

$$\lim_{x\to a} \frac{f'(x)}{g'(x)} = L,$$

then

$$\lim_{x\to a} \frac{f(x)}{g(x)} = L.$$

This makes the evaluation of many limits quite easy.

Example. Find

$$\lim_{x\to 1} \frac{x^3 - 3x + 2}{x^3 - x^2 - x + 1}.$$

Solution. We differentiate the numerator and denominator.

$$\lim_{x\to 1} (x^3 - 3x + 2) = \lim_{x\to 1} (x^3 - x^2 - x + 1) = 0.$$

However,

$$\lim_{x\to 1} \frac{3x^2 - 3}{3x^2 - 2x - 1},$$

is still indeterminate. Therefore, we differentiate the numerator and denominator again to find that

$$\lim_{x\to 1} \frac{6x}{6x - 2} = \frac{6}{4} = \frac{3}{2}.$$

Hence by two applications of L'Hospital's rule we have

$$\lim_{x\to 1} \frac{x^3 - 3x + 2}{x^3 - x^2 - x + 1} = \frac{3}{2}.$$

There is another form of the rule which is useful. We need an extended definition of limit. If given a real number M, there exists a neighborhood U of x_0 such that for all $x \in U$, $f(x) > M$, then we say that

$$\lim_{x\to x_0} f(x) = \infty.$$

As we have warned before, ∞ is not a point, but merely a way of indicating that a quantity increases without bound.

If given a real number M, there exists a real number N such that for all $x > N$, $f(x) > M$, then we say

$$\lim_{x \to +\infty} f(x) = \infty.$$

$\lim_{x \to -\infty} f(x) = \infty$ is defined analogously.

These limits do not exist in the usual sense; when we say in the future that

$$\lim_{x \to a} f(x)$$

exists, we shall mean, as we have in the past, that there is a real number L with

$$\lim_{x \to a} f(x) = L,$$

and similarly for $\lim_{x \to \pm\infty} f(x)$.

L'Hospital's Rule, second part: If $f(x)$ and $g(x)$ are differentiable,

$$\lim_{x \to a} f(x) = \lim_{x \to a} g(x) = \infty \qquad \text{and} \qquad \lim_{x \to a} \frac{f'(x)}{g'(x)} = L,$$

then

$$\lim_{x \to a} \frac{f(x)}{g(x)} = L.$$

The two parts of the rule are also valid if a in $\lim_{x \to a}$ is replaced by $+\infty$ or $-\infty$.

Example. Find

$$\lim_{x \to +\infty} \frac{2x - 1}{3x^2 - 2x}.$$

Solution. The hypotheses of the second part of the rule are satisfied. Therefore we differentiate to get

$$\lim_{x \to +\infty} \frac{2}{6x - 2} = 0,$$

so

$$\lim_{x \to +\infty} \frac{2x - 1}{3x^2 - 2x} = 0.$$

Warning. To apply the rule *both* numerator and denominator must have 0 (or both have ∞) as their limit. Otherwise conventional methods of evaluation of the limit should be used. The expressions $0/0$ and ∞/∞ are called *indeterminate forms.* The value of an indeterminate form is not known until it is evaluated in a particular case.

Note that in both versions of the rule we must know whether the functions are differentiable and what their derivatives are. Thus, for example, L'Hospital's rule is of no help in calculating the limit of the difference quotient

$$\frac{f(x) - f(x_0)}{x - x_0},$$

which appears in the definition of the derivative of $f(x)$.

PROBLEMS

1. Given $f(x) = x^3$, $a = 0$, $b = 2$, find for what $c \in [a, b,]$

$$f'(c) = \frac{f(b) - f(\mathrm{a})}{b - a}.$$

2. Given $f(x) = 2x^2 - 6x + 3$, $a = -1$, $b = 2$, find for what $c \in [a, b]$,

$$f'(c) = \frac{f(b) - f(\mathrm{a})}{b - a}.$$

3. Given $f(x) = (2x - 1)/(x + 2)$, $a = -1$, $b = 1$, find for what $c \in [a, b]$,

$$f'(c) = \frac{f(b) - f(a)}{b - a}.$$

4. Given $f(x) = 2x^3 + 3$, $a = 1$, $b = 2$, find for what $c \in [a, b]$,

$$f'(c) = \frac{f(b) - f(a)}{b - a}.$$

5. Given $f(x) = 3x^2 - 4x + 5$, $a = -1$, $b = 1$, find for what $c \in [a, b]$,

$$f'(c) = \frac{f(b) - f(a)}{b - a}.$$

6. Suppose the motion of a particle is given by

$$s(t) = -16t^2 - 80t + 100.$$

Find at what point between $t = 0$ and $t = 100$ the velocity is 1520.

7. Use L'Hospital's Rule to find the following limits:

a) $\lim_{x \to 3} \frac{x^3 - 2x^2 - 2x - 3}{x^2 - 9}$.

b) $\lim_{x \to +\infty} \frac{2x^3 - x^2 + 3x + 1}{3x^3 + 2x^2 - x - 1}$.

c) $\lim_{x \to 1} \frac{x^4 - 2x^2 + 1}{3x^3 - 3x^2 + x - 1}$.

d) $\lim_{x \to -1} \frac{4x^3 + 4x^2 - 2x - 2}{x^4 - x^3 - x^2 + x}$.

8. Evaluate the following limits:

a) $\lim_{x\to 2} \dfrac{2x^3 - 2x^2 + x - 10}{4x - 8}$.

b) $\lim_{x\to 0} \dfrac{x^4 + 4x^3 - 2x^2 + x}{x^3 - 4x^2 + x}$.

c) $\lim_{x\to 1} \dfrac{x^3 + 5x^2 - 4x - 2}{2x^3 - 6x + 4}$.

6.4 CURVE SKETCHING

In Chapter 2 we discussed briefly the graphing of linear and quadratic equations. With the notions of limit and derivative available we are able to obtain more information to facilitate sketching the graphs of certain functions. The usual purpose in sketching a curve is to obtain a general idea of how the dependent variable (graphed on the vertical axis) behaves with respect to the independent variable (graphed on the horizontal axis); thus we endeavor to obtain a good approximation of the shape of the graph without plotting a large number of points. If $y = f(x)$ we want to know such things as: For what range of values of x is y an increasing function of x? Where does it achieve its maximum value?

Most of the methods described here work only for differentiable or at best continuous functions, but even if we want to examine the behavior of some function which is not continuous we can frequently approximate it by a continuous function and so use these methods.

A. Intercepts

It is usually useful to determine where the graph crosses the coordinate axes. In general this is simply the problem discussed in Chapter 2 of finding the solution set of an equation. We summarize the techniques.

In the linear and quadratic cases:

1) $$y = ax + b.$$

2) $$y = ax^2 + bx + c.$$

3) $$x = ay^2 + by + c.$$

we have a general method for finding the value of one variable when the other is zero, namely (see Fig. 6.18)

1) $x = 0,\ y = b,$ $\qquad y = 0,\ x = -b/a.$

2) $x = 0,\ y = c,$ $\qquad y = 0,\ x = (-b \pm \sqrt{b^2 - 4ac})/2a.$

3) $x = 0,\ y = (-b \pm \sqrt{b^2 - 4ac})/2a,$ $\qquad y = 0,\ x = c.$

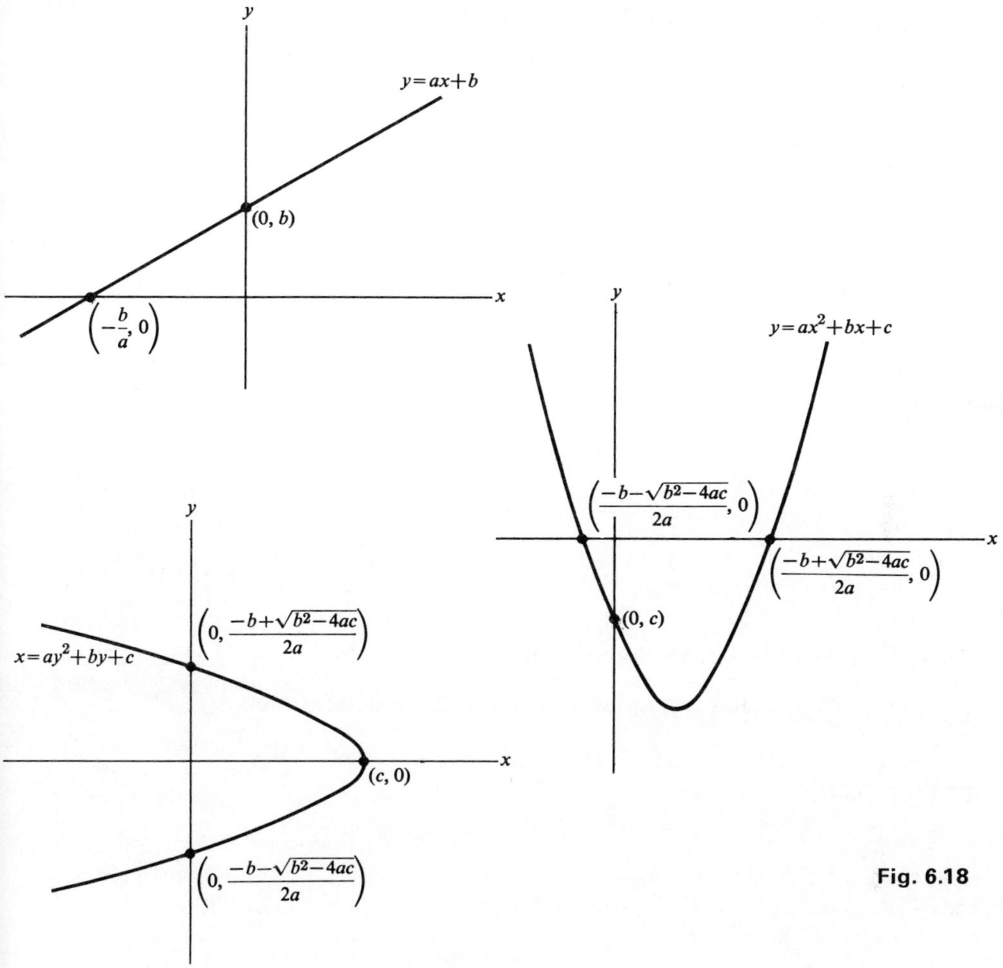

Fig. 6.18

Some equations of higher degree are disguised linear or quadratic equations and we can apply the above formulas. For example,

$$x^3 - 5 = y,$$

is linear in x^3 so that for

$$y = 0, \qquad x = \sqrt[3]{5},$$

and if $x = 0$, $y = 5$ (Fig. 6.19).

$3x^4 + 4x^2 - 4 = y$ is quadratic in x^2 so $y = 0$ implies that

$$x^2 = \frac{-4 \pm \sqrt{16 - 4(3)(-4)}}{6} = -2, \tfrac{2}{3}.$$

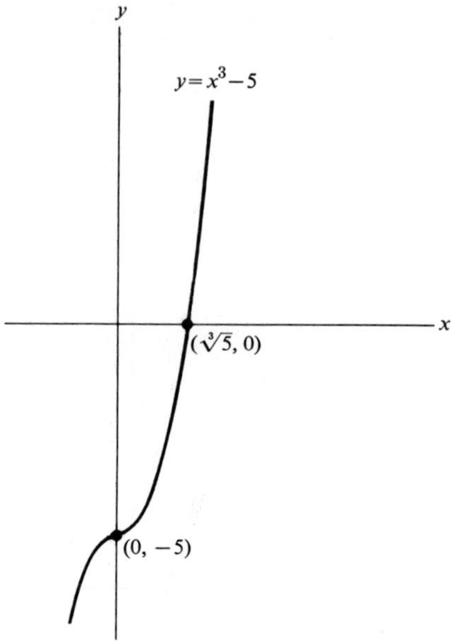

Fig. 6.19

Then $x^2 = -2$ has no real solutions, but $x^2 = \frac{2}{3}$ yields $x = \pm\sqrt{\frac{2}{3}}$ (Fig. 6.20).

There are formulas giving the solution of the general cubic

$$y = ax^3 + bx^2 + cx + d,$$

and the quartic

$$y = ax^4 + bx^3 + cx^2 + dx + e,$$

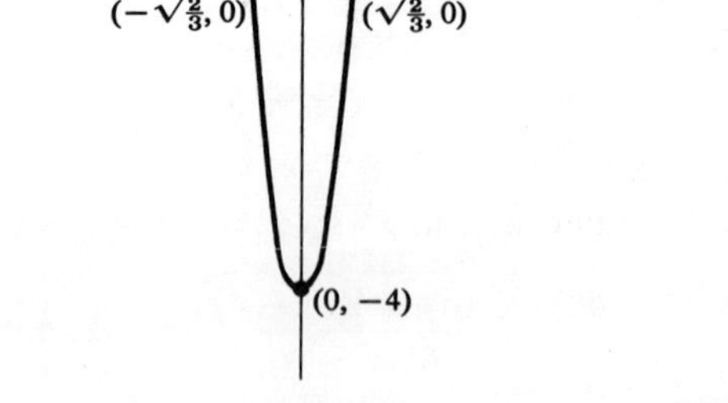

Fig. 6.20

(and similarly with the roles of x and y reversed), but we shall not give them, because, not only are they complicated, but they are practically useless for actually obtaining roots. (For 5th degree and higher there is no such general solution.)

However, we observe that by substituting values for x we can locate the intercepts approximately: if one value for x, say x_0, gives a positive value for y and another, say x_1, gives a negative, then, since a polynomial is continuous, the graph must cross the x-axis between x_0 and x_1.

Example. Let

$$y = 2x^3 + x^2 - 8x - 4.$$

We have the following table:

x	-2	-1	0	1	2	3
y	0	3	-4	-9	0	35

Thus we have located two intercepts, $x = 2, -2$ (Fig. 6.21).

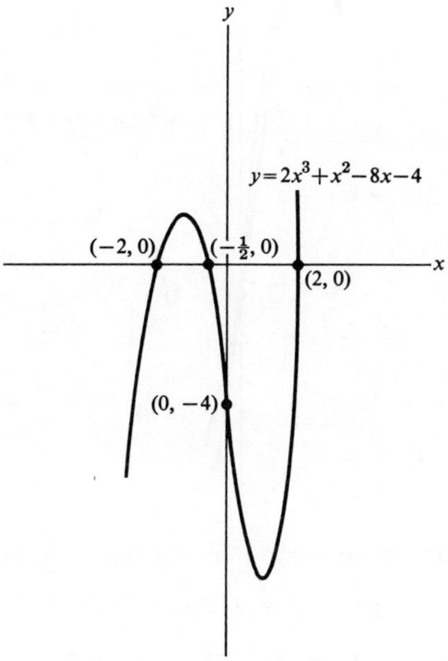

Fig. 6.21

Complex solutions occur in pairs, so we know that there is a third real solution. From the sign changes in y we determine that it is between $x = 0$ and $x = -1$. It is clear that as x increases above $x = 3$, y remains positive, and as x decreases below $x = -2$, y becomes (and remains) negative. Hence there are no more solutions. We knew this anyway since a third-degree equation has only three solutions. The solution between 0 and -1 can be located more precisely by the substitution of other values between -1 and 0 for x and the computation of corresponding values for y or by approximating the cubic equation in the interval $[-1, 0]$ by the straight line joining $(-1, 3)$ and $(0, -4)$. The slope of this line is

$$\frac{3 - (-4)}{-1 - 0} = -7.$$

Its y-intercept is -4, so the equation of the straight line is $y = -7x - 4$. Its x-intercept is $-4/7$ (Fig. 6.22).

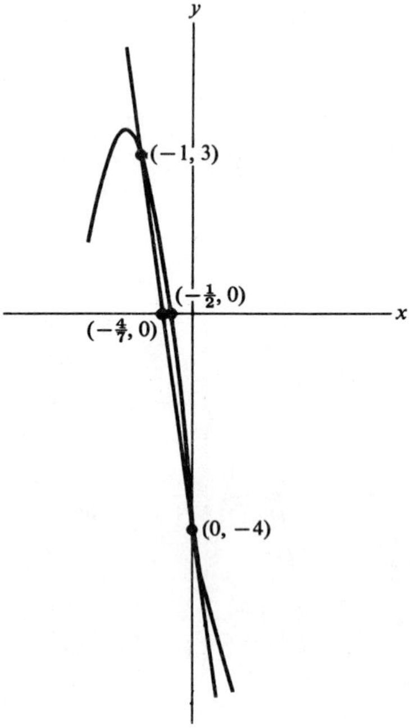

Fig. 6.22

In this particular case, we observe that the cubic can be factored as

$$(x - 2)(x + 2)(2x + 1),$$

to give us the intercept $x = -1/2$ as a check on the accuracy of our approximation. However, most cubics are not so amenable to factorization. See Chapter 2 for a more detailed discussion of roots of polynomials.

Warning. We emphasize that only the real solutions of a polynomial equation

$$y = a_n x^n + \cdots + a_1 x + a_0,$$

appear as the x-intercepts. For example, in the cubic

$$y = x^3 + x^2 + x + 1,$$

the roots are $x = \pm i, -1$, but its only x-intercept is -1 (Fig. 6.23). Thus the fundamental theorem of algebra gives us only the *maximum* number of intercepts (but the precise number of roots). However, as a guide to determining whether all intercepts have been located we note again that nonreal roots occur in pairs. Thus, for example, a fifth-degree equation has 1, 3, or 5 real roots and hence 1, 3, or 5 x-intercepts.

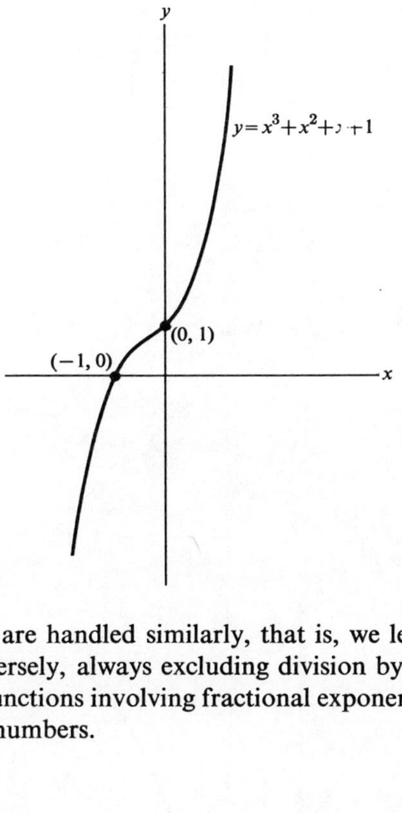

Fig. 6.23

Rational functions are handled similarly, that is, we let $x = 0$ and ask what happens to y and conversely, always excluding division by zero in our computations. With algebraic functions involving fractional exponents, the caveat concerns even roots of negative numbers.

Examples

1. Find the intercepts of

$$y = \frac{2x^2 - 1}{x - 4}.$$

(Fig. 6.24.)

Solution. When $x = 0$, $y = \frac{1}{4}$. When $y = 0$,

$$2x^2 - 1 = 0 \quad \text{and} \quad x = \pm\frac{1}{\sqrt{2}}.$$

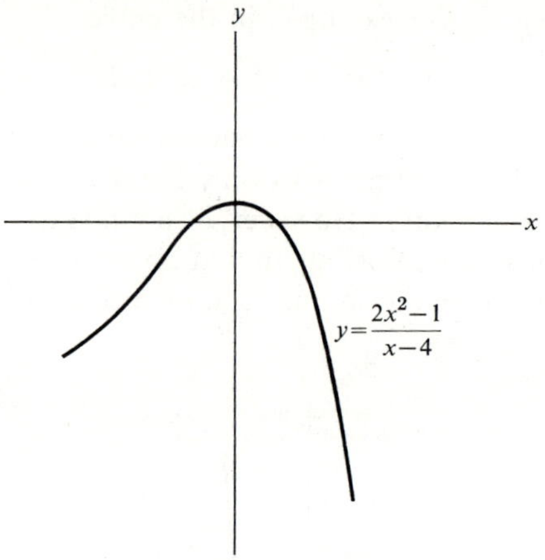

Fig. 6.24

2. Find the intercepts of

$$y = \sqrt{3x^2 - 5x + 2}.$$

(Fig. 6.25.)

Solution. When $x = 0$, $y = \sqrt{2}$. When $y = 0$,

$$\sqrt{3x^2 - 5x + 2} = 0,$$

so

$$3x^2 - 5x + 2 = 0$$

$$(3x - 2)(x - 1) = 0$$

$$x = \tfrac{2}{3}, 1.$$

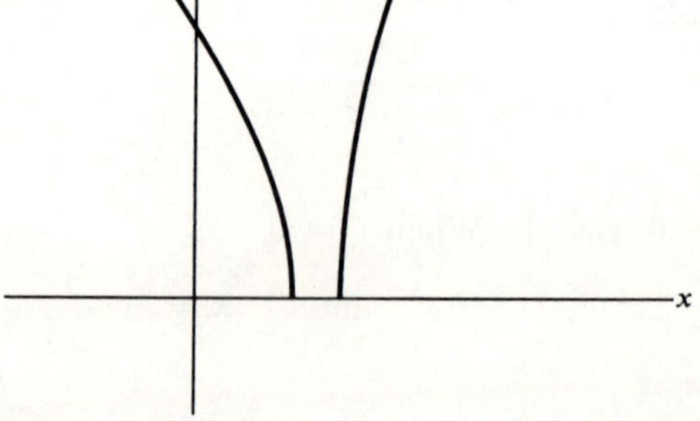

Fig. 6.25

3. Find the intercepts of

$$y = \frac{1}{\sqrt{x^2 + x - 1}}.$$

(Fig. 6.26.)

Solution. When $x = 0$, y is not real, so there is no y-intercept. When $y = 0$, we have $0 = 1$, which is impossible; so there is also no x-intercept.

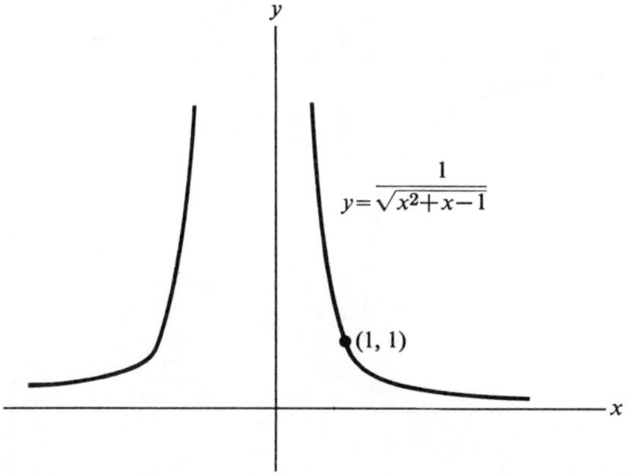

Fig. 6.26

B. Asymptotes

We observe that (see Fig. 6.27)

$$xy = 1,$$

has neither x- nor y-intercepts, but that as x becomes large, y approaches zero, and as x approaches zero, y becomes arbitrarily large (and similarly for negative values). We symbolize this information by

$$\begin{aligned} y &\to 0 \quad &\text{as} \quad x &\to +\infty \\ y &\to 0 \quad &\text{as} \quad x &\to -\infty \\ y &\to +\infty \quad &\text{as} \quad x &\to 0^+ \\ y &\to -\infty \quad &\text{as} \quad x &\to 0^-, \end{aligned}$$

where the last two symbols indicate one-sided approaches to 0.

In this case the x and y axes are *asymptotes*, that is, limiting values. Note again that ∞ is not a number; the symbol is simply shorthand for "arbitrarily large."

Another example of such behavior (see Fig. 6.28) is the function

$$y = \frac{x - 2}{(x + 1)(x + 2)}.$$

As x approaches -1 from the left the denominator becomes small and is negative. Since the numerator is also negative, y becomes arbitrarily large and positive.

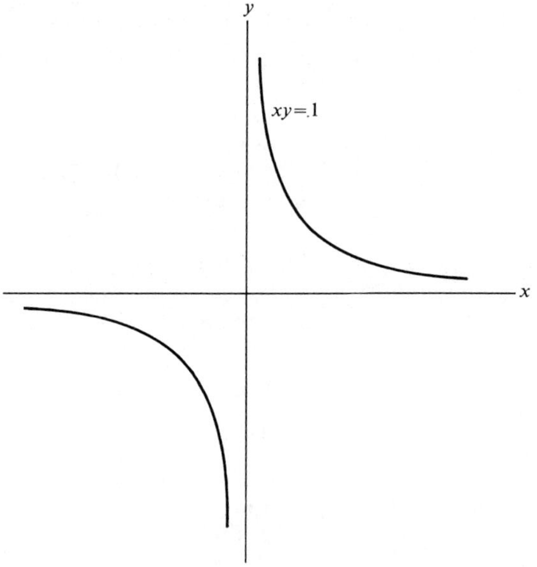

Fig. 6.27

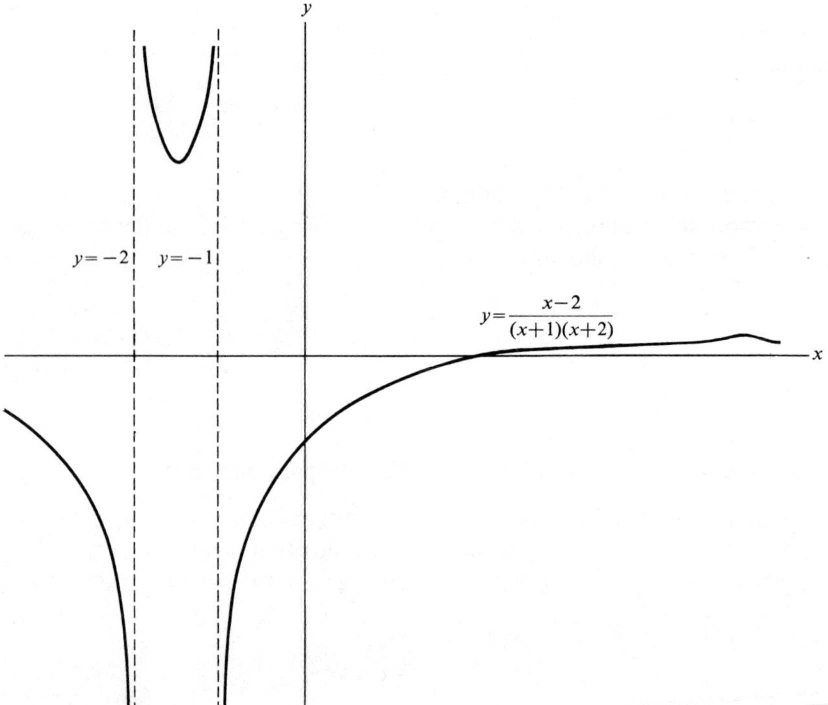

Fig. 6.28

As x approaches -1 from the right, the denominator is positive and the numerator negative, so that y becomes arbitrarily large in absolute value, but is negative. The behavior is similar at $x = -2$, so that there are two vertical asymptotes.

Using L'Hospital's rule, we find that $\lim_{x\to+\infty} y = 0$, and similarly for $x \to -\infty$. We have a y-intercept at -1 and when $y = 0, x = 2$ for the x-intercept. It is useful to check the behavior of $y = f(x)$ for very large positive or negative values of x even if no asymptotes occur.

Warning. Parabolas never have asymptotes. For example (Fig. 6.29a)

$$y = x^2 + 2,$$

simply becomes large as $x \to \pm\infty$. But (see Fig. 6.29b),

$$y^2 - x^2 = 4,$$

the upper branch of which behaves roughly as does $x^2 + 2$, becomes large and has the lines $x = y$ and $x = -y$ as asymptotes. This is because $\pm x$ are actually limiting values for

$$y = \pm\sqrt{4 + x^2},$$

as $x \to \pm\infty$.

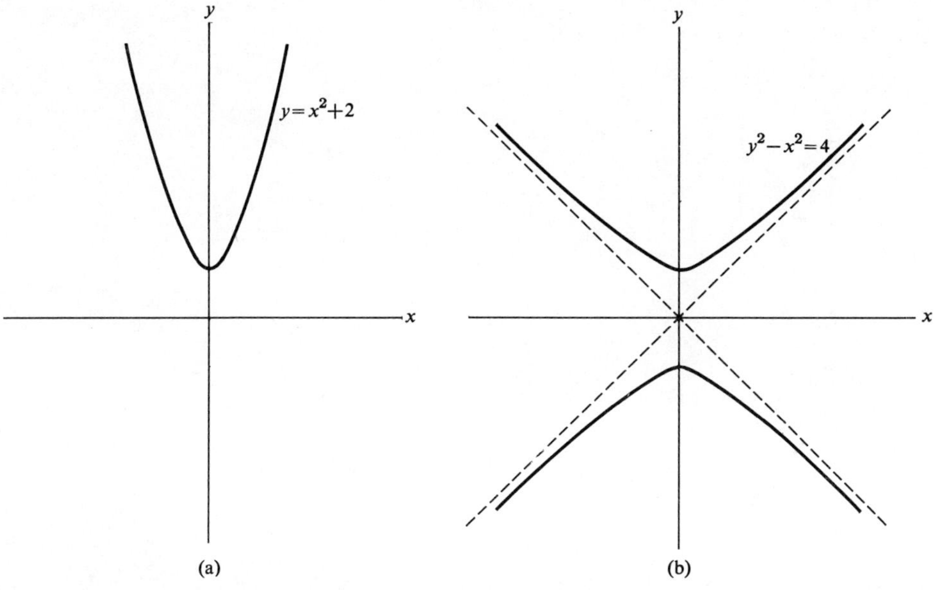

Fig. 6.29

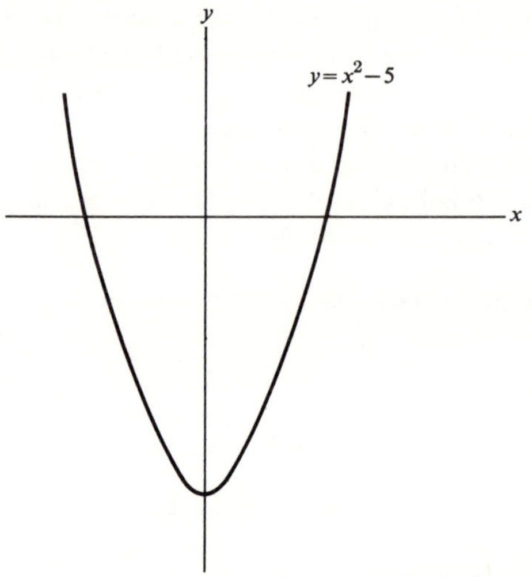

Fig. 6.30

C. Symmetry

A function $y = f(x)$ is said to be *symmetric* with respect to x if $f(x) = f(-x)$ for all x in its domain. For example, if $f(x) = x^2 - 5$ then $f(-x) = (-x)^2 - 5 = x^2 - 5 = f(x)$ and $f(x)$ is symmetric with respect to x. In fact any function having only even powers of x is symmetric with respect to x. We observe the graph of $x^2 - 5 = y$ in Fig. 6.30.

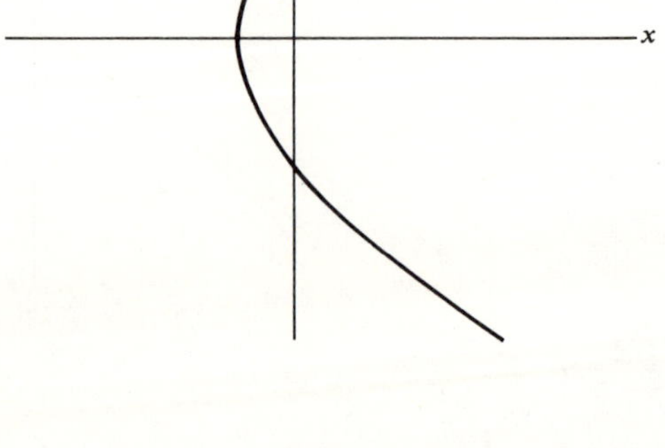

Fig. 6.31

The graph is symmetric about the y-axis in the ordinary sense of the word "symmetric." The same is true of the graph of any function symmetric with respect to x. We sometimes call such functions "symmetric with respect to the y-axis."

It is clear that if a function remains unchanged when y is replaced by $-y$ we can speak of symmetry with respect to y or with respect to the x-axis. For example,

$$y^2 = 4x + 3,$$

is symmetric with respect to y (Fig. 6.31).

A similar sort of symmetry with respect to any line parallel to either of the axes can be considered. For example,

$$y = x^2 - 5x + 4,$$

is symmetric about the line $x = \frac{5}{2}$ (Fig. 6.32). In fact, any parabola is symmetric about a line through its vertex perpendicular to the tangent line there.

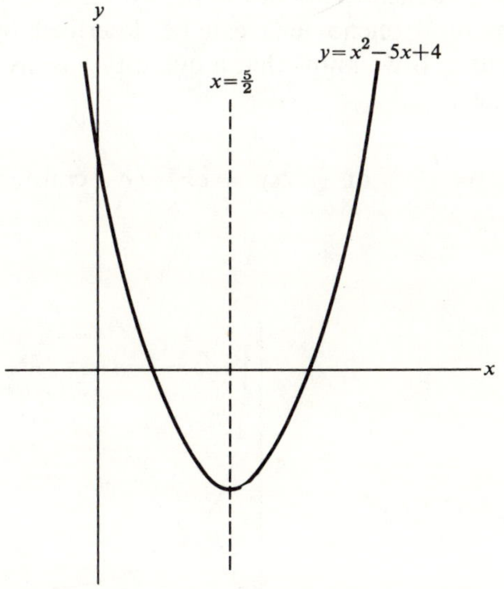

Fig. 6.32

It is clear that symmetry in x and y can occur simultaneously (Fig. 6.33). For example, in

$$4x^2 + y^2 = 9.$$

Another type of symmetry occurs when a function is such that x and y may be interchanged without altering the function. For example, $xy = 1$ or $x^2 + y^2 = 2$ (Figs. 6.34 and 6.35). Here the geometric symmetry is about the line $y = x$.

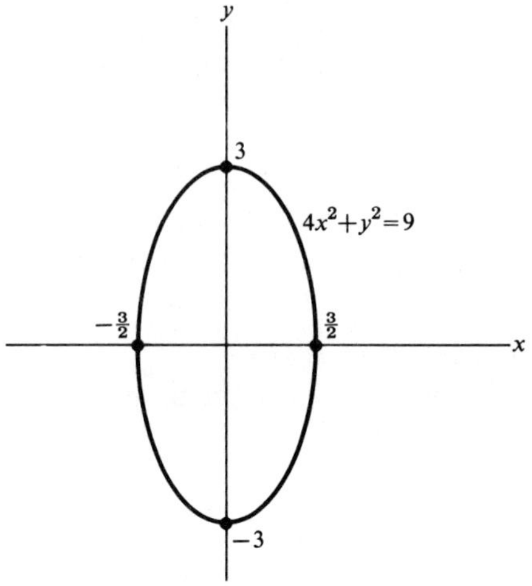

Fig. 6.33

The function $xy = 1$ mentioned in the preceding sections is an example of a *hyperbola.* Many natural phenomena can be described by an equation which represents such a figure. If we know that a quantity y is inversely proportional to a quantity x, we write

$$y = \frac{c}{x} \qquad \text{or} \qquad xy = c, \qquad c \text{ a constant.}$$

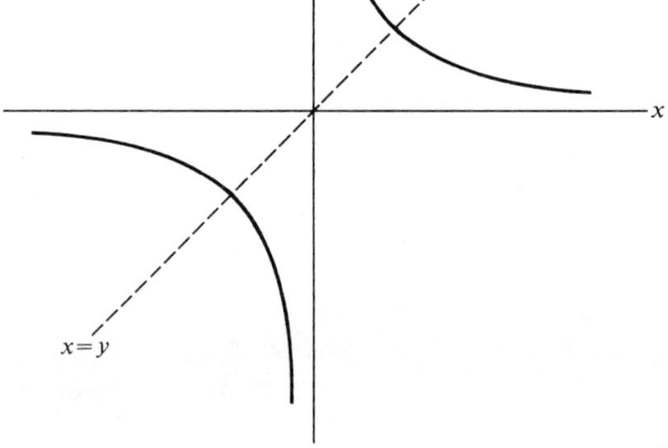

Fig. 6.34

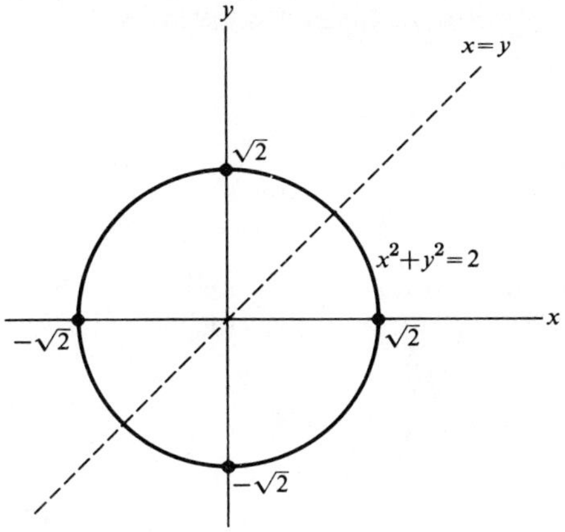

Fig. 6.35

Sketching such a curve is easy if we observe that the points $(\sqrt{c}, \sqrt{c})$ and $(-\sqrt{c}, -\sqrt{c})$ must lie on it and that the axes are asymptotes.

Example. A demand law may be given in the form

$$p = \frac{c}{d},$$

where p is the price, d the demand, and c a constant. The graph of this equation is a hyperbola, but since it has no intercepts it cannot be a true description of an economic situation for very small values of d or p. We usually expect demand laws to be valid only throughout a restricted domain in any case. Our demand model is better written as

$$p = \frac{c}{d}, \qquad (0 < a \leq d \leq b)$$

(Fig. 6.36).

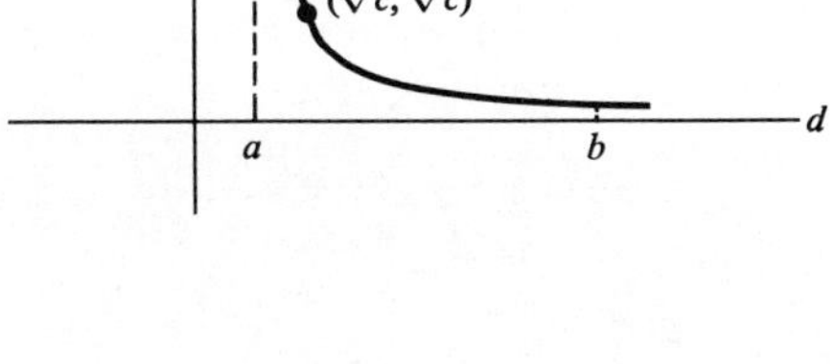

Fig. 6.36

A more general form of demand law is given by

$$(d - h)(p - k) = c, \qquad (h, k, c \text{ constants}).$$

By rewriting this as

$$p - k = \frac{c}{d - h},$$

(Fig. 6.37) we see that as $d \to h^+$, $p \to +\infty$ and as $p \to k^+$, $d \to +\infty$, p and d are never negative so that $d = h$ and $p = k$ are asymptotes.

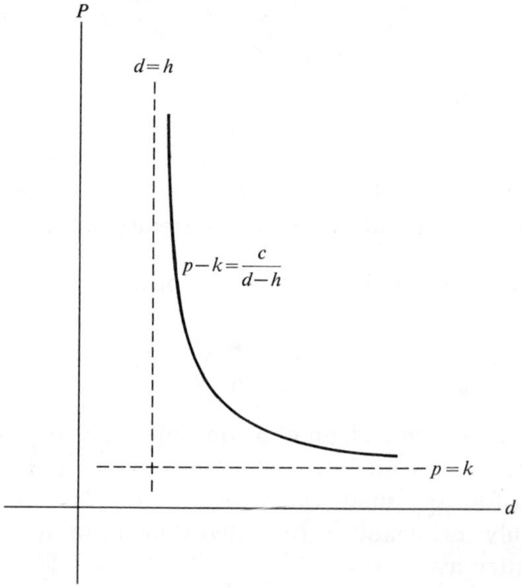

Fig. 6.37

Equations of the form

$$\frac{x^2}{a^2} - \frac{y^2}{b^2} = 1, \qquad -\frac{x^2}{a^2} + \frac{y^2}{b^2} = 1,$$

(a, b constants) also represent hyperbolas (Figs. 6.38 and 6.39). How to find the asymptotes is somewhat more difficult. We study the case

$$\frac{x^2}{a^2} - \frac{y^2}{b^2} = 1;$$

the other involves only a reversal of the roles of the variables. We have x-intercepts of $\pm a$. Values of x in the interval $[-a, a]$ are excluded since they would produce a nonreal value for y. Outside this interval, since

$$y = \pm b\sqrt{x^2/a^2 - 1},$$

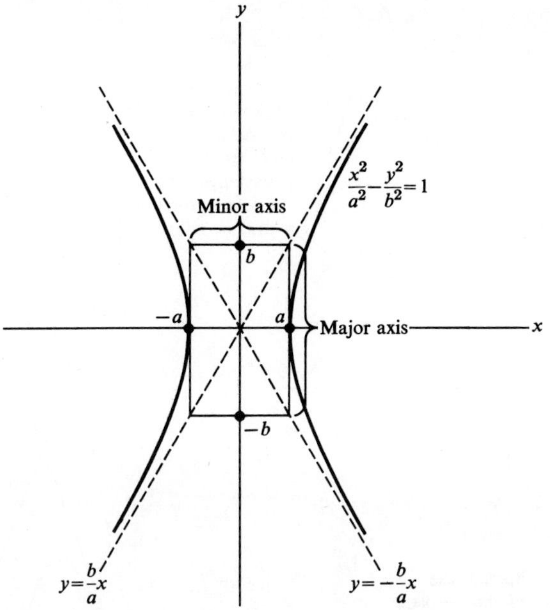

Fig. 6.38

as $|x|$ increases so does $|y|$. For very large values of $|x|$, $|y|$ is very close to, but less than, $b|x|/a$. In fact, although we shall not prove it,

$$y = \frac{b}{a}x \qquad \text{and} \qquad y = -\frac{b}{a}x,$$

are asymptotes. To sketch this curve we draw the asymptotes and the intercepts and rough in the rest. We call the larger of $2a$ and $2b$ the *major axis* and the other the *minor axis*. A hyperbola is the set of all points in which the difference in distances from two fixed points is constant.

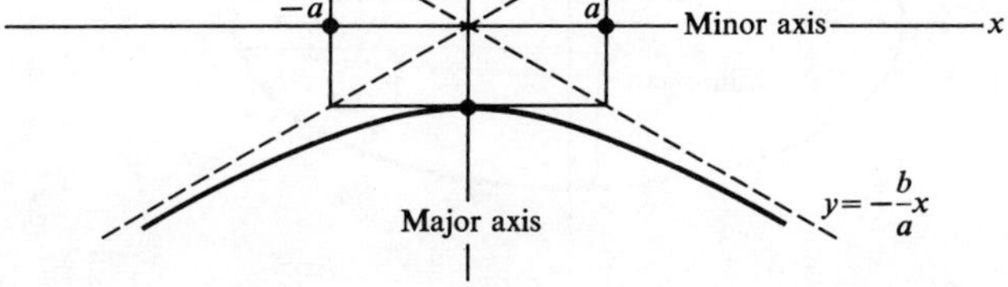

Fig. 6.39

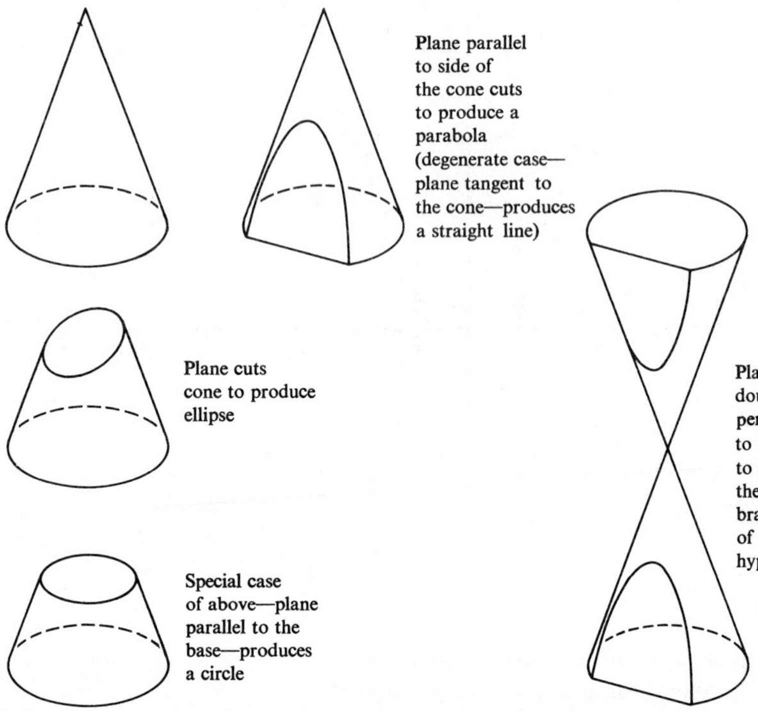

Fig. 6.40

D. Conic Sections

Straight lines, parabolas, and hyperbolas are part of a family of curves called *conic sections* (see Fig. 6.40). Another member of the family is the ellipse. An *ellipse*, centered at the origin, is an equation of the form

$$\frac{x^2}{a^2} + \frac{y^2}{b^2} = 1. \tag{6.2}$$

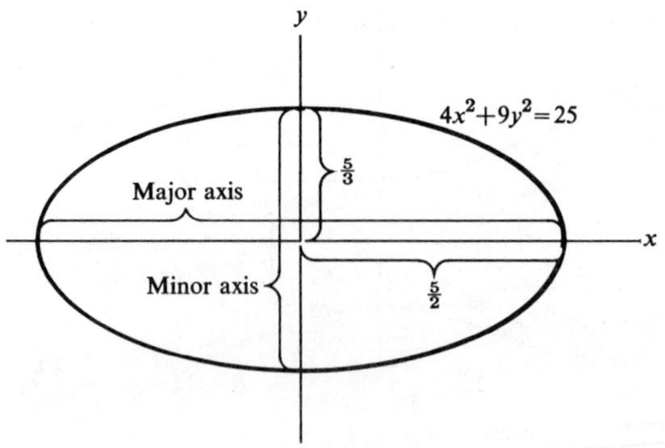

Fig. 6.41

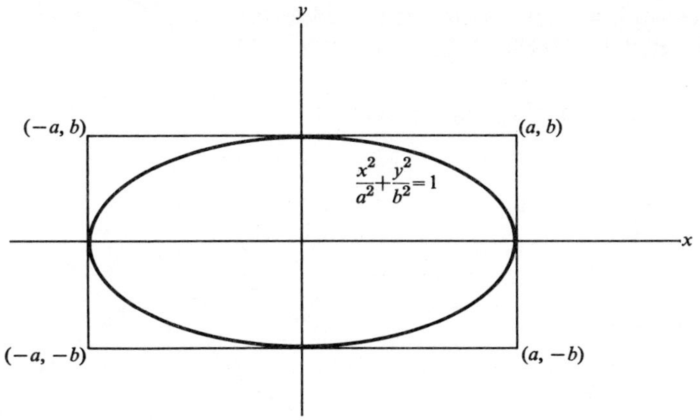

Fig. 6.42

For example, Fig. 6.33 represents an ellipse, as does

$$4x^2 + 9y^2 = 25. \tag{6.3}$$

(See Fig. 6.41.)

In standard form, Eq. (6.3) is

$$\frac{x^2}{25/4} + \frac{y^2}{25/9} = 1.$$

The ellipse (Eq. (6.2)) has four intercepts, $x = \pm a$, $y = \pm b$. These also represent maximum and minimum values for the variables, since $|x| > a$ makes y the square root of a negative number and similarly for $|y| > b$. Thus the entire graph lies within the rectangle whose vertices are $(\pm a, \pm b)$ (Fig. 6.42). We can draw the rectangle and rough in the sketch of the curve. An ellipse is the set of all points the sum of whose distances from two points, called the *foci*, is constant (Fig. 6.43). A circle is a special case of an ellipse which occurs when the coefficients of x^2 and y^2 are equal; this happens when the foci are identical, namely the center of the circle.

Each conic section has special properties. We recall that a parabola is the set of points equidistant from a fixed point, the focus, and a fixed line, the directrix. Parabolic curves occur naturally; for example, a parabola is the path traveled by an object thrown or fired into the air (neglecting air resistance). Parabolic reflectors

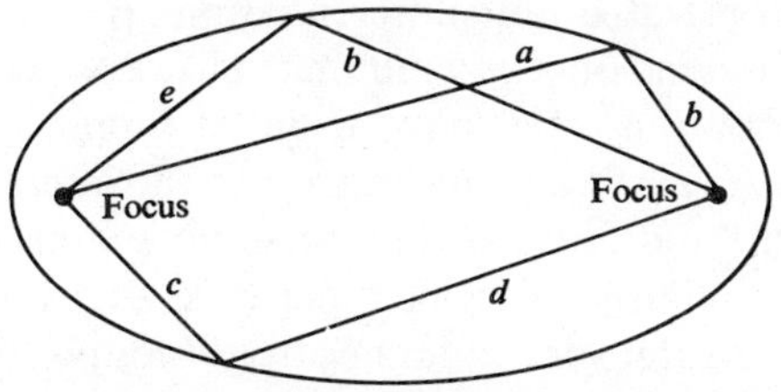

$a+b=c+d=e+f$, etc.

Fig. 6.43

are useful because if a light source is placed at the focus the rays hit the reflector and are reflected in parallel beams (see Fig. 6.44). Conversely a solar furnace consists of a parabolic mirror which focuses the sun's parallel rays at a small spot at the focus, producing intense heat.

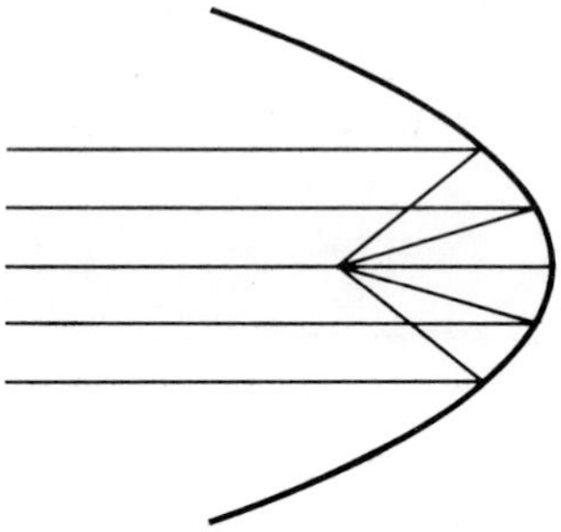

Fig. 6.44

The properties of the ellipse have led to the building of elliptical structures to obtain special acoustical effects. The slightest sound produced at one focus will reflect from the walls to the other focus, making it audible even though the spots are quite far apart (see Fig. 6.45).

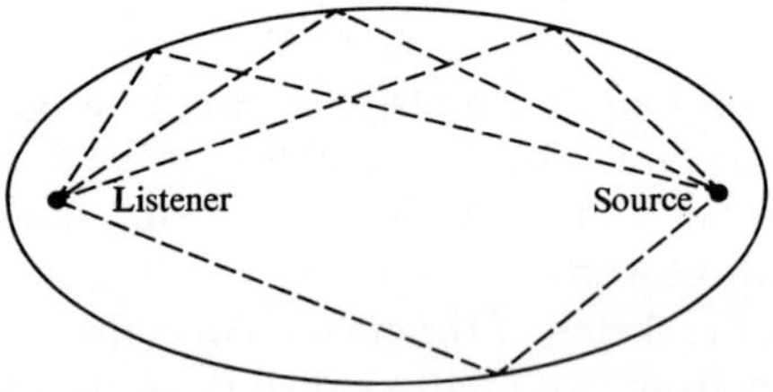

Fig. 6.45

The problem of conic sections not centered at the origin is more complicated, but one can always transform the equations into one of the standard forms above by a change of axes. We discuss the general problem of transformation of coordinates in Chapter 11.

E. Derivatives

We have already demonstrated how differentiating gives us the slope of a tangent line and how we may find maxima and minima. However, computing the derivative gives additional information. We have from Theorem 5 that $f'(x) > 0$ for all $x \in (a, b)$ implies that $f(x)$ is strictly increasing on the interval (a, b). In particular, a function is one-to-one and thus has an inverse on an interval where its derivative is strictly positive. Similarly, a strictly negative derivative indicates a strictly decreasing function. Moreover, a nonnegative (nonpositive) derivative tells us that a function is nondecreasing (nonincreasing).

Example.

$$f(x) = 3x^3 - 5x^2 + x - 2$$

$$f'(x) = 9x^2 - 10x + 1.$$

To see more clearly how the derivative behaves we write it as

$$f'(x) = (9x - 1)(x - 1).$$

$f'(x)$ is positive whenever the factors are both positive or both negative. Thus $f'(x) > 0$ if $x > 1$ or if $x < \frac{1}{9}$ and $f(x)$ is increasing for those values of x. On the other hand $f'(x) < 0$ if $\frac{1}{9} < x < 1$, and the function is decreasing for those values of x. Moreover, $f'(x) = 0$ at $x = \frac{1}{9}$, 1, the critical points. We have

$$f''(x) = 18x - 10$$

so $f''(\frac{1}{9}) = -8$, and $f''(1) = 8$. Therefore $x = \frac{1}{9}$ is a local maximum and $x = 1$ is a local minimum. We also observe that as $x \to +\infty, f(x) \to +\infty$ and as $x \to -\infty$, $f(x) \to -\infty$ and that when $x = 0$, $f(x) = -2$, the y-intercept. Thus we can sketch the curve illustrated in Fig. 6.46.

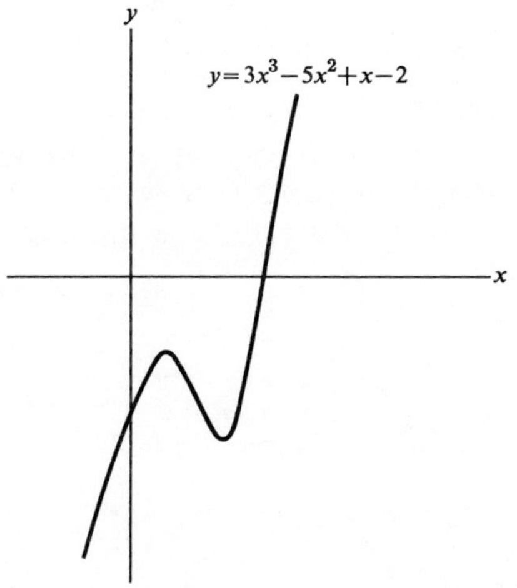

Fig. 6.46

We have already used second derivatives to discover the nature of critical points. They provide additional information about the shape of a curve. We say that a curve is *concave upward* on an interval $[a, b]$ if at every point in this interval the curve is above the tangent line to the curve. If the curve is below the tangent line at every point, we say that it is *concave downward* (see Fig. 6.47). Recall that the equation of the tangent line at $(x_0, f(x_0))$ is given by

$$y - f(x_0) = f'(x_0)(x - x_0),$$

from which we have

$$y = f(x_0) + f'(x_0)(x - x_0).$$

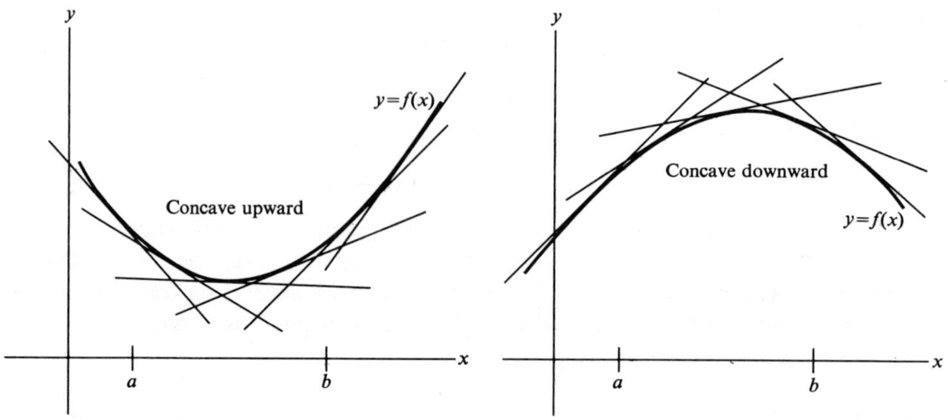

Fig. 6.47

Now if the curve is concave upward, $f(x)$ must be greater than or equal to y. That is,

$$f(x) \geq f(x_0) + f'(x_0)(x - x_0).$$

It is possible to show, using the MVT, that this is the case if and only if $f''(x) > 0$ for all $x \in (a, b)$. This is intuitively clear if we observe in the sketch that the derivative $f'(x)$ is increasing on the concave upward curve and hence its derivative, $f''(x)$, is positive. Analogously, the curve is concave downward on (a, b) if and only if $f''(x) < 0$ for all $x \in (a, b)$.

Example.

$$f(x) = x^2 - 2x + 4$$

$$f'(x) = 2x - 2$$

$$f''(x) = 2.$$

We have that there is a minimum at $x = 1$ and a y-intercept at 4 (see Fig. 6.48). Moreover, there is no x-intercept since the roots of $x^2 - 2x + 4$ are not real. From the second derivative we see that the slope of the tangent is always increasing from large negative values to large positive values and the curve is always concave upward.

If a point $x_0 \in (a, b)$ is such that for all $x \in (a, x_0)$, $f''(x) > 0$ and for all $x \in (x_0, b)$, $f''(x) < 0$ (or conversely), then x_0 is said to be a *point of inflection* for $f(x)$. That is, at a point of inflection a function changes from concave upward to concave downward. If the second derivative is continuous at a point of inflection x_0, then by the intermediate value theorem we must have that $f''(x_0) = 0$. However, as we see below, it is not the case that if $f''(x_0) = 0$, then x_0 is a point of inflection.

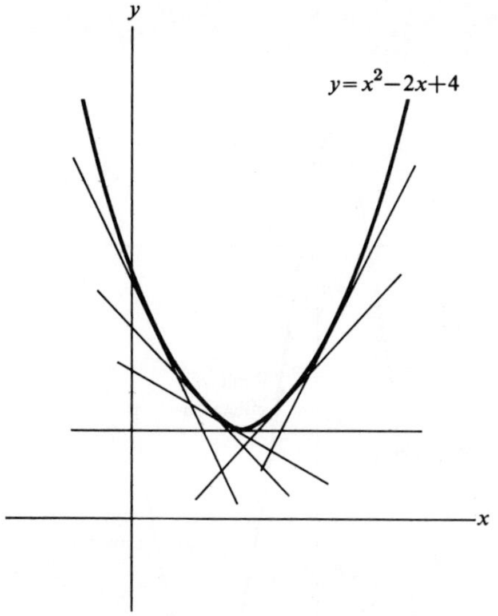

Fig. 6.48

Examples

1. If

$$f(x) = x^3,$$

then 0 is both a critical point and a point of inflection.

2. If

$$f(x) = x^4,$$

then $f''(0) = 0$, but $f''(x) = 12x^2$ is always nonnegative, so the function does not change its concavity at $x = 0$ and so 0 is not a point of inflection.

3. If

$$f(x) = x^3 - 3x^2 - 4x + 2,$$

then $f'(x) = 3x^2 - 6x - 4$ and $f''(x) = 6x - 6$. Now for $x < 1, f''(x) < 0$ and for $x > 1, f''(x) > 0$. Thus 1 is a point of inflection (see Fig. 6.49). We note also that $f''(1) = 0$, but that 1 is not a critical point of $f(x)$ since $f'(1) \neq 0$.

F. Parametric Equations

If we are so fortunate as to be given two variables, one expressed as a function of the other, we now have available various techniques enabling us to sketch a curve which gives us a good idea of the behavior of the dependent variable with respect to the independent. However, we frequently encounter equations of the form

$$\begin{aligned} x &= f(t) \\ y &= g(t). \end{aligned} \qquad (*)$$

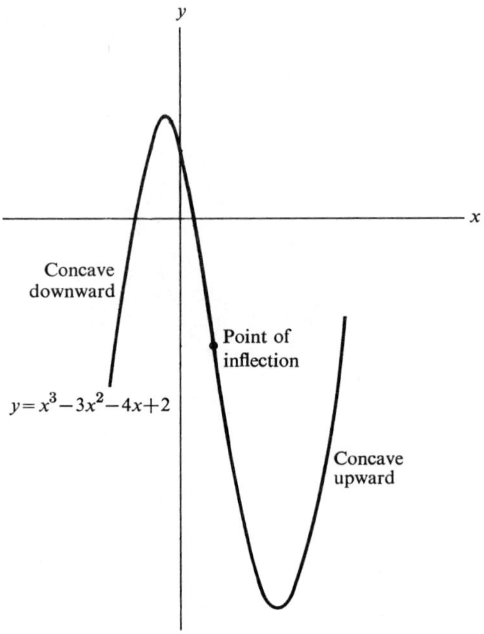

Fig. 6.49

That is, both x and y are expressed as functions of another variable. Usually such a variable is denoted by t since frequently in applications it represents time. We call (*) a *parametric equation with parameter t*, or a curve in *parametric form*.

When x and y are given by a parametric equation, can we sketch a graph of y as a function of x? In some cases it is clear that a simple substitution reduces a parametric representation to an ordinary equation giving y as a function of x. For example, if

$$x = t - 2$$

$$y = 3t^2 - 2t + 1,$$

then

$$t = x + 2.$$

Thus

$$y = 3(x + 2)^2 - 2(x + 2) + 1$$

$$= 3x^2 + 10x + 9,$$

a simple parabola, is easily sketched by the methods of this section (see Fig. 6.50).

On the other hand, if

$$x = t^3$$

$$y = t^2,$$

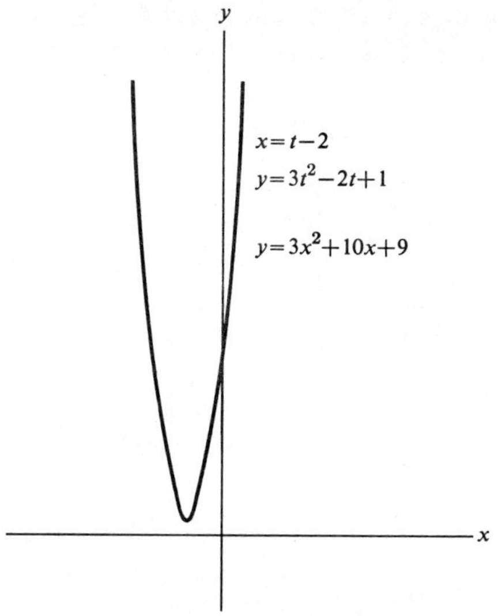

Fig. 6.50

it is easier to proceed directly. We find that

$$\frac{dx}{dt} = 3t^2,$$

which is nonnegative for all values of t. Thus x is always nondecreasing. Also

$$\frac{dy}{dt} = 2t,$$

so y decreases for negative t and increases for positive t. Moreover $y = t^2$ is always nonegative. To graph a parametric equation, we assign values to t and compute the corresponding values at x and y as shown in Fig. 6.51.

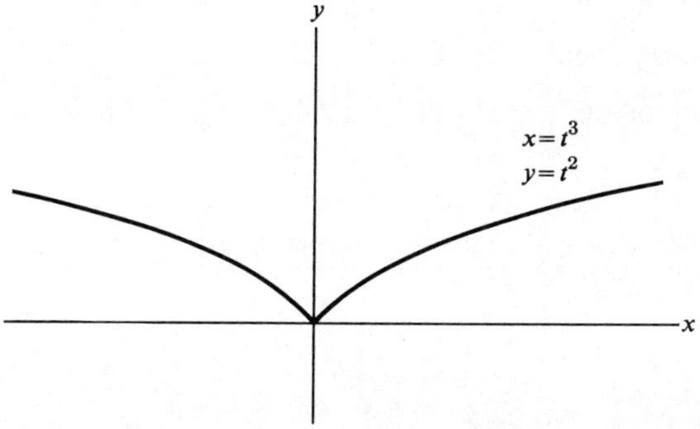

Fig. 6.51

Warning. In the graph of a parametric equation, there is no "t-axis." In other words, the values we assign to the parameter do not appear on the graph.

SUMMARY

We summarize our guide lines for sketching the graph of a function:

1. Check for intercepts.
2. Observe the behavior as each of the variables becomes very large.
3. Look for symmetry.
4. Take the derivative to calculate possible maxima or minima and to observe regions of increase or decrease.
5. Differentiate again for more information on the shape of the curve.

PROBLEMS

1. Find the horizontal and vertical asymptotes of the following functions:

   a) $f(x) = \dfrac{2x - 1}{x + 4}$.
   b) $f(x) = 3/2x$.
   c) $f(x) = \dfrac{1}{(x + 2)(3x - 4)}$.
   d) $f(x) = \dfrac{2x^2 + 1}{x - 4}$.
   e) $f(x) = \dfrac{2x}{x^2 + 1}$.
   f) $f(x) = \dfrac{2}{x^2 - x - 6}$.

2. Describe any symmetry in the graphs of the following:
   a) $y^2 + 4x^2 = 16$.
   b) $xy = 4$.
   c) $y = 2x^2 - 5x - 3$.
   d) $x^2 - 3y^2 = 10$.
   e) $4y^2 + x^2 = 1$.
   f) $2xy = 1$.
   g) $y = 5x^2 - 10x + 4$.
   h) $x^2 - y^2 = 2$.

3. Classify the conic sections of Problem 2.

4. On what subsets of the domain of each function is the function (i) increasing, (ii) decreasing?

   a) $f(x) = 15$.
   b) $f(x) = 5x - 4$.
   c) $f(x) = 2x^2 + 8x - 15$.
   d) $f(x) = 6x^3 - 4x^2 + 2x - 5$.
   e) $f(x) = \dfrac{1}{3x - 1}$.
   f) $f(x) = 3/2x^2$.

5. Describe the concavity of the graphs of the following functions:

a) $f(x) = x^{1/3}$.
b) $f(x) = 2x^3 + 5x^2 - 8x + 1$.
c) $f(x) = \dfrac{2x}{x + 1}$.
d) $f(x) = 4x^3 - 9x^2$.
e) $f(x) = x^3 - 4x^2$.

6. Sketch the graphs of the following functions:

a) $x = 2t^2 - t + 1,$ $y = 3t + 1.$
b) $x = t^3 + 1,$ $y = 2t^2.$
c) $x = 1/t,$ $y = 2t + 1.$
d) $x = t^2 - 1,$ $y = \sqrt{t}.$
e) $x = 4t - 3,$ $y = t + 4.$
f) $x = t^2 + 1,$ $y = t^4 - 4.$
g) $x = 2t - 1,$ $y = \sqrt{2t}.$
h) $x = 2t - 1,$ $y = t^2 - t - 2.$

7. Sketch the following curves:

a) $f(x) = x^3 + 2x^2 - 3x - 2$.
b) $f(x) = -x^3 + 2x^2 - x + 1$.
c) $f(x) = \dfrac{2x + 3}{3x - 2}$.
d) $f(x) = \dfrac{x^2 + x + 7}{\sqrt{2x + 1}}$.
e) $f(x) = x + 1/x$.
f) $f(x) = x\sqrt{9 - x}$.
g) $f(x) = \dfrac{x}{x + 2}$.
h) $f(x) = x^4 + 4x^3 + 6x^2$.
i) $f(x) = \dfrac{2x}{x^2 + 4}$.
j) $f(x) = (x + 2)(x - 3)^3$.

8. Describe the regions of increase and decrease and the concavity of the following functions:

a) $f(x) = x^{2/3}$.
b) $f(x) = 15x$.
c) $f(x) = \dfrac{1}{x - 1}$.
d) $f(x) = 2x^2 - x^3$.
e) $f(x) = \dfrac{x}{2x + 1}$.
f) $f(x) = 2x^2 - 5x + 2$.
g) $f(x) = -x^3 - 2x^2 + 1$.

6.5 APPLICATIONS

Example 1.—Revenue. If we are given the average revenue as a function of the output

$$AR = f(Q),$$

then the total revenue R is

$$R = f(Q)Q.$$

The marginal revenue is defined to be (whenever f is a differentiable function)

$$MR = \frac{dR}{dQ},$$

which is equal to

$$\frac{dR}{dQ} = f(Q) \cdot 1 + Q \cdot f'(Q) = f(Q) + Qf(Q).$$

This gives us

$$MR - AR = Qf'(Q),$$

that is, the marginal revenue and the average revenue differ by $Qf'(Q)$. Now Q is always positive and $f'(Q)$, being the slope of the curve

$$AR = f(Q),$$

is usually known to be negative. Thus, their product is usually negative and the average revenue is usually greater than the marginal revenue.

Let us find the marginal revenue if the average revenue is given by

$$AR = 63 + \frac{1}{Q}.$$

Solution. The revenue is

$$R = Q\left(63 + \frac{1}{Q}\right) = 63Q + 1,$$

and

$$MR = \frac{dR}{dQ} = 63.$$

Example 2. Suppose we are given the total revenue as a function of the output

$$R = f(Q),$$

and the output as a function of the labor input

$$Q = g(L).$$

Then the total revenue may be expressed as a function of the labor input

$$R = f(g(L)).$$

Since R is now represented as a composite function we can find dR/dL by using the chain rule

$$\frac{dR}{dL} = \frac{dR}{dQ}\frac{dQ}{dL},$$

provided that f and g are differentiable.
In the expression for dR/dL, dR/dQ is the marginal revenue function and

$$\frac{dQ}{dL} = MPP_L,$$

is the marginal physical product-of-labor function. The expression dR/dL is the marginal revenue product-of-labor function, denoted by MRP_L. Then the chain rule is the well-known economic result

$$MRP_L = MR \cdot MPP_L.$$

In particular, suppose that

$$R = 3Q^2 - 4Q + 15,$$

and

$$Q = 16L^2 - 10.$$

Then

$$\frac{dR}{dL} = \frac{dR}{dQ} \cdot \frac{dQ}{dL} = (6Q - 4)(32L),$$

or in terms of L

$$\frac{dR}{dL} = (6(16L^2 - 10) - 4)(32L) = 3072L^3 - 2048L$$

so that if the input is 10 hours of labor, the marginal revenue product-of-labor is 3,051,520.

Example 3—Income Tax. Suppose that Mr. Smith has deductions of \$6,000 plus 2 percent of his income. Suppose, moreover, that he learns from the Federal tax-rate schedule that if his net taxable income is between \$25,000 and \$35,000 then his tax is \$3700 plus 31 percent of the excess over \$25,000. He is interested in knowing how the percentage of his total income paid in Federal income tax varies as his income increases in the \$25,000 to \$35,000 range.

Solution. We have

$$\begin{aligned} \text{tax} = T(I) &= 3700 + 0.31(I - 25{,}000 - 6{,}000 - 0.02I) \\ &= 3700 + 0.3038I - 9610 \\ &= 0.3038I - 5910. \end{aligned}$$

Now the percentage of income paid as tax is

$$P = \frac{T(I)}{I} = \frac{0.3038I - 5910}{I} = 0.3038 - \frac{5910}{I}.$$

We want the rate of change of this percentage, so we calculate:

$$\frac{dP}{dI} = \frac{5910}{I^2}.$$

Example 4—Enzymes. The Schutz-Borisoff law with regard to the action of enzymes such as pepsin or rennin is given by

$$x = K\sqrt{cat},$$

where c is the concentration of enzyme, a the initial concentration of the substrate (for example, albumen or milk), t the time elapsed, x the amount transformed, and K a constant. Find the rate at which the transformation takes place.

Solution. We have

$$\frac{dx}{dt} = K\tfrac{1}{2}(cat)^{-1/2}(ca) = \frac{K\sqrt{ca}}{2\sqrt{t}}.$$

Since the rate is inversely proportional to $\sqrt{t}$, the transformation is rapid at first (not defined, however, for $t = 0$) and slows down with increasing values of t.

Example 5—Hydrolysis. When ethyl acetate is hydrolyzed in the presence of acetic acid as a catalyst the following reaction occurs:

$$\underset{\text{ethyl acetate}}{CH_3COOC_2H_5} + H_2O \rightarrow \underset{\text{acetic acid}}{CH_3COOH} + C_2H_5OH.$$

As the reaction progresses, the amount of acetic acid available to act as a catalyst increases, and so the rate of reaction increases. Simultaneously, since some of it is being hydrolyzed, the active mass of the ester decreases, causing a deacceleration in the reaction velocity. When is the velocity at maximum?

Solution. Let

a moles/liter = initial concentration of acetic acid

b moles/liter = initial concentration of ester

x = number of moles hydrolysed after time t, producing x moles of acetic acid.

Then at time t there are

$a + x$ moles of acid

$b - x$ moles of ester

and

$$v = k(a + x)(b - x), \qquad \text{where } k \text{ is a constant.}$$

Differentiating we obtain

$$\frac{dv}{dt} = -k(a + x) + k(b - x) = -2kx - ka + kb.$$

When

$$\frac{dv}{dt} = -2kx - ka + kb = 0,$$

$$x = \tfrac{1}{2}(b - a).$$

Thus the velocity is at a maximum when there are

$$a + x = a + \tfrac{1}{2}(b - a) = \tfrac{1}{2}(a + b)$$

liters of acetic acid.

Example 5—Related Rates. Suppose oil is being pumped into the triangular tank shown in Fig. 6.52 at a rate of 15 cu ft per minute. If the depth of the tank is 50 ft, the width at the top is 40 ft, and the length is 100 ft, at what rate is the depth of oil in the tank increasing?

The volume of oil in the tank is given by

$$V = \tfrac{1}{2}bh \cdot 100.$$

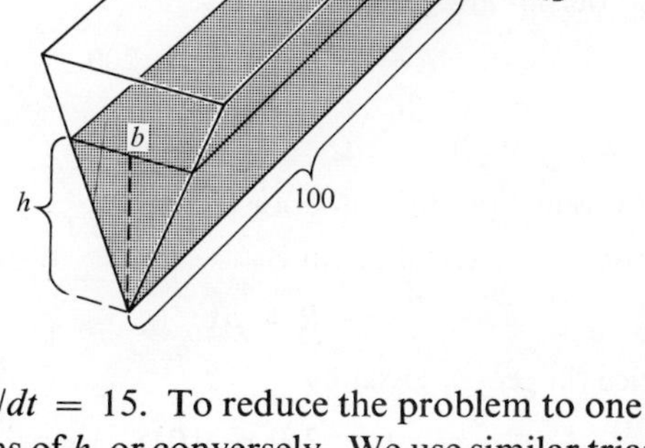

Fig. 6.52

Also, we know that $dV/dt = 15$. To reduce the problem to one that we can handle, we need to put b in terms of h, or conversely. We use similar triangles (see Fig. 6.53). Looking at the end of the tank, we see that the triangle representing the part already filled is similar to the triangle representing the entire end of the tank. Thus

$$\frac{40}{50} = \frac{b}{h} \quad \text{or} \quad b = \tfrac{4}{5}h.$$

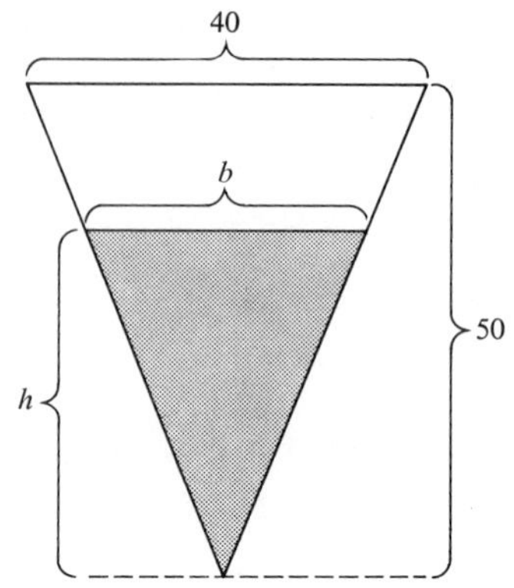

Fig. 6.53

Now we have

$$V = \tfrac{1}{2}(\tfrac{4}{5}h \cdot h) \cdot 100$$
$$= 40h^2.$$

Using implicit differentiation,

$$\frac{dV}{dt} = 80h \frac{dh}{dt}$$

$$\frac{dh}{dt} = \frac{dV/dt}{80h} = \frac{15}{80h}.$$

Thus, for example, at a depth of 2 ft the depth is increasing at a rate of $\frac{15}{160}$ ft per minute.

PROBLEMS

1. If the revenue and output are given by

$$R = 3Q^3 - 2Q + 200,$$

and

$$Q = L + 2,$$

find the marginal revenue product-of-labor (L).

2. The revenue function for a certain item is

$$R = px,$$

where p is the price (in cents), given by

$$p = 250 - 0.01x,$$

and x the number of items sold daily. If the cost is

$$c(x) = 75x + 15{,}000,$$

and the profit is revenue minus cost, find the number of items which should be produced to maximize the profit.

3. Suppose the government assesses a 9 cent added-value tax to the item described in Problem 2. How much of the tax should be passed on to the consumer? Why? Compare profits before and after the imposition of the tax.

4. Show that if $R(x)$ and $c(x)$ are the revenue and cost functions respectively, with marginal revenue $R'(x)$ and marginal cost $c'(x)$,then the profit

$$P(x) = R(x) - c(x),$$

is at a maximum when marginal revenue is equal to marginal cost.

5. Miss Jones has deductions of \$1500 plus 3 percent of her income. For net taxable income between \$10,000 and \$12,000 the tax is \$976 plus 13 percent of the excess over \$10,000. Find how the percentage of her total income paid in Federal income tax varies if the net taxable income is in the \$10,000 to \$12,000 range.

6. The fixed costs (land, architect's fees, etc.) for a building are \$1,500,000. The cost of erecting the first floor is \$400,000, of erecting the second floor \$450,000, and in general the nth floor costs

$$400{,}000 + (n - 1)50{,}000.$$

If the minimum height is 5 stories and the maximum 50 stories and the annual profit from each floor is \$75,000, how many stories should be constructed to maximize the annual rate of return?

$$\text{rate of return} = \frac{\text{net income}}{\text{total investment}}.$$

7. At noon a boat starts from A and sails south at the rate of 16 mph. At 1 p.m. another boat starts from A and sails west at the rate of 12 mph. How fast is the distance between the ships increasing at 3 p.m. the same day?

8. If a stone dropped in a pool sends out concentric ripples and the outer ripple increases in diameter at the rate of 5 ft/sec, how fast is the area of the disturbed surface increasing after 10 sec?

9. Water is being pumped into a cylindrical tank of radius 10 ft at the rate of 50 ft^3 per minute. How fast is the water level rising when the water is 5 ft deep?

10. A man 6 ft tall walks at the rate of 4 ft per sec toward a street light 20 ft above the ground. At what rate is the tip of his shadow moving? At what rate is the length of his shadow changing when he is 12 ft from the light?

11. When air expands adiabatically the pressure p and volume v satisfy

$$pv^{1.4} = k,$$

k a constant. When the pressure is 100 lb/in^2, and the volume is 50 in^3 and is decreasing at the rate of 5 in^3 per sec, how rapidly is the pressure changing?

12. If the revenue and output are given by

$$R = 4Q^3 - 2Q^2 + 300 \qquad \text{and} \qquad Q = \frac{L}{2} + 1,$$

find the marginal revenue product-of-labor.

13. The revenue function for a certain item is

$$R = px,$$

where p is the price, given by

$$p = 1000 - 0.005x,$$

and x the number of items sold daily. If the cost is

$$c(x) = 5x^2 + 3x + 10{,}000,$$

find the maximum profit.

14. The fixed costs for an apartment building are \$1,000,000; the cost of erecting the nth floor is

$$250{,}000 + (n - 1)30{,}000.$$

The minimum height is three stories and the maximum height is ten stories. If the annual profit from each floor is \$50,000, how many floors should be built to maximize the annual rate of return?

15. A ladder 15 ft long leans against a wall. The bottom of the ladder slides away from the wall at a rate of 1 ft per sec. At what rate is the top of the ladder moving down the wall when the bottom is 2 ft from the wall?

16. A swimming pool is 20 ft wide, 60 ft long, 5 ft deep at one end and 9 ft deep at the other, with the rate of depth constantly increasing. If water is pumped in at the rate of 10 ft^3 per sec, how fast is the water level rising when the water is 2 ft deep at the shallow end?

17. A man standing on an overpass drops a manhole cover at the instant when a car traveling 44 ft per sec is 50 ft from the overpass moving towards it. If the overpass is 25 ft high, how fast is the distance between the cover and the car decreasing $\frac{1}{2}$ second later? Will the cover hit the car?

18. A helicopter takes off vertically at 88 ft per sec from a 750 ft high building. Simultaneously a man walks out of the building at the rate of 8.8 ft per sec. How fast is the distance between the helicopter and the man increasing 2 sec after takeoff?

7 / INTEGRATION

7.1 AREA UNDER A PARABOLA

Consider the problem of finding the area under a parabola with vertex at the origin (Fig. 7.1). The first difficulty which arises is that we lack a definition of area. We define the area of a triangle to be

$$A = \tfrac{1}{2}bh,$$

where b is its base and h its altitude. From this we can define the area of any polygon by breaking it up into triangles (see Fig. 7.2).

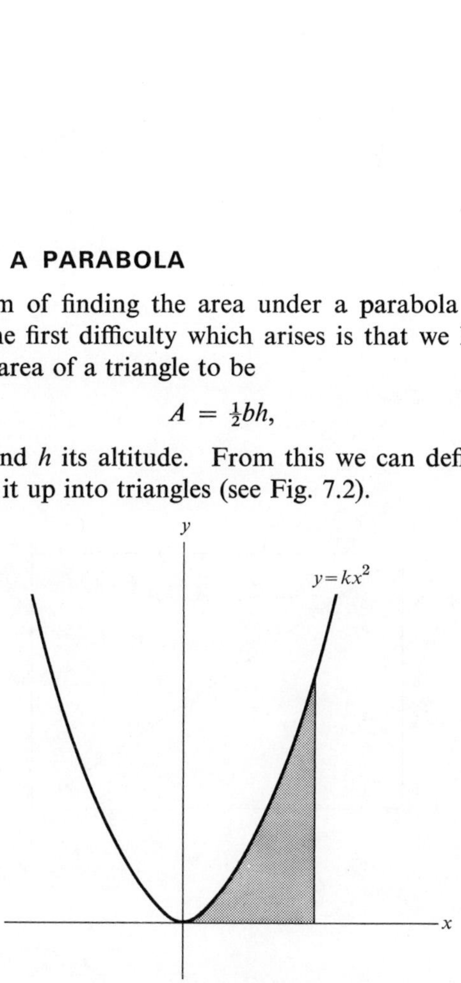

Fig. 7.1

Special cases give us the area of a rectangle

$$A = lw,$$

where l is the length and w is the width (Fig. 7.3), and the area of a trapezoid (Fig. 7.4) is

$$\begin{aligned} A &= \tfrac{1}{2}bh + \tfrac{1}{2}b'h \\ &= \tfrac{1}{2}h(b + b'). \end{aligned}$$

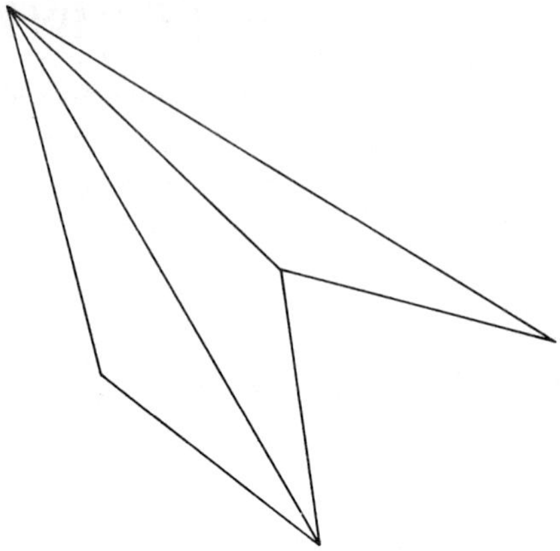

Fig. 7.2

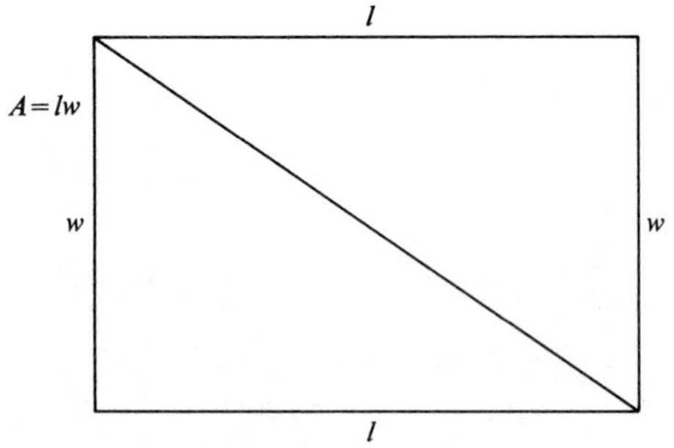

Fig. 7.3

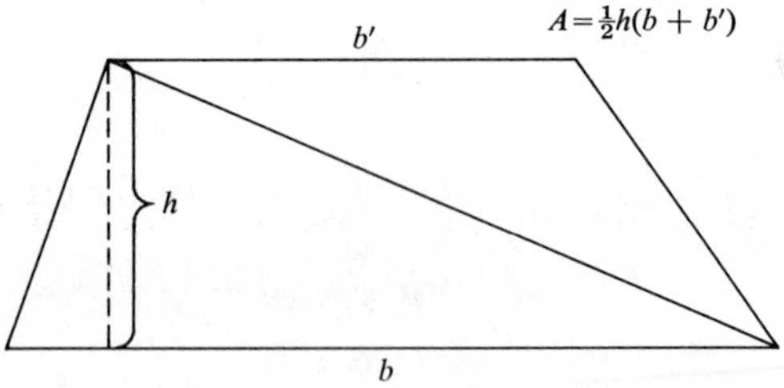

Fig. 7.4

However, we cannot subdivide the area under a parabola into triangles. We can get a series of approximations by using rectangles of decreasing width (see Fig. 7.5).

Suppose we accept an intuitive notion of area in the case of the parabola. Is it possible, by means of a limiting process, to find the area using these approximations? In other words, do these approximations somehow converge to a number which we can reasonably call the area? The answer, which was known to Archimedes in the third century B.C., is Yes.

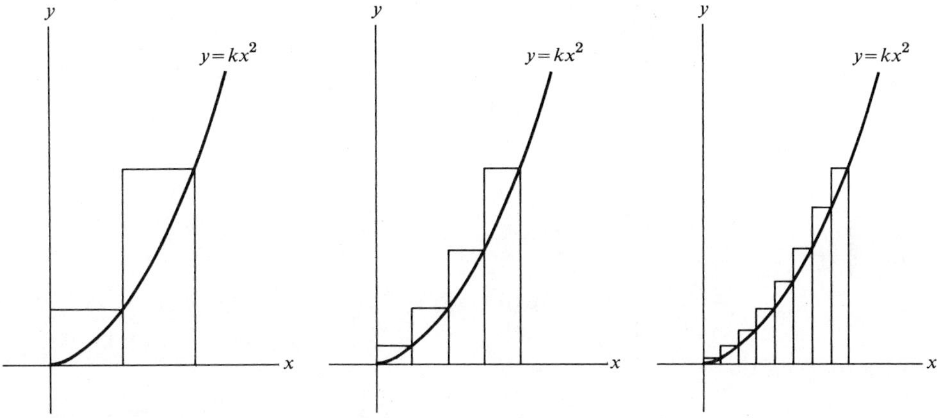

Fig. 7.5

Suppose it is the area under $y = kx^2$ from 0 to a that we wish to compute. We subdivide the interval $[0, a]$ into n equal subintervals

$$[x_0, x_1], [x_1, x_2], \ldots, [x_{n-1}, x_n], \qquad (a = x_n, 0 = x_0),$$

and construct n rectangles (Fig. 7.6), each of width a/n and length $f(x_i) = kx_i^2$, $i = 1, \ldots, n$. Let us call the sum of the areas of the n rectangles S_n. Then

$$S_n = \sum_{i=1}^{n} \frac{a}{n} kx_i^2 = \frac{ak}{n} \sum_{i=1}^{n} x_i^2.$$

But $x_i = ia/n$, so this sum is equal to

$$\frac{ak}{n}\left[\left(\frac{a}{n}\right)^2 + \left(\frac{2a}{n}\right)^2 + \cdots + \left(\frac{na}{n}\right)^2\right] = \frac{a^3k}{n^3}[1 + 2^2 + \cdots + n^2]$$

$$= \frac{a^3k}{n^3} \sum_{i=1}^{n} i^2.$$

From Chapter 1, we have that the sum of the squares of the first n integers is

$$\frac{n}{6}(n + 1)(2n + 1).$$

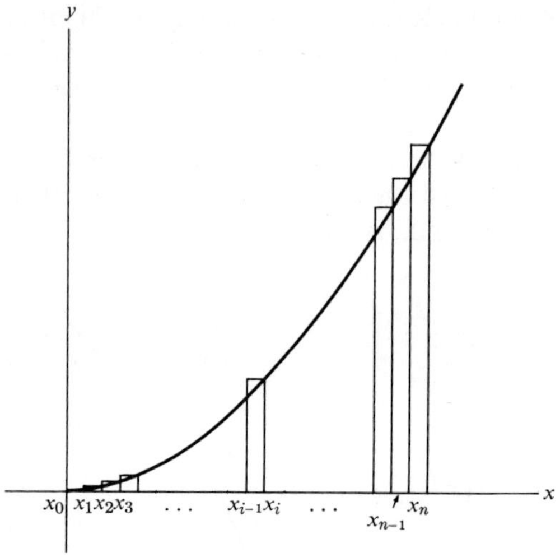

Fig. 7.6

Therefore

$$S_n = \frac{a^3k}{n^3} \cdot \frac{n}{6}(n+1)(2n+1) = \frac{a^3k}{6n^3}(2n^3 + 3n^2 + n).$$

Let us rewrite this as

$$\frac{a^3k}{6}\left(2 + \frac{3}{n} + \frac{1}{n^2}\right),$$

in order to observe that as $n \to \infty$,

$$S_n \to \frac{a^3k}{3}.$$

That is,

$$\lim_{n\to\infty} S_n = \frac{a^3k}{3},$$

since for sufficiently large n, $(3/n) + (1/n^2)$ is arbitrarily small.

Now suppose we start over again and make our approximation by means of rectangles interior to the parabolic area. Then each rectangle has area

$$\frac{ak}{n} f(x_{i-1}) = \frac{ak}{n} x_{i-1}^2 = \frac{ak}{n}\left(\frac{a^2}{n^2}(i-1)^2\right).$$

The sum of the areas of these rectangles in Fig. 7.7 is

$$S_n' = \frac{a^3k}{n^3}\sum_{i=1}^{n}(i-1)^2 = \frac{a^3k}{n^3}\sum_{i=0}^{n-1} i^2$$

$$= \frac{a^3k}{n^3}\,\frac{(n-1)(n-1+1)(2(n-1)+1)}{6}$$

$$= \frac{a^3k}{6}\left(2 - \frac{3}{n} + \frac{1}{n^2}\right).$$

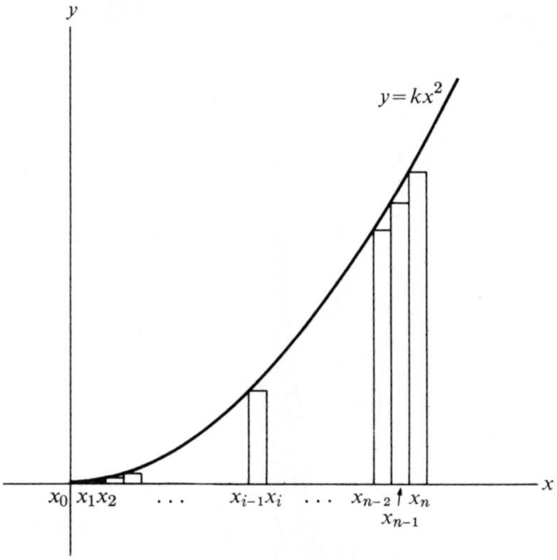

Fig. 7.7

Therefore

$$\lim_{n\to\infty} S_n' = \frac{a^3k}{3}.$$

Thus it seems reasonable to say that the area is $a^3k/3$. In fact, we adopt the following definition. If a region R of the plane is such that

1) there is a sequence of polygonal regions $P_1, \ldots, P_n$, containing R, with area of $P_i = S_i$, such that

$$\lim_{n\to\infty} S_n = L,$$

2) there is a sequence of polygonal regions $P_1', \ldots, P_n'$, contained in R such that

$$\lim_{n\to\infty} S_n' = L, \text{ where area of } P_i' = S_i',$$

then we say that L is the *area* of R.

Example. Find the area under $y = 2x^2$ between 0 and 4 (Fig. 7.8).

Solution. We have the general formula

$$A = \frac{a^3k}{3};$$

in this example $k = 2$ and $a = 4$ so

$$A = \frac{4^3 \cdot 2}{3} = \frac{128}{3}.$$

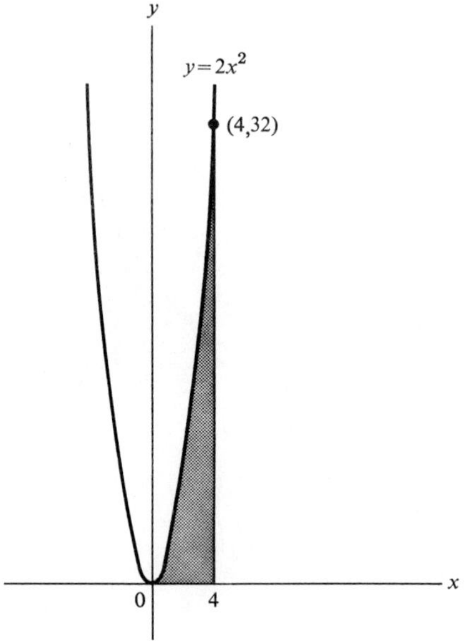

Fig. 7.8

It is easy to extend our method. Suppose we are interested in the area under a parabola (Fig. 7.9)

$$y = kx^2, \qquad (k > 0),$$

between a and b, $0 < a < b$.

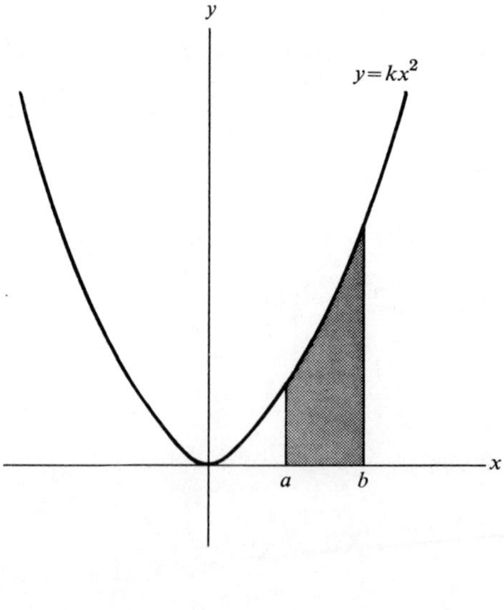

Fig. 7.9

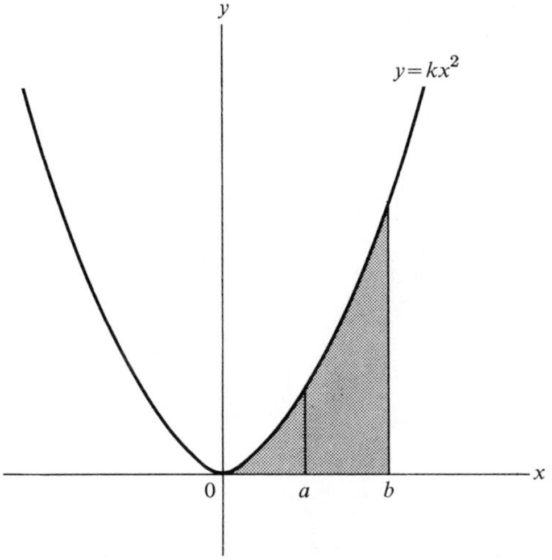

Fig. 7.10

We merely find that the area under the curve (Fig. 7.10) from 0 to b is

$$A_1 = \frac{b^3k}{3},$$

and the area under the curve from 0 to a (see Fig. 7.11) is

$$A_2 = \frac{a^3k}{3},$$

and subtract to get the desired result:

$$A = A_1 - A_2 = \frac{k}{3}(b^3 - a^3).$$

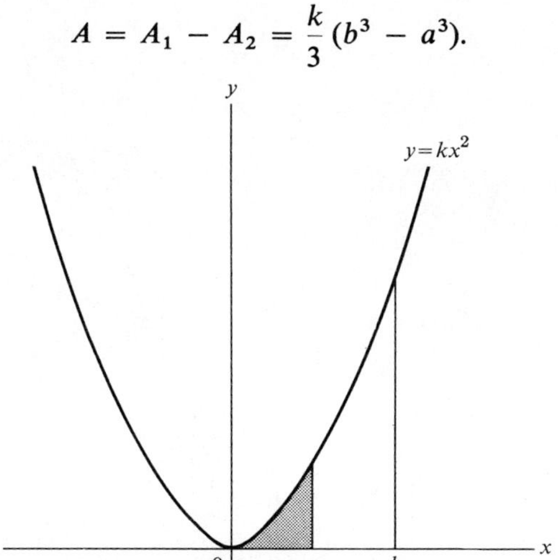

Fig. 7.11

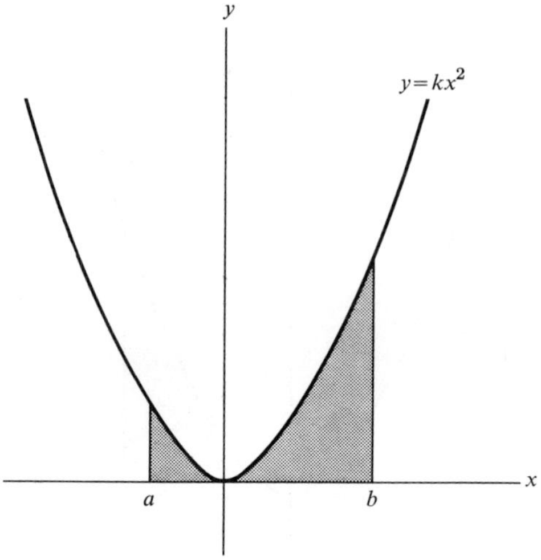

Fig. 7.12

Now suppose we want the area under the parabola $y = kx^2$ from a to b where $a < 0$, $a < b$, $b > 0$ (Fig. 7.12). We observe that the curve is symmetric about the y-axis so that the area from a to 0 should be the same as the area from 0 to $-a$ (note that if $a < 0$, then $-a > 0$) (see Fig. 7.13a), which we know to be

$$A_1 = \frac{(-a)^3 k}{3} = \frac{-a^3 k}{3}.$$

The area from 0 to b in Fig. 7.13b is as above:

$$A_2 = \frac{b^3 k}{3}.$$

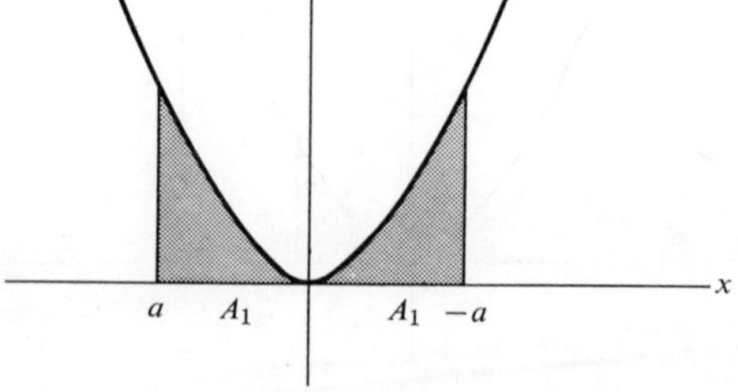

Fig. 7.13(a)

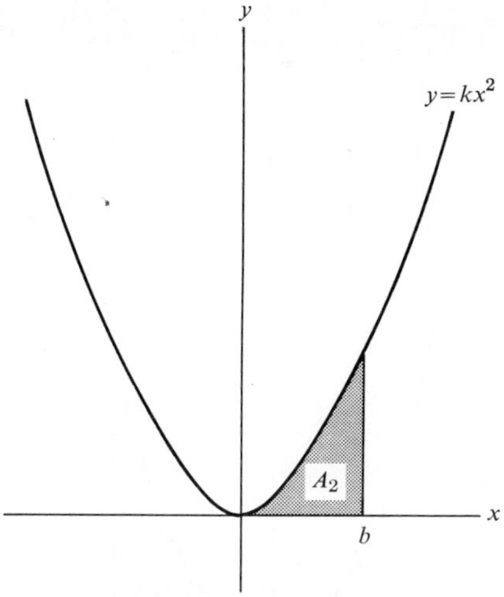

Fig. 7.13(b)

We add the two areas for the desired result as follows:

$$A = A_1 + A_2 = \frac{-a^3k}{3} + \frac{b^3k}{3} = \frac{k}{3}(b^3 - a^3).$$

Comparing the last two cases we see that the same formula holds. The reader can verify that it is also valid if $a < b < 0$. Thus we have

The area under the parabola $y = kx^2$, $(k > 0)$ from a to b $(a < b)$ is given by

$$A = \frac{k}{3}(b^3 - a^3).$$

Examples

1. Find the area under $y = x^2/2$ from -1 to 3 (Fig. 7.14).

Solution. We have

$$A = \frac{k}{3}(b^3 - a^3) = \tfrac{1}{6}(3^3 - (-1)^3) = \tfrac{14}{3}.$$

2. Find the area under the parabola $y = 2x^2$ from -2 to -1 (Fig. 7.15).

Solution. We have

$$A = \frac{k}{3}(b^3 - a^3) = \tfrac{2}{3}((-1)^3 - (-2)^3) = \tfrac{14}{3}.$$

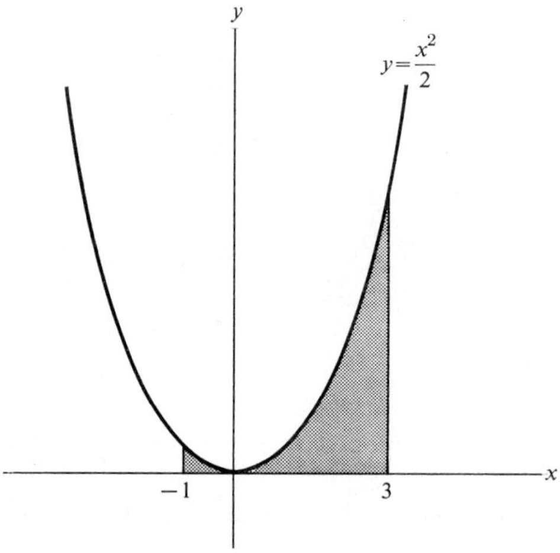

Fig. 7.14

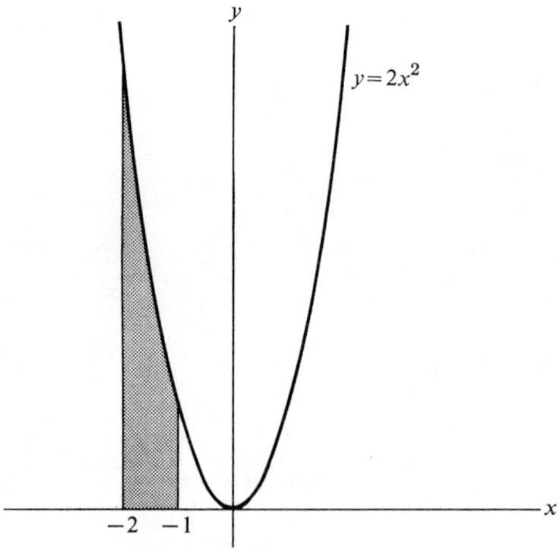

Fig. 7.15

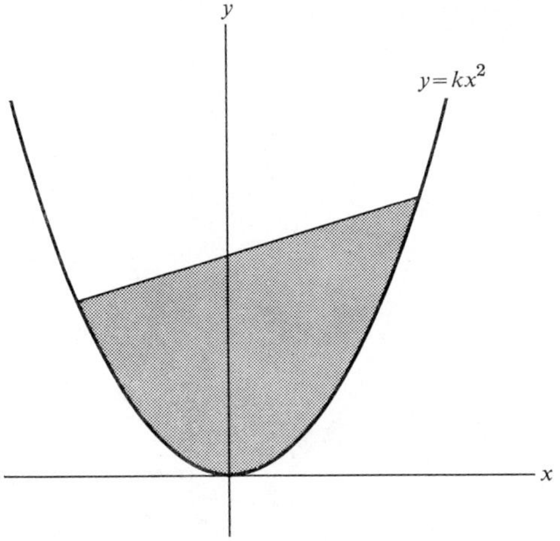

Fig. 7.16

The problem which Archimedes solved was not in the form in which we have given it. He was concerned with the area of a *parabolic sector*—the region between a parabola and a chord of the parabola (see Fig. 7.16). It is an equivalent problem, however, for to obtain the area of the shaded region in Fig. 7.16, we drop perpendiculars to the x-axis to form a trapezoid (see Fig. 7.17). We compute the area of this trapezoid and subtract from it the area under the curve, calculated as above.

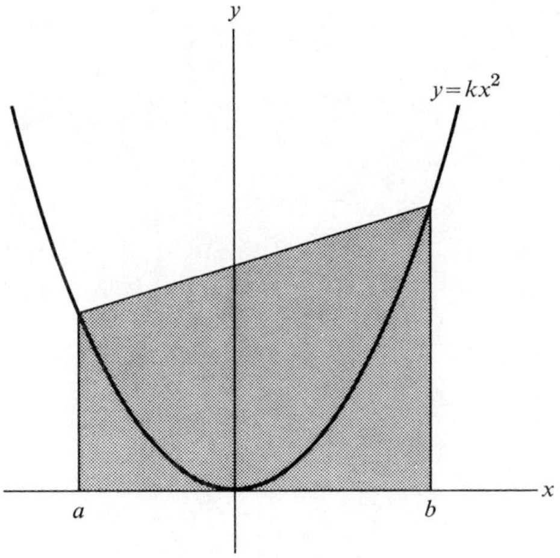

Fig. 7.17

Example. Find the area of the parabolic sector cut by $x + 6y - 4 = 0$ from the parabola $y = x^2/2$ (Fig. 7.18).

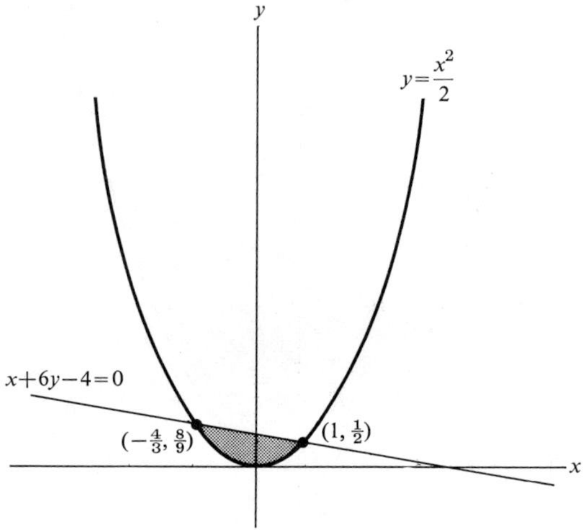

Fig. 7.18

Solution. First we must find the intersection of the straight line and the parabola. Solving the linear equation for y, we have

$$y = \frac{4 - x}{6}.$$

Substituting this in the quadratic, we obtain

$$\frac{4 - x}{6} = \frac{x^2}{2}$$

$$3x^2 + x - 4 = 0$$

$$(3x + 4)(x - 1) = 0$$

$$x = \frac{-4}{3}, 1.$$

Thus the intersections are at $(-4/3, \frac{8}{9})$ and $(1, \frac{1}{2})$. The area of the trapezoid (Fig. 7.19) bounded by $x + 6y - 4 = 0$, the x-axis and the lines $x = -4/3$, $x = 1$ is

$$\tfrac{1}{2}h(b + b') = \tfrac{1}{2}(\tfrac{7}{3})(\tfrac{8}{9} + \tfrac{1}{2})$$

$$= \tfrac{175}{108}.$$

The area under the parabola is

$$\frac{k}{3}(b^3 - a^3) = \frac{1}{3} \cdot \frac{1}{2}\left(1^3 - \left(\frac{-4}{3}\right)^3\right)$$

$$= \frac{91}{162}.$$

Thus the area of the parabolic sector is $\frac{175}{108} - \frac{91}{162} = \frac{343}{324}$.

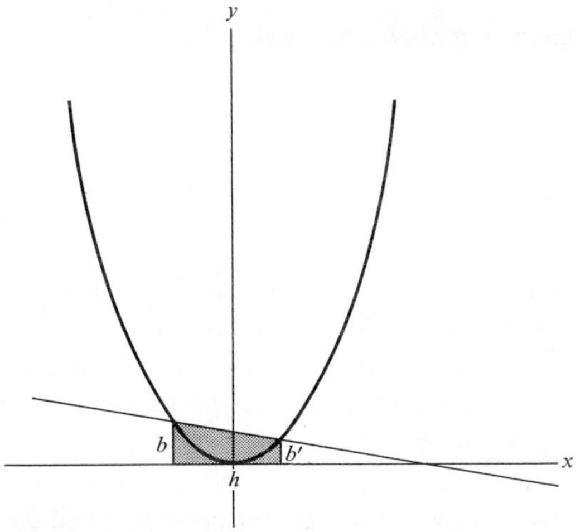

Fig. 7.19

Now we can extend the method of this section to other parabolas whose vertex is on the positive y-axis.

Example. Find the area under the parabola (Fig. 7.20)

$$y = 4x^2 + 2,$$

between -1 and 2.

Solution. We compute first the area A_1 of the region between the curve and a line through the vertex parallel to the x-axis. To do this we simply think of the x-axis as being translated up 2 units. Then

$$A_1 = \tfrac{1}{3}4(2^3 - (-1)^3) = 12.$$

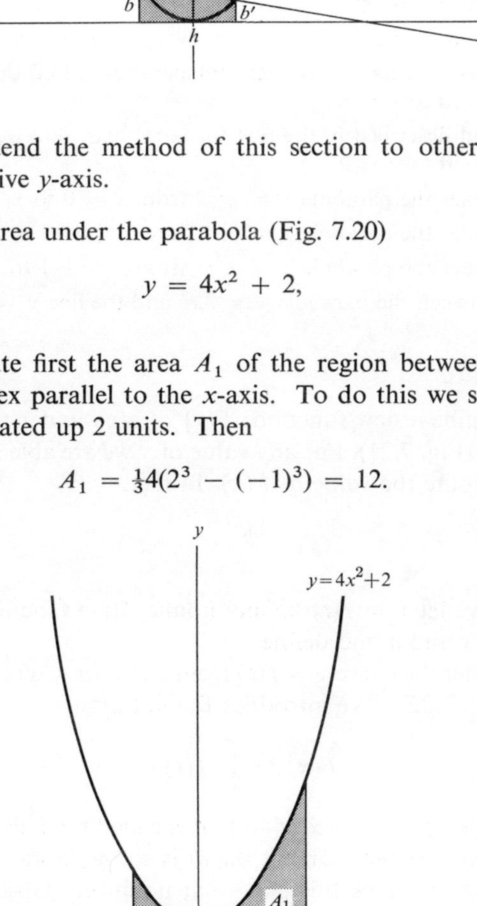

Fig. 7.20

The remaining region is simply a rectangle. Its area is given by

$$A_2 = lw = 2 \cdot 3 = 6.$$

Thus the total area is

$$A = A_1 + A_2 = 12 + 6 = 18.$$

PROBLEMS

1. Find the area under the parabola $y = 4x^2$ from $x = 0$ to $x = 5$.
2. Find the area under the parabola $y = 3x^2$ from $x = -1$ to $x = 2$.
3. Find the area between the parabola $y = 2x^2/3$ and the line $y = 2x$.
4. Find the area under the parabola $y = \frac{1}{10}x^2$ between $x = 10$ and $x = 100$.
5. Find the area between the parabola $y = 4 - x^2$ and the x-axis.
6. Find the area bounded by the curves $x = 5y^2$, $y = 1$ and $x = 0$.
7. Using the method described in the text for parabolas, find the area under the curve $y = 5x$ from $x = 0$ to $x = 4$.
8. Using the method described in the text for parabolas, find the area under the curve $y = x^3$ from $x = 0$ to $x = 2$.
9. Find the area under the parabola $y = x^2/2$ from $x = 0$ to $x = 4$.
10. Find the area under the parabola $y = 2x^2$ from $x = 1$ to $x = 2$.
11. Find the area under the parabola $y = x^2/3$ from $x = -1$ to $x = 5$.
12. Find the area between the parabola $y = 3x^2$ and the line $y = 2x + 1$.

7.2 THE INTEGRAL

Suppose that we define a new function: $F(x)$ = area under the parabola $y = kt^2$ from a to x $(a < x)$ (Fig. 7.21). For any value of x we are able, by the method of the last section, to compute the value of $F(x)$. In fact,

$$F(x) = \frac{k}{3}(x^3 - a^3).$$

Now suppose we let $y = f(t)$ be any nonnegative function continuous on an interval containing x and a and define

$F(x)$ = area under the curve $y = f(t)$ from a to x $(a \leq x)$ (assuming an intuitive notion of area) (Fig. 7.22). We introduce the notation

$$F(x) = \int_a^x f(t)\,dt,$$

to be read: the *integral from a to x of f(t)*. a is called the *lower limit* of integration and x the *upper limit*. At the moment the dt is simply notational. In Section 5 it will become clear why we pick this particular notation. Observe that

$$F(a) = \int_a^a f(t)\,dt = 0.$$

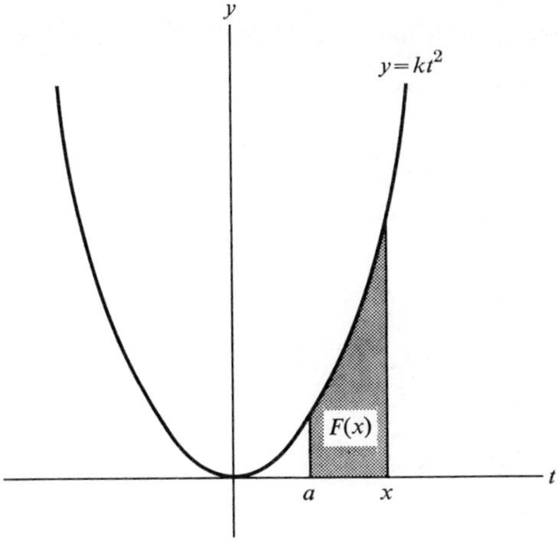

Fig. 7.21

There is no need to be alarmed about the apparent shift in variables. We use $y = f(t)$ instead of $y = f(x)$ because we want to reserve x for use in the definition of $F(x)$. When we are actually computing the value of a function, it is to x that we shall be assigning a value, in conformity with our previous usage. The t is really a dummy variable which is used only to express the function whose graph is used in the definition of $F(x)$. In this notation the area under a parabola $y = kt^2$ is written as

$$F(x) = \int_a^x kt^2\,dt = \frac{k}{3}(x^3 - a^3).$$

It is a function of the upper limit; for a particular value, say $x = b$, we come up with a number for the function value, namely the area $F(b) = (k/3)(b^3 - a^3)$.

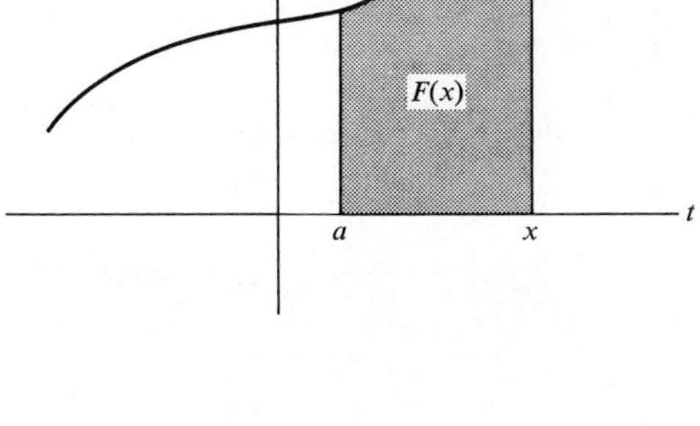

Fig. 7.22

Example. Find the area under the parabola $y = 4x^2$ from $x = 1$ to $x = 2$.

Solution. The area formula is

$$F(x) = \int_a^x 4t^2\,dt,$$

with $a = 1$, $x = 2$, so we have

$$F(2) = \int_1^2 4t^2\,dt = \tfrac{4}{3}(2^3 - 1^3) = \tfrac{28}{3}.$$

So far we are able to compute the value of $F(x) = \int_a^x b(t)\,dt$ only for very special functions $f(t)$:

1) The parabolas of the last section.

2) The case where the graph of $y = f(t)$ is a straight line. For if $y = mt + b$ (see Fig. 7.23), the area under the curve is simply a trapezoid.

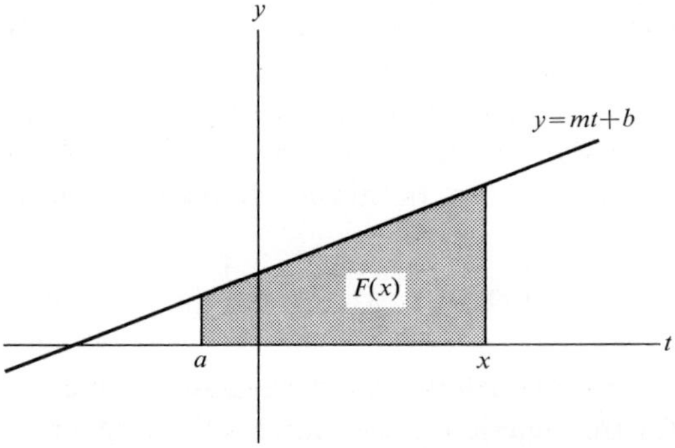

Fig. 7.23

Example. Find the area between the line $y = 3x + 1$ and the x-axis from $x = 0$ to $x = 1$ in Fig. 7.24.

Solution. When $x = 0$, $y = 1$, and when $x = 1$, $y = 4$. Thus the two bases of the trapezoid are 1 and 4 and the height is 1. The area function is

$$F(x) = \int_0^x (3t + 1)\,dt.$$

We have

$$F(1) = \int_0^1 (3t + 1)\,dt = \tfrac{1}{2}h(b + b') = \tfrac{1}{2} \cdot 1(1 + 4) = \tfrac{5}{2}.$$

We examine the method used for parabolas to see whether it can be easily generalized. We observe that a crucial step involved the formula

$$\sum_{i=1}^{n} i^2 = \frac{n}{6}(n + 1)(2n + 1).$$

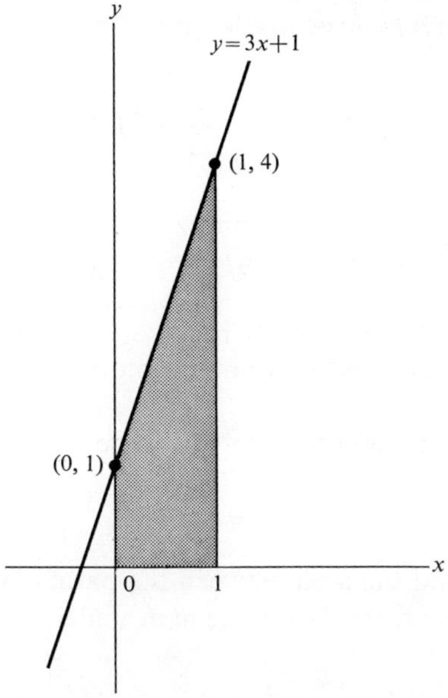

Fig. 7.24

We would need a similar summation formula for more complicated functions. For example (Fig. 7.25), if

$$y = t^4 + t + 1,$$

were the function under consideration, we would need to know a closed-form expression for

$$\sum_{i=1}^{n} i^4 \quad \text{and for} \quad \sum_{i=1}^{n} i,$$

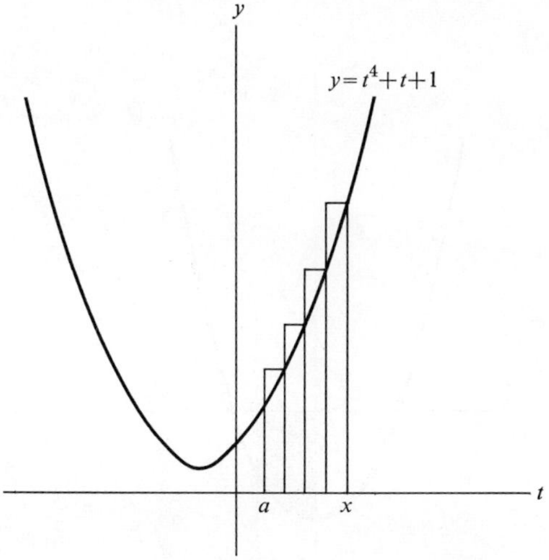

Fig. 7.25

since a division into rectangles would produce as the sum of the areas of the rectangles:

$$S_n = \sum_{i=1}^{n} \frac{x}{n} f(t_i) = \sum_{i=1}^{n} \frac{x}{n} \left[\left(\frac{ix}{n}\right)^4 + \frac{ix}{n} + 1 \right]$$

$$= \frac{x^5}{n^5} \sum_{i=1}^{n} i^4 + \frac{x^2}{n^2} \sum_{i=1}^{n} i + \frac{x}{n} \sum_{i=1}^{n} 1.$$

Such methods rapidly get out of hand, so we shall try a different approach, using differentiation. First let us examine more closely the functions defined in terms of intergrals.

Can we extend the definition to cover negative functions? For example, suppose that

$$y = x^2 - 2,$$

and that we want to find the area between the parabola (Fig. 7.26), the x-axis, the y-axis, and $x = 1$. The formula for the area under $y = x^2$ from 0 to 1 gives us

$$A = \frac{k}{3}(b^3 - a^3) = \tfrac{1}{3}(1^3) = \tfrac{1}{3}.$$

On the other hand, when we had a parabola with vertex $(c, 0)$, $c > 0$, we added to the area as computed above the area of a rectangle c by $b - a$ to get the area under the curve. An analogous procedure for $c = -2$ and $b - a = 1$ gives us a rectangular area of -2, so adding the two numbers gives

$$A = -2 + \tfrac{1}{3} = \frac{-5}{3}.$$

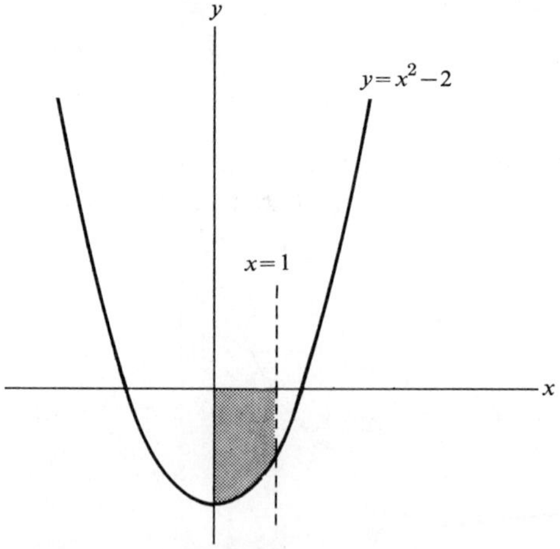

Fig. 7.26

The notion of negative area may appear alarming, but it turns out to be a useful convention. We shall agree that *areas lying below the x-axis are negative.* Then we remove the restriction placed on $f(t)$ in the definition at the beginning of this section and say: If $f(t)$ is continuous on an interval containing x and a then we define

$$F(x) = \int_a^x f(t)\,dt = \begin{cases}\text{the area between the graph of } f(t), \text{ the } t\text{-axis,} \\ t = a, \text{ and } t = x. \qquad (x \geq a).\end{cases}$$

Thus if the graph of $f(t)$ is as illustrated in Fig. 7.27, we have

$$F(x) = \int_a^x f(t)\,dt = A_1 + A_2 + A_3,$$

where A_1 and A_3 are negative and A_2 is positive.

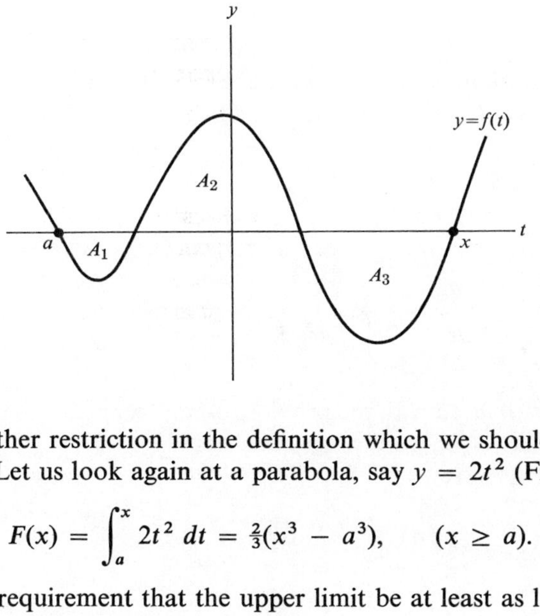

Fig. 7.27

There is another restriction in the definition which we should like to remove, namely $x \geq a$. Let us look again at a parabola, say $y = 2t^2$ (Fig. 7.28).

$$F(x) = \int_a^x 2t^2\,dt = \tfrac{2}{3}(x^3 - a^3), \qquad (x \geq a).$$

If we ignore the requirement that the upper limit be at least as large as the lower limit, we would, by a mechanical application of the formula, define a new function $F_1(x)$ by:

$$F_1(x) = \int_x^a 2t^2\,dt = \tfrac{2}{3}(a^3 - x^3) = -F(x).$$

In other words, the reversal of the limits changes the sign. We can think of this as computing area from right to left instead of from left to right. It is useful to make the definition

$$\int_x^a f(t)\,dt = -\int_a^x f(t)\,dt.$$

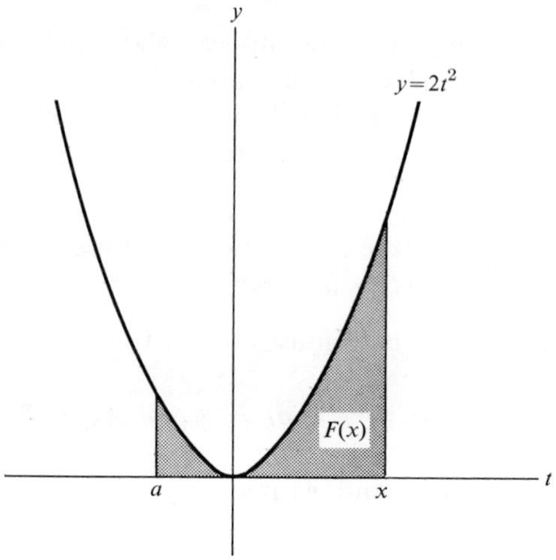

Fig. 7.28

Now we can deal with integrals whose lower limit is the larger simply by computing the integrals with the limits reversed and adding a minus sign.

Example. Find

$$\int_1^0 3t^2\,dt.$$

Solution

$$\int_1^0 3t^2\,dt = -\int_0^1 3t^2\,dt = \frac{-3}{3}(1^3 - 0^3) = -1.$$

However, we are still faced with the problem of not being able to compute the value of the integral in other than a few special cases.

PROBLEMS

1. Compute

a) $\int_0^{-1} 2t^2\,dt.$ b) $\int_2^1 \frac{t^2}{3}\,dt.$ c) $\int_1^6 (2 - t^2)\,dt.$

d) $\int_0^2 -3t^2\,dt.$ e) $\int_{-1}^1 \frac{-t^2}{2}\,dt.$ f) $\int_0^1 (t^2 + 2)\,dt.$

g) $\int_0^5 (1 - t^2)\,dt.$ h) $\int_0^2 3t^2\,dt.$ i) $\int_3^1 \frac{t^2}{2}\,dt.$

j) $\int_1^2 (3 - t^2)\,dt.$ k) $\int_0^1 -2t^2\,dt.$

2. Find the areas described:
   a) between $y = x^2 - 1$ and the x-axis.
   b) between $y = 2x^2 - 1$ and the x-axis from $x = -1$ to $x = 2$.
   c) between $y = -x^2 - 2$ and the x-axis from $x = 0$ to $x = 1$.
   d) between $y = x^2 - 4$ and $y = 2x - 1$.
   e) between $y = 2x^2 - 1$ and the x-axis.
   f) between $y = -2 - x^2$ and the x-axis from $x = 1$ to $x = 2$.

7.3 THE DERIVATIVE OF THE INTEGRAL

We turn to differentiation as our tool for evaluating a function

$$F(x) = \int_a^x f(t)\, dt.$$

Since $F(x)$ is a perfectly good function of x, we have

$$F'(x_0) = \lim_{x \to x_0} \frac{F(x) - F(x_0)}{x - x},$$

if the limit exists. We have

$$\lim_{x \to x_0} \frac{F(x) - F(x_0)}{x - x_0} = \lim_{x \to x_0} \frac{\int_a^x f(t)\, dt - \int_a^{x_0} f(t)\, dt}{x - x_0}.$$

Assume for the moment that $x > x_0$.

We observe that the numerator

$$\int_a^x f(t)\, dt - \int_a^{x_0} f(t)\, dt,$$

is the difference of two areas and is in fact the area shown in Fig. 7.29 under $f(t)$ between x_0 and x. Hence we write

$$\lim_{x \to x_0} \frac{F(x) - F(x_0)}{x - x_0} = \lim_{x \to x_0} \frac{\int_{x_0}^x f(t)\, dt}{x - x_0}.$$

The numerator represents the shaded area in Fig. 7.29. If $f(x) > f(x_0)$, as in our figure, then

$$f(x_0)(x - x_0) < \int_{x_0}^x f(t)\, dt < f(x)(x - x_0),$$

with the inequalities reversed if $f(x_0) > f(x)$.
From this we have

$$f(x_0) < \frac{\int_{x_0}^x f(t)\, dt}{x - x_0} < f(x), \qquad \text{if } f(x) > f(x_0)$$

or

$$f(x_0) > \frac{\int_{x_0}^x f(t)\, dt}{x - x_0} > f(x), \qquad \text{if } f(x) < f(x_0).$$

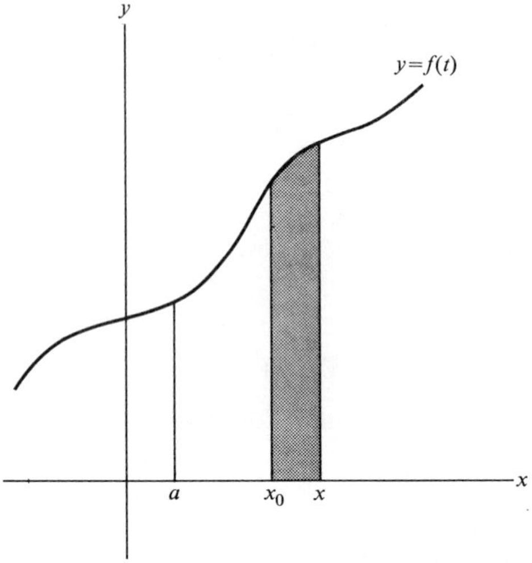

Fig. 7.29

The situation is the same for $x < x_0$. Moreover, since $f(x)$ is continuous

$$\lim_{x\to x_0} f(x) = f(x_0) = \lim_{x\to x_0} f(x_0).$$

We see that as $x \to x_0$,

$$\frac{\int_{x_0}^{x} f(t)\, dt}{x - x_0}$$

is squeezed between two quantities having the same limit, namely $f(x_0)$. This gives us that

$$F'(x_0) = \lim_{x\to x_0} \frac{\int_{x_0}^{x} f(t)\, dt}{x - x_0} = f(x_0).$$

We state this result as a theorem.

Theorem 1. If $f(t)$ is continuous on an interval containing x and a and

$$F(x) = \int_a^x f(t)\, dt,$$

then $F(x)$ is differentiable and

$$F'(x) = f(x).$$

Examples

1. If

$$F(x) = \int_0^x 5t^2\, dt \qquad \text{then} \qquad F'(x) = 5x^2.$$

2. If

$$F(x) = \int_{-1}^{x} \frac{t^2 + 1}{2t + 4}\, dt \qquad \text{then} \qquad F'(x) = \frac{x^2 + 1}{2x + 4},$$

for $x \neq -2$.

3. Suppose

$$F(x) = \int_0^x \sqrt{1 - t^2}\, dt.$$

Then for $|x| \leq 1$,

$$F'(x) = \sqrt{1 - x^2}.$$

Warning. One must not be led astray by the thought of differentiation to the extent that he differentiates the function *under* the integral sign. For instance, in Example 1, $F'(x) = 5x^2$, not $10x$.

Although we still have not evaluated the function $F(x)$ defined by an integral, we do know what its derivative is. Any function which is differentiable has a uniquely determined derivative; the converse does not hold. For example, if

$$f(x) = 2x^2 + x + 1,$$

and

$$g(x) = 2x^2 + x - 3,$$

then

$$f'(x) = g'(x) = 4x + 1.$$

We do notice that while $f(x)$ and $g(x)$ are not equal, they do differ only by a constant. This is a general phenomenon.

Theorem 2. If $F'(x) = G'(x)$ for all x, then

$$F(x) - G(x) = k, \quad \text{a constant.}$$

Proof. Define a new function

$$H(x) = F(x) - G(x).$$

We can write

$$H'(x) = 0.$$

Therefore, by Theorem 6.5,

$$H(x) = F(x) - G(x)$$

is a constant.

Corollary. If $F(x)$ and $G(x)$ are differentiable functions such that

1) for all x, $F'(x) = G'(x)$, and
2) There exists b such that $F(b) = G(b)$, then $F(x) = G(x)$ for all x.

Proof. From Theorem 2,

$$F(x) - G(x) = k,$$

k a constant. But

$$F(b) - G(b) = 0,$$

so $k = 0$ and $F(x) = G(x)$ for all x.

We apply Theorem 2 to evaluate functions defined by integrals.

Examples

1. Let

$$F(x) = \int_a^x 2t^3\, dt.$$

Then $F'(x) = 2x^3$.

Now we ask: What function did we differentiate in order to get $2x^3$? In differentiating polynomials we reduce exponent by 1—hence in the reverse process, which we call *anti-differentiation* or *integration*, we must raise the exponent by 1. Moreover, in differentiation we multiply by the exponent so here we must divide. In general, if

$$F'(x) = a_n x^n \qquad \text{then} \qquad F(x) = \frac{a_n x^{n+1}}{n+1} + k.$$

Thus in our example, since

$$F'(x) = 2x^3,$$

we know from Theorem 2 that

$$F(x) = \frac{x^4}{2} + k,$$

where k is some constant. To evaluate k we apply the corollary to Theorem 2. We know that

$$F(a) = \int_a^a f(t)\, dt = 0.$$

Thus if we find k such that $(x^4/2) + k$ evaluated at a is 0, we have the right function.

We must have

$$\frac{a^4}{2} + k = 0 \qquad \text{so} \qquad k = \frac{-a^4}{2}.$$

Therefore

$$F(x) = \frac{x^4}{2} - \frac{a^4}{2}.$$

2. Find the area under the curve $x^2 - 2x + 1$ from 0 to 2 (Fig. 7.30).

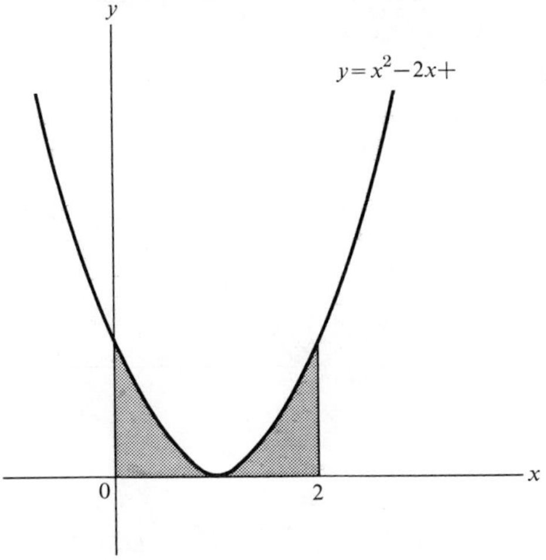

Fig. 7.30

Solution. Let

$$F(x) = \int_0^x (t^2 - 2t + 1)\, dt.$$

Then

$$F'(x) = x^2 - 2x + 1 \quad \text{and} \quad F(x) = \frac{x^3}{3} - x^2 + x + k.$$

$F(0) = 0$ so that $k = 0$. Thus

$$F(x) = \frac{x^3}{3} - x^2 + x,$$

and $F(2) = \frac{8}{3} - 4 + 2 = \frac{2}{3}$, the desired area.

3. Find the area between the curves $y = x^2$ and $y = \sqrt{x}$ in Fig. 7.31.

Solution. The curves intersect at (0, 0) and (1, 1). If we find the area under $y = \sqrt{x}$ and from it subtract the area under $y = x^2$, we shall have the desired area.

Define

$$F(x) = \int_0^x \sqrt{t}\, dt, \qquad G(x) = \int_0^x t^2\, dt.$$

Then

$$F'(x) = x^{1/2}, \qquad G'(x) = x^2,$$

and

$$F(x) = \tfrac{2}{3}x^{3/2} + k_1, \qquad G(x) = \frac{x^3}{3} + k_2.$$

$$F(0) = 0 = k_1, \qquad G(0) = 0 = k_2.$$

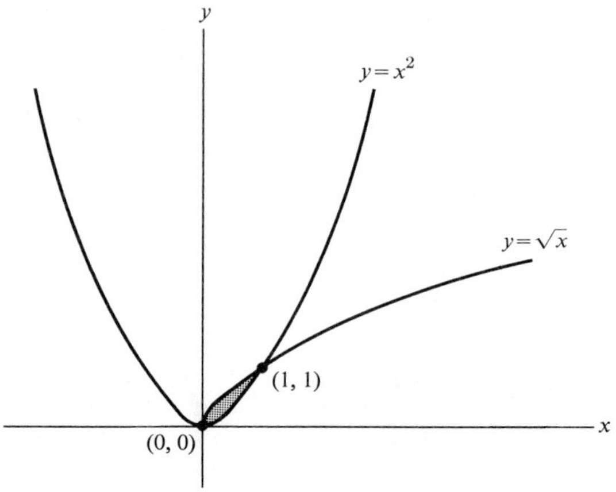

Fig. 7.31

Therefore,

$$F(x) = \tfrac{2}{3}x^{3/2}, \qquad G(x) = \frac{x^3}{3}.$$

The desired area is

$$F(1) - G(1) = \tfrac{2}{3} - \tfrac{1}{3} = \tfrac{1}{3}.$$

To summarize our method:

If

$$F(x) = \int_a^x f(t)\, dt,$$

then $F'(x) = f(x)$. Therefore, if $f(x) = a_n x^n$, then

$$F(x) = \frac{a_n x^{n+1}}{n+1} + k.$$

To evaluate k, set $F(a) = 0$ and solve for k.

PROBLEMS

1. Find a function whose derivative is each of the following:
   a) $x^4 - x^3 + 2x - 5$. b) $-1/2x^2$.
   c) $4t^5 - 5t^3 + t - 4$. d) $4x^{1/3}$.
   e) $1/\sqrt{7y}$. f) $\sqrt{2x - 5}$.
   g) $4x^2 - 2x^3 + x - 1$. h) $2/x^3$.
2. Find the area under the curve $y = 17x^2$ between $x = 2$ and $x = 5$.
3. Find $F'(x)$.

   a) $F(x) = \int_0^x 3t^2\, dt$. b) $F(x) = \int_x^1 (2t + 1)\, dt$.

c) $F(x) = \int_2^x \sqrt{1 + t}\, dt.$

d) $F(x) = \int_0^x 3t^4\, dt.$

e) $F(x) = \int_x^a (4t - 2)\, dt.$

f) $F(x) = \int_2^x \sqrt{t - 2}\, dt.$

4. If $F(x) = \int_{-1}^x (t^3 - t^2 + 2t + 1)\, dt$, find $F'(2)$.

5. If $F(x) = \int_0^x \sqrt{1 + \sqrt{1 - t^2}}\, dt$, find $F'(1)$.

6. Find $F'(2)$ if $F(x) = \int_1^x \frac{1}{t}\, dt.$

7. Find the area under the curve $y = x^{15}$ from $x = 0$ to $x = 1$.

8. Evaluate

a) $\int_{-2}^{-1} x^2\, dx.$

b) $\int_{-1}^{-3} x^3\, dx.$

c) $\int_1^2 (x^3 - 4x^2)\, dx.$

d) $\int_0^1 \sqrt{t}\, dt.$

e) $\int_{-1}^1 3x^2\, dx.$

f) $\int_{-1}^1 (3x^5 - 5x^3 + 2x)\, dx.$

g) $\int_{-1}^1 (5x^7 + 6x^6 + 4x^2 + 1)\, dx.$

9. If $F(x) = \int_1^{x^2} (1 + t^2)^{1/2}\, dt$, find $F'(x)$.

7.4 THE FUNDAMENTAL THEOREM OF CALCULUS

In the last section we evaluated a function

$$F(x) = \int_a^x f(t)\, dt,$$

by determining a function, say $G(x)$, whose derivative was the same as that of $F(x)$, and then calculating by what constant k this function and $F(x)$ differed. That is, we found a k such that

$$G(x) + k = F(x).$$

It would be nice to have a general expression for this constant k. Let us look at an example. If

$$F(x) = \int_1^x (3t^2 + 2)\, dt,$$

then $G(x) = x^3 + 2x$ is such that $G'(x) = F'(x) = 3x^2 + 2$ so that $F(x) = x^3 + 2x + k$. $F(1) = 0$, so $k = -1 - 2 = -3$. But $G(1) = 3$, so that in this case $k = -G(1)$ and $F(x) = G(x) - G(1)$. That is, we find that if we take the function $G(x)$ without constant term whose derivative is the same as that of $F(x)$ and subtract the value of $G(x)$ at the lower limit, we obtain $F(x)$. Moreover, if we want the value of the function $F(x)$ for a particular value of x, say b, we have

$$F(b) = \int_1^b (3t^2 + 2)\, dt = G(b) - G(1).$$

Our very first area problem

$$\begin{aligned} F(b) &= \text{area under the parabola } kx^2 \text{ from } a \text{ to } b \\ &= \int_a^b kt^2\, dt = \frac{k}{3}(b^3 - a^3) \end{aligned}$$

is also solved by this procedure. If we let

$$G(x) = \frac{k}{3}x^3,$$

then $G'(x) = F'(x) = kx^2$ and

$$F(b) = \frac{k}{3}(b^3 - a^3) = G(b) - G(a).$$

Are these special cases of a general theorem? The answer is given by

Theorem 3. *Fundamental Theorem of Calculus.* If $f(t)$ is a continuous function defined on an interval containing a and b, then

$$\int_a^b f(t)\, dt = F(b) - F(a),$$

where $F(t)$ is any function whose derivative is $f(t)$.

Proof. Let

$$G(x) = \int_a^x f(t)\, dt.$$

Let $H(x)$ be a function without constant term whose derivative is $f(x)$. Then

$$G(x) = H(x) + k.$$

But $G(a) = 0$ by definition, so

$$G(a) = H(a) + k,$$

and $k = -H(a)$. Thus

$$G(x) = H(x) - H(a),$$

and

$$G(b) = \int_a^b f(t)\, dt = H(b) - H(a).$$

Now let $F(x)$ be any function whose derivative is $f(x)$. Then

$$F(x) = H(x) + c,$$

where c is a constant. Now,

$$\begin{aligned} F(b) - F(a) &= H(b) + c - (H(a) + c) \\ &= H(b) - H(a) \\ &= \int_a^b f(t)\, dt, \end{aligned}$$

as required.

A natural question arises. In the statement of the theorem we have "any function whose derivative is $f(t)$"—why not *the* function whose constant term is zero? The answer is found in the third from last line of the proof where the constant c which comes from $F(b)$ is cancelled by the c from $F(a)$. Note also that what the fundamental theorem tells us is how to evaluate the constant which arises in the integration, namely the constant is just the value of $F(x)$ at the lower limit.

We call any function whose derivative is $f(x)$ an *antiderivative* of $f(x)$. (Note the use of "an" instead of "the" because there is no unique antiderivative.)

$\int_a^b f(t)\, dt$ is called a *definite integral*; $f(t)$ is the *integrand*. Note that the definite integral is a real number, while

$$F(x) = \int_a^x f(t)\, dt,$$

is a function of x.

Examples

1. Evaluate

$$\int_2^3 (3s^3 - 4s + 1)\, ds.$$

Solution

$$F(s) = \frac{3s^4}{4} - 2s^2 + s,$$

is a function such that

$$F'(s) = 3s^3 - 4s + 1.$$

Therefore

$$\int_2^3 (3s^3 - 4s + 1)\, ds = F(3) - F(2) = \tfrac{159}{4}.$$

If $F(x)$ is an antiderivative of $f(x)$, we usually write

$$\int_a^b f(x)\, dx = F(x)\,|_a^b = F(b) - F(a).$$

2. Find the area under the curve

$$y = x^4 - 5,$$

between -1 and 2.

Solution. Since the area from -1 to x is given by

$$F(x) = \int_{-1}^{x} (t^4 - 5)\, dt,$$

we write the area we want to find as

$$A = \int_{-1}^{2} (x^4 - 5)\, dx.$$

Then an antiderivative of $x^4 - 5$ is $(x^5/5) - 5x$ so

$$\int_{-1}^{2} (x^4 - 5)\, dx = \frac{x^5}{5} - 5x \Big|_{-1}^{2} = \frac{-42}{5}.$$

Warning. We must be certain that the function under the integral sign is continuous over the interval between the limits of integration in order to apply the fundamental theorem. For example, we cannot use it to evaluate

$$\int_{-1}^{1} \frac{1}{x^2}\, dx,$$

because $1/x^2$ is not continuous at the origin.

It is also important to observe that success in applying the fundamental theorem depends upon finding some function whose derivative is the integrand. This is a straightforward process for polynomials, but in general there are no formulas enabling us to deal with all rational functions or all roots.

We are not yet able to evaluate

$$\int_{a}^{b} \sqrt{x^2 + 1}\, dx,$$

because $\sqrt{x^2 + 1}$ is not the derivative of any function we have as yet encountered. Note, however, that

$$\int_{a}^{b} \sqrt{x^2 + 1}\, x\, dx,$$

is evaluated by observing that the derivative of $x^2 + 1$ is $2x$ and rewriting the integral as

$$\tfrac{1}{2} \int_{a}^{b} \sqrt{x^2 + 1}\, 2x\, dx,$$

so that it becomes clear that $\sqrt{x^2 + 1}\, 2x$ is the derivative of

$$\frac{(x^2 + 1)^{3/2}}{\frac{3}{2}}.$$

Then we evaluate the integral as

$$\tfrac{1}{2} \cdot \tfrac{2}{3}(x^2 + 1)^{3/2}\Big|_a^b = \tfrac{1}{3}[(b^2 + 1)^{3/2} - (a^2 + 1)^{3/2}].$$

This type of problem introduces a big stumbling block for the neophyte investigator who asks: What happened to the x in

$$\int_a^b \sqrt{x^2 + 1}\, x\, dx?$$

The answer is that the x is simply part of the function $\sqrt{x^2 + 1}\, x$ whose antiderivative is $\frac{1}{3}(x^2 + 1)^{3/2}$. If in doubt when computing antiderivatives, all we need to do is to differentiate and see whether we obtain the original function.

It is useful to think of quantities such as $x^2 + 1$ in the above example as new variables and to reverse, in a sense, the chain rule for differentiation. For example, let

$$u(x) = x^2 + 1 \qquad \text{so that} \qquad \frac{du}{dx} = 2x.$$

Then we write

$$\sqrt{x^2 + 1}\, x\, dx,$$

as

$$\tfrac{1}{2}\sqrt{x^2 + 1}\, 2x\, dx,$$

which is

$$\tfrac{1}{2}u^{1/2}\, du, \qquad \left(du = \frac{du}{dx}\, dx\right).$$

However, we cannot legitimately write

$$\int_a^b \sqrt{x^2 + 1}\, x\, dx = \tfrac{1}{2}\int_a^b u^{1/2}\, du,$$

since the limits a and b are for the variable x and not for the new variable u. However, since u is a function of x, the limits on u are given by $u(a)$ and $u(b)$. In general we have

Theorem 4. (Change of Variables Theorem). If u is a differentiable function of x and f is a continuous function of u, then

$$\int_a^b f(u(x))\, \frac{du}{dx}\, dx = \int_{u(a)}^{u(b)} f(u)\, du.$$

Proof

$$\int_{u(a)}^{u(b)} f(u)\, du = F(u)\Big|_{u(a)}^{u(b)} = F(u(b)) - F(u(a)),$$

where $F(u)$ is an antiderivative of $f(u)$. But by the chain rule,

$$\frac{dF}{dx}(u(x)) = f(u(x))\frac{du}{dx},$$

so

$$\int_a^b f(u(x))\frac{du}{dx}\,dx = F(u(x))\Big|_a^b = F(u(b)) - F(u(a))$$

$$= \int_{u(a)}^{u(b)} f(u)\,du,$$

as required.

We now have a supplementary integration formula. If

$$F'(x) = f(x) \qquad \text{and} \qquad f(x)\,dx = a_n u^n\,du,$$

u a differentiable function of x, then

$$F(x) = \frac{a_n u^{n+1}}{n+1} + k.$$

In particular,

$$\int_a^b f(x)\,dx = \frac{a_n(u(b))^{n+1}}{n+1} - \frac{a_n(u(a))^{n+1}}{n+1}.$$

The hit-and-miss method of recognizing the integrand as the derivative of some function can be supplemented by the techniques of integration discussed in Chapter 12, but the problem of integration is intrinsically more difficult than differentiation.

Examples

1. Evaluate

$$\int_0^1 \frac{x}{(x^2+2)^2}\,dx.$$

Solution. Let $u = x^2 + 2$. Then $du/dx = 2x$.

$$\int_0^1 \frac{x}{(x^2+2)^2}\,dx = \frac{1}{2}\int_0^1 \frac{2x}{(x^2+2)^2}\,dx = \frac{1}{2}\int_2^3 u^{-2}\,du = -\frac{u^{-1}}{2}\Big|_2^3 = \frac{1}{12}.$$

2. Evaluate

$$\int_1^3 \frac{x+2}{\sqrt{x^2+4x}}\,dx.$$

Solution. Let $u = x^2 + 4x$. Then $du/dx = 2x + 4$. Therefore,

$$\int_1^3 \frac{x+2}{\sqrt{x^2+4x}}\,dx = \frac{1}{2}\int_1^3 \frac{2x+4}{\sqrt{x^2+4x}}\,dx = \frac{1}{2}\int_5^{21} \frac{du}{\sqrt{u}} = u^{1/2}\Big|_5^{21} = \sqrt{21} - \sqrt{5}.$$

So far we have just used integration for the calculation of area. However, just as differentiation has many applications, integration provides solutions to a variety of problems.

Suppose the velocity of a particle is given by

$$v = t^2 - 4t + 1.$$

We want to find an equation representing its motion. We know that $v = ds/dt$, so we ask: What function has a derivative of $t^2 - 4t + 1$? The answer is: Any function of the form

$$s = \frac{t^3}{3} - 2t^2 + t + k,$$

k a constant. Thus our original information is not adequate to determine the motion equation. However, suppose we are given that the initial value of s is 2 (when $t = 0$). Then

$$2 = s(0) = k,$$

so

$$s = \frac{t^3}{3} - 2t^2 + t + 2.$$

3. Sometimes we must integrate twice. It has been experimentally determined that the acceleration of a freely falling body due to gravity is a constant, g. (We neglect air resistance and assume that the height from which it falls is negligible compared to the earth's diameter.) In the metric system $g = -980$ cm per second per second and in the British system, $g = -32$ ft per second per second (negative because the acceleration is directed downward). Find how far a ball falls in two seconds if it is dropped from a height of 100 ft.

Solution. We have

$$a = \frac{dv}{dt} = -32.$$

Any antiderivative of -32 is given by

$$v = -32t + k.$$

Since the body falls freely it has zero initial velocity, so when $t = 0$, $v = 0$. This yields $k = 0$ and

$$v = -32t.$$

Integrating again produces

$$s = -16t^2 + k.$$

When $t = 0$, $s = 100$ so $k = 100$. The motion equation is

$$s = -16t^2 + 100.$$

When $t = 2$, $s = -16 \cdot 4 + 100 = 36$. Since s represents the height above ground, the ball falls 64 ft in two seconds.

4. If the marginal revenue is given by

$$\frac{dR}{dQ} = 9 - 3Q,$$

discuss the revenue curve.

Solution. We want an expression for the revenue R whose derivative will be the marginal revenue. We find that

$$R = 9Q - \tfrac{3}{2}Q^2 + k.$$

To evaluate k we observe that when $Q = 0$ we must have $R = 0$ from the definitions; therefore $k = 0$. The revenue curve is the graph of

$$R = 9Q - \tfrac{3}{2}Q^2,$$

restricted to the domain [0, 6] by the fact that we cannot have negative revenue (see Fig. 7.32). The revenue function has a maximum where

$$\frac{dR}{dQ} = 9 - 3Q = 0, \qquad Q = 3.$$

The equation for the unit price is given by

$$\frac{R}{Q} = \frac{9Q - \frac{3}{2}Q^2}{Q} = 9 - \tfrac{3}{2}Q.$$

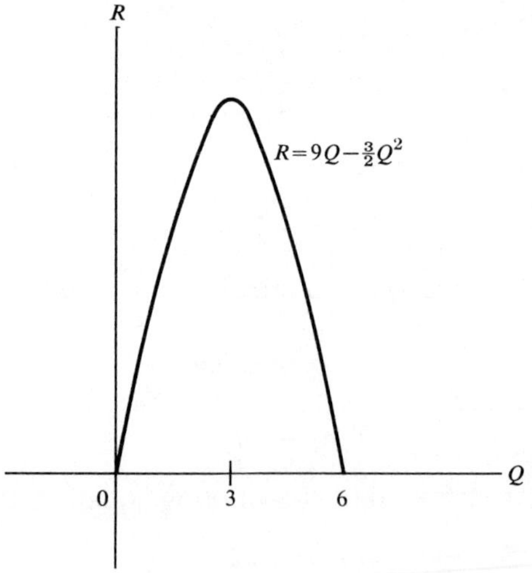

Fig. 7.32

5. If the temperature T at time t is given by

$$T = 49 + 5t - 2t^2,$$

find the average temperature over the period from $t = 0$ to $t = 24$.

Solution. The first difficulty we encounter is in giving a meaning to "average." If we have a variable with a finite number of values, say $T_1 = 5$, $T_2 = 10$, $T_3 = 8$; to find the average we take the sum of the numbers and divide by the number of summands:

$$T_{\text{average}} = \frac{T_1 + T_2 + T_3}{3} = \frac{23}{3}.$$

However, in this problem, T is given as a continuous function of a real variable, so no finite summation process will work. Our solution of the area problem for parabolas gives us a hint that integration is somehow an infinite analog of summation, so we make the following definition:

If $y = f(x)$ is continuous on an interval containing a and b, the *average value* of y on $[a, b]$ is given by

$$y_{\text{av}} = \frac{1}{b - a} \int_a^b f(x)\, dx.$$

Intuitively, the definition makes sense, since the area of the rectangle of base $b - a$ and height y_{av} shown in Fig. 7.33 has the same area as the area $\int_a^b f(x)\, dx$ under the curve.

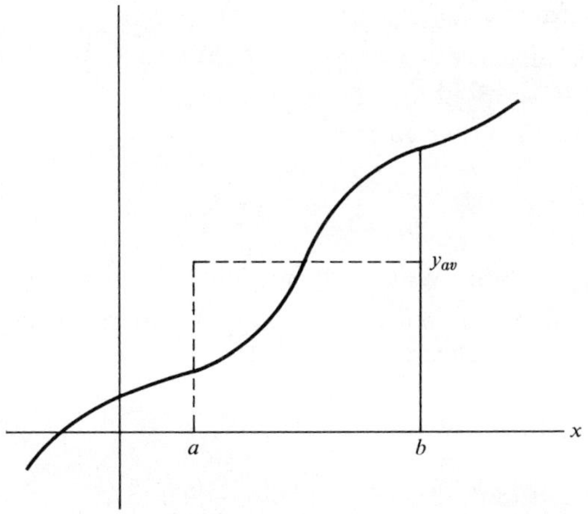

Fig. 7.33

To return to our problem, we have

$$\begin{aligned} T_{\text{av}} &= \frac{1}{24 - 0} \int_0^{24} (49 + 5t - 2t^2)\, dt \\ &= \tfrac{1}{24}(49t + \tfrac{5}{2}t^2 - \tfrac{2}{3}t^3)\Big|_0^{24} \\ &= -275. \end{aligned}$$

PROBLEMS

1. Evaluate

a) $\displaystyle\int_{-1}^{1} (x^5 - 4x^4)\, dx.$

b) $\displaystyle\int_3^5 \sqrt{x - 3}\, dx.$

c) $\displaystyle\int_0^2 x(x^2 + 1)^{1/2}\, dx.$

d) $\displaystyle\int_{-1}^{0} \frac{t^3}{(t^4 + 1)^2}\, dt.$

e) $\displaystyle\int_a^b \frac{2x}{\sqrt{5x^2 + 2}}\, dx.$

f) $\displaystyle\int_0^2 \frac{3t^2 + t}{(2t^3 + t^2 - 4)^{1/3}}\, dt.$

2. Find the area under the curve
$$y = x^{4/5} - 5x + 1$$
from $x = 1$ to $x = 2$.

3. If the velocity of a particle is given by $ds/dt = v = 7t - 6$, find its equation of motion if $s = 5$ when $t = 1$.

4. If the velocity of a particle is given by $ds/dt = v = 2t + 5$, find its equation of motion if $s = 10$ when $t = 0$.

5. If the acceleration of a particle is given by $a = dv/dt = 4t - 1$, find its equation of motion if its initial velocity is 40 and if $s = 500$ when $t = 1$.

6. If the acceleration of a particle is given by $a = 2t - 5$, and when $t = 0$, $v = 16$, and $s = 12$, find the equation of motion.

7. If the velocity of a particle is given by $v = 200 - t + 3t^2$, find the average velocity in the period from $t = 0$ to $t = 25$.

8. If the marginal revenue is given by
$$\frac{dR}{dQ} = 54 - 7Q,$$
find the revenue function. What is the maximum value for the revenue?

9. It is possible to find the definite integral of functions which are not continuous. Develop a method to evaluate
$$\int_0^5 [x]\, dx,$$
where $[x]$ is the greatest integer less than or equal to x.

10. Find the area between the two given curves:

a) $y = x^2$, $y = 2x + 3$. b) $y = x^2$, $y = x^3$.

c) $y = \dfrac{1}{\sqrt{2x - 1}}$, $y = 2x^2$, between $x = 1$ and $x = 2$.

11. The amount of capital stock y is related to investment I by the equation

$$\frac{dy}{dt} = I(t).$$

If the rate of net investment at time t is given (in thousands of dollars per year) by

$$I(t) = 3t^{1/3} + 7,$$

find the amount of capital stock formed in the first ten years of the investment.

12. If $y = x^3 - 4x + 1$, find the average value of y on $[0, 2]$.

13. If the marginal revenue is given by

$$\frac{dR}{dQ} = \frac{39 - Q}{2},$$

find the revenue function. What value of Q maximizes the revenue?

14. Find the area between $y = x^2$ and $y = 4 - x^2$.

15. If a car's brakes provide constant deceleration of 36 ft per second per second, and if the car must be stopped within 50 ft of applying the brakes, what is the maximum allowable velocity for the car?

7.5 THE INDEFINITE INTEGRAL

Let us return to the definition of derivative. If $y = f(x)$ is differentiable at x_0, we recall that

$$f'(x_0) = \lim_{x \to x_0} \frac{f(x) - f(x_0)}{x - x_0} = \lim_{\Delta x \to 0} \frac{f(x_0 + \Delta x) - f(x_0)}{\Delta x} = \lim_{\Delta x \to 0} \frac{\Delta f}{\Delta x},$$

where Δx is the increment in x, Δf the increment in $f(x)$, and $f'(x_0)$ is the slope of the tangent to the curve $f(x)$ at $(x_0, f(x_0))$. Let us look at a point (x, y) on this tangent line (Fig. 7.34). Since $(x_0, f(x_0))$ is also on this line, we know that

$$\frac{y - f(x_0)}{x - x_0} = f'(x_0).$$

But $x - x_0$ is Δx, so

$$y - f(x_0) = f'(x_0)\,\Delta x.$$

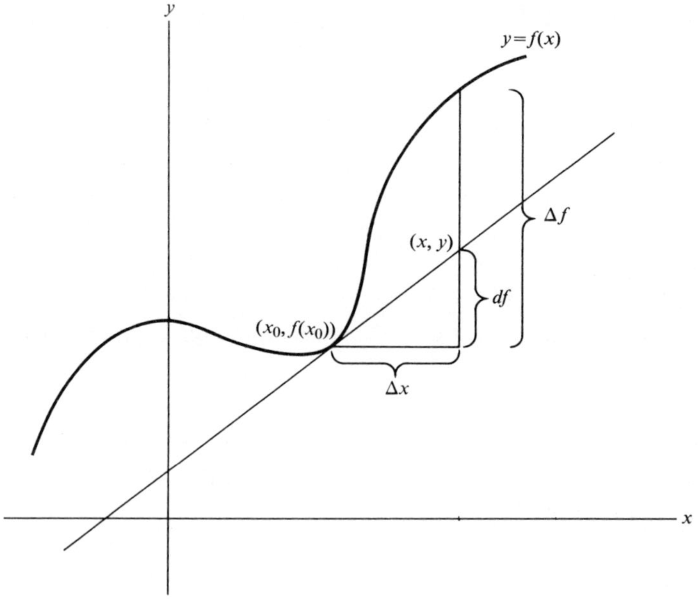

Fig. 7.34

This quantity is called the *differential* of $f(x)$ at x_0 and is denoted by df. Now Δx is a variable and since x_0 is fixed, $f(x_0)$ is also. Therefore

$$df = f'(x_0)\,\Delta x,$$

defines the differential as a function of the variable Δx. Note that df approximates Δf for small values of Δx. To conform to standard notation, we replace the variable Δx by dx, remembering that as usual the name we give to a variable is not important, so long as no confusion arises. Then we write

$$df = f'(x)\,dx,$$

as the definition of the differential of f; an alternative form is

$$\frac{df}{dx} = f'(x).$$

Since df is really a function of the variable dx, we are justified in treating df/dx as a quotient, as we did, for example, in the chain rule.

We now introduce another use of the integral notation. We write

$$\int f(x)\,dx,$$

to represent the class of all functions whose differential is $f(x)\,dx$; equivalently, the class of all functions whose derivative is $f(x)$. Since if $F(x)$ is any element of this class, then any other element can be expressed as $F(x) + C$, for some constant C, we write

$$\int f(x)\,dx = F(x) + C,$$

where $F(x)$ is a function such that $F'(x) = f(x)$ and C is a constant. $\int f(x)\,dx$ is called the *indefinite integral*. It is neither a function (as $\int_a^x f(t)\,dt$ is) nor a real number (as $\int_a^b f(x)\,dx$ is), but a class of functions.

We have already discussed obtaining $F(x)$ by reversing the process of differentiation. We summarize this process by means of the following formula.

$$\boxed{\int u^a\,du = \frac{u^{a+1}}{a+1} + C,}$$

where a is a real number, $a \neq -1$, C a constant, and du the differential of the function u. We postpone the discussion of the evaluation of integrals which do not fit into this pattern until Chapters 8 and 12.

Examples

1. Find

$$\int (3x^2 - 1)x\,dx.$$

Solution. $u = 3x^2 - 1$, $du = 6x\,dx$ so that

$$\int (3x^2 - 1)x\,dx = \frac{1}{6}\frac{(3x^2 - 1)^2}{2} + C.$$

2. Find

$$\int 2\,dx.$$

Solution. To apply the formula directly we would have to think of 2 as $2x^0$, but it is pretty evident, without resorting to this, that

$$\int 2\,dx = 2x + C.$$

3. Find

$$\int \frac{1}{x+1}\,dx.$$

Solution. So far we are out of luck. $a = -1$ is excluded from the formula above for the good reason that we don't know how to differentiate anything that ends up with a negative one as exponent. Chapter 8 handles this sort of problem.

4. Find

$$\int \frac{2x}{\sqrt{x^2 + 4}}\, dx.$$

Solution. Here $u = x^2 + 4$, $du = 2x\, dx$ so

$$\int \frac{2x}{\sqrt{x^2 + 4}} = 2\sqrt{x^2 + 4} + C.$$

We list without proof some useful properties of the integral. For continuous functions $f(x)$, $g(x)$, k a real number,

1. $$\int (f(x) + g(x))\, dx = \int f(x)\, dx + \int g(x)\, dx;$$

2. $$\int kf(x)\, dx = k \int f(x)\, dx.$$

Definite integrals have additive and multiplicative properties analogous to these:

$$\int_a^b (f(x) + g(x))\, dx = \int_a^b f(x)\, dx + \int_a^b g(x)\, dx,$$

and

$$\int_a^b kf(x)\, dx = k \int_a^b f(x)\, dx.$$

In addition they have the following property, which we have used without referring to it explicitly. If $f(x)$ is continuous on an interval containing a and c and $a \le b \le c$, then

$$\int_a^b f(x)\, dx + \int_b^c f(x)\, dx = \int_a^c f(x)\, dx.$$

Warning. It is *not* the case that for all continuous functions $f(x)$, $g(x)$,

$$\int (f(x)g(x))\, dx = \left(\int f(x)\, dx\right)\left(\int g(x)\, dx\right).$$

In fact, we have such an equality only if at least one of $f(x)$, $g(x)$ is a constant function.

PROBLEMS

1. Integrate:

a) $\int (3x + 2)\, dx.$

b) $\int (3x^3 - 2x^2 + 4x - 1)\, dx.$

c) $\int (2x - 5)^{17}\, dx.$

d) $\int x(3x^2 - 4)^{15}\, dx.$

e) $\int \sqrt{2x + 3}\, dx.$ f) $\int x\sqrt{25x^2 - 4}\, dx.$

g) $\int (x^2 - 1/\sqrt{x})\, dx.$ h) $\int \frac{y}{\sqrt{3y^2 + 1}}\, dy.$

i) $\int \frac{t + 2}{(t^2 + 4t - 3)^2}\, dt.$ j) $\int (12 - t)^{2/3}\, dt.$

k) $\int (3x^3 + 2x^2 - 4)\, dx.$ l) $\int \frac{1}{(2x - 1)^2}\, dx.$

m) $\int \frac{t^2}{(t^3 - 4)^5}\, dt.$ n) $\int \frac{1}{\sqrt{1 - 2x}}\, dx.$

2. A *differential equation* is an equation involving differentials. One solves such an equation in its simplest form by integrating both sides. Solve the following differential equations:

a) $dy = (x^2 + 2)\, dx.$ b) $\frac{dy}{x} = \frac{dx}{y}.$

c) $\frac{dy}{dx} = \frac{1}{\sqrt{x}} + \sqrt{x}.$ d) $\frac{dy}{dx} = \sqrt{xy}.$

e) $\frac{dy}{dx} = 2xy^3.$ f) $dy = \sqrt{z - 5}\, dz.$

g) $\frac{ds}{dt} = (2t + 1)^3.$ h) $\frac{dy}{dx} = -2xy^2.$

i) $dy = \sqrt{2 - x}\, dx.$ j) $\frac{dy}{dx} = \frac{x^2}{y^2}.$

k) $dy = \sqrt{1 + y}\, dx.$

3. Find y if $dy/dx = 4x^2 - 5x + 3$ and $y = 2$ when $x = 0$.

4. Find y if

$$\frac{dy}{dx} = \frac{1}{(2x + 1)^2} \quad \text{and} \quad y = 1 \text{ when } x = 1.$$

5. The differential approximates the increment in the function by

$$\Delta f \sim df = f'(x_0)\, \Delta x.$$

For example, $\sqrt{36.5}$ is approximated by letting $f(x) = \sqrt{x}$, $x_0 = 36$ and $\Delta x = 0.5$. Then

$$f'(x) = \frac{1}{2\sqrt{x}},$$

$$f'(36) = \frac{1}{2\sqrt{36}} = \frac{1}{12}.$$

Thus

$$\begin{aligned}\Delta f &= f(x_0 + x) - f(x_0)\\ &= \sqrt{36.5} - \sqrt{36}\\ &\sim df = f'(x_0)\,\Delta x = \tfrac{1}{12}\cdot 0.5 = \tfrac{1}{24}.\end{aligned}$$

Finally,

$$\sqrt{36.5} = f(x_0) + \Delta f \sim 6\tfrac{1}{24}.$$

Approximate

$$\sqrt{101}.$$

7.6 APPLICATIONS, IMPROPER INTEGRALS

In this section we look at two integration problems. The first requires only a straightforward application of the techniques of this chapter, while for the second we shall introduce the notion of improper integrals.

1. Consumers' Surplus. Suppose that we are given a demand curve (price as a function of demand) and that the market demand d_0 and the corresponding price p_0 are determined in some way, either under pure competition or under monopoly. There are consumers who would have been willing to pay more than the market price p_0. Under certain economic assumptions, the consumer gain is represented (in Fig. 7.35) by the area between the line $p = p_0$ and the demand curve. This amount is called the *consumers' surplus.*

If the demand curve is $p = 25 - d^2/10{,}000$ and the market demand is 250, then the market price p_0 is \$18.75. The consumers' surplus is given by

$$\begin{aligned}\text{C.S.} &= \int_0^{250} 25 - \frac{d^2}{10{,}000}\,dd - \int_0^{250} 18.75\,dd\\ &= \int_0^{250} \left(6.25 - \frac{d^2}{10{,}000}\right) dd\\ &= 6.25d - \frac{d^3}{30{,}000}\bigg|_0^{250}\\ &= 1041 \quad \text{(approximately).}\end{aligned}$$

Now suppose we have a condition of pure competition in which market equilibrium is determined as the intersection of the demand and supply curves, shown in Fig. 7.36. Then the amount supplied (which is the same as the market demand) is d_0 and the price is p_0. But there are producers who would have been willing to supply the commodity at a price less than p_0. The area between the supply curve and the line $p = p_0$ represents their gain and is called the *producers' surplus.*

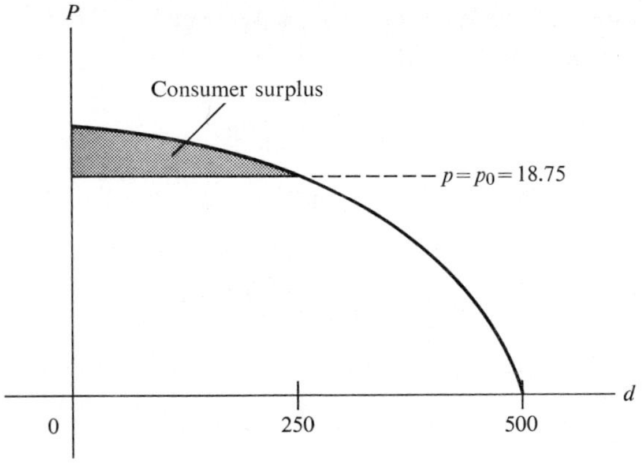

Fig. 7.35

Suppose $p = 0.0003\, d^2$ is the supply curve. Then its intersection with the demand curve above is at (250, 18.75) and the producers' surplus is given by

$$\text{P.S.} = \int_0^{250} 18.75\, dd - \int_0^{250} 0.0003\, d^2\, dd$$

$$= \int_0^{250} (18.75 - 0.0003\, d^2)\, dd$$

$$= 18.75\, d - 0.0001\, d^3 \Big|_0^{250}$$

$$= 3125.$$

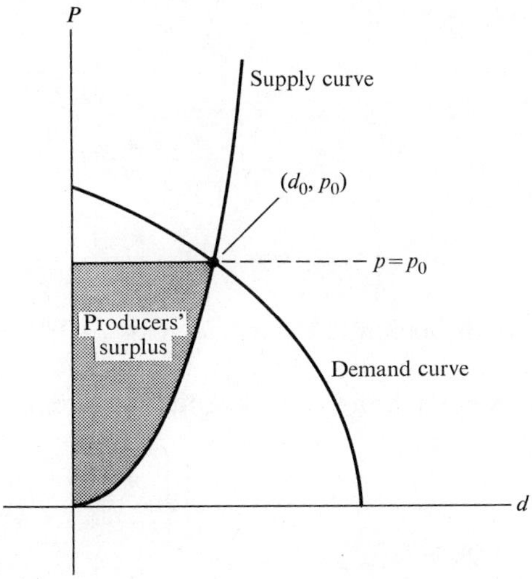

Fig. 7.36

2. ***Improper Integrals.*** If as $x \to +\infty$, $y = f(x)$ has the x-axis as its asymptote, it seems reasonable to ask whether for large values of x, the functions

$$\int_0^x f(t)\,dt,$$

might not tend to a finite limit. If they do, it would make sense to denote this limit by

$$\int_0^{+\infty} f(t)\,dt.$$

We make the following definitions. If $y = f(x)$ is continuous for all $x \geq b$, then

$$\int_b^{+\infty} f(t)\,dt = \lim_{x \to +\infty} \int_b^x f(t)\,dt,$$

if the limit exists.

If $y = f(x)$ is continuous for all $x \leq b$, then

$$\int_{-\infty}^b f(t)\,dt = \lim_{x \to -\infty} \int_x^b f(t)\,dt,$$

if the limit exists. In both cases if the limit exists, the integral is said to be *convergent*.

If $y = f(x)$ is continuous for all x,

$$\int_{-\infty}^{+\infty} f(t)\,dt = \int_{-\infty}^0 f(t)\,dt + \int_0^{-\infty} f(t)\,dt,$$

if the right-hand side is defined.

The integrals of this definition are called *improper integrals*. We note that since an integral with a variable as an upper or lower limit defines a function, we are talking about the existence of the limit of a function.

Examples

1. Find

$$\int_1^{+\infty} \frac{1}{x^2}\,dx.$$

Solution. (Fig. 7.37).

$$\lim_{x \to +\infty} \int_1^x \frac{1}{t^2}\,dt = \lim_{x \to +\infty} \left[\frac{-1}{x} - (-1)\right] = 1.$$

2. The limit in the definition may not exist. This can happen in two ways—either the value of the function becomes arbitrarily large, or the values fluctuate in some manner. An example of the first case is:

$$\int_0^{+\infty} x\,dx = \lim_{x \to +\infty} \int_0^x t\,dt = \lim_{x \to +\infty} \frac{x^2}{2},$$

which does not exist.

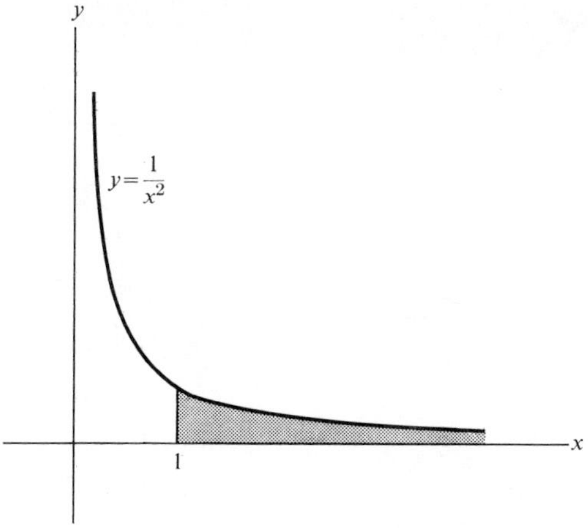

Fig. 7.37

To illustrate the second consider the function $y = f(x)$ whose graph is shown in Fig. 7.38. We have

$$\int_0^1 f(x)\,dx = \tfrac{1}{2}, \qquad \int_0^2 f(x)\,dx = 0, \qquad \int_0^3 f(x)\,dx = \tfrac{1}{2},$$

and in general

$$\int_0^n f(x)\,dx = \begin{cases} \tfrac{1}{2}, & \text{if } n \text{ is odd} \\ 0, & \text{if } n \text{ is even.} \end{cases}$$

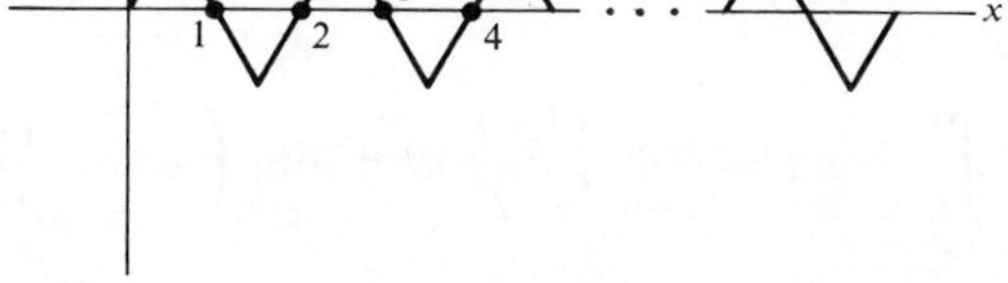

Fig. 7.38

There is still another sort of improper integral. We recall that we have always required that the integrand be continuous. Integration should, however, be possible in other cases. For example, we should be able to use integration to find the area under the graph of the *greatest integer function* (see Fig. 7.39):

$$y = [x] = \text{greatest integer (less than or equal to } x).$$

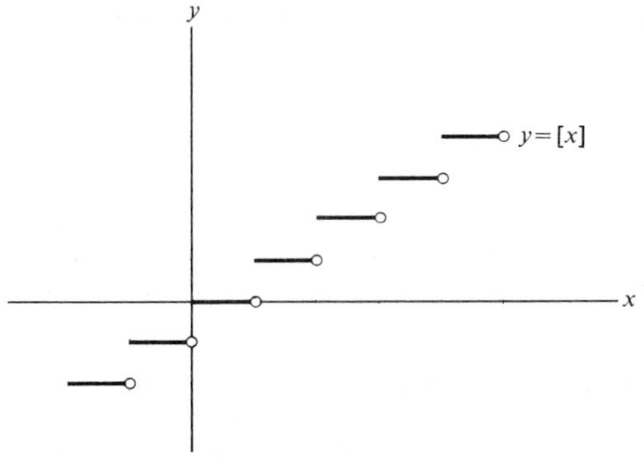

Fig. 7.39

Let us look at a more general problem. Suppose $f(t)$ is continuous for all t such that $a < t \leq b$. Then define

$$\int_a^b f(t)\,dt = \lim_{x \to a} \int_x^b f(t)\,dt,$$

if this limit exists. If $f(t)$ is continuous for all t such that $a \leq t < b$, define

$$\int_a^b f(t)\,dt = \lim_{x \to b} \int_a^x f(t)\,dt,$$

if this limit exists. Again we say that the integral is *convergent* if the limit exists.

Suppose we want to evaluate

$$\int_0^1 \frac{1}{\sqrt{x}}\,dx.$$

We write

$$\int_0^1 \frac{1}{\sqrt{x}}\,dx = \lim_{x \to 0} \int_x^1 \frac{1}{\sqrt{t}}\,dt = \lim_{x \to 0} (2 - 2\sqrt{x}) = 2.$$

On the other hand,

$$\int_0^1 \frac{1}{x^2}\,dx = \lim_{x \to 0} \int_x^1 \frac{1}{t^2}\,dt = \lim_{x \to 0} \left(-1 + \frac{1}{x}\right)$$

so, since this limit does not exist, the integral is not convergent.

3. *Probability*. A *probability distribution* over the nonnegative real numbers is a function F such that

 1) $F(0) = 0$.
 2) F is nondecreasing, that is, $x_1 < x_2$ implies that $F(x_1) \leq F(x_2)$.
 3) $\lim_{x \to \infty} F(x) = 1$.

A *probability measure* over the nonnegative real numbers is a continuous function f such that

1) $f(x) \geq 0$ for each $x \geq 0$.
2) $\int_0^{+\infty} f(x)\,dx = 1$.

We compare this with the finite probability measures of Chapter 3. The value $f(x)$ is the weight assigned to the point x and $\int_0^{+\infty} f(x)\,dx$ corresponds to the sum of the weights. The difference here is, of course, that there are an infinite number of points to which a weight is assigned.

Given a probability measure we can define a probability distribution by

$$F(x) = \int_0^x f(t)\,dt.$$

This gives the probability of the set $[0, x]$ of outcomes.

In actuarial work it is important to be able to predict how long someone will live. Suppose the probability of a person dying at age t is given by

$$f(t) = \frac{1}{2(1+t)^{3/2}}.$$

The form of this function is in practice determined on the basis of statistics concerning the percentage of people who have previously died at various ages.

For $f(t)$ as above, $f(t) \geq 0$ for all $t \geq 0$. Also,

$$\int_0^{+\infty} \frac{1}{2(1+t)^{3/2}}\,dt = \lim_{x\to+\infty} \int_0^x \frac{1}{2(1+t)^{3/2}}\,dt = 1.$$

To find the probability of the subject living to more than 65, we compute the probability of his dying at 65 or before:

$$\begin{aligned} F(65) &= \int_0^{65} \frac{1}{2(1+t)^{3/2}}\,dt \\ &= \frac{-2}{2}\left[(1+t)^{-1/2}\right]\Big|_0^{65} \\ &= 1 - \frac{1}{\sqrt{66}}, \end{aligned}$$

which is approximately 0.87. Thus the probability of living to be more than 65 is $1 - 0.87 = 0.13$.

PROBLEMS

1. If the demand curve is

$$p = 47 - \frac{d^3}{150{,}000},$$

the market demand 75 and the market price 44.19, determine the consumers' surplus.

2. If the demand curve is as in Problem 1, there is a condition of pure competition and the supply curve is given by

$$p = 0.0002\, d^3,$$

determine the producers' surplus.

3. A *frequency distribution* is a function f from a set of values of a variable to a set of real numbers whose elements designate the frequency of occurrence of various values. If the domain of a frequency distribution is an interval on the real line, it is *continuous.* If the domain is a finite subset of the set of positive integers, it is *discrete.* In the latter case the *arithmetic mean* of the frequency distribution is given by

$$\bar{x} = \frac{\sum_{i=1}^{n} x_i f(x_i)}{\sum_{i=1}^{n} f(x_i)},$$

where x_i is the ith value of the variable and $f(x_i)$ the frequency with which it occurs. The definition in the continuous case makes use of the definite integral

$$\bar{x} = \frac{\int_a^b x\, f(x)\, dx}{\int_a^b f(x)\, dx},$$

where $f(x)$ is the frequency distribution defined on $[a, b]$. If $f(x) = -x^2 + 16{,}000x + 432{,}000{,}000$ is the frequency distribution of incomes between \$5000 and \$15,000, find the arithmetic mean of the incomes.

4. Determine whether each of the following integrals is convergent, and if it is convergent, evaluate it.

a) $\displaystyle\int_1^{\infty} -\frac{1}{x^2}\, dx.$

b) $\displaystyle\int_0^1 \frac{dx}{\sqrt{1-x}}.$

c) $\displaystyle\int_0^2 \frac{x}{\sqrt{4-x^2}}\, dx.$

d) $\displaystyle\int_0^1 \frac{2}{x^2}\, dx.$

e) $\displaystyle\int_{-1}^1 \frac{dx}{x^3}.$

f) $\displaystyle\int_{-2}^0 (x+1)^{-1/3}\, dx.$

g) $\displaystyle\int_{-\infty}^{\infty} \frac{dx}{x^3}.$

h) $\displaystyle\int_0^{\infty} x(x^2+1)^{-2}\, dx.$

i) $\displaystyle\int_0^1 \frac{x}{\sqrt{1-x^2}}\, dx.$

j) $\displaystyle\int_0^{\infty} \frac{1}{x^{5/4}}\, dx.$

k) $\displaystyle\int_0^{\infty} \frac{x}{(x^2+1)^3}\, dx.$

5. Show that $F(x) = x/(1 + x)$ is a probability distribution.

6. Show that $f(t) = (t + 256)^{-5/4}$ is a probability measure. If $f(t)$ is the probability that a light bulb will fail in t days, find the probability of a light bulb lasting 1000 days.

7. A probability measure is given by

$$f(t) = \frac{2}{3(t+8)^{4/3}}.$$

Find the probability of $t \leq 10$.

8 / TRIGONOMETRIC, LOGARITHMIC, AND EXPONENTIAL FUNCTIONS

In this chapter we shall define the sine and cosine, first as functions of angles and then as functions of real numbers. From them we shall derive the other trigonometric functions and some useful identities. The treatment of elementary trigonometry is brief, but self-contained. However, many of the proofs which depend only on plane geometry are omitted. The last part of the chapter is an introduction to logarithmic and exponential functions and a study of some of their applications.

8.1 TRIGONOMETRIC FUNCTIONS

The most elementary definition of the trigonometric functions employs a right triangle (see Fig. 8.1):

$$\text{the sine of angle } A = \frac{\text{opposite side}}{\text{hypotenuse}} = \frac{a}{c}$$

$$\text{the cosine of angle } A = \frac{\text{adjacent side}}{\text{hypotenuse}} = \frac{b}{c}.$$

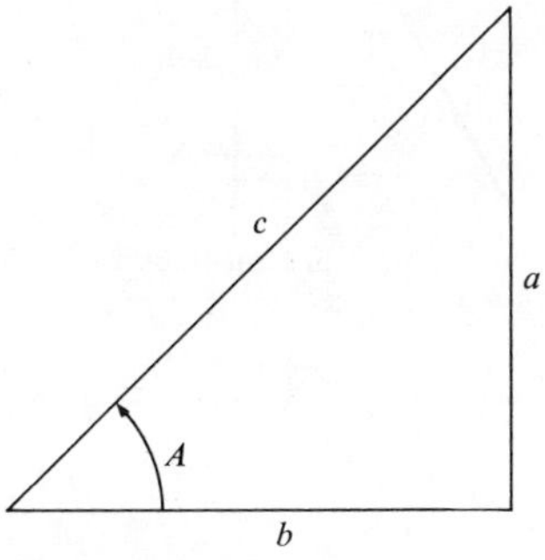

Fig. 8.1

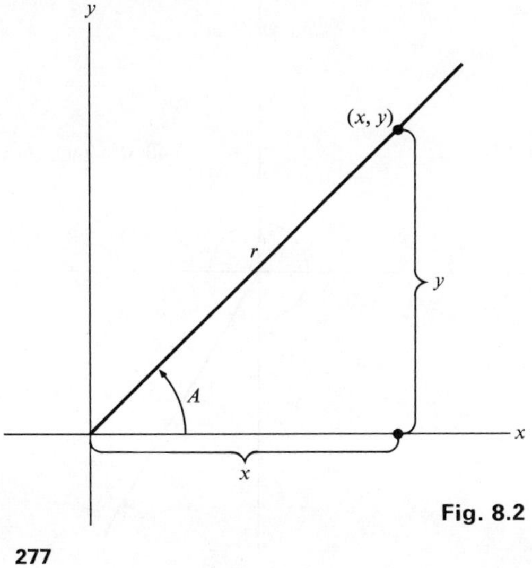

Fig. 8.2

Moving from plane geometry to the use of coordinates, we select any point (x, y) on a line (called a *ray*) making an angle A with the x-axis at the origin (see Fig. 8.2). Then

$$\text{sine } A = \frac{y}{r} = \frac{y}{\sqrt{x^2 + y^2}}$$

while

$$\text{cosine } A = \frac{x}{r} = \frac{x}{\sqrt{x^2 + y^2}}.$$

It does not matter which point on the ray we choose since by using similar triangles we can show that the defining ratios are the same. We can also define the sine and cosine for angles greater than 90 degrees (see Fig. 8.3) by

$$\text{sine } A = \frac{y}{r} \quad \text{and} \quad \text{cosine } A = \frac{x}{r},$$

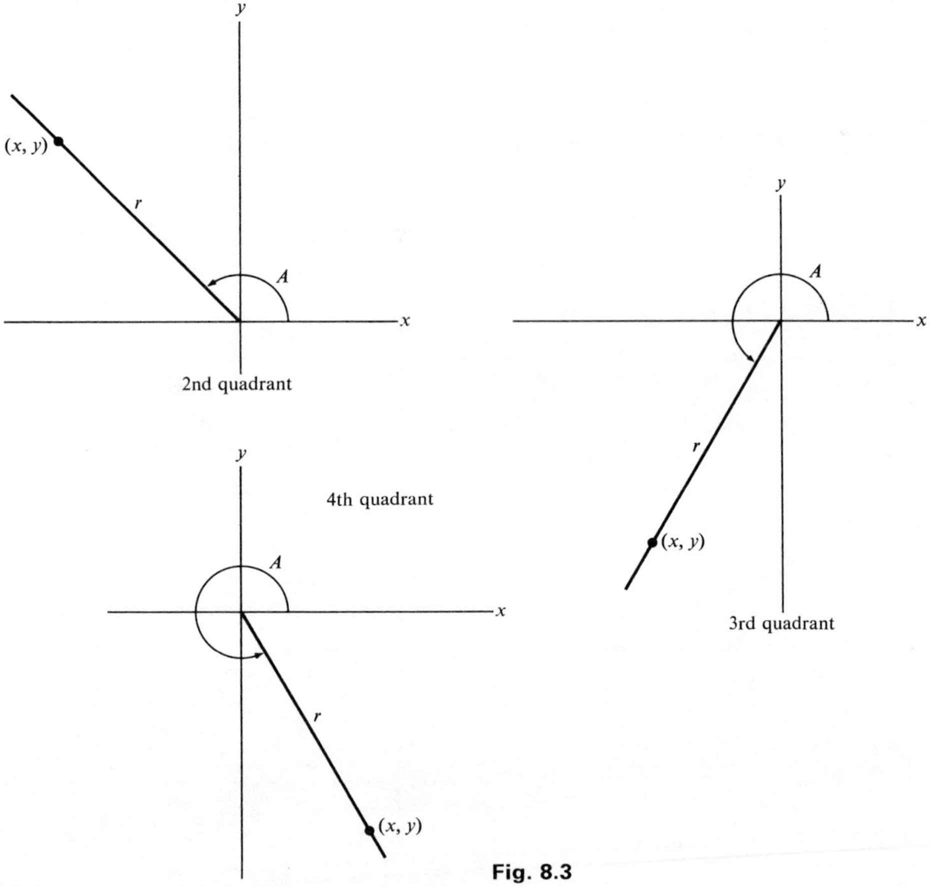

Fig. 8.3

where r is the distance of (x, y) from the origin and is always positive. We write sin and cos for short.

It is not necessary to consider angles greater than 360 degrees since we obtain an angle with the same sine and cosine by subtracting multiples of 360 (Fig. 8.4).

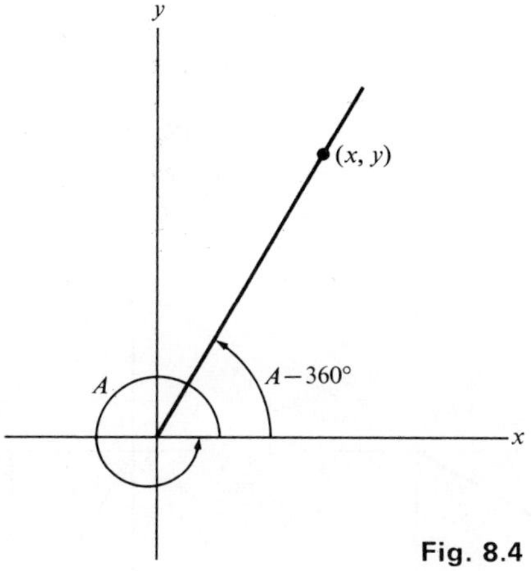

Fig. 8.4

We also define

$$\text{tangent } A = \tan A = \frac{y}{x} = \frac{\sin A}{\cos A}$$

$$\text{cotangent } A = \operatorname{ctn} A = \frac{x}{y} = \frac{\cos A}{\sin A} = \frac{1}{\tan A}$$

$$\text{secant } A = \sec A = \frac{r}{x} = \frac{1}{\cos A}$$

$$\text{cosecant } A = \csc A = \frac{r}{y} = \frac{1}{\sin A}.$$

Directly from the definitions we find that the sine and the cosecant are positive in the first and second quadrants and negative in the third and fourth (Fig. 8.5); the cosine and the secant are positive in the first and fourth quadrants and negative in the second and third; the tangent and the cotangent are positive in the first and third quadrants and negative in the second and fourth.

To define the sine, cosine, etc. of a real number, we introduce a new unit of measurement. A *radian* is the angle subtended by an arc of length 1 in a circle of radius 1 (Fig. 8.6). Since the circumference of a circle is $2\pi r$, this means that the whole circle subtends an angle of 2π radians. We list radian equivalents of common

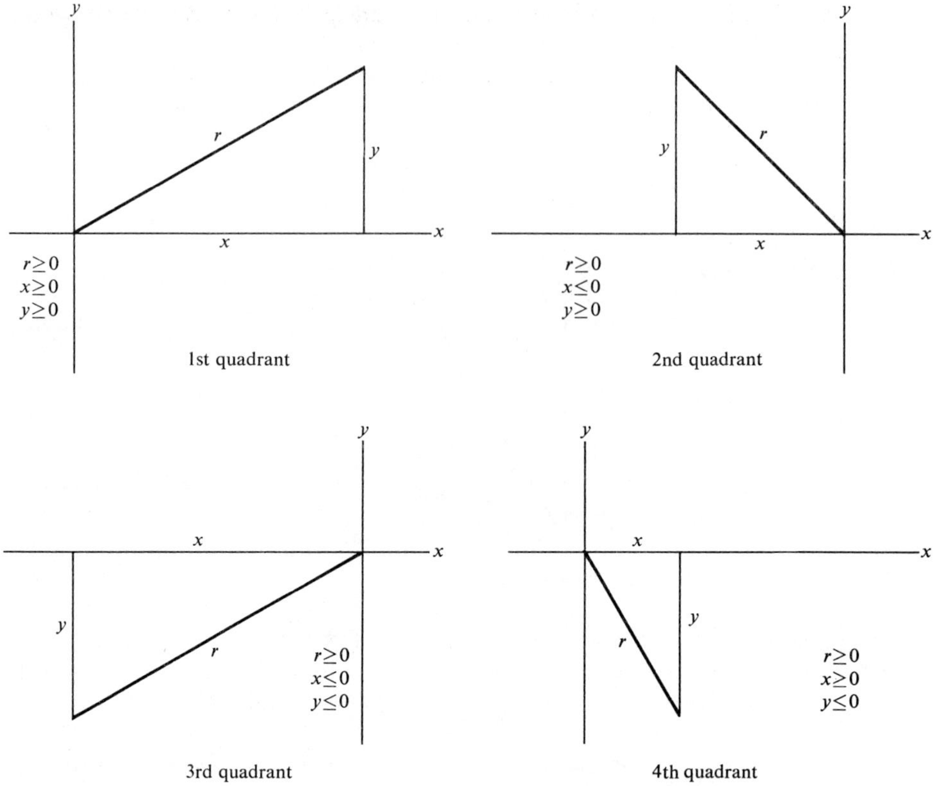
y
r
y
x
r≥0
x≥0
y≥0
1st quadrant
y
r
y
x
r≥0
x≤0
y≥0
2nd quadrant
y
x
y
r
r≥0
x≤0
y≤0
3rd quadrant
y
x
y
r
r≥0
x≥0
y≤0
4th quadrant

Fig. 8.5

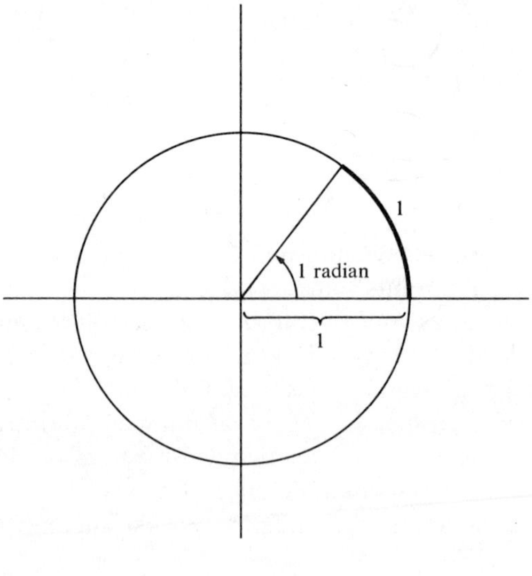
1
1 radian
1

Fig. 8.6

angles together with the sines and cosines of these angles. (The other trigonometric functions are readily computed from this information.)

Angle in degrees	Angle in radians	Sine	Cosine
0	0	0	1
30	$\pi/6$	$1/2$	$\sqrt{3}/2$
45	$\pi/4$	$1/\sqrt{2}$	$1/\sqrt{2}$
60	$\pi/3$	$\sqrt{3}/2$	$1/2$
90	$\pi/2$	1	0
180	π	0	-1
270	$3\pi/2$	-1	0
360	2π	0	1

This table should be committed to memory if it is not already known to the reader. We shall be using radian measurement since it is the more useful one for our purposes.

As yet we have no means of computing the trigonometric functions for other than a few special angles. Thus if the information is needed, we resort to any standard book of mathematical tables, or to the tables in the back of this book.

Now we are ready to define the sine as a function of real numbers: to the real number x we assign the number which is the sine of x radians. We denote this function by $\sin x$. Its domain is all the real numbers and its image is $[-1, 1]$.

We define the functions $\cos x$, $\tan x$, $\text{ctn}\, x$, $\sec x$, and $\csc x$ similarly. It is left to the reader to determine the domain and image of these functions.

We list without proof the following useful identities for all real numbers x and y:

$$\sin x \equiv \cos\left((\pi/2) - x\right)$$
$$\cos x \equiv \sin\left((\pi/2) - x\right)$$
$$\left.\begin{aligned}\sin x &\equiv \sin(x + 2k\pi)\\ \cos x &\equiv \cos(x + 2k\pi)\end{aligned}\right\} k \text{ an integer}$$
$$\sin x \equiv -\sin(-x)$$
$$\cos x \equiv \cos(-x)$$
$$\sin^2 x + \cos^2 x \equiv 1, \qquad \text{where } \sin^2 x \text{ means } (\sin x)^2$$
$$1 + \tan^2 x \equiv \sec^2 x$$
$$1 + \text{ctn}^2\, x \equiv \csc^2 x$$
$$\left.\begin{aligned}\sin(A + B) &\equiv \sin A \cos B + \cos A \sin B\\ \cos(A + B) &\equiv \cos A \cos B - \sin A \sin B\end{aligned}\right\} \text{addition formulas}$$

As special cases of the above two addition formulas, we have the double-angle formulas:

$$\sin 2A \equiv 2 \sin A \cos A$$

$$\cos 2A \equiv \cos^2 A - \sin^2 A.$$

From these two we derive the half-angle formulas:

$$\sin \frac{A}{2} \equiv \sqrt{\frac{1 - \cos A}{2}}$$

$$\cos \frac{A}{2} \equiv \sqrt{\frac{1 + \cos A}{2}}.$$

Remember that the use of the term "identity" indicates that the relations expressed hold for *all* values of x for which the relevant functions are defined. We can also consider trigonometric equations, which we solve as we do algebraic equations, by finding for what values of x they are true. For example, if

$$\sin x = \cos x,$$

then

$$\frac{\sin x}{\cos x} = 1, \qquad \text{so that} \qquad \tan x = 1$$

and $x = \pi/4 + k\pi$, where k is any integer.

It is handy to be familiar with the graphs of the trigonometric functions shown in Fig. 8.7.

We can see from their graphs that none of these is one-to-one on its entire domain. If we want to define an inverse for one of these functions, it will have to be under some restrictions. All of the trigonometric functions are periodic—that is, their values go through a cycle and then repeat the cycle at regular intervals. More precisely, a function is *periodic* with period p if $f(x + kp) = f(x)$ for each integer k. The sine, cosine, secant, and cosecant have period 2π, whereas the tangent and cotangent have period π.

To define an inverse for the function $\sin x$, we restrict its domain to $-\pi/2 \leq x \leq \pi/2$. In this interval it is one-to-one, so that its inverse exists; we call the inverse function the *arcsin*. That is, if

$$y = \sin x, \qquad \frac{-\pi}{2} \leq x \leq \frac{\pi}{2}$$

then

$$x = \arcsin y;$$

the image of the arcsin function is $[-\pi/2, \pi/2]$ and its domain is $[-1, 1]$.

The arccos and arctan are defined analogously, except that the image of the arccos is taken to be $[0, \pi]$. The inverses of the other functions are rarely used. Some authors write $\arcsin x$ as $\sin^{-1} x$, but we shall avoid this practice because of

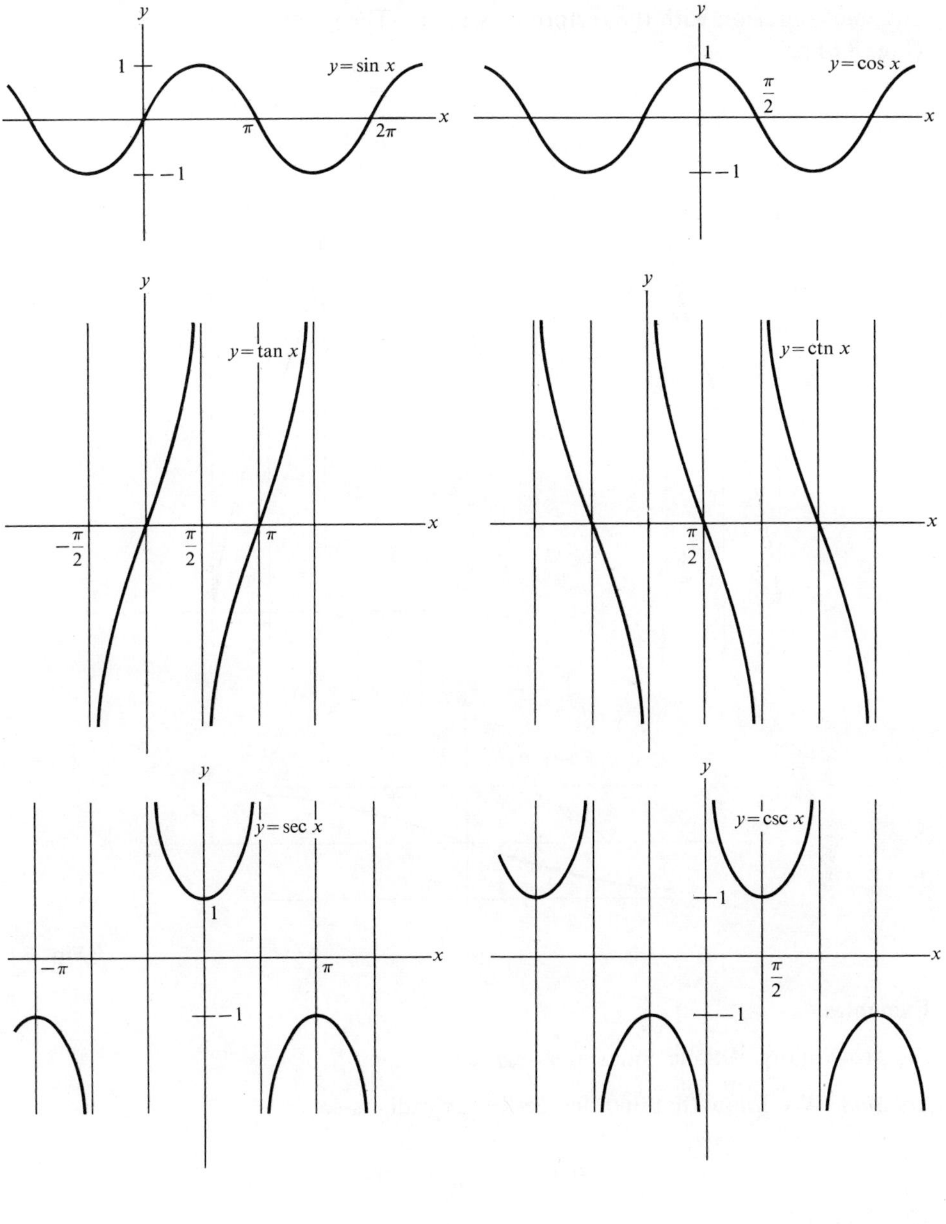
y
1
−1
π
2π
x
y = sin x
y
1
π/2
−1
x
y = cos x
y
−π/2
π/2
π
x
y = tan x
y
π/2
x
y = ctn x
y
1
−π
π
−1
x
y = sec x
y
1
π/2
−1
x
y = csc x

Fig. 8.7

possible confusion with the reciprocal $1/\sin x$. The graphs of the inverse functions (Fig. 8.8) are:

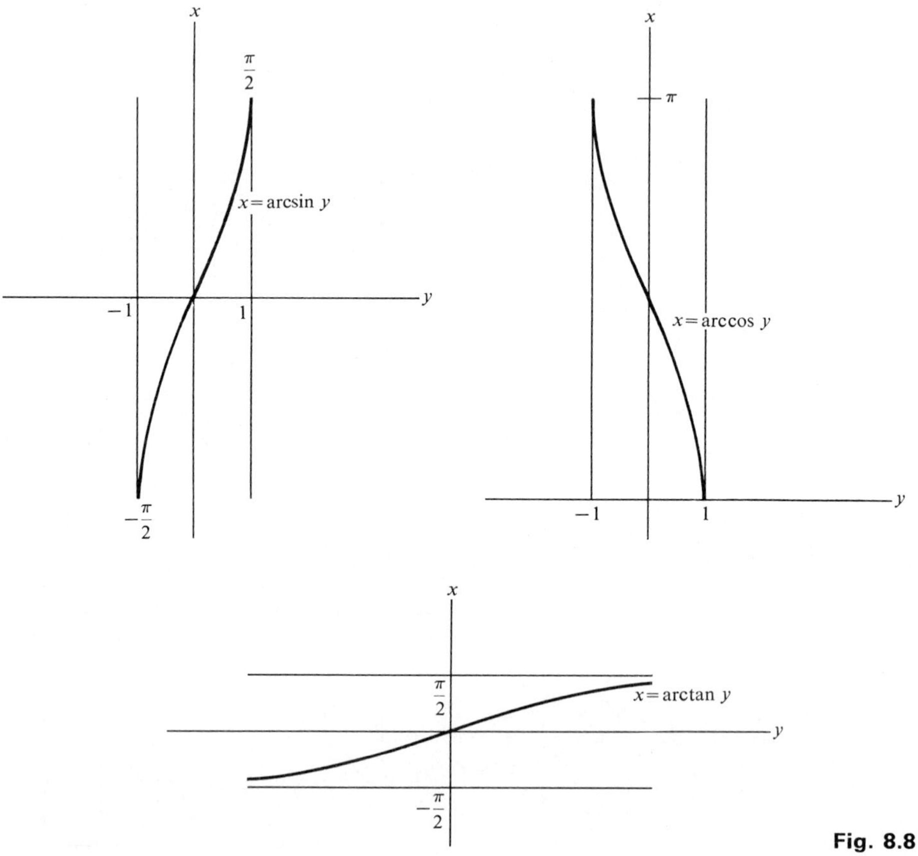

Fig. 8.8

Examples

1. How many radians equal one degree?

Solution. We know that 360 degrees $= 2\pi$ radians so

$$1 \text{ degree} = \frac{\pi}{180} \text{ radians.}$$

2. Find $\sin \pi/12$.

Solution

$$\sin \frac{\pi}{12} = \sin \frac{\pi/6}{2} = \sqrt{\frac{1 - \cos \pi/6}{2}} = \sqrt{\frac{1 - \sqrt{3}/2}{2}}.$$

3. Find $\sin 5\pi/12$.

Solution

$$\begin{aligned}\sin\frac{5\pi}{12} &= \sin\left(\frac{\pi}{6}+\frac{\pi}{4}\right)\\ &= \sin\frac{\pi}{6}\cos\frac{\pi}{4}+\cos\frac{\pi}{6}\sin\frac{\pi}{4}\\ &= \left(\frac{1}{2}\right)\left(\frac{1}{\sqrt{2}}\right)+\left(\frac{\sqrt{3}}{2}\right)\left(\frac{1}{\sqrt{2}}\right)\\ &= \frac{1+\sqrt{3}}{2\sqrt{2}}.\end{aligned}$$

4. Find $\arcsin \frac{1}{2}$.

Solution. The real number $\pi/6$ is in the interval $[0, \pi]$ and $\sin \pi/6 = \frac{1}{2}$. Therefore, $\arcsin \frac{1}{2} = \pi/6$.

5. Find $\arctan(-1)$.

Solution. The real number $-\pi/4$ is in the interval $[-\pi/2, \pi/2]$ and $\tan -\pi/4 = -1$. Therefore, $\arctan(-1) = -\pi/4$.

PROBLEMS

1. Convert to radians:

a) 5 degrees.	b) 540 degrees.	c) 3 degrees.
d) 20 degrees.	e) 75 degrees.	f) 3600 degrees.
g) 10 degrees.	h) 630 degrees.	i) 160 degrees.

2. Convert to degrees:

a) 1 radian.	b) $5\pi/12$ radians.	c) $3\pi/2$ radians.
d) 10π radians.	e) $\pi/8$ radians.	f) 2 radians.
g) $7\pi/12$ radians.	h) $3\pi/8$ radians.	i) $11\pi/12$ radians.

3. Solve the trigonometric equations:

a) $\sin x = \sin 2x$.	b) $\sin x = 2 \sin x$.
c) $2 \sin x = \sin 2x$.	d) $\sin x = \tan x$.
e) $\cos x = \sin x \tan x$.	f) $2 \sin^2 x - \sin x - 1 = 0$.

4. Derive a formula for $\sin 3x$ in terms of $\sin x$.

5. Find (without using tables)

a) $\cos \pi/8$.	b) $\tan 5\pi/12$.	c) $\sin 3\pi$.
d) $\cos \pi/12$.	e) $\sin 5\pi/6$.	f) $\text{ctn}\, 3\pi/4$.
g) $\sin \pi/8$.	h) $\cos 5\pi/12$.	i) $\tan 3\pi/4$.

6. Sketch the graph of
   a) $y = 3 \sin x$.
   b) $y = \sin 3x$.
   c) $y = \frac{1}{3} \sin x$.
   d) $y = -\sin \dfrac{x}{3}$.
   e) $y = 2 \tan \dfrac{x}{2}$.
   f) $y = 2 \cos (-3x)$.
   g) $y = \sin 2x$.
   h) $y = \frac{1}{2} \cos x$.
   i) $y = \tan (x + (\pi/3))$.
7. Find:
   a) $\arcsin -1$.
   b) $\arctan 1$.
   c) $\arccos \sqrt{3}/2$.
   d) $\arctan (\tan \pi/3)$.
   e) $\sin (\arcsin -\sqrt{3}/2)$.
   f) $\cos (\arcsin \sqrt{2}/2)$.
   g) $\arcsin -1/2$.
   h) $\arccos \sqrt{2}/2$.
   i) $\arctan -1/\sqrt{3}$.
   j) $\cos (\arccos \frac{1}{3})$.

8.2 DERIVATIVES OF TRIGONOMETRIC FUNCTIONS

We begin by deriving the derivative of the sine function. We use the fact that $\cos x$ is a continuous function—a fact which seems apparent from the graph, but which we have not proved.

Theorem 1. $\sin x$ is a differentiable function and

$$\frac{d \sin x}{dx} = \cos x.$$

Proof. If the derivative exists, it is equal to

$$\begin{aligned}\lim_{\Delta x \to 0} \frac{\sin (x + \Delta x) - \sin x}{\Delta x} &= \lim_{\Delta x \to 0} \frac{\sin x \cos \Delta x + \cos x \sin \Delta x - \sin x}{\Delta x} \\ &= \lim_{\Delta x \to 0} \frac{\sin x}{\Delta x} (\cos \Delta x - 1) + \lim_{\Delta x \to 0} \cos x \frac{\sin \Delta x}{\Delta x} \\ &= \sin x \lim_{\Delta x \to 0} \frac{\cos \Delta x - 1}{\Delta x} + \cos x \lim_{\Delta x \to 0} \frac{\sin \Delta x}{\Delta x}. \qquad (8.1)\end{aligned}$$

Thus the proof is reduced to the problem of showing that

$$\lim_{\Delta x \to 0} \frac{\cos \Delta x - 1}{\Delta x} \quad \text{and} \quad \lim_{\Delta x \to 0} \frac{\sin \Delta x}{\Delta x}.$$

exist and computing them.

Taking the second limit first, we draw a circle of radius 1 and an angle of Δx radians (Fig. 8.9). We consider $\Delta x > 0$, since if $\Delta x < 0$, $\dfrac{\sin \Delta x}{\Delta x}$ has the same sign as for $\Delta x > 0$. Now the area of the sector of the circle OAD satisfies the inequalities:

area of triangle OAC $<$ area of OAD $<$ area of triangle OBD.

We have that $\sin \Delta x = s/1 = s$, $\cos \Delta x = a/1 = a$, and $\tan \Delta x = t/1 = t = \sin \Delta x/\cos \Delta x$. The area of OAC is given by

$$\frac{sa}{2} = \frac{\sin \Delta x \cos \Delta x}{2}$$

and that of OBD by

$$\tfrac{1}{2} \cdot 1 \cdot t = \frac{\sin \Delta x}{\cos \Delta x}.$$

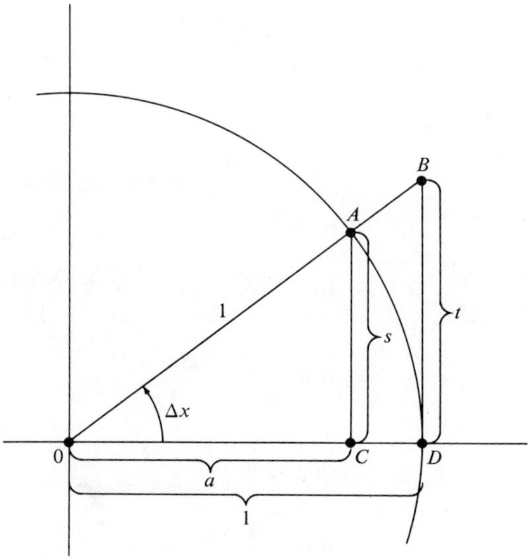

Fig. 8.9

The area of a sector of a circle subtended by an arc of A radians is $A/2\pi$ times the area of the circle. Thus the area of the sector subtended by the arc Δx is $(\Delta x/2\pi)\pi = \Delta x/2$. Substituting in the inequalities we obtain

$$\frac{\sin \Delta x \cos \Delta x}{2} < \frac{\Delta x}{2} < \frac{\sin \Delta x}{2 \cos \Delta x},$$

$$\sin \Delta x \cos \Delta x < \Delta x < \frac{\sin \Delta x}{\cos \Delta x}.$$

The first inequality yields, upon division by Δx and by $\cos \Delta x$ (which are both positive),

$$\frac{\sin \Delta x}{\Delta x} < \frac{1}{\cos \Delta x}.$$

We multiply the second inequality by $\cos \Delta x$ and divide by Δx to obtain

$$\cos \Delta x < \frac{\sin \Delta x}{\Delta x}.$$

Putting these together we have

$$\cos \Delta x < \frac{\sin \Delta x}{\Delta x} < \frac{1}{\cos \Delta x}.$$

Now assuming that $\cos x$ is a continuous function so that $\lim_{\Delta x \to 0} \cos \Delta x = \cos 0 = 1$, we get that $\dfrac{\sin \Delta x}{\Delta x}$ is squeezed between two quantities which have 1 as their limit. Thus

$$\lim_{\Delta x \to 0} \frac{\sin \Delta x}{\Delta x}$$

must be 1.

It remains to evaluate

$$\lim_{\Delta x \to 0} \frac{\cos \Delta x - 1}{\Delta x}.$$

We rewrite this as

$$\begin{aligned}
\lim_{\Delta x \to 0} \frac{\cos \Delta x - 1}{\Delta x} &= \lim_{\Delta x \to 0} \frac{(\cos \Delta x - 1)(\cos \Delta x + 1)}{\Delta x\,(\cos \Delta x + 1)} \\
&= \lim_{\Delta x \to 0} \frac{\cos^2 \Delta x - 1}{\Delta x\,(\cos \Delta x + 1)} \\
&= \lim_{\Delta x \to 0} \frac{-\sin^2 \Delta x}{\Delta x\,(\cos \Delta x + 1)} \\
&= \lim_{\Delta x \to 0} \frac{\sin \Delta x}{\Delta x} \lim_{\Delta x \to 0} \sin \Delta x \cdot \frac{1}{\cos \Delta x + 1} \\
&= 1 \cdot 0 \\
&= 0.
\end{aligned}$$

We substitute these limits in (8.1) to obtain

$$\begin{aligned}
\frac{d \sin x}{dx} &= \lim_{\Delta x \to 0} \frac{\sin (x + \Delta x) - \sin x}{\Delta x} \\
&= \sin x \cdot 0 + \cos x \cdot 1 \\
&= \cos x.
\end{aligned}$$

Theorem 2. cos x is differentiable and

$$\frac{d \cos x}{dx} = -\sin x.$$

Proof

$$\begin{aligned}\lim_{\Delta x \to 0} \frac{\cos (x + \Delta x) - \cos x}{\Delta x} &= \lim_{\Delta x \to 0} \frac{\cos x \cos \Delta x - \sin x \sin \Delta x - \cos x}{\Delta x} \\ &= \lim_{\Delta x \to 0} \frac{\cos x(\cos \Delta x - 1)}{\Delta x} - \lim_{\Delta x \to 0} \frac{\sin x \sin \Delta x}{\Delta x} \\ &= \cos x \lim_{\Delta x \to 0} \frac{\cos \Delta x - 1}{\Delta x} - \sin x \lim_{\Delta x \to 0} \frac{\sin \Delta x}{\Delta x} \\ &= \cos x \cdot 0 - \sin x \cdot 1 \\ &= -\sin x.\end{aligned}$$

Theorem 3. tan x is differentiable except at

$$x = \frac{(2k + 1)\pi}{2},$$

k an integer and

$$\frac{d \tan x}{dx} = \sec^2 x.$$

Proof. We write tan x as sin x/cos x and apply the rule for differentiation of a quotient.

$$\begin{aligned}\frac{d \tan x}{dx} &= \frac{d(\sin x/\cos x)}{dx} \\ &= \frac{\cos x \cos x - \sin x(-\sin x)}{\cos^2 x} \\ &= \frac{\cos^2 x + \sin^2 x}{\cos^2 x} \\ &= \frac{1}{\cos^2 x} \\ &= \sec^2 x.\end{aligned}$$

We list the derivatives of the other trigonometric functions, but leave the proof to the reader:

$$\frac{d \operatorname{ctn} x}{dx} = -\csc^2 x. \qquad \frac{d \csc x}{dx} = -\csc x \operatorname{ctn} x. \qquad \frac{d \sec x}{dx} = \sec x \tan x.$$

We can also apply the chain rule to obtain:

If u is a differentiable function of x, then

$$\frac{d \sin u}{dx} = \cos u \frac{du}{dx} \qquad \frac{d \cos u}{dx} = -\sin u \frac{du}{dx}$$

$$\frac{d \tan u}{dx} = \sec^2 u \frac{du}{dx} \qquad \frac{d \operatorname{ctn} u}{dx} = -\csc^2 u \frac{du}{dx}$$

$$\frac{d \sec u}{dx} = \sec u \tan u \frac{du}{dx} \qquad \frac{d \csc u}{dx} = -\csc u \operatorname{ctn} u \frac{du}{dx}.$$

Examples

1. Find

$$\frac{d(\sin (2x^2 + 2))}{dx}.$$

Solution. Let $u = 2x^2 + 2$; then $du/dx = 4x$.

$$\frac{d(\sin (2x^2 + 2))}{dx} = \frac{d \sin u}{dx} = \cos u \frac{du}{dx} = 4x \cos (2x^2 + 2).$$

2. Find

$$\frac{d \cos 1/x}{dx}, \qquad x \neq 0.$$

Solution. Let $u = 1/x$; then $du/dx = -1/x^2$.

$$\frac{d \cos 1/x}{dx} = \frac{d \cos u}{dx} = -\sin u \frac{du}{dx} = \frac{1}{x^2} \sin \frac{1}{x}.$$

3. Find the derivative of $(\tan 3x + \sin x^2)^3$.

Solution. Let $v = 3x$, $w = x^2$, and $u = \tan v + \sin w$. Then

$$\frac{dv}{dx} = 3,$$

$$\frac{dw}{dx} = 2x,$$

$$\frac{du}{dx} = \sec^2 v \frac{dv}{dx} + \cos w \frac{dw}{dx}$$

$$= 3 \sec^2 3x + 2x \cos x^2.$$

Finally

$$\frac{d(\tan 3x + \sin x^2)^3}{dx} = \frac{du^3}{dx} = 3u^2 \frac{du}{dx}$$

$$= 3(\tan 3x + \sin x^2)^2(3 \sec^2 3x + 2x \cos x^2).$$

Warning. Observe the difference between

$$\sin^2 x = (\sin x)^2 \qquad \text{and} \qquad \sin x^2.$$

To compute the derivative of $\sin^2 x$, we let $u = \sin x$. Then

$$\frac{d \sin^2 x}{dx} = \frac{du^2}{dx} = 2u \frac{du}{dx} = 2 \sin x \cos x.$$

On the other hand,

$$\frac{d \sin x^2}{dx} = 2x \cos x^2.$$

4. Find the derivative of

$$\frac{\cos (3x + 1)}{\sin (x^3 - x)}.$$

Solution. Let $u = 3x + 1$, $v = x^3 - x$. Then

$$\frac{du}{dx} = 3, \qquad \frac{dv}{dx} = 3x^2 - 1.$$

Thus

$$\frac{d(\cos (3x + 1)/\sin (x^3 - x))}{dx}$$

$$= \frac{d(\cos u/\sin v)}{dx} = \frac{\sin v\,(-\sin u)\, du/dx - (\cos u)(\cos v)\, dv/dx}{\sin^2 v}$$

$$= \frac{-3 \sin (x^3 - x) \sin (3x + 1) - (3x^2 - 1) \cos (3x + 1) \cos (x^3 - x)}{\sin^2 (x^3 - x)}.$$

We learned in Chapter 5 to determine the derivative of the inverse of a function from the knowledge of the derivative of the function. Suppose $y = \sin x$ so that $x = \arcsin y$. Then

$$\frac{dy}{dx} = \cos x \qquad \text{and} \qquad \frac{dx}{dy} = \frac{1}{dy/dx} = \frac{1}{\cos x}.$$

But

$$\cos x = \sqrt{1 - \sin^2 x} = \sqrt{1 - y^2}; \qquad \frac{dx}{dy} = \frac{1}{\sqrt{1 - y^2}}.$$

Similarly, if $y = \cos x$ so that $x = \arccos y$,

$$\frac{dy}{dx} = -\sin x \qquad \text{and} \qquad \frac{dx}{dy} = \frac{1}{dy/dx} = \frac{-1}{\sin x}.$$

But

$$\sin x = \sqrt{1 - \cos^2 x} = \sqrt{1 - y^2},$$

so

$$\frac{dx}{dy} = \frac{-1}{\sqrt{1 - y^2}}.$$

We also find the derivative of the arctan function: If $y = \tan x$, then $x = \arctan y$.

$$\frac{dy}{dx} = \sec^2 x \qquad \text{and so} \qquad \frac{dx}{dy} = \frac{1}{1 + \tan^2 x} = \frac{1}{1 + y^2}.$$

Application of the chain rule leads to the following formulas:

$$\frac{d \arcsin u}{dx} = \frac{du/dx}{\sqrt{1 - u^2}}, \qquad \frac{d \arccos u}{dx} = \frac{-du/dx}{\sqrt{1 - u^2}}, \qquad \frac{d \arctan u}{dx} = \frac{du/dx}{1 + u^2},$$

where u is a differentiable function of x.

Examples

1. If $y = \arctan 2x$, find dy/dx.

Solution. Let $u = 2x$; then $du/dx = 2$.

$$\frac{dy}{dx} = \frac{d \arctan u}{dx} = \frac{1}{1 + u^2} \cdot \frac{du}{dx} = \frac{2}{1 + (2x)^2} = \frac{2}{1 + 4x^2}.$$

Warning. Be sure when making the substitution for u to square all of u and not just x.

2. If $f(x) = \arcsin x^2$, find $f'(x)$.

Solution. Let $u = x^2$; then $du/dx = 2x$.

$$f'(x) = \frac{df}{du}\frac{du}{dx} = \frac{du/dx}{\sqrt{1 - u^2}} = \frac{2x}{\sqrt{1 - x^4}}.$$

3. If $y = (\arccos (3x + 1))^3$, find dy/dx.

Solution. Let $v = 3x + 1$, $u = \arccos v$. Then

$$\frac{dv}{dx} = 3, \qquad \frac{du}{dx} = \frac{-1}{\sqrt{1 - v^2}}\frac{dv}{dx}.$$

We have

$$\frac{dy}{dx} = 3u^2 \frac{du}{dx} = 3(\arccos(3x + 1))^2 \frac{-3}{\sqrt{1 - (3x + 1)^2}}$$

$$= \frac{-9(\arccos(3x + 1))^2}{\sqrt{-9x^2 - 6x}}.$$

4. A man is walking along the shore at a rate of 3 mph. A lighthouse is situated two miles offshore from the point where he begins walking. The beacon from the lighthouse follows him as he walks. At what rate is the light turning when he has covered one mile?

Solution. The first step is to draw a diagram (see Fig. 8.10): If s represents the man's position, then $ds/dt = 3$. We also have that

$$\tan \theta = \frac{s}{2} \quad \text{or} \quad \theta = \arctan \frac{s}{2}.$$

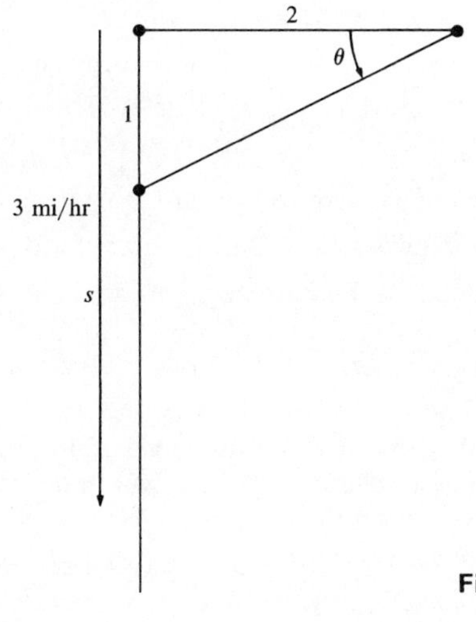

Fig. 8.10

The rate of revolution is given by

$$\frac{d\theta}{dt} = \frac{1}{1 + (s/2)^2} \cdot \frac{1}{2} \cdot \frac{ds}{dt} = \frac{2}{4 + s^2} \cdot 3 = \frac{6}{4 + s^2}.$$

Therefore when $s = 1$, $d\theta/dt = \frac{6}{5}$ radians per hour.

PROBLEMS

1. Find dy/dx.

   a) $y = \sin 3x$.
   b) $y = 3 \sin x$.
   c) $y = \cos(3x^2 + 1)$.
   d) $y = \tan(1/x)$.
   e) $\tan \dfrac{x}{y} = 1$.
   f) $y = (\sin 3x)^2$.
   g) $y = (1 + \tan 2x)^{1/3}$.
   h) $\sin^2 y + \dfrac{1}{\cos^2 x} = 1$.
   i) $y = \sqrt{2 + \cos 2x}$.
   j) $y = 2 \sin x$.
   k) $y = \sin \dfrac{x}{2}$.
   l) $y = \tan x^2$.
   m) $y = \cos \sqrt{x}$.
   n) $y = \sqrt{1 - \tan x}$.

2. Find dy/dx.

   a) $y = \arcsin \dfrac{x}{2}$.
   b) $y = 3 \arctan \dfrac{x}{3}$.
   c) $y = 2 \arccos 2x$.
   d) $y = \arctan x^2$.
   e) $y = \arcsin \dfrac{x - 1}{x + 1}$.
   f) $y = x \arcsin x + \sqrt{1 - x^2}$.
   g) $y = \arcsin x^2$.
   h) $y = x \arctan \sqrt{x}$.

3. Find the equation of the line tangent to $y = \tan x$ at $x = \pi/3$.
4. Graph $y = \sin x - \cos x$ for $0 \leq x \leq 2\pi$. Find the local maxima and minima.
5. Graph the following functions. Discuss maxima, minima, concavity, and points of inflection.
   a) $y = x/2 + \sin x,\ 0 \leq x \leq 2\pi$.
   b) $y = x - \sin x,\ 0 \leq x \leq 2\pi$.
   c) $y = \tan x + 1,\ -\pi \leq x \leq \pi$.
6. A 50-ft ladder is propped against a wall at an angle of 60°. If it starts to slip and the angle changes at the rate of $\pi/180$ radians/sec, at what rate is the bottom of the ladder moving away from the wall when the angle is 30°?
7. A balloon leaving the ground 1 km from an observer rises at the rate of 150 m/min. How fast is the angle of elevation of the observer's line of sight increasing when the balloon is 1 km high?
8. A man is walking along a road at the rate of 5ft/sec. A searchlight on the ground 40 ft from the road is kept trained on him. At what rate is the searchlight revolving when the man is 30 ft away from the point on the road nearest the light?
9. A ship is traveling 15 knots on a course parallel to shore and one nautical mile from it. The light from a lighthouse is kept trained on the ship. At what rate is the light revolving when the ship is directly opposite the lighthouse? When the ship is two miles from the lighthouse?

8.3 INTEGRATION OF TRIGONOMETRIC FUNCTIONS

We list the integration formulas which are obtained from the differentiation formulas of the last section:

$$\int \cos u \, du = \sin u + C$$

$$\int \sin u \, du = -\cos u + C$$

$$\int \sec^2 u \, du = \tan u + C$$

$$\int \csc^2 u \, du = -\text{ctn}\, u + C$$

$$\int \sec u \tan u \, du = \sec u + C$$

$$\int \csc u \,\text{ctn}\, u \, du = -\csc u + C$$

$$\int \frac{1}{\sqrt{1 - u^2}} \, du = \arcsin u + C$$

$$\int \frac{-1}{\sqrt{1 - u^2}} \, du = \arccos u + C$$

$$\int \frac{1}{1 + u^2} \, du = \arctan u + C.$$

The reader will notice that this list does not include the integrals of all elementary trigonometric functions, e.g., $\int \tan u \, du$ is missing. We need more techniques, and indeed, new functions, to handle some of these. This just emphasizes that integration is basically more difficult than differentiation—$\int f(x)\, dx$, $f(x)$ continuous, always exists, but we may not be able to evaluate it.

Examples

1. Find

$$\int \cos 2x \, dx.$$

Solution. We have

$$\int \cos 2x \, dx = \tfrac{1}{2} \int 2 \cos 2x \, dx = \tfrac{1}{2} \sin 2x + C.$$

2. Find

$$\int x \sin 3x^2 \, dx.$$

Solution. Let $u = 3x^2$, $du = 6x\,dx$. Then

$$\int x \sin 3x^2 \, dx = \tfrac{1}{6} \int 6x \sin 3x^2 \, dx = -\tfrac{1}{6} \cos 3x^2 + C.$$

3. Find

$$\int \cos^2 x \sin x \, dx.$$

Solution. Let $u = \cos x$, $du = -\sin x \, dx$. Then

$$\int \cos^2 x \sin x = -\int u^2 \, du = \frac{-u^3}{3} + C = \frac{-\cos^3 x}{3} + C.$$

4. Find

$$\int \frac{3}{\cos^2 x} \, dx.$$

Solution. The integral is equal to $3 \int \sec^2 x \, dx = 3 \tan x + C$.

5. Find

$$\int \frac{\cos x \, dx}{\sqrt{1 + \sin x}}.$$

Solution. Let $u = \sin x$, $du = \cos x \, dx$. Then

$$\int \frac{\cos x \, dx}{\sqrt{1 + \sin x}} = \int \frac{du}{\sqrt{1 + u}} = 2(1 + u)^{1/2} + C = 2\sqrt{1 + \sin x} + C.$$

6. Find

$$\int \sin^2 x \, dx.$$

Solution. As it stands this seems hopeless. However, we recall the half-angle formula

$$\sin \frac{x}{2} = \sqrt{\frac{1 - \cos x}{2}}.$$

We write $\sin^2 x$ as

$$\frac{1 - \cos 2x}{2}$$

and

$$\int \sin^2 x \, dx = \tfrac{1}{2} \int (1 - \cos 2x) \, dx = \tfrac{1}{4}(2x - \sin 2x) + C.$$

7. Find

$$\int \frac{1}{1 + 3x^2}\, dx.$$

Solution. Let $u = \sqrt{3}x$, $du = \sqrt{3}\, dx$. Then

$$\int \frac{1}{1 + 3x^2}\, dx = \frac{1}{\sqrt{3}} \int \frac{\sqrt{3}\, dx}{1 + 3x^2} = \frac{1}{\sqrt{3}} \int \frac{du}{1 + u^2} = \frac{1}{\sqrt{3}} \arctan u$$

$$= \frac{1}{\sqrt{3}} \arctan \sqrt{3}x.$$

Warning. If the integral represents an arctan or an arcsin function, u^2 appears in the denominator, either in $\sqrt{1 - u^2}$ or in $(1 + u^2)$, and no u appears in the numerator. If there is a u (e.g., $\int u/\sqrt{1 - u^2}\, du$) in the numerator we can use the formula of Chapter 7:

$$\int u^a\, du = \frac{u^{a+1}}{a + 1} + C(n \neq -1),$$

or other means (see Section 5).

8. Evaluate

$$\int \frac{x\, dx}{\sqrt{1 - 2x^2}}.$$

Solution. Let $u = 1 - 2x^2$; then $du = -4x\, dx$.

$$\int \frac{x\, dx}{\sqrt{1 - 2x^2}} = \frac{-1}{4} \int \frac{-4x\, dx}{\sqrt{1 - 2x^2}} = \frac{-1}{4} \int u^{-1/2}\, du$$

$$= \frac{-1}{4} 2u^{1/2} + C = -\frac{\sqrt{1 - 2x^2}}{2} + C.$$

When in doubt, check the answer by differentiating to obtain the integrand.

9. The procedure called nitrogen washout is used to determine the volume of a lung. In normal respiration the air which is drawn into the lungs is 80 percent nitrogen. The concentration in the lungs is also, for practical purposes, 80 percent nitrogen. (Almost no nitrogen is absorbed by the blood.) Therefore the volume of the lungs is the volume of nitrogen in the lungs divided by 0.80. The only remaining question is: How do we measure the volume of nitrogen?

If the subject is given nitrogen-free air to breathe, the lungs will gradually expire all the initially contained nitrogen. We assume that we have a method for collecting the expired air, and measuring it and its nitrogen concentration. If we add up the amount of nitrogen in successive breaths, the sum will give us the total volume of nitrogen originally in the lungs.

For uniformity suppose each sample has a volume of 1 ml and we record the concentrations; we then graph the concentration as one coordinate and the number of each successive sample as the other. Thus, for example, after 4 ml have been measured the volume of nitrogen expired (see Fig. 8.11) is

$$0.80(1 \text{ ml}) + 0.79(1 \text{ ml}) + 0.75(1 \text{ ml}) + 0.74(1 \text{ ml}) = 3.08 \text{ ml}.$$

Suppose we determine—the means of doing so is not our present concern—that if we connect these points by a smooth curve, this curve is closely approximated by the graph of

$$N = \frac{32{,}000}{t^2 + 40{,}000}.$$

Then the volume of nitrogen after t ml is the area under the curve from 0 to t. The amount of nitrogen in a sample is zero (to two decimal places) after 2600 ml have been expired, so we have

$$\text{volume of nitrogen} = \int_0^{2600} \frac{32{,}000}{t^2 + 40{,}000}\, dt.$$

The volume of the lungs is then this number divided by 0.8. We must evaluate:

$$\int_0^{2600} \frac{32{,}000}{t^2 + 40{,}000}\, dt.$$

Rewriting the integrand as

$$\frac{32{,}000}{40{,}000((t/200)^2 + 1)}$$

and letting $u = t/200$ so that $du = dt/200$, we have

$$\int_0^{2600} \frac{32{,}000}{t^2 + 40{,}000}\, dt = \frac{32{,}000}{40{,}000} \cdot 200 \int_0^{2600} \frac{dt/200}{((t/200)^2 + 1)}$$

$$= 160 \int_0^{13} \frac{du}{u^2 + 1}$$

$$= 160 \arctan u\Big|_0^{13}$$

$$= 160\,(\arctan 13 - \arctan 0)$$

$$= 160(1.493) \quad \text{(by use of a table of trigonometric functions)}$$

$$= 238.88 \text{ ml}.$$

Volume of lungs $= 238.88/0.8 = 298.6$ ml.

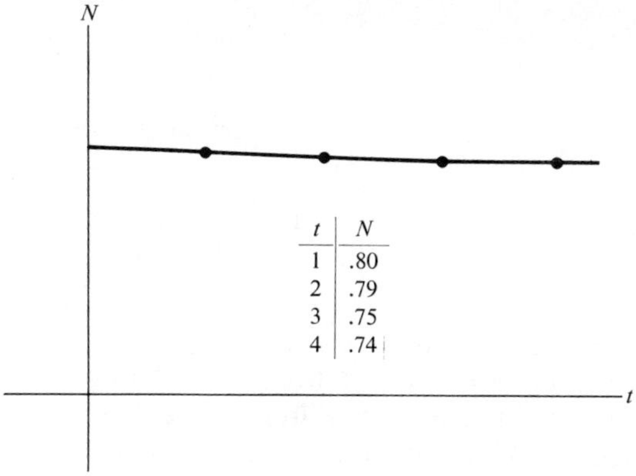

Fig. 8.11

PROBLEMS

1. Evaluate the definite integrals:

a) $\int_0^{\pi} 2 \sin x\, dx.$

b) $\int_0^{2\pi} \frac{\sin x \cos x}{2}\, dx.$

c) $\int_{\pi/3}^{\pi/2} \tan \frac{x}{2} \sec \frac{x}{2}\, dx.$

d) $\int_{\pi/4}^{\pi/2} \frac{\cos x}{\sin^2 x}\, dx.$

e) $\int_{\sqrt{2}}^{2/\sqrt{3}} \frac{x\, dx}{\sqrt{x^2 - 1}}.$

f) $\int_0^{1/2} \frac{dx}{\sqrt{1 - 2x^2}}.$

g) $\int_{-1}^{1} \frac{dx}{1 + x^2}.$

h) $\int_0^{\pi/2} \cos 3x\, dx.$

i) $\int_0^{1/2} \frac{x}{\sqrt{2 - x^2}}\, dx.$

j) $\int_{\pi/6}^{\pi/4} \operatorname{ctn} 2x \csc 2x\, dx.$

k) $\int_{\pi/6}^{\pi/3} \sin^2 x \cos x\, dx.$

l) $\int_{\sqrt{2}}^{2} \frac{1}{\sqrt{1 - (x^2/4)}}\, dx.$

m) $\int_0^{2} \frac{1}{9 + x^2}\, dx.$

2. Integrate:

a) $\int \sin 4x\, dx.$

b) $\int \tan^4 x \sec^2 x\, dx.$

c) $\int \sin 2x \cos 2x\, dx.$

d) $\int x \cos x^2\, dx.$

e) $\int \frac{\cos x - \sin x}{(\sin x + \cos x)^3}\, dx.$

f) $\int \frac{dx}{\sqrt{1 - 4x^2}}.$

g) $\displaystyle\int \frac{dx}{4 + x^2}$.

h) $\displaystyle\int \tan^3 x \sec^2 x\, dx$.

i) $\displaystyle\int \frac{x}{(1 + x^2)^2}\, dx$.

j) $\displaystyle\int x^2 \sin 3x^3\, dx$.

k) $\displaystyle\int \tan x \sec^2 x\, dx$.

l) $\displaystyle\int \frac{1}{\sqrt{1 - 9x^2}}\, dx$.

3. Find the indicated areas:
   a) between the x-axis and $y = \cos x$ from $x = 0$ to $x = \pi$.
   b) between the x-axis and $y = 3 \sin x/2$ from $x = \pi/3$ to $x = \pi/2$.
   c) between the x-axis and $y = x + \sin x$ from $x = 0$ to $x = 2\pi$.
   d) between $y = 1/\sqrt{1 - x^2}$ and the x-axis from $x = 0$ to $x = 1/2$.
   e) between $y = 1/(1 + 4x^2)$ and the x-axis from $x = -1$ to $x = 1$.
   f) between the x-axis and $y = \sin 2x$ from $x = 0$ to $x = \pi/2$.
   g) between the x-axis and $y = 3/\sqrt{1 - x^2}$ from $x = -1/2$ to $x = 1/2$.
   h) between the x-axis and $y = 1/(1 + x^2)$ from $x = 0$ to $x = \sqrt{3}$.
   i) between the x-axis and $y = x + \cos x$ from $x = 0$ to $x = \pi$.

8.4 LOGARITHMIC AND EXPONENTIAL FUNCTIONS AND THEIR DERIVATIVES

We are familiar with exponents which are rational numbers. If c is a real number, we define

$$c^{m/n} = (c^m)^{1/n}, \qquad n, m \text{ integers}, \qquad n \neq 0.$$

These rational exponents obey certain laws such as:

$$(c^a)^b = c^{ab}$$

$$c^a c^b = c^{a+b},$$

for a real number c and a and b rational numbers. We have remarked previously that we should like to be able to use irrational numbers as exponents; for example: Can we give a reasonable meaning to 2^π? In this section we shall in fact define functions

$$f(x) = c^x \qquad (c > 0 \text{ a real number}),$$

whose domain is the set of all real numbers.

We approach the problem of defining such functions, called *exponential functions*, through a back door. The concept of logarithm should be a familiar one: if

$$a^x = b > 0,$$

then $\log_a b = x$, with a as the base of the logarithm. *Common logarithms* are logarithms with base 10. Such logarithms are very useful for computational purposes because our number system has 10 as its base; this means that the logarithms of powers of 10 are integers:

$$\log_{10} 1 = 0, \qquad \log_{10} 10 = 1, \qquad \log_{10} 100 = 2, \qquad \log_{10} 0.1 = -1, \quad \text{etc.}$$

Moreover, one need have at hand only the logarithms for the real numbers between 1 and 10 since, for example, the logarithm (in base 10) of 971 is $\log_{10} 9.71 + \log_{10} 100$. The efficiency of logarithms for computational purposes is due to the fact that they reduce multiplication problems to addition (this is the principle upon which slide rules are constructed), and the raising of powers to multiplication. That $\log ab = \log a + \log b$ and $\log a^b = b \log a$ follows directly from the definition. For example, to compute

$$\frac{(1873 \cdot 2154 \cdot 5.6)^3}{(643 \cdot 971)^{1/2}}.$$

we use a table to find the logarithms (base 10) of 1.873, 2.154, 5.6, 6.43, and 9.71. Then, letting x be the desired answer we have

$$\log x = 3\,(\log 1.873 + 3 + \log 2.154 + 3 + \log 5.6)$$
$$-\tfrac{1}{2}\,(\log 6.43 + 2 + \log 9.71 + 2).$$

The final step is to use a table to find what number has this quantity as its common logarithm. (For details on the use of logarithms, see Appendix B.)

However, it is not for the purpose of computation that we use logarithms in this chapter. We define a function, called the *natural logarithm*: for $x > 0$,

$$\log x = \text{area between the graph of } 1/t \text{ and the } t\text{-axis from 1 to } x$$

$$\boxed{\log x = \int_1^x \frac{1}{t}\,dt.}$$

(Some authors use ln to denote natural logarithms.)
Then by Theorem 1 of Chapter 7

$$\frac{d \log x}{dx} = \frac{1}{x}.$$

Moreover, by the chain rule, if u is a differentiable function of x, then

$$\frac{d \log u}{dx} = \frac{1}{u}\frac{du}{dx}.$$

Let us see whether this function has the useful properties that we would expect of logarithms.

Theorem 4. For any real numbers $x, y \geq 0$,

$$\log xy = \log x + \log y.$$

Proof. Consider $\log xy$ as a function of x, with y a constant. Then

$$\frac{d \log xy}{dx} = \frac{y}{xy} = \frac{1}{x} = \frac{d \log x}{dx}.$$

Since $\log xy$ and $\log x$ have the same derivative (with respect to x), they differ by a constant. That is,

$$\log xy = \log x + C.$$

This holds for all $x > 0$ and in particular for $x = 1$, so $\log y = 0 + C$. From this we get

$$\log xy = \log x + \log y,$$

as required.

Warning. Remember that the logarithm of a product is the sum of the logarithms of the factors. The logarithm of a sum cannot in general be simplified—it is *not* the case that for all $x, y > 0$, $\log (x + y) = \log x + \log y$.

From Theorem 4 there follows by induction:

Theorem 5. For a a rational number, $x > 0$,

$$\log x^a = a \log x.$$

Although we still do not have a definition of x^a for a irrational we define $\log x^a$ to be $a \log x$ for all real a.

Note that since $\log x$ is defined only on the positive real numbers, if $u = f(x)$, $\log u$ is defined only for those values of x making $u = f(x) > 0$. For example, $\log (x - 1)$ is defined only for $x - 1 > 0$, that is, for $x > 1$. However,

$$\frac{d}{dx} \log |u| = \frac{d}{dx} \log u = \frac{1}{u} \frac{du}{dx}, \qquad u > 0$$

$$\frac{d}{dx} \log |u| = \frac{d}{dx} \log (-u) = \frac{1}{-u} \left(-\frac{du}{dx}\right) = \frac{1}{u} \frac{du}{dx}, \qquad u < 0.$$

Examples

1. If $f(x) = \log |3x + 1|$, find $f(x)$.

Solution. Let $u = 3x + 1$. Then

$$\frac{d \log |3x + 1|}{dx} = \frac{d \log |u|}{dx} = \frac{1}{u} \frac{du}{dx} = \frac{3}{3x + 1}.$$

2. Compute

$$\frac{d \log (\sqrt{x} + 1/\sqrt{x})}{dx}.$$

Solution. Let $u = \sqrt{x} + 1/\sqrt{x}$. Then $du/dx = 1/2\sqrt{x} - 1/2x^{3/2}$ and

$$\frac{d \log(\sqrt{x} + 1/\sqrt{x})}{dx} = \frac{d \log u}{dx} = \frac{1}{u}\frac{du}{dx}$$

$$= \frac{1}{\sqrt{x} + 1/\sqrt{x}}\left(\frac{1}{2\sqrt{x}} - \frac{1}{2x^{3/2}}\right).$$

3. If

$$y = \log\sqrt{\frac{2x^2 + 1}{3x^3 - 1}}, \qquad \text{find } y'.$$

Solution. We write

$$\log\sqrt{\frac{2x^2 + 1}{3x^3 - 1}} = \tfrac{1}{2}(\log(2x^2 + 1) - \log(3x^3 - 1)).$$

Then

$$y' = \frac{1}{2}\left(\frac{4x}{2x^2 + 1} - \frac{9x^2}{3x^3 - 1}\right).$$

4. Find

$$\lim_{x \to +\infty} \frac{\log x}{x}.$$

Solution. Applying L'Hospital's rule yields

$$\lim_{x \to +\infty} \frac{\log x}{x} = \lim_{x \to +\infty} \frac{1/x}{1} = 0.$$

Let us examine the graph shown in Fig. 8.12:

$$y = \log x \qquad (x > 0).$$

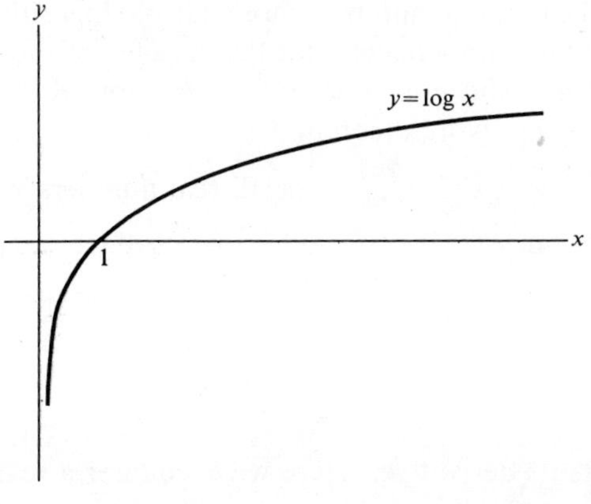

Fig. 8.12

Since the derivative $1/x$ is always positive, the function is strictly increasing. The second derivative

$$y'' = \frac{-1}{x^2}$$

is always negative. There are no critical points nor points of inflection. There is an x-intercept at $x = 1$, but no y-intercept. As $x \to +\infty$, $y \to +\infty$ and as $x \to 0$, $y \to -\infty$.

Since $\log x$ is a strictly increasing function, it is one-to-one and has an inverse. Let us call the inverse function *exp*. That is, exp is the function defined by

$$\exp x = y \qquad \text{if and only if} \qquad \log y = x.$$

This new function also has some nice properties.

Theorem 6.

$$\exp(x + y) = (\exp x)(\exp y).$$

Proof. Let

$$\exp x = a, \qquad \exp y = b.$$

From the definition of exp we have

$$\log a = x, \qquad \log b = y.$$

Then by Theorem 4,

$$\log ab = \log a + \log b = x + y.$$

Returning to the definition of exp, we have

$$\exp(x + y) = ab = (\exp x)(\exp y),$$

as required.

As a consequence of Theorem 5 and the definition of exp, we have

Theorem 7. For a real number a,

$$(\exp x)^a = \exp xa.$$

We define e to be that real number whose natural logarithm is 1, i.e., $\log e = 1$. Geometrically, e is the number such that the area under the curve $y = 1/x$ shown in Fig. 8.13 between 1 and e is equal to 1. We note that since $\exp(1) = e$ by definition, we get, by applying theorem 7

$$\exp x = e^x, \qquad \text{for all real numbers } x.$$

What can we say about e^x? We already know its algebraic properties, such as

$$e^{x+y} = e^x e^y$$

$$(e^x)^r = e^{xr}$$

$$e^0 = 1,$$

so next we compute its derivative. Here we encounter a really amazing fact!

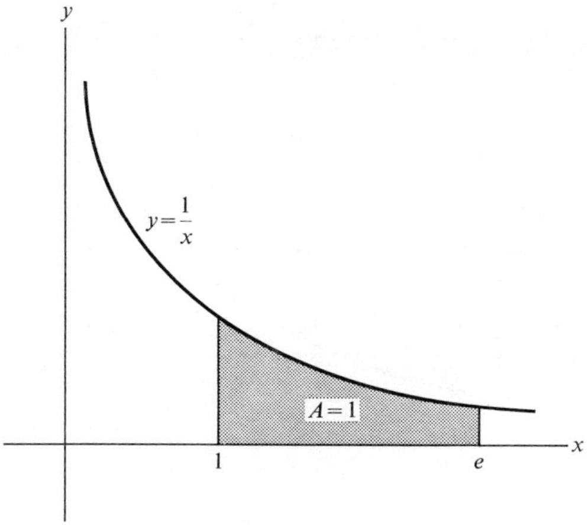

Fig. 8.13

Theorem 8. e^x is differentiable and $de^x/dx = e^x$. That is, the derivative of the function is that function itself.

Proof. If $e^x = y$, then $\log y = x$, because they are inverse functions. Therefore

$$\frac{dx}{dy} = \frac{d \log y}{dy} = \frac{1}{y}$$

yields

$$\frac{de^x}{dx} = \frac{dy}{dx} = \frac{1}{dx/dy} = \frac{1}{1/y} = y = e^x.$$

Let us examine the graph of e^x (Fig. 8.14). Since it is the inverse of the log function, whose domain is the positive real numbers, its image is the positive real numbers; that is, $e^x > 0$, for all x. On the other hand, its domain is the set of all real numbers. Its derivative, e^x, is always positive, so it is strictly increasing. Use of the chain rule yields: If u is a differentiable function of x, then

$$\frac{de^u}{dx} = e^u \frac{du}{dx}.$$

Examples

1. Find

$$de^{x^2}/dx.$$

Solution. Let

$$u = x^2, \qquad \frac{du}{dx} = 2x$$

and

$$\frac{de^{x^2}}{dx} = \frac{de^u}{dx} = e^u \frac{du}{dx} = 2xe^{x^2}.$$

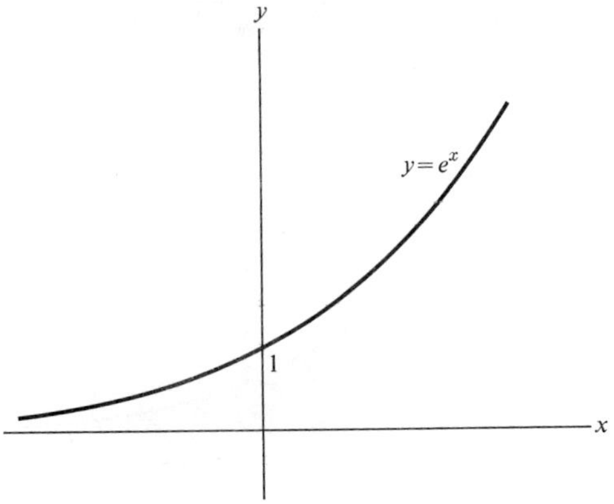

Fig. 8.14

2. Find the derivative of

$$\frac{e^{\sin x}}{e^{3x}}.$$

Solution. First observe that

$$\frac{e^{\sin x}}{e^{3x}} = e^{\sin x - 3x}.$$

Thus let $u = \sin x - 3x$ and $du/dx = \cos x - 3$. Then

$$\frac{d(e^{\sin x}/e^{3x})}{dx} = \frac{de^u}{dx} = e^u \frac{du}{dx} = e^{\sin x - 3x}(\cos x - 3).$$

3. Find

$$\lim_{x \to +\infty} \frac{x}{e^x}.$$

Solution. Applying L'Hospital's rule gives us

$$\lim_{x \to +\infty} \frac{1}{e^x} = 0.$$

4. Find the derivative of $(\log e^{1/x})(\log \sin x)$.

Solution. First we note that $\log e^{1/x} = 1/x$. Therefore, we want to compute the derivative of $1/x \log \sin x$. This is a product so we have

$$\frac{1}{x}\frac{d(\log \sin x)}{dx} + \frac{-1}{x^2}\log \sin x = \frac{1}{x}\frac{1}{\sin x}\cos x - \frac{1}{x^2}\log \sin x$$

$$= \frac{\operatorname{ctn} x}{x} - \frac{\log \sin x}{x^2}.$$

Warning. While

$$\log xy = \log x + \log y,$$

and

$$\log x^a = a \log x,$$

it is *not* the case that for all x and y

$$(\log x)(\log y) = \log xy,$$

nor is it true that for all x and y

$$(\log x)(\log y) = \log (x + y).$$

Finally, it is also *not* the case that

$$(\log x)(\log y) = \log y^x.$$

To summarize, be careful to apply the addition and multiplication rules for logarithms correctly.

We can also define other exponential functions:

$$f(x) = c^x, \qquad c \text{ any real number } > 0,$$

by

$$c^x = \exp (x \log c) = e^{x \log c}.$$

Thus, for example, we can finally give a meaning to 2^π:

$$2^\pi = e^{\pi \log 2}.$$

Using our previous knowledge of log and exp we see that these functions obey the expected rules:

$$c^{x+y} = c^x c^y$$

$$(c^x)^a = c^{ax}.$$

The derivative of an exponential function is given by

$$\frac{dc^x}{dx} = \frac{de^{x \log c}}{dx} = e^{x \log c}(\log c) = c^x \log c.$$

We are also able to define the *power function*:

$$f(x) = x^a = e^{a \log x},$$

for any real number a, $x > 0$. It is left to the reader to verify that this coincides with our previous definition for a rational number a.

Now we can prove that the differentiation formula of Chapter 5

$$\frac{dx^a}{dx} = ax^{a-1},$$

for any real number $a \neq -1$, is valid for $x > 0$. We have

$$x^a = e^{a \log x}.$$

Letting $u = a \log x$, we have

$$\frac{du}{dx} = \frac{a}{x}.$$

Then

$$\begin{aligned}\frac{dx^a}{dx} &= e^u \frac{du}{dx} = e^{a \log x} \frac{a}{x} \\ &= x^a \cdot \frac{a}{x} \\ &= ax^{a-1}.\end{aligned}$$

We state this result as a theorem:

Theorem 8.

$$\frac{dx^a}{dx} = ax^{a-1} \quad \text{for} \quad x > 0,\ a \neq 0.$$

PROBLEMS

1. Find dy/dx.
   a) $y = \log(1 + x^2)$.
   b) $y = 3 \log |6 - 2x^3|$.
   c) $y = \log \left| \dfrac{x - 1}{2x + 2} \right|$.
   d) $y = \log 2x^3 - 5$.
   e) $y = e^{2x}$.
   f) $y = e^{-x^2+1}$.
   g) $y = \sqrt{e^x}$.
   h) $y = \dfrac{1 - e^x}{1 + e^x}$.
   i) $y = \sin e^x$.
   j) $y = \log \cos x^2$.
   k) $y = e^{\log x}$.
   l) $y = e^{x^2 \log x}$.
   m) $y = x(\log x)^2$.
   n) $y = 2^{x^2}$.
   o) $y = \log 2x^2$.
   p) $y = \sin \log x$.
   q) $y = \sqrt{\log 2x}$.
   r) $y = e^{\sin 2x}$.
   s) $y = x \log |x|$.
   t) $y = e^{1/x}$.
2. Use L'Hospital's rule to evaluate the following limits:
   a) $\lim_{x \to 1} \dfrac{e^x - e}{x^2 - 1}$.
   b) $\lim_{x \to \infty} x^{1/x}$.
   (*Hint:* consider the log of the function.)

c) $\lim_{x\to+\infty} x \sin \frac{1}{x}$.

d) $\lim_{x\to+\infty} \frac{x}{e^x}$.

e) $\lim_{x\to 0} \frac{\sin x^2}{x}$.

f) $\lim_{x\to 0} \frac{x}{e^x - 1}$.

g) $\lim_{x\to\infty} \frac{x^n}{e^x}$.

3. Graph xe^x. Discuss maxima, minima, concavity, and points of inflection.
4. Psychologists claim that the ability of a child to memorize during his first four years is given by

$$f(x) = x \log x + 1.$$

At what age during this period is a child's ability to memorize at its minimum?

5. In Hull theory it is assumed that the habit strength is given in terms of the number x of repetitions by

$$H(x) = 100(1 - e^{-ix}),$$

i a positive constant. Find the rate of change of $H(x)$.

8.5 INTEGRATION OF EXPONENTIAL AND LOGARITHMIC FUNCTIONS

The title of this section is in a sense a misnomer. While we can integrate exponential functions very easily ($\int e^u\, du = e^u + C$), we are not able to evaluate $\int \log u\, du$. This integral will be treated in Chapter 12. However, we are able to handle integrals leading to the logarithm function

$$\int \frac{1}{u}\, du = \log |u| + C.$$

The key to mastery of integration of any functions is practice. We give examples of some types of problems which are likely to arise.

Examples

1. Evaluate

$$\int_0^2 e^{-5x}\, dx.$$

Solution. Let $u = -5x$, $du = -5\, dx$. Then

$$\int_0^2 e^{-5x}\, dx = \frac{-1}{5}\int_0^2 e^{-5x}(-5)\, dx = \frac{-1}{5}\int_0^{-10} e^u\, du$$

$$= \left.\frac{-e^u}{5}\right|_0^{-10}$$

$$= \frac{-1}{5e^{10}} + \frac{1}{5}.$$

2. Find

$$\int \frac{e^x}{e^{2x}+1}\,dx.$$

Solution. Let $u = e^x$, $du = e^x\,dx$. Then

$$\int \frac{e^x\,dx}{e^{2x}+1} = \int \frac{1}{u^2+1}\,du = \arctan u + C = \arctan e^x + C.$$

3. Find

$$\int \frac{e^{\sqrt{x}}}{\sqrt{x}}\,dx.$$

Solution. Let $u = \sqrt{x}$, $du = 1/2\sqrt{x}\,dx$. Then

$$\int \frac{e^{\sqrt{x}}}{\sqrt{x}}\,dx = 2\int e^u\,du = 2e^u + C = 2e^{\sqrt{x}} + C.$$

4. Find

$$\int \frac{x}{x^2+1}\,dx.$$

Solution. Let $u = x^2 + 1$, $du = 2x\,dx$. Then

$$\int \frac{x\,dx}{x^2+1} = \frac{1}{2}\int \frac{1}{u}\,du = \tfrac{1}{2}\log u + C = \tfrac{1}{2}\log(x^2+1) + C.$$

5. Find

$$\int \tan x\,dx.$$

Solution. We write $\tan x = \sin x/\cos x$ and let $u = \cos x$. Then $du = -\sin x\,dx$ and

$$\int \tan x\,dx = -\int \frac{du}{u} = -\log u + C = -\log\cos x + C.$$

6. Find

$$\int \frac{x^2}{3x-1}\,dx.$$

Solution. Dividing $3x - 1$ into x^2, we obtain $(x/3) + \frac{1}{9} + 1/9(3x-1)$, therefore,

$$\begin{aligned}\int \frac{x^2}{3x-1}\,dx &= \int\left(\frac{x}{3} + \frac{1}{9} + \frac{1}{9(3x-1)}\right)dx\\ &= \int \frac{x\,dx}{3} + \int \frac{dx}{9} + \int \frac{dx}{9(3x-1)}\\ &= \frac{x^2}{6} + \frac{x}{9} + \frac{1}{27}\log|3x-1| + C.\end{aligned}$$

7. Find $d2^x/dx$.

Solution

$$\frac{d2^x}{dx} = 2^x \log 2.$$

8. If $dy/dx = xe^{x^2}$, find y.

Solution. This is an example of a *differential equation.* To solve it, we write it as

$$dy = xe^{x^2}\,dx,$$

and integrate both sides of the equation:

$$y + C_1 = \frac{e^{x^2}}{2} + C_2.$$

Combining the arbitrary constants,

$$y = \frac{e^{x^2}}{2} + C.$$

We summarize the differentiation and integration formulas for trigonometric, exponential, and logarithmic functions.

SUMMARY

$$\frac{d \sin u}{dx} = \cos u \frac{du}{dx} \qquad \frac{d \cos u}{dx} = -\sin u \frac{du}{dx}$$

$$\frac{d \tan u}{dx} = \sec^2 u \frac{du}{dx} \qquad \frac{d \operatorname{ctn} u}{dx} = -\csc^2 u \frac{du}{dx}$$

$$\frac{d \sec u}{dx} = \sec u \tan u \frac{du}{dx} \qquad \frac{d \csc u}{dx} = -\csc u \operatorname{ctn} u \frac{du}{dx}$$

$$\frac{d \arcsin u}{dx} = \frac{du/dx}{\sqrt{1 - u^2}} \qquad \frac{d \arccos u}{dx} = \frac{-du/dx}{\sqrt{1 - u^2}}$$

$$\frac{d \arctan u}{dx} = \frac{du/dx}{1 + u^2}$$

$$\frac{d \log u}{dx} = \frac{du/dx}{u} \qquad \frac{de^u}{dx} = e^u \frac{du}{dx}$$

$$\frac{dc^u}{dx} = c^u \log c \frac{du}{dx} \qquad \frac{du^a}{dx} = au^{a-1} \frac{du}{dx}$$

$$\int \sin u \, du = -\cos u + C \qquad \int \cos u \, du = \sin u + C$$

$$\int \sec^2 u\, du = \tan u + C \qquad \int \csc^2 u\, du = -\text{ctn}\, u + C$$

$$\int \sec u \tan u\, du = \sec u + C \qquad \int \csc u\, \text{ctn}\, u\, du = -\csc u + C$$

$$\int \frac{du}{\sqrt{1 - u^2}} = \text{acrsin}\, u + C \qquad \int \frac{du}{1 + u^2} = \arctan u + C$$

$$\int \frac{du}{u} = \log |u| + C \qquad \int e^u\, du = e^u + C.$$

. . . .

PROBLEMS

1. Evaluate the integrals:

a) $\displaystyle\int e^{3x}\, dx.$

b) $\displaystyle\int_e^{e^2} \frac{dx}{x \log x}.$

c) $\displaystyle\int \frac{e^x}{1 + 2e^x}\, dx.$

d) $\displaystyle\int e^{\cos x + 1} \sin x\, dx.$

e) $\displaystyle\int \frac{dx}{e^{2x}}.$

f) $\displaystyle\int_0^1 3^x\, dx.$

g) $\displaystyle\int \frac{x^2}{2x^3 - 4}\, dx.$

h) $\displaystyle\int \frac{\sin x}{1 + 2\cos x}\, dx.$

i) $\displaystyle\int \frac{x}{1 + 2x^2}\, dx.$

j) $\displaystyle\int_0^1 e^{x/2}\, dx.$

k) $\displaystyle\int \frac{x}{1 + x}\, dx.$

l) $\displaystyle\int_0^{\sqrt{\pi}} \frac{x \sin x^2}{2 \cos x^2}\, dx.$

m) $\displaystyle\int_{-1}^1 \frac{x}{(1 + 3x^2)^2}\, dx.$

n) $\displaystyle\int \frac{e^x}{\sqrt{1 - 4e^x}}\, dx.$

2. The population of a certain state is increasing at the rate of 3 percent per year. If this rate continues, how many years will it take the population to double?

3. In the inversion of raw sugar, the rate of change of the amount of raw sugar present varies as the amount of raw sugar remaining. If after one hour, 100 pounds has been reduced to 95 pounds, how much raw sugar remains at the end of ten hours?

4. The population of a certain city increases at a rate of 10 percent per year. If this rate continues, how many years will it take the population to double?

5. The growth of bacteria is proportional to the number of bacteria present. If after 10 hours 1000 have increased to 2000, how many will be present at the end of 100 hours?

6. Find the area of the prescribed regions:
   a) between the curves $y = e^{2x}$ and $y = e^x$ from $x = 0$ to $x = 2$.
   b) between $y = (\log x)/x$ from $x = 1$ to $x = 10$.
   c) between $y = e^{2x}$ and the x axis from $x = -100$ to $x = -1$.
   d) between $y = x$ and $y = e^x$ from $x = 0$ to $x = 1$.
   e) between $y = (\log x)/x$ and $y = xe^{x^2}$ from $x = 1$ to $x = 2$.
7. Find the average value of e^x over the interval $[1, 2]$.
8. Show that $f(x) = 2/\pi(1 + x^2)$ is a probability measure. Find its associated probability distribution.
9. Find the average value of $1/x$ from $x = 1$ to $x = 10$.

8.6 APPLICATIONS

In many situations the rate of change of some quantity is proportional to that quantity:

$$\frac{dQ}{dt} = kQ, \qquad k \text{ a constant.}$$

The solution to this differential equation is found by writing it as

$$\frac{dQ}{Q} = k\,dt$$

and integrating both sides

$$\log Q = kt + C.$$

Let us denote $Q(0)$ by Q_0. Then

$$\log Q_0 = C,$$

and

$$\log Q = kt + \log Q_0.$$

But kt is the logarithm of e^{kt}, so we can write

$$\log Q = \log e^{kt} + \log Q_0.$$

From this equation, we have

$$Q = Q_0 e^{kt}.$$

If we need a numerical answer for certain values of Q_0, k, and t, we can use a table of natural logarithms, or a table listing powers of e.

1. Radioactive decay. The rate of decay of a radioactive substance is proportional to the amount of the substance present. The half-life of radium is 1600 years (i.e., half of a given amount of radium will be gone after 1600 years). If we start with 100 grams of radium, after how many years will there be 20 grams remaining?

Solution. We have

$$Q = Q_0 e^{kt} = 100e^{kt}.$$

To evaluate k we use the fact that at $t = 1600$, $Q = Q_0/2 = 50$:

$$50 = 100e^{1600k}$$
$$0.5 = e^{1600k}.$$

From the exponential table in the back of the book

$$1600k = -0.69, \qquad k = -0.00043.$$

Now we want to know what t is when Q is 20. We have

$$20 = 100e^{-0.00043t}$$
$$0.2 = e^{-0.00043t}$$
$$-0.00043t = 1.6 \quad \text{(from the table)}$$
$$t = 3721 \text{ years.}$$

The dating of very old organic material depends upon measuring the amount of carbon 14 (a radioactive isotope of carbon) left in the substance, and then using the method above with k computed using the half-life of carbon 14.

When the constant k in the equation

$$Q = Q_0e^{kt},$$

is negative, as in the example above, this is an *exponential decay* equation. If k is positive, the equation represents *exponential growth.*

2. Exponential growth—bacteria. Under ideal circumstances the rate of growth of bacteria is proportional to the number of bacteria present. If the initial number of bacteria is 1000 and 10 hours later 6000 are present, find the number 15 hours after the first count.

Solution. Using the exponential growth equation

$$Q = Q_0e^{kt},$$

we have

$$6000 = 1000e^{10k}$$
$$6 = e^{10k}.$$

We obtain $10k = 1.74$ from the table, giving $k = 0.174$. Thus

$$Q = 1000e^{0.174t}.$$

If

$$t = 15, \qquad Q = 1000e^{2.61}, \qquad Q = 13{,}606.$$

3. Exponential growth—population. Under "normal" conditions the rate of change of population is considered to be proportional to the population. If a city had a population of 50,000 in 1960 and 100,000 in 1970, what, assuming "normal" conditions, will its population be in 1980?

Solution. Again $Q = Q_0e^{kt}$. We have

$$\begin{aligned} 100{,}000 &= 50{,}000e^{10k} \\ 2 &= e^{10k} \\ 10k &= 0.69 \\ k &= 0.069. \end{aligned}$$

Thus

$$Q = 50{,}000e^{0.069t}.$$

If $t = 20$,

$$\begin{aligned} Q &= 50{,}000e^{1.38} \\ &= 200{,}000 \text{ (approximately)}. \end{aligned}$$

4. Exponential growth—compound interest. In Chapter 2 we used the formula for principal invested at compound interest:

$$P = P_0(1 + tr)^n,$$

where P_0 is the initial principal, r is the rate per year, t the number of years per compounding period, and n the number of compounding periods.

We rewrite this as

$$P = P_0\left[\left(1 + \frac{1}{m}\right)^m\right]^{ry}$$

where $m = 1/tr$ and y = number of years = nt. Now as $t \to 0$, that is, as the compounding periods get shorter, $m \to +\infty$. If

$$P = P_0 \lim_{m\to+\infty}\left[\left(1 + \frac{1}{m}\right)^m\right]^{ry}$$

the interest is said to be *compounded continuously*.

Let us look at

$$\lim_{m\to+\infty}\left(1 + \frac{1}{m}\right)^m.$$

The (natural) logarithm of $(1 + 1/m)^m$ is given by $m \log(1 + 1/m)$. We want to find the limit of this logarithm as $m \to +\infty$. In order to compute this by L'Hospital's rule we write

$$\begin{aligned} \lim_{m\to+\infty} m \log\left(1 + \frac{1}{m}\right) &= \lim_{m\to+\infty} \frac{\log(1 + 1/m)}{1/m} \\ &= \lim_{m\to+\infty} \frac{(1/(1 + 1/m))(-1/m^2)}{-1/m^2}. \\ &= 1. \end{aligned}$$

Thus since the log of e is 1 it seems reasonable to suppose, although we have not proved it, that

$$\lim_{m \to +\infty} \left(1 + \frac{1}{m}\right)^m = e.$$

In fact, $\lim_{m \to \infty} (1 + 1/m)^m$ is often taken as the definition of e and our definition is derived as a property.

Thus P_0 invested at r percent interest compounded continuously for y years yields

$$P = P_0 e^{ry}.$$

If interest is compounded daily, the length of the compounding period is sufficiently short that the formula for continuous compounding is usually used as an adequate approximation.

We can also obtain the above formula by observing that with interest being compounded continuously, the principal is proportional to the rate of change

$$\frac{dP}{dy} = rP.$$

In other words, at any time y the money is increasing at rate of the interest rate times the principal *at that time*. From this differential equation we obtain the same formula,

$$P = P_0 e^{ry}.$$

Example. Suppose \$1000 is invested at 5 percent interest compounded continuously. What is the amount at the end of one year? Compare this with the amount if interest is compounded quarterly and with the amount at simple interest.

Solution. For continuous compounding

$$\begin{aligned} P &= P_0 e^{ry} \\ &= 1000e^{0.05} \\ &= 1051.30. \end{aligned}$$

For quarterly compounding

$$\begin{aligned} P &= P_0(1 + tr)^n \\ &= 1000(1 + 0.05(\tfrac{1}{4}))^4 \\ &= 1050.94. \end{aligned}$$

For simple interest (one "compounding period")

$$\begin{aligned} P &= P_0(1 + 0.05) \\ &= 1050. \end{aligned}$$

5. Elasticity of demand. Economists recognize that the demand for some commodities is more sensitive to price changes than the demand for others. Hence it may be useful to consider the ratio of the relative change in demand to the relative change in price, called the *elasticity of demand.* If the demand x is regarded as a function of the price p, then the elasticity of demand with respect to price is given by

$$\frac{E_x}{E_p} = \lim_{p\to 0} \frac{-dx/x}{dp/p} = -\frac{p}{x}\frac{dx}{dp}.$$

Note that E_x/E_p is a function of x.

Example. Consider the demand function

$$p = e^{-x} \qquad (0 \leq x \leq 8).$$

as shown in Fig. 8.15. Find the elasticity of demand at $x = 0$, 4, and 8.

$$\frac{dp}{dx} = -e^{-x}, \qquad \frac{dx}{dp} = \frac{1}{dp/dx} = -e^{x}, \qquad \frac{E_x}{E_p} = -\frac{e^{-x}}{x}\cdot(-e^{x}) = 1/x.$$

At $x = 0$, the elasticity is undefined.
At $x = 4$, the elasticity is 1/4.
At $x = 8$, the elasticity is 1/8.

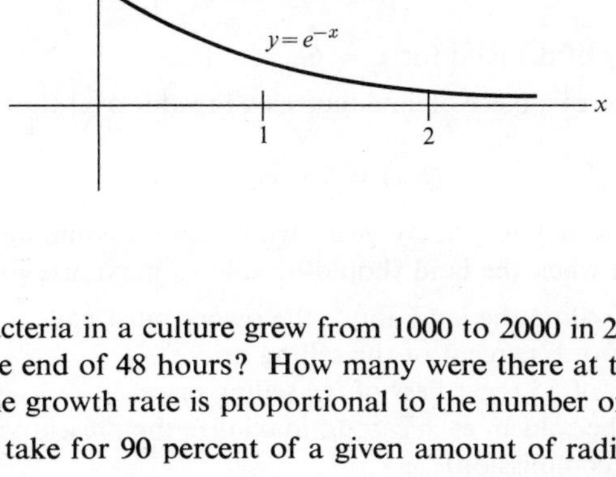

Fig. 8.15

PROBLEMS

1. The number of bacteria in a culture grew from 1000 to 2000 in 24 hours. How many will there be at the end of 48 hours? How many were there at the end of 12 hours? (Assuming that the growth rate is proportional to the number of bacteria present.)
2. How long does it take for 90 percent of a given amount of radium to disappear?
3. The half-life of carbon 14 is 5570 years. How old is a specimen of a plant whose carbon 14 content is 20 percent of that of a present-day live specimen?
4. How long does it take for $\frac{1}{4}$ of a given amount of radium to disappear? $\frac{3}{4}$?

5. How long does it take \$100 invested at 6 percent compounded continuously to increase to \$150?
6. How old is a specimen whose carbon 14 content is 10 percent of that of a present-day live specimen?
7. How long does it take \$100 invested at 5 percent compounded continuously to double?
8. Approximately how long does it take for money invested at 5.5 percent compounded daily to triple? (Use continuous compounding to approximate daily compounding.)
9. If the demand function is given by

$$p = 7e^{-x/2} + 42,$$

find the elasticity of demand for $x = 2$.

10. The present value P of a sum of money $Q(y)$ to be received y years from now is the amount that must be invested now, at the annual rate of interest r, compounded m times a year, in order to produce $Q(y)$ at the end of y years. Thus

$$Q(y) = P\left[\left(1 + \frac{1}{m}\right)^m\right]^{ry},$$

or if the interest is compounded continuously

$$Q(y) = Pe^{ry}.$$

Therefore

$$P = Q(y)[(1 + 1/m)^m]^{-ry},$$

or

$$P = Q(y)e^{-ry}.$$

Find the present value of \$100 received in 15 years, assuming 5 percent interest compounded (a) quarterly; (b) continuously.

11. If a demand function is given by

$$p = 17e^{-x/3} + 51,$$

find the elasticity of demand for $x = 6$.

12. Suppose the value of a piece of land increases in value over the next 20 years according to the formula:

$$Q(y) = 15 + 2y - 0.1y^2,$$

where $Q(y)$ is the selling price y years from now. Assuming 6 percent continuous interest, find out when the land should be sold to maximize its present value.

13. Suppose that in selling the land above the owner must pay
a) a commission of 6 percent of the selling price;
b) a commission of \$5 regardless of the selling price.
When should it be sold in each case to maximize the present value of the net revenue (selling price − commission)?

14. Suppose that the construction cost of a certain facility is given by

$$C(t) = kt^{1/2},$$

where t is durability in years and k is a constant. Suppose moreover that Q is the annual rate of net savings resulting from the facilities. Then, assuming a constant rate r of interest, the present value of future savings to be realized over x years is

$$P = \int_0^x Qe^{-rt}\, dt.$$

The optimal life of the facility is that number of years which maximizes the ratio of the present value of future savings to cost. That is, the value of x for which

$$R(x) = \frac{\int_0^x Qe^{-rt}\, dt}{kx^{1/2}}.$$

is a maximum. Find the optimal life.

15. Find the present value of \$100 to be received in 10 years assuming 6 percent interest compounded continuously.
16. Suppose a chain letter asks that it be sent on to four people. Suppose, moreover, that everyone complies and that it takes two days to deliver a letter. How many people will the letter have reached at the end of 30 days?

9 / FUNCTIONS OF SEVERAL VARIABLES

In Chapter 4 we talked of y as a function of x. However, in applications one quantity may be a function of several others. For example, the cost of an item may be determined by transportation, labor, and materials. We may write this relation as

$$C = f(t, l, m).$$

Since changes in any of these alter the cost, if we want to minimize this quantity we would need to find the derivative with respect to all of these.

9.1 PARTIAL DIFFERENTIATION

If each n-tuple of values for $x_1, \ldots, x_n$ determines a unique value for y, we say that $y = f(x_1, \ldots, x_n)$ is a *function of n variables*. We define the ith *partial derivative* of $f(x_1, \ldots, x_n)$ to be

$$\lim_{h \to 0} \frac{f(x_1, \ldots, x_i + h, \ldots, x_n) - f(x_1, \ldots, x_n)}{h}$$

if this limit exists. Note that the definition requires that the function be defined at $(x_1, \ldots, x_i + h, \ldots, x_n)$ for small values of h. We denote the ith partial derivative of f by

$$\frac{\partial f}{\partial x_i} \quad \text{or} \quad f_i(x_1, \ldots, x_n) \quad \text{or} \quad D_i f.$$

What we are really doing is asking for the derivative of f as a function of x_i alone, with the other variables held constant; indeed, that is precisely how partial derivatives are computed—all variables except one are treated as constants. Each partial derivative is itself a function.

Examples

1. If

$$f(x, y) = x^3y - 2xy^2,$$

find

$$\frac{\partial f}{\partial x} \quad \text{and} \quad \frac{\partial f}{\partial y}.$$

Solution. Treating y and x in turn as constants yields:

$$\frac{\partial f}{\partial x} = 3x^2y - 2y^2; \qquad \frac{\partial f}{\partial y} = x^3 - 4xy.$$

2. If $f(x, y, z) = xyz$, find the third partial derivative of f at $(1, -1, 2)$.

Solution

$$D_3 f(x, y, z) = xy$$

$$D_3 f(1, -1, 2) = -1$$

3. If the output Y is expressed as a function of the labor N and capital K,

$$Y = f(N, K),$$

then the changes in Y which result from changes in N and K are represented by the first and second partial derivatives, respectively.

The calculation of higher-order partial derivatives proceeds by iteration of the differentiation process. That is,

$$\frac{\partial^2 f(x, y)}{\partial x^2} = \frac{\partial}{\partial x}\left(\frac{\partial f(x, y)}{\partial x}\right),$$

and in general

$$\frac{\partial^n f(x_1, \ldots, x_m)}{\partial x_i^n} = \frac{\partial}{\partial x_i}\left(\frac{\partial^{n-1} f(x_1, \ldots, x_m)}{\partial x_i^{n-1}}\right).$$

However, we may also compute *mixed partials*:

$$\frac{\partial^2 f(x, y)}{\partial y\, \partial x} = \frac{\partial}{\partial y}\left(\frac{\partial f(x, y)}{\partial x}\right), \qquad \frac{\partial^2 f(x, y)}{\partial x\, \partial y} = \frac{\partial}{\partial x}\left(\frac{\partial f(x, y)}{\partial y}\right);$$

and in general

$$\frac{\partial^n f(x_1, \ldots, x_m)}{\partial x_{i_1}, \ldots, \partial x_{i_n}} = \frac{\partial}{\partial x_{i_1}}\left(\frac{\partial^{n-1} f(x_1, \ldots, x_m)}{\partial x_{i_2}, \ldots, \partial x_{i_n}}\right),$$

where $x_{i_1}, \ldots, x_{i_n}$ is some collection of the $x_1, \ldots, x_n$ (repetition allowed). The mixed partial above is of *order n*.

The question which arises is whether the order of differentiation is crucial. That is, can we expect that

$$\frac{\partial^2 f}{\partial x\, \partial y} = \frac{\partial^2 f}{\partial y\, \partial x}\,?$$

The answer is that if all partials of $f(x_1, \ldots, x_n)$ exist and are continuous, then the computation of mixed partials is independent of the order of differentiation.

Examples

1. Compute all second order partials of $f(x, y) = x^2 + 2xy^2$.

Solution

$$\frac{\partial f}{\partial x} = 2x + 2y^2, \qquad \frac{\partial f}{\partial y} = 4xy.$$

Then

$$\frac{\partial^2 f}{\partial y\, \partial x} = 4y = \frac{\partial^2 f}{\partial x\, \partial y} \quad \text{and} \quad \frac{\partial^2 f}{\partial x^2} = 2, \qquad \frac{\partial^2 f}{\partial y^2} = 4x.$$

2. If

$$f(x, y, z) = 2x^3 + 2xyz + z^3,$$

compute

$$\frac{\partial^3 f}{\partial x^3} \quad \text{and} \quad \frac{\partial^3 f}{\partial x\, \partial y\, \partial z}.$$

Solution

$$\frac{\partial f}{\partial x} = 6x^2 + 2yz \qquad \frac{\partial^2 f}{\partial x_2} = 12x \qquad \frac{\partial^3 f}{\partial x^3} = 12.$$

Since

$$\frac{\partial^3 f}{\partial x\, \partial y\, \partial z} = \frac{\partial^3 f}{\partial y\, \partial z\, \partial x},$$

we compute

$$\frac{\partial^2 f}{\partial z\, \partial x} = \frac{\partial}{\partial z}\left(\frac{\partial f}{\partial x}\right) = 2y.$$

Finally,

$$\frac{\partial^3 f}{\partial y\, \partial z\, \partial x} = \frac{\partial}{\partial y}\left(\frac{\partial^2 f}{\partial z\, \partial x}\right) = 2$$

The chain rule for partial differentiation is slightly more complicated.

Suppose f is differentiable as a function of each of $x_1, \ldots, x_n$ and each x_i is differentiable as a function of each of $u_1, \ldots, u_m$. Then

$$\frac{\partial f}{\partial u_i} = \frac{\partial f}{\partial x_1}\frac{\partial x_1}{\partial u_i} + \cdots + \frac{\partial f}{\partial x_n}\frac{\partial x_n}{\partial u_i}.$$

Example. If

$$f(x, y, z) = xyz + 2x \quad \text{and} \quad x = 3u, \quad y = uv^2, \quad z = u + v,$$

then

$$\frac{\partial f}{\partial u} = \frac{\partial f}{\partial x}\frac{\partial x}{\partial u} + \frac{\partial f}{\partial y}\frac{\partial y}{\partial u} + \frac{\partial f}{\partial z}\frac{\partial z}{\partial u}$$

$$= (yz + 2)(3) + (xz)v^2 + (xy)(1)$$

$$= (uv^2(u + v) + 2)(3) + (3u)(u + v)(v^2) + (3u)(uv^2)$$

$$= 3u^2v^2 + 3uv^3 + 6 + 3u^2v^2 + 3uv^3 + 3u^2v^2$$

$$= 9u^2v^2 + 6uv^3 + 6.$$

PROBLEMS

1. If $f(x, y, z) = 3x^4 - yz + z^2x$, find $\partial f/\partial x$, $\partial f/\partial y$, $\partial f/\partial z$, and $\partial^3 f/\partial z^3$.
2. If $f(x, y) = \arcsin y/x$, find $\partial f/\partial y\,(1, 1)$.
3. If $f(x, y) = \log y^x$, find $\partial f/\partial x$, $\partial^2 f/\partial y^2$, and $\partial^2 f/\partial x\,\partial y$.
4. If $f(x, y, z) = e^x + e^y + e^{z^2}$, find $\partial^3 f/\partial x\,\partial y\,\partial z$.
5. If $f(x, y) = xy/(x^2 + y^2)$, find $f_1(1, -2)$.
6. If $f(x, y, z) = xyz^2 + yzx^2$, find $f_3(0, 1, 2)$.
7. If $f(x, y) = \int_x^y t^2\,dt$, find $\partial f/\partial x$ and $\partial f/\partial y$.
8. If $f(x, y, z) = \sin xy + x\cos z$, find f_1, f_2, and f_3.
9. If $f(x, y, z) = x^2z - y + 2xy^2$, find $\partial^2 f/\partial x^2$ and $\partial^2 f/\partial x\,\partial y$.
10. If $f(x, y) = e^{x^2}/y$, find all second partials.
11. If $f(x, y, z) = 3x^2e^y + 2z\tan x$, find $\partial^2 f/\partial x\,\partial y$ and $\partial^2 f/\partial y^2$.
12. If $f(x, y) = e^x \sin y$, find $f_2(1, \pi/4)$.
13. If $f(x, y, z) = 4xz - 3z^2 - y^2/2$, find $f_2(0, 1, 2)$.
14. If $f(x, y, z) = ze^{x+y}$, find $f_2(1, -1, 1)$.
15. If $f(x, y) = \int_y^x (t + 2)\,dt$, find $\partial f/\partial x$ and $\partial f/\partial y$.

9.2 MAXIMA AND MINIMA

If all the partial derivatives of a function f of n variables are 0 at $(x_1, \ldots, x_n)$, $(x_1, \ldots, x_n)$ is a *critical point* of f. Local maxima and minima and maxima and minima are defined analogously to the case of functions of a single variable.

Example

1. Find the critical points of

$$f(x, y) = 3x^2 - 2xy + 2y^2 - 5x.$$

Solution

$$\frac{\partial f}{\partial x} = 6x - 2y - 5,$$

$$\frac{\partial f}{\partial y} = -2x + 4y.$$

Setting the partials equal to zero yields the system

$$6x - 2y - 5 = 0$$

$$-2x + 4y = 0,$$

whose solution is $x = 1$, $y = \frac{1}{2}$. Therefore, $(1, \frac{1}{2})$ is the only critical point.

2. Find the critical points of

$$f(x, y, z) = 2x^2 - y^2 + 4z.$$

Solution

$$\frac{\partial f}{\partial x} = 4x, \qquad \frac{\partial f}{\partial y} = -2y, \qquad \frac{\partial f}{\partial z} = 4.$$

The partials can never all be zero so there are no critical points.

3. Find the critical points of

$$f(x, y) = x^2y - 5y^2 - 20y.$$

Solution

$$\frac{\partial f}{\partial x} = 2xy, \qquad \frac{\partial f}{\partial y} = x^2 - 10y - 20.$$

If $2xy = 0$, either $x = 0$ or $y = 0$. If $x = 0$, and

$$\frac{\partial f}{\partial y} = x^2 - 10y - 20 = 0,$$

then

$$y = -2.$$

If $y = 0$,

$$x^2 = 20 \quad \text{and} \quad x = \pm 2\sqrt{5}.$$

Therefore the critical points are $(0, -2)$, $(2\sqrt{5}, 0)$, $(-2\sqrt{5}, 0)$.

To connect critical points with local maxima and minima, we need to define the n-dimensional analogue of open and closed intervals. Recall that a neighborhood of a point x on the real line is an open interval (a, b) with $x \in (a, b)$.

For purposes of generalization we consider only those neighborhoods of a point which are centered at that point, that is, those of the form

$$(x - \delta, x + \delta),$$

δ the *radius* of the neighborhood. Observe that

$$(x - \delta, x + \delta) = \{y \mid |x - y| < \delta\}.$$

On the real line $|x - y|$ is the distance between the points x and y. In the plane the distance between (x_1, x_2) and (y_1, y_2) is

$$\sqrt{(x_1 - y_1)^2 + (x_2 - y_2)^2} \qquad \text{(see Fig. 9.1).}$$

In general, if $X = (x_1, \ldots, x_n)$ and $Y = (y_1, \ldots, y_n)$ then the *distance* between X and Y is defined to be

$$d(X, Y) = \sqrt{(x_1 - y_1)^2 + \cdots + (x_n - y_n)^2}.$$

A *neighborhood* of $X = (x_1, \ldots, x_n)$ is the set of all $Y = (y_1, \ldots, y_n)$ such that $d(X, Y) < \delta$, for some positive real number δ:

$$N_\delta(X) = \{Y \mid d(X, Y) < \delta\}.$$

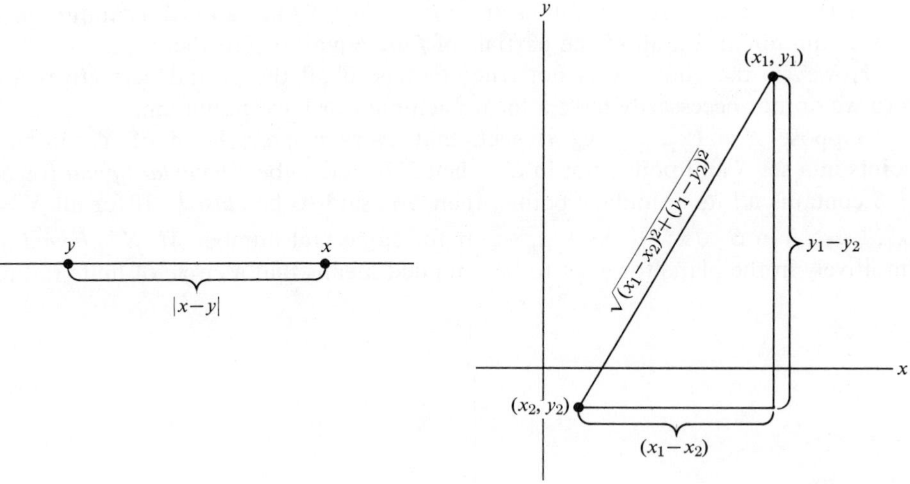

Fig. 9.1

If D is a set of n-tuples of real numbers such that for each $X \in D$ there is a neighborhood U_X of X such that $U_X \subset D$, then D is said to be an *open* set. Intuitively, in the plane a set D is open if at each of its points (x_1, x_2) it is possible to draw a circle with center (x_1, x_2) which is completely contained in D (Fig. 9.2).

Example. $D = \{(x, y) \mid x^2 + y^2 < 1\}$ is open, but $D' = \{(x, y) \mid x^2 + y^2 \leq 1\}$ is not, since it is not possible to surround the points on the circumference by circles lying entirely in D' (see Fig. 9.3).

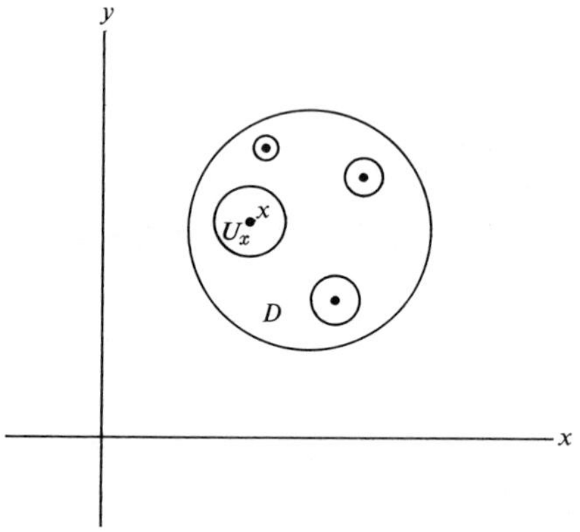

Fig. 9.2

We state without proof:

Theorem 1. If $f(x_1, \ldots, x_n)$ is defined on an open set U and all of its partial derivatives exist and are continuous at $X_0 \in U$, then if f has a local maximum or a local minimum at X_0, all of the partials of f are equal to zero there.

However, the converse is not true; that is, if all the partial derivatives are zero we do not necessarily have a local maximum or local minimum.

Suppose $X = (x_1, \ldots, x_n)$ is such that every neighborhood of X contains points in a set S and points not in S. Then X is said to be a *boundary point* for S. If S contains all its boundary points, then S is said to be *closed*. If for all $X = (x_1, \ldots, x_n)$ in S, $\sqrt{x_1^2 + \cdots + x_n^2} < M$ for some real number M, S is *bounded*. Intuitively in the plane for a set to be bounded means that a circle of finite radius

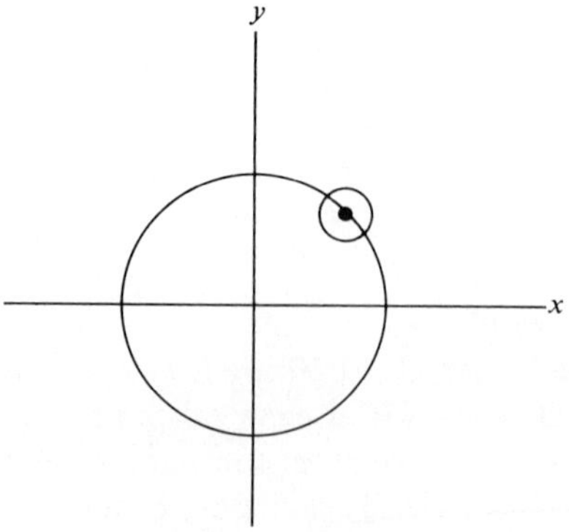

Fig. 9.3

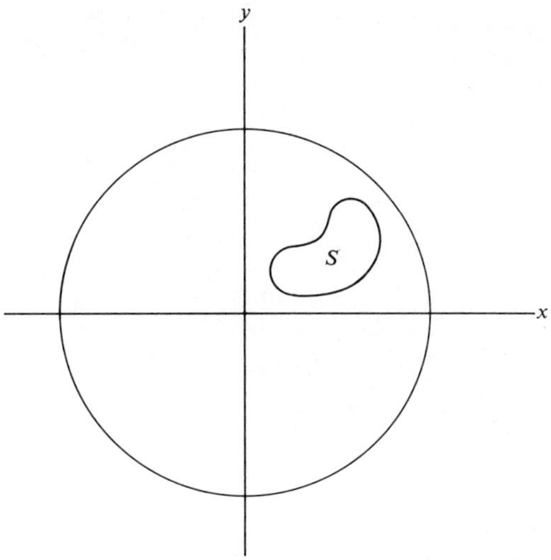

Fig. 9.4

(see Fig. 9.4), center at the origin, can be drawn so as to contain S. On the real line, a bounded set can be contained in an interval of the form $(-M, M)$.

Examples

1. Let $S = \{(x, y) \mid x^2 + y^2 \leq 1\}$. The boundary points of S are those points on the unit circle, i.e., (x, y) such that $x^2 + y^2 = 1$. They are all contained in S, so S is closed. S is also bounded.

Warning. All sets which are not open are not necessarily closed.

$$S = \{(x, y) \mid x < 2 \quad \text{and} \quad y = 0\},$$

is neither open nor closed (Fig. 9.5). It is not open because any neighborhood of, for example, $(1, 0) \in S$ must contain points outside of S. On the other hand, $(2, 0)$ is a boundary point, but is not in S, so S is not closed.

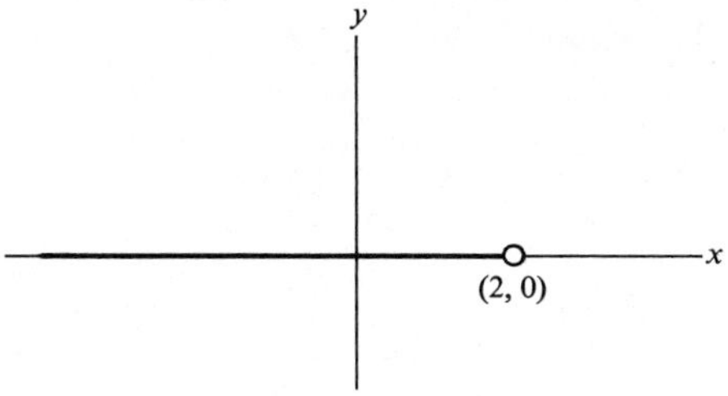

Fig. 9.5

2. Consider $\{(x, y) \mid x \leq y\}$. The boundary points of S are those points on the line $x = y$ (see Fig. 9.6). Hence S is closed, but not bounded.

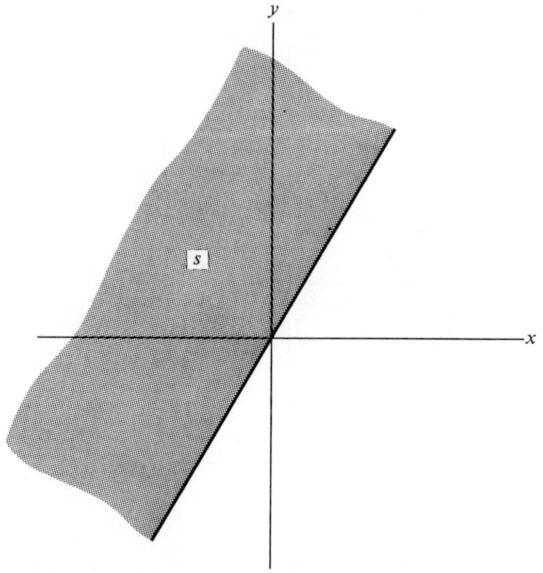

Fig. 9.6

Theorem 2. If $f(x_1, \ldots, x_n)$ is defined on a closed and bounded set S, then there exist X_0, Y_0 in S such that

$$f(X_0) \geq f(X) \geq f(Y_0),$$

for all $X \in S$.

Thus we see that the closed and bounded set plays the role of the closed interval. From Theorem 2 we know that on a closed and bounded set a function has a maximum and a minimum; by Theorem 1 if they are not at boundary points they are critical points. Therefore the procedure for finding maxima and minima on closed and bounded sets is straightforward: Find the critical points, compute the value of the function at the critical points and on the boundary and compare the values.

Example. Let $f(x, y, z) = xyz$ be given on the region defined by $|x| \leq 1$, $|y| \leq 1$, $|z| \leq 1$—a cube centered at the origin, of side 2. Find the critical points, maxima and minima.

Solution

$$f_1 = yz$$
$$f_2 = xz$$
$$f_3 = xy.$$

All partials are 0 whenever $x = y = 0$, $x = z = 0$, or $y = z = 0$, i.e., on the coordinate axes. None of these critical points is a local maximum or a local minimum since at each point P on any of the axes the value of f is zero, but in

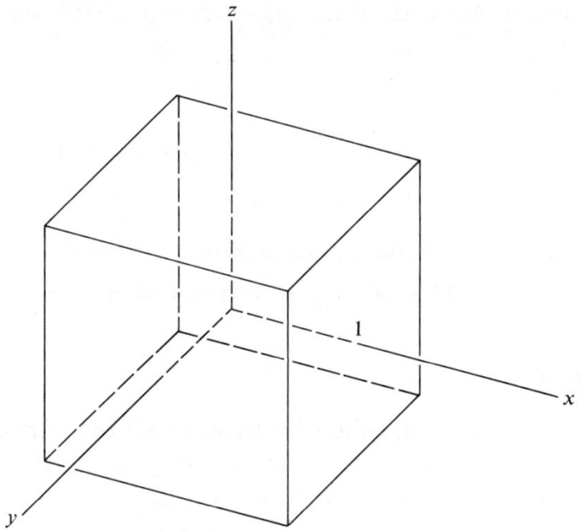

Fig. 9.7

every neighborhood of P the function $f(x, y, z) = xyz$ takes on positive and negative values. However, at the corners (which are boundary points) the value of f is -1, the minimum, or $+1$, the maximum.

PROBLEMS

1. Find the critical points of the following functions:
   a) $f(x, y) = x^2 - 4xy + y^2 - 4x + 3y$.
   b) $f(x, y, z) = 2y + 3x - x^2 - y^2 - z$.
   c) $f(x, y) = x \sin y + y$.
   d) $f(x, y, z) = xy + xz + yz$.
   e) $f(x, y, z) = x^2 + y^2 + z^2 - 4$.
   f) $f(x, y) = \sin xy$.
   g) $f(x, y) = 3x^2 - 4y^2 + 5xy$.
   h) $f(x, y, z) = x^2 + 4y^2 + z^2$.
   i) $f(x, y, z) = xy - yz + 3y^2$.
2. Determine which of the following sets are open, which are closed, and which are bounded.
   a) $\{(x, y) \mid x \geq 2, y < 1\}$.
   b) $\{(x, y) \mid x^2 + 4y^2 \geq 16\}$.
   c) $\{(x, y) \mid x < 1, y < 1\}$.
   d) $\{(x, y) \mid 0 \leq x \leq 2, y = x\}$.
   e) $\{(x, y) \mid 2 < (1/y) < 4, 0 < x < 2\}$.
   f) $\{(x, y) \mid x < 2, y \geq 3\}$.
   g) $(x, y) \mid 4x^2 + y^2 \leq 1\}$.
   h) $\{(x, y) \mid y \leq 2x^2\}$.
   i) $\{(x, y) \mid |x| < 1, y = 2x\}$.
3. Find the local maxima and minima (if any) of the following functions.
   a) $f(x, y) = x^2 + xy^3 - 2y$.
   b) $f(x, y) = 5x + 12y + \dfrac{7}{x} - \dfrac{8}{y}$.
   c) $f(x, y) = 10\,xy - 5x^2y - 2xy^2$.
   d) $f(x, y, z) = \dfrac{(xyz)^{1/2}}{x + y + z + 1}$.
   e) $f(x, y) = x^3 + y^3 + 1$.
   f) $f(x, y) = 3x^2 - 2y^2$.
   g) $f(x, y, z) = \sqrt{x + y + z^2}$.
   h) $f(x, y, z) = \sin xy - z^2$.

4. Find the maximum value of $f(x, y, z) = x^2 + y^2 + z^2$ on the cube $|x| \leq 1$, $|y| \leq 1$, $|z| \leq 1$.
5. If the cost function and the revenue function for a firm producing two quantities are given by
$$R = p_1q_1 + p_2q_2 \qquad (p_i = \text{price}, \quad q_i = \text{quantity}),$$
and
$$C = 4q_1^2 + 3q_1q_2 + 4q_2^2,$$
find when the profit is maximized, assuming fixed prices.
6. Find the distance between the origin and the plane $2x + y - 2z = 3$.

9.3 INTEGRATION

Partial differentiation is accomplished by treating all the variables except one as constants; the analogue is iterated integration—integration of a function of several variables by holding all variables except one constant.

The integral
$$\int_c^d f(x, y)\, dy$$
means the integral of $f(x, y)$ obtained by holding x fixed and integrating $f(x, y)$ as a function of y. For example,
$$\int_{-1}^2 x^2y^3\, dy = \frac{x^2y^4}{4}\bigg|_{y=-1}^{y=2} = \frac{15}{4}x^2.$$
Such an integration yields a function of x which can in turn be integrated with respect to x:
$$\int_a^b \int_c^d f(x, y)\, dy\, dx. \tag{9.1}$$
For example,
$$\int_0^1 \int_1^2 x^2y^3\, dy\, dx = \int_0^1 \tfrac{15}{4} x^2\, dx = \tfrac{15}{4} \cdot \frac{x^3}{3}\bigg|_0^1 = \tfrac{15}{12}.$$
Usually we write (9.1) as
$$\int_a^b dx \int_c^d f(x, y)\, dy,$$
to avoid confusion about the order of integration.

In general, we define an iterated integral
$$\int_{a_1}^{b_1} dx_1, \ldots, \int_{a_n}^{b_n} f(x_1, \ldots, x_n)\, dx_n$$
by integrating the function with respect to each variable successively from right to left.

Examples

1. Evaluate the integral of $f(x, y) = x + y^2$ on the rectangle $0 \leq x \leq 1$, $0 \leq y \leq 2$.

Solution. We have

$$\int_0^1 dx \int_0^2 (x + y^2)\, dy = \int_0^1 xy + \frac{y^3}{3}\Big|_0^2 dx$$

$$= \int_0^1 \left(2x + \frac{8}{3}\right) dx$$

$$= x^2 + \frac{8x}{3}\Big|_0^1$$

$$= 1 + \frac{8}{3} = \frac{11}{3}.$$

2. It is also possible to have limits which are functions of the variables being held constant; that is, we may integrate over a region which is not a rectangle. For example, evaluate the integral of $f(x, y) = x + y$ on the quarter circle

$$\{(x, y) \mid x^2 + y^2 \leq 1, x, y \geq 0\}.$$

Solution

$$\int_0^1 dx \int_0^{\sqrt{1-x^2}} (x + y)\, dy = \int_0^1 xy + \frac{y^2}{2}\Big|_0^{\sqrt{1-x^2}} dx$$

$$= \int_0^1 \left(x\sqrt{1 - x^2} + \frac{1 - x^2}{2}\right) dx$$

$$= \frac{-(1 - x^2)^{3/2}}{3} + \frac{x}{2} - \frac{x^3}{6}\Big|_0^1$$

$$= \frac{2}{3}.$$

We recall that in Chapter 7 the notion of integration of a function of one variable was first introduced by computing the area under a parabola as the limit of the sum of the areas of rectangles (see Fig. 9.8):

$$\int_a^b f(x)\, dx = \lim_{n \to \infty} S_n,$$

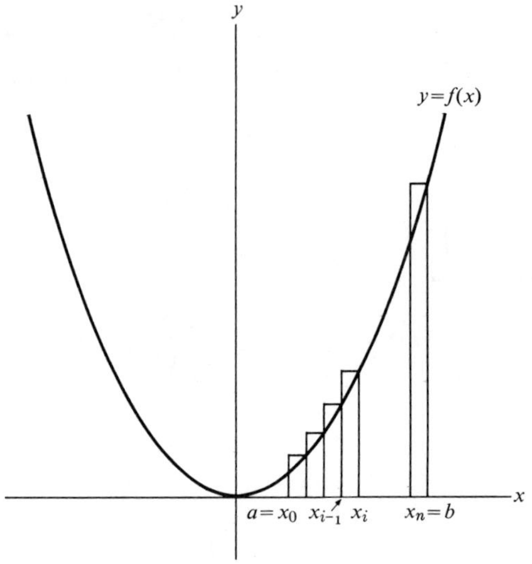

Fig. 9.8

where

$$S_n = \sum_{i=1}^{n} f(x_i)\,\Delta x_i, \qquad (\Delta x_i = x_i - x_{i-1}).$$

We should be able to generalize this notion to functions of several variables by extending the concept of area to R^n.

A subset D of R^n consisting of all $X = (x_1, \ldots, x_n)$ such that

$$a_i \le x_i \le b_i, \qquad i = 1, \ldots, n,$$

is called a *coordinate rectangle* (see Fig. 9.9). Then the volume of D is given by

$$V(D) = (b_1 - a_1)\cdots(b_n - a_n).$$

Subdividing each of the intervals $[a_i, b_i]$ results in subdividing the coordinate rectangle into smaller coordinate rectangles, say $D_1, \ldots, D_r$—the more the subdivisions of the intervals, the larger the number of rectangles. Such a subdivision is called a *grid*. The maximum of the lengths of the sides of the rectangles $D_1, \ldots, D_r$ of a grid is called the *mesh* of the grid.

Now let f be a function of n variables, G be a grid on a rectangle D in the domain of f, $D_1, \ldots, D_r$ be the rectangles of the grid, $m(G)$ be the mesh of the grid. Form the sum

$$\sum_{i=1}^{r} f(X_i)V(D_i),$$

where $X_i = (x_{i_1}, \ldots, x_{i_n})$ is in D_i. Then we define the *multiple integral:*

$$\int_D f\,dV = \lim_{m(G)\to 0} \sum_{i=1}^{r} f(X_i)V(D_i).$$

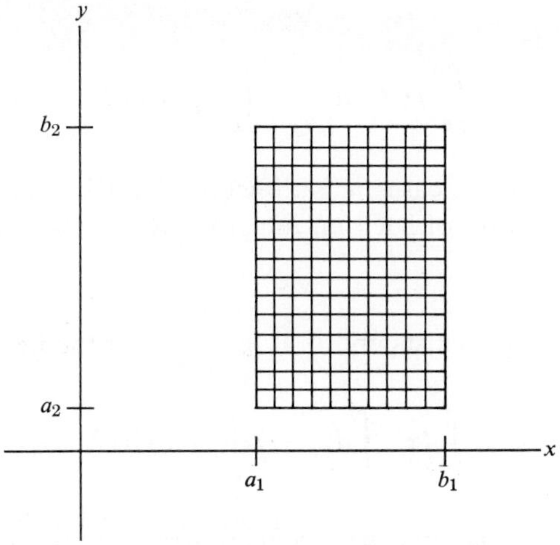

Fig. 9.9

if the limit exists. If the limit does exist, we say that f is *integrable* on D. We shall not prove any existence theorems, but simply remark that if f is continuous on D, then the limit does exist and is independent of the choice of a particular $X_i \in D_i$.

There is also the problem of whether it is possible to define an integral

$$\int_B f(X)\, dV$$

on a set B which is not necessarily a rectangle. If B is a sufficiently well-behaved set (Fig. 9.10), then we may define an integral over it—for example, if B is bounded and has a boundary which consists of finitely many smooth segments.

The definition of the integral of a function over such a set B is a generalization of the case when B is a rectangle. We put a grid on B as if B were a rectangle and do not worry about the odd shapes near the boundary. Just as before we define

$$\int_B f\, dV = \lim_{m(G)\to 0} \sum_{i=1}^{n} f(X_i)V(D_i),$$

where $X_i \in D_i$, provided the limit exists. It can be shown (with considerable difficulty) that the limit exists for continuous functions f and nice sets B.

Examples of "sufficiently well-behaved" subsets of R^2

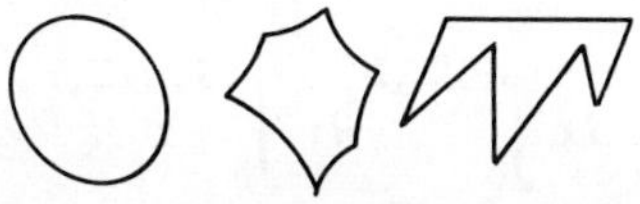

Fig. 9.10

Then we define

$$\int_B f(X)\, dV = \int_D f_B(X)\, dV.$$

Unfortunately, in most cases it is difficult, if not impossible, to evaluate the limit used in the definition of such an integral. But a theorem, which we do not prove, bails us out:

Theorem 3. If $\int_B f(X)\, dV$ exists and if at least one of the iterated integrals

$$\int dx_{i_1} \int dx_{i_2} \cdots \int f(X)\, dx_{i_n}$$

exists (where $x_{i_1}, \ldots, x_{i_n}$ is some arrangement of $x_1, \ldots, x_n$, $X = (x_1, \ldots, x_n)$ and each of the integrals is a definite integral—it is too unwieldy to write in the limits), then all of the iterated integrals exist and are equal to

$$\int_B f(X)\, dV.$$

Thus if we know that f and B are sufficiently regular that the multiple integral and one of the iterated integrals exist we can pick the easiest iterated integral to evaluate, knowing that the value thus obtained is the value of the multiple integral (and of the other iterated integrals).

The greatest difficulty in applying Theorem 3 is determining the limits of integration in each of the integrals. The general method is to start with the first integral, determine the limits on it, and proceed in order with the other integrals. At each stage the limits will be functions (possibly constant functions) of the variables used in the preceding integrals. We illustrate the method.

Examples

1. Let $f(x, y, z) = xyz$ and

$$B = \{(x, y, z) \mid x^2 + y^2 + z^2 \le 4, \quad y, z \ge 0\} \qquad \text{(Fig. 9.11).}$$

f is defined and continuous on B which is bounded and has a smooth boundary. Also $|f(X)| < 8$ for all $X \in B$. Therefore $\int_B f(X)\, dV$ exists and is equal to

$$\int_{-2}^{2} dx \int_0^{\sqrt{4-x^2}} dy \int_0^{\sqrt{4-x^2-y^2}} xyz\, dz.$$

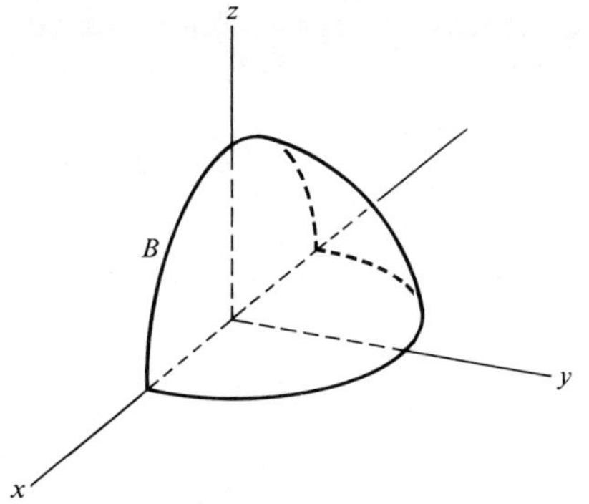

Fig. 9.11

Evaluating this iterated integral yields

$$\int_B f(X)\,dV = \frac{1}{2}\int_{-2}^{2} dx \int_0^{\sqrt{4-x^2}} xy(4 - x^2 - y^2)\,dy$$
$$= \frac{1}{2}\int_{-2}^{2} x\left(2(4 - x^2) - \frac{x^2}{2}(4 - x^2) - \frac{(4 - x^2)^2}{4}\right)dx$$
$$= 0.$$

2. The theorem telling us that the value of the integral is (under certain conditions) independent of the order of integration is very useful in situations such as the following: Find

$$\int_B e^{y^2}\,dV,$$

where

$$B = \{(x, y) \mid 0 \le x \le 1, \quad y < x\} \qquad \text{(Fig. 9.12).}$$

Solution. We can set this up as

$$\int_B e^{y^2}\,dV = \int_0^1 dx \int_0^x e^{y^2}\,dy.$$

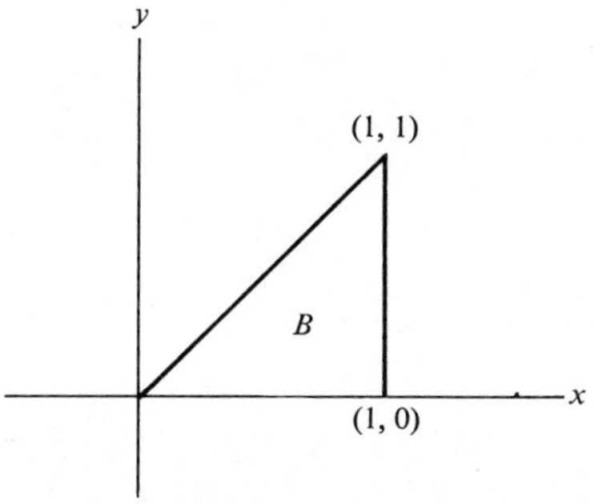

Fig. 9.12

However, e^{y^2} is not the derivative of any elementary function so we cannot evaluate the integral easily. Let us rewrite the integral:

$$\begin{aligned}\int_B e^{y^2}\,dV &= \int_0^1 dy \int_0^y e^{y^2}\,dx.\\ &= \int_0^1 e^{y^2} y\,dy\\ &= \frac{1}{2} e^{y^2}\Big|_0^1\\ &= \frac{e-1}{2}.\end{aligned}$$

Multiple integrals enjoy the usual properties of integrals.

$$\int_B kf(X)\,dV = k\int_B f(X)\,dV$$

$$\int_B (f(X) + g(X))\,dV = \int_B f(X)\,dV + \int_B g(X)\,dV.$$

$$\int_{B_1 \cup B_2} f(X)\,dV = \int_{B_1} f(X)\,dV + \int_{B_2} f(X)\,dV,$$

if $B_1 \cap B_2$ is empty, where in each case if the right-hand side exists, then the left-hand side exists and they are equal.

PROBLEMS

1. Evaluate the integrals:

a) $\int_0^3 dy \int_0^2 y\,dx.$

b) $\int_0^3 dy \int_{y^2}^y x\,dx.$

c) $\int_{-1}^1 dy \int_y^{2y} (x + y)\,dx.$

d) $\int_{-1}^0 dx \int_1^{1-x} \frac{x}{y^2}\,dy.$

e) $\int_{-1}^1 dx \int_0^1 (x^3 + 3x^2y - y^3 + 2xy)\,dy.$

f) $\int_1^2 dx \int_{x^2}^{x^3} \frac{1}{y}\,dy.$

g) $\int_0^{\sqrt{\pi}} dx \int_{x^2}^{2x^2} x\cos y\,dy.$

h) $\int_0^1 dx \int_0^2 dy \int_y^{2y} (x^2 + y^2 + z^2)\,dz.$

i) $\int_0^1 dy \int_0^2 e^{x+y}\,dx.$

j) $\int_{-1}^1 dx \int_1^y x^3y\,dx.$

k) $\int_1^2 dy \int_1^2 \frac{1}{xy}\, dx.$ l) $\int_{-1}^{1} dx \int_{x^2}^{x} y^2\, dy.$

2. Evaluate the double integral:

a) $\int_R (x^2 + y^2)\, dV,\ R = \{(x, y) \mid y^2 \le x \le 2, 0 \le y \le 2\}.$

b) $\int_R \frac{x}{\sqrt{1 - y^2}}\, dV,\ R = \{(x, y) \mid \sqrt{y} \le x \le \frac{1}{2}, 0 \le y \le \frac{1}{2}\}.$

c) $\int_R x^3 y\, dV,\ R = \{(x, y) \mid x^2 + y^2 \le 1\}.$

d) $\int_R (x^2 - y^2)\, dV,\ R = \{(x, y) \mid 0 \le x \le 1, x \le y \le x^2\}.$

e) $\int_R x^2 y^2\, dV,\ R = \{(x, y) \mid y \le x \le 1, 0 \le y \le 1\}.$

f) $\int_R e^x\, dV,\ R = \{(x, y) \mid 1 \le x \le 2y, 0 \le y \le 1\}.$

g) $\int_R y \sin x\, dV,\ R = \{(x, y) \mid y^2 \le x \le 2y^2, 0 \le y \le \sqrt{\pi}\}.$

h) $\int_R \frac{y}{\sqrt{1 - x^2}}\, dV,\ R = \{(x, y) \mid 0 \le x \le \frac{1}{2}, -1 \le y \le 1\}.$

3. Find the volume of the subsets of 3-dimensional space bounded by
a) $y = 0, z = 0, y = 2x, x = 1, x + 2y - z = 1.$
b) $y = 0, z = 0, y = x, x + y = 1, x + y = 1, x + y + z = 2.$

9.4 APPLICATIONS

1. Curve fitting. The number of points required to determine a curve which passes through given points is equal to the number of constants in the equation of the curve:

A straight line $y = ax + b$ is determined by two points.

A parabola $y = ax^2 + bx + c$, is determined by three points, etc.

In particular, given any three points we may find exactly one parabola passing through them.

Example. Consider the demand law

$$y = \frac{4}{x + 1} - 1 \qquad (3 \ge x \ge 0).$$

This is, of course, not the equation of a parabola. However, (0, 3), (1, 1), and (3, 0) lie on this curve and we can find a parabola on which they also lie. Each point (x, y) on a parabola must satisfy

$$y = ax^2 + bx + c,$$

so we have

$$3 = c$$

$$1 = a + b + c$$

$$0 = 9a + 3b + c.$$

From these equations we obtain $a = \frac{1}{2}$, $b = -5/2$, $c = 3$. Therefore the equation of the desired parabola is

$$y = \frac{x^2}{2} - \frac{5x}{2} + 3$$

This curve may fairly accurately approximate the original demand curve near the three points they have in common (see Fig. 9.13), but it is far from accurate away from them—in particular it is negative between $x = 2$ and $x = 3$ and the demand function is never negative.

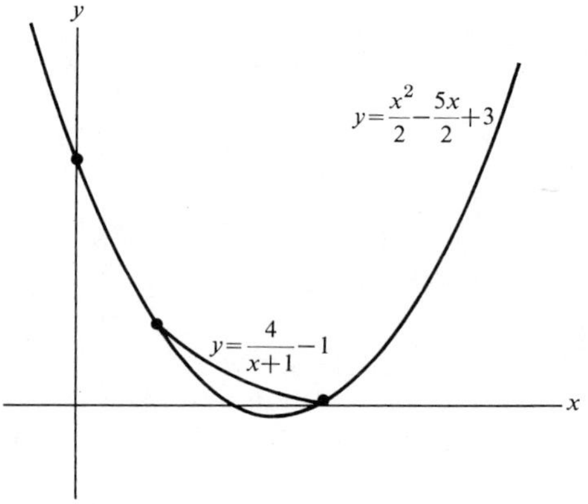

Fig. 9.13

What if we want to approximate a curve through more points than there are constants in the equation of the curve? For example, suppose we want a straight line "approximately" through three points, or a parabola "approximately" through four points. It very likely will be unsatisfactory just to pick a subset consisting of the number of points needed to determine the curve and to proceed as above.

There are several better ways to carry out the construction which approximates given data, a process known as "curve-fitting." We use the method of *least squares.* If we have a set of n numbers $\{x_1, \ldots, x_n\}$ to be replaced by a single number $\bar{x}$, we first form the differences

$$\bar{x} - x_i, \qquad i = 1, \ldots, n.$$

Let

$$S = \sum_{i=1}^{n} (\bar{x} - x_i)^2$$

$$= n\bar{x}^2 - \left(2 \sum_{i=1}^{n} x_i\right) \bar{x} + \sum_{i=1}^{n} x_i^2.$$

We want to minimize S. S is a function of the single variable $\bar{x}$, so we find

$$\frac{dS}{d\bar{x}} = 2n\bar{x} - 2 \sum_{i=1}^{n} x_i = 0; \qquad \frac{d^2S}{d\bar{x}^2} = 2n > 0,$$

so we have a minimum. Thus

$$\bar{x} = \frac{\sum_{i=1}^{n} x_i}{n}$$

is the desired value of $\bar{x}$.

We see that the arithmetic average of the x_i minimizes S. We say the $\bar{x}$ is the *best fit to the data in the sense of least squares.*

Let us now see how this is used in curve fitting. We have a set of data $\{(x_1, y_1), \ldots, (x_n, y_n)\}$; for example, the y_i might represent the observed temperatures at the times x_i. We want to find the straight line $y = mx + b$ which best (in the sense of least squares) approximates this data. Such a line is called a *line of regression.*

For a given x_i we have an estimate of y from the equation of the line:

$$y = mx_i + b.$$

Then we seek values for b and m so as to minimize

$$S = \sum_{i=1}^{n} (b + mx_i - y_i)^2.$$

Necessary conditions are

$$\frac{\partial S}{\partial b} = 0 = \frac{\partial S}{\partial m},$$

which lead to

$$2 \sum_{i=1}^{n} (b + mx_i - y_i) = 0$$

and

$$2 \sum_{i=1}^{n} (b + mx_i - y_i)x_i = 0.$$

We can rewrite these as

$$\left(\sum_{i=1}^{n} x_i\right) m + nb = \sum_{i=1}^{n} y_i \tag{9.2}$$

and

$$\left(\sum_{i=1}^{n} x_i^2\right) m + \left(\sum_{i=1}^{n} x_i\right) b = \sum_{i=1}^{n} x_i y_i. \tag{9.3}$$

No ambiguity arises since the critical value obtained from these conditions can only be a minimum.

Example. Find the straight line which is the best least squares fit to the data

x	y
1	2
2	3
3	4
4	5
5	7
6	9
7	12

(See Fig. 9.14.)

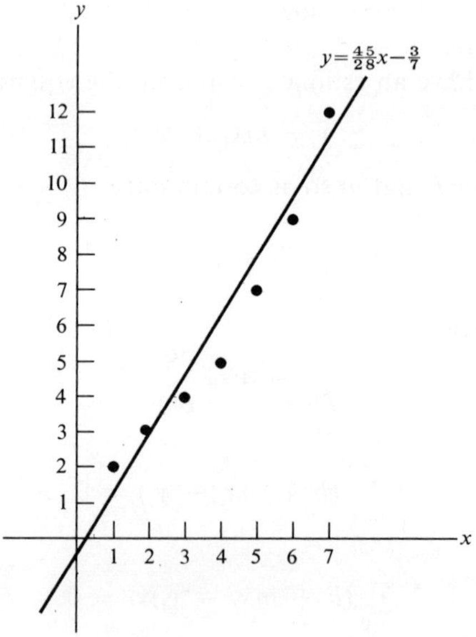

Fig. 9.14

Solution. Substituting the values from the table in (9.2) and (9.3), we obtain

$$28m + 7b = 42$$

$$140m + 28b = 213.$$

The solution to this system is $m = \frac{45}{28}$, $b = -3/7$. Hence the desired line is

$$y = \tfrac{45}{18}x - \tfrac{3}{7}.$$

If we are given a set of data which does not appear to have a linear trend, we might try to fit it to a parabola:

$$y = ax^2 + bx + c.$$

The method is precisely the same, but it is the function

$$S = \sum_{i=1}^{n} (c + bx_i + ax_i^2 - y_i)^2$$

which we want to minimize.

The necessary conditions are

$$\frac{\partial S}{\partial c} = \frac{\partial S}{\partial b} = \frac{\partial S}{\partial a} = 0,$$

which lead to

$$\left(\sum_{i=1}^{n} x_i^2\right) a + \left(\sum_{i=1}^{n} x_i\right) b + nc = \sum_{i=1}^{n} y_i, \tag{9.4}$$

$$\left(\sum_{i=1}^{n} x_i^3\right) a + \left(\sum_{i=1}^{n} x_i^2\right) b + \left(\sum_{i=1}^{n} x_i\right) c = \sum_{i=1}^{n} x_i y_i, \tag{9.5}$$

and

$$\left(\sum_{i=1}^{n} x_i^4\right) a + \left(\sum_{i=1}^{n} x_i^3\right) b + \left(\sum_{i=1}^{n} x_i^2\right) c = \sum_{i=1}^{n} x_i^2 y_i. \tag{9.6}$$

Example. Fit a parabola of the form $y = ax^2 + bx + c$ to the data

x	y
0	1
1	2
2	3
3	5
4	8

(See Fig. 9.15.)

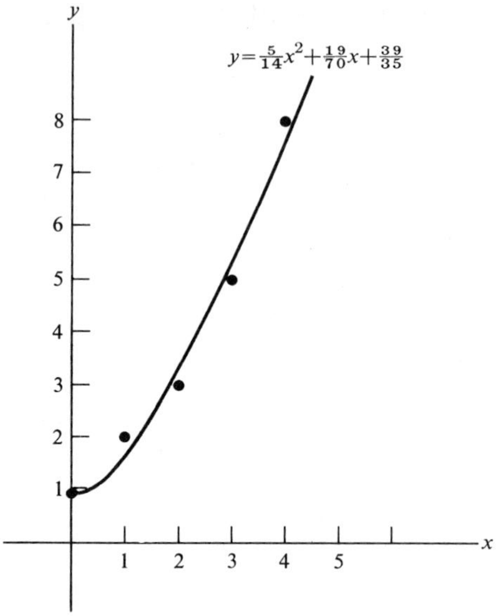

Fig. 9.15

Solution. From (4), (5), and (6) we have

$$30a + 10b + 5c = 19$$
$$100a + 30b + 10c = 55$$
$$354a + 100b + 30c = 187.$$

Solving simultaneously yields

$$a = \tfrac{5}{14}, \qquad b = \tfrac{19}{70}, \qquad c = \tfrac{39}{35}.$$

The best parabola to which to fit the data is

$$y = \tfrac{5}{14}x^2 + \tfrac{19}{70}x + \tfrac{30}{35}.$$

Other types of curves may also be used for the fitting of data. For example, an exponential curve

$$y = ae^{bx},$$

frequently seems to approximate the data best in such things as population growth (see Chapters 8 and 14). The principle involved is the same: computation of differences, differentiation of the sum of the squares of these differences, and calculation of the constants from the system of equations produced by setting the partial derivatives equal to zero.

2. *Probability and density.* A function p of n variables which is integrable on a set D is called a *probability density* on D if

1) $p(X) \geq 0$, for all $X \in D$;
2) $\int_D p(X)\, dV = 1$.

If E is an experiment with possible outcomes $(x_1, \ldots, x_n)$ distributed according to the density p, then the *probability* that the outcome lies in a set B is defined by

$$P[E \in B] = \int_B p(X)\, dV.$$

Example. If $p(X)$ is defined on

$$D = \{(x, y) \mid x^2 + y^2 \leq 1, \quad x, y \geq 0\},$$

by

$$p(X) = c(2x + xy),$$

find a value for c which makes p a probability density on D.

Solution. We have

$$\begin{aligned}\int_D p(X)\, dV &= c \int_0^1 dy \int_0^{\sqrt{1-y^2}} (2x + xy)\, dx \\ &= c \int_0^1 \left(1 - y^2 + \frac{y - y^3}{2}\right) dy \\ &= c \left[y - \frac{y^3}{3} + \frac{y^2}{4} - \frac{y^4}{8}\right]_0^1 \\ &= c\,\frac{19}{24}.\end{aligned}$$

In order for the integral to equal 1, it is necessary that $c = \frac{24}{19}$.

Now suppose that the outcomes of an experiment E are distributed according to this probability density. Compute the probability that the y-value for E is no more than $\frac{1}{2}$.

Solution. We want to integrate $p(X)$ over

$$B = \{(x, y) \mid x^2 + y^2 \leq 1, \quad x \geq 0, \quad 0 \leq y \leq \tfrac{1}{2}\}.$$

$$\begin{aligned}\int_B p(X)\, dV &= \frac{24}{19} \int_0^{1/2} dy \int_0^{\sqrt{1-y^2}} (2x + xy)\, dx \\ &= \frac{24}{19} \left[y - \frac{y^3}{3} + \frac{y^2}{4} - \frac{y^4}{8}\right]_0^{1/2} \\ &= 0.648 \quad \text{(approximately)}.\end{aligned}$$

3. ***Volume.*** A lawn 200 ft × 20 ft is to be watered with a moving sprinkler. The sprinkler moves down the center of the lawn as indicated in Fig. 9.16. At the same time it sprays two jets of water perpendicular to the direction in which it is moving. The number of gallons per square foot falling on a point P on the lawn which is y ft up or down from the center line is given by

$$\frac{1}{100}\left(1-\frac{x}{200}\right)\left(1-\frac{y^2}{100}\right),$$

where x is the distance the sprinkler has traveled along the center line when it sprays the given point P. Find the total number of gallons of water used to spray the lawn in one transversal by the sprinkler.

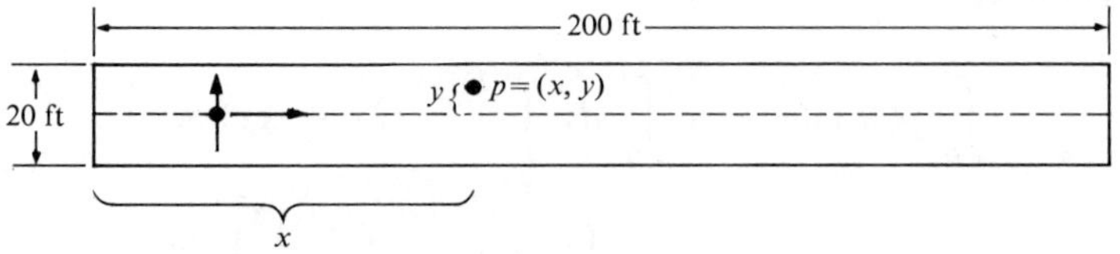

Fig. 9.16

Solution. The total volume of water is given by

$$\begin{aligned}
T &= \frac{1}{100}\int_0^{200} dx \int_{-10}^{10}\left(1-\frac{x}{200}\right)\left(1-\frac{y^2}{100}\right) dy \\
&= \frac{1}{100}\int_0^{200} dx \int_{-10}^{10}\left(1-\frac{x}{200}-\frac{y^2}{100}+\frac{xy^2}{200(100)}\right) dy \\
&= \frac{1}{100}\int_0^{200}\left[y-\frac{xy}{200}-\frac{y^3}{300}+\frac{xy^3}{60000}\right]_{-10}^{10} dx \\
&= \frac{1}{100}\int_0^{200}\left(20-\frac{x}{10}-\frac{20}{3}+\frac{x}{30}\right) dx \\
&= \frac{1}{100}\int_0^{200}\left(\frac{40}{3}-\frac{x}{15}\right) dx \\
&= \frac{1}{100}\left[\frac{40}{3}x-\frac{x^2}{30}\right]_0^{200} \\
&= \frac{40}{3}.
\end{aligned}$$

PROBLEMS

1. Find the straight line which is the best least squares fit to the data.

a)

x	y
0	2
1	4
2	5
3	8
4	9
5	12

b)

x	y
1	3
2	3
3	3.5
4	4
5	4.5
6	5
7	5.5

c)

x	y
0	3
1	2
2	1
3	0.5
4	0
5	−0.5
6	−1

d)

x	y
1	0
2	2
3	4
4	5
5	7
6	8
7	10

2. Fit a parabola $y = ax^2 + bx + c$ to the data.

a)

x	y
0	2
1	1
2	0
3	1
4	2
5	3
6	5

b)

x	y
0	0
1	1
2	3
3	5
4	4
5	3
6	2

c)

x	y
1	0
2	2
3	3
4	4
5	2
6	1

d)

x	y
0	−2
1	−1
2	0
3	1
4	3
5	4

3. Show that

$$p(X) = \tfrac{3}{2}(x^2 + y^2),$$

defines a probability density on

$$D = \{(x, y) \mid 0 \leq x \leq 1, \quad 0 \leq y \leq 1\}.$$

Find the probability that $x \leq y/2$.

4. The *average value* of $f(x_1, \ldots, x_n)$ on S is defined to be

$$\frac{\int_S f(x_1, \ldots, x_n)\, dV}{\int_S dV}.$$

Find the average value of $x^2 - 2xy + y^2$ on $S = \{(x, y) \mid x^2 + y^2 \leq 1, y \geq 0\}$.

5. A gas obeys the law $pv = RT$ (p = pressure, v = volume, T = temperature, R a constant). When $v = 100$ cu in., $p = 50$ lb/in.^{2}, $dv/dt = -5$ cu ft/sec, and $dp/dt = 7.5$ lb/in.^{2}/sec. Find dT/dt.

6. Find the average value of:
a) e^{x+y} on $S = \{(x, y) \mid 0 \leq x \leq 1, 0 \leq y \leq 1\}$.
b) $x^2 - 2y^2 + 4$ on $S = \{(x, y) \mid 0 \leq x \leq 1, x^2 \leq y \leq x\}$.
c) x/y on $S = \{(x, y) \mid 1 \leq x \leq 2, 1 \leq y \leq 2\}$.
d) $y \sin xy$ on $S = \{(x, y) \mid 0 \leq x \leq 1, 0 \leq y \leq \pi/2\}$.

10 / LINEAR EQUATIONS AND MATRICES

In Chapter 2 we discussed systems of two or three linear equations in two or three variables, such as:

$$\begin{aligned} x + y + 2z &= 3 \\ 2x - y + 2z &= 0 \\ -x + 2y - z &= 1 \end{aligned} \tag{10.1}$$

We now want to consider systems of any finite number of linear equations in any finite number of variables and to develop an efficient method of solution.

10.1 LINEAR EQUATIONS

A system of *m linear equations in n variables* (an $m \times n$ system) is an array of the form

$$\begin{aligned} a_{11}x_1 + a_{12}x_2 + \cdots + a_{1n}x_n &= k_1 \\ a_{21}x_1 + a_{22}x_2 + \cdots + a_{2n}x_n &= k_2 \\ \vdots \qquad & \\ a_{m1}x_1 + a_{m2}x_2 + \cdots + a_{mn}x_n &= k_m \end{aligned} , \tag{10.2}$$

where the a_{ij} and the k_i are real numbers and the x_j are variables.

Examples

1. (10.1) is a system of three linear equations in three variables.
2. $$\begin{aligned} x + z &= 2 \\ y + w &= 0, \end{aligned}$$

   is a system of two linear equations in four variables.
3. $$\begin{aligned} x^2 + y^3 &= 5 \\ x - 4y &= 0, \end{aligned}$$

   is not a system of linear equations since the first equation has powers of the variables greater than one.

4. Suppose three liters of a solution which is 40 percent alcohol are required. All that are available are two solutions, one of which is 10 percent alcohol and the other 90 percent. How much of each should be used to obtain the desired solution?

Solution. We set up a system of two linear equations in two variables. Let

$$x = \text{amount of 10 percent solution}$$

$$y = \text{amount of 90 percent solution.}$$

The amount of solution required is three liters and the amount of alcohol required is

$$0.40(3) = 1.2 \text{ liters.}$$

We have

$$\begin{aligned} x + y &= 3 \\ 0.1x + 0.9y &= 1.2, \end{aligned}$$

a system of two equations in two variables, from which $x = \frac{15}{8}$, $y = \frac{9}{8}$.

If in the system (10.2), $k_i = 0$, $i = 1, \ldots, n$, then the system is said to be *homogeneous*. If some $k_j \neq 0$, the system is *nonhomogeneous*. An ordered n-tuple of real numbers $(b_1, \ldots, b_n)$ is said to be a *solution* of (10.2) if

$$\begin{aligned} a_{11}b_1 + a_{12}b_2 + \cdots + a_{1n}b_n &= k_1 \\ a_{21}b_1 + a_{22}b_2 + \cdots + a_{2n}b_n &= k_2 \\ &\vdots \\ a_{m1}b_1 + a_{m2}b_2 + \cdots + a_{mn}b_n &= k_m. \end{aligned}$$

Our problem is to find such n-tuples, if any exist. The collection of all solutions is called the *solution set of the system*.

We recall that for two or three equations the usual method of solution is that of elimination of variables. For example, in Eq. (10.1),

1) $$x + y + 2z = 3$$

2) $$2x - y + 2z = 0$$

3) $$-x + 2y - z = 1$$

we add (1) and (3) to obtain

3′) $$3y + z = 4.$$

Then we multiply (1) by 2 to obtain

1′) $$2x + 2y + 4z = 6,$$

from which we subtract (2) to obtain

1″) $$3y + 2z = 6.$$

Now (3′) and (1″) comprise a system of two equations in two unknowns. We subtract (1″) from (3′) to get a "system" of one equation:

$$3'') \qquad -z = -2,$$

from which we see that $z = 2$.

Then we substitute this value for z in (3′) or (1″) to obtain a value for y, $y = \frac{2}{3}$. Finally we substitute for y and z in one of the original equations to find a value for x, $x = -5/3$. Thus the solution is $(-5/3, \frac{2}{3}, 2)$.

If the equations of a system of linear equations are obtained from the equations of another system by multiplying the original equations by constants and adding them to each other, the resulting equations are said to be *linear combinations* of the original equations. The equations of the systems (1′), (2), (3); (1″), (2), (3); (1″), (2), (3′), and (1″), (2), and (3″) above are linear combinations of the equations of (10.1).

The success of the elimination method described above depends upon the fact that a system of linear combinations of equations of the original system has the same solution set as the original equations. Systems having the same solution sets are said to be *equivalent*. We should like to formulate the elimination method so as to generalize it to an $m \times n$ system and to simplify the computations involved.

Warning. Not every system of linear equations has a solution and indeed not all systems having a solution have a unique solution. Determining the existence and nature of the solution will be discussed in the following section.

PROBLEMS

1. Solve the following systems of linear equations:

a) $\begin{aligned} 3x - 4y &= 1 \\ 2x + 3y &= -2. \end{aligned}$

b) $\begin{aligned} -4x + y &= 0 \\ 7x - 2y &= 0. \end{aligned}$

c) $\begin{aligned} x + y + 2z &= 2 \\ 2x - 2y + z &= 0 \\ 2x - z &= 1. \end{aligned}$

d) $\begin{aligned} 2x + y &= 0 \\ y + 4z &= 1 \\ w + z &= 2 \\ 2x - 2z &= 0. \end{aligned}$

e) $\begin{aligned} 7x + 7y + 7z &= 7 \\ -x - y - z &= -1 \\ 2x + 2y + 2z &= 2. \end{aligned}$

f) $\begin{aligned} 3x - 4y + z &= 2 \\ -x + y - z &= 3 \\ 6x - 8y + 2z &= -1. \end{aligned}$

g) $\begin{aligned} 2x + 3y &= 0 \\ x - y &= 1. \end{aligned}$

h) $\begin{aligned} -x + y - z &= 1 \\ 2x + 4z &= 0. \\ 3x + y + z &= -2. \end{aligned}$

i) $\begin{aligned} z + w &= 5 \\ 2z - 5w &= -2. \end{aligned}$

j) $\begin{aligned} 2x - 2y + z &= 2 \\ 3x - y + z &= 4 \\ -x + 2y - z &= 1. \end{aligned}$

2. If it is required to produce one liter of 50 percent sulfuric acid solution from a 20 percent solution and a 95 percent solution, how much of each should be used?
3. Three types of trucks are available for rental. Obsolescent Objects has three types of products to transport. The load chart is:

Truck	Product 1	2	3
A	2	1	1
B	1	2	1
C	0	2	1

   How many trucks should be rented to transport exactly 11 of product 1, 9 of product 2, and 7 of product 3? (Assume each truck fully loaded.)
4. Find the intersection of the demand curve $y = 15 - 3p$ and the supply curve $y = -6 + 5p$. (The intersection is the market price under pure competition.)
5. Find the intersection of the demand curve $y = 20 - 2p$ and the supply curve $y = -10 + 6p$.
6. If it is required to produce one liter of 40 percent alcohol solution from a 10 percent solution and a 90 percent solution, how much of each should be used?

10.2 MATRICES AND GAUSS ELIMINATION

The problem of finding the solution set of a system of linear equations is considerably simplified from a notational point of view by the observation that the variables serve no useful purpose except as placeholders. Thus one could write the system (10.2) of the last section as

$$\begin{pmatrix} a_{11} & a_{12} & \cdots & a_{1n} \\ a_{21} & a_{22} & \cdots & a_{2n} \\ \vdots & & & \\ a_{m1} & a_{m2} & \cdots & a_{mn} \end{pmatrix} \begin{pmatrix} x_1 \\ x_2 \\ \vdots \\ x_n \end{pmatrix} = \begin{pmatrix} k_1 \\ k_2 \\ \vdots \\ k_m \end{pmatrix}. \tag{10.3}$$

The array

$$\begin{pmatrix} a_{11} & a_{12} & \cdots & a_{1n} \\ a_{21} & a_{22} & \cdots & a_{2n} \\ \vdots & & & \\ a_{m1} & & \cdots & a_{mn} \end{pmatrix}$$

is denoted by (a_{ij}) or by A, and called the *coefficient matrix* of the system (10.3). In general an array which consists of entries of real numbers in horizontal *rows* and vertical *columns* is called a *matrix*. An $m \times n$ *matrix* has m rows and n columns. In (10.3) A is an $m \times n$ matrix, $X = (x_j)$ is an $n \times 1$ matrix and $K = (k_i)$ is an $m \times 1$ matrix. $m \times n$ is the *dimension* of the matrix; if the matrix is square ($m = n$) it is said to be of dimension n. We can write (10.3) as

$$AX = K.$$

We form an augmented matrix for this equation by adding K as the $(n + 1)$st column of the matrix A:

$$\left(\begin{array}{cccc|c} a_{11} & a_{12} & \cdots & a_{1n} & k_1 \\ a_{21} & a_{22} & \cdots & a_{2n} & k_2 \\ \vdots & & & & \\ a_{m1} & a_{m2} & \cdots & a_{mn} & k_m \end{array}\right).$$

We write this matrix as $(A \mid K)$. We then perform the operations such as we did in the elimination process above to the augmented matrix to obtain a matrix representing a system whose solution is evident. Let us apply this method to (10.1). The augmented matrix of (10.1) is

$$\left(\begin{array}{ccc|c} 1 & 1 & 2 & 3 \\ 2 & -1 & 2 & 0 \\ -1 & 2 & -1 & 1 \end{array}\right).$$

The operations produce the following:

$$\left(\begin{array}{ccc|c} 1 & 1 & 2 & 3 \\ 2 & -1 & 2 & 0 \\ 0 & 3 & 1 & 4 \end{array}\right) \qquad \text{Replace third row by third row plus first row.}$$

$$\left(\begin{array}{ccc|c} 2 & 2 & 4 & 6 \\ 2 & -1 & 2 & 0 \\ 0 & 3 & 1 & 4 \end{array}\right) \qquad \text{Multiply first row by 2.}$$

$$\left(\begin{array}{ccc|c} 0 & 3 & 2 & 6 \\ 2 & -1 & 2 & 0 \\ 0 & 3 & 1 & 4 \end{array}\right) \qquad \text{Replace first row by first row minus the second row.}$$

$$\left(\begin{array}{ccc|c} 0 & 3 & 2 & 6 \\ 2 & -1 & 2 & 0 \\ 0 & 0 & -1 & -2 \end{array}\right) \qquad \text{Replace third row by third row minus first row.}$$

where the last matrix is the augmented matrix for a system including $-z = -2$, which we solve for z. Then we substitute the value for z to solve for x and y.

The operations we may perform, corresponding to the linear combinations of equations, are called *elementary row operations*:

1) multiplication of a row of a matrix by a constant $c \neq 0$;

2) replacement of the ith row by the ith row plus a constant $c \neq 0$ times the jth row, $i \neq j$;

3) interchange of two rows.

A matrix B obtained from a matrix A by a finite number of elementary row operations is said to be *row-equivalent* to A. We state without proof a basic result concerning these operations.

Theorem 1. If $(A \mid K)$ and $(B \mid Y)$ are row-equivalent augmented matrices, then the systems of linear equations

$$AX = K \quad \text{and} \quad BX = Y,$$

have the same solution set.

We must now develop a systematic method for reducing a given augmented matrix to something manageable. The method we choose is called the *Gauss elimination* method; it is by no means the only method for solving a system of linear equations (see Appendix C for another), but it has the advantage of being efficiently adaptable to computer solutions.

Some definitions are in order. The *determinant* of a matrix is a real-number associated with each $n \times n$ matrix as follows:

If the ith row and jth column of an $n \times n$ $(n \geq 2)$ matrix are deleted, the resulting submatrix is denoted by M_{ij}. The definition of determinant proceeds by induction.

If $n = 1$, the determinant of an $n \times n$ matrix A is its single entry.

If $n \geq 2$, the determinant of A is given by

$$\det A = a_{i1}A_{i1} + a_{i2}A_{i2} + \cdots + a_{in}A_{in},$$

where $A_{ij} = (-1)^{i+j} \det M_{ij}$. Note that if $n = 2$,

$$\det \begin{pmatrix} a & b \\ c & d \end{pmatrix} = ad - bc.$$

We state the following theorem without proof.

Theorem 2. If A is an $n \times n$ matrix, its determinant does not depend on the choice of the row i and

$$\det A = a_{1j}A_{1j} + \cdots + a_{nj}A_{nj},$$

independent of the choice of j.

Theorem 2 allows us to obtain the determinant of a matrix by expanding on any row or column.

Example. Find the determinant of

$$A = \begin{pmatrix} 1 & 2 & -3 & 1 \\ 0 & -1 & 4 & 0 \\ 5 & -1 & 2 & 3 \\ -3 & 1 & 2 & 0 \end{pmatrix}.$$

Solution. We choose to use the second row of A for our computations because of the presence of the two zeros. We have

$$\det A = 0(-1)\det\begin{pmatrix} 2 & -3 & 1 \\ -1 & 2 & 3 \\ 1 & 2 & 0 \end{pmatrix} + (-1)(1)\det\begin{pmatrix} 1 & -3 & 1 \\ 5 & 2 & 3 \\ -3 & 2 & 0 \end{pmatrix}$$

$$+ 4(-1)\det\begin{pmatrix} 1 & 2 & 1 \\ 5 & -1 & 3 \\ -3 & 1 & 0 \end{pmatrix} + 0(1)\det\begin{pmatrix} 1 & 2 & -3 \\ 5 & -1 & 2 \\ -3 & 1 & 2 \end{pmatrix}.$$

We again pick rows with zeros, the third in each case, to get

$$\det A = (-1)\left[(-3)\det\begin{pmatrix} -3 & 1 \\ 2 & 3 \end{pmatrix} + (-1)(2)\det\begin{pmatrix} 1 & 1 \\ 5 & 3 \end{pmatrix} + (1)(0)\det\begin{pmatrix} 1 & -3 \\ 5 & 2 \end{pmatrix}\right]$$

$$+ (-4)\left[(1)(-3)\det\begin{pmatrix} 2 & 1 \\ -1 & 3 \end{pmatrix} + (-1)(1)\det\begin{pmatrix} 1 & 1 \\ 5 & 3 \end{pmatrix} + (1)(0)\det\begin{pmatrix} 1 & 2 \\ 5 & -1 \end{pmatrix}\right]$$

$$= 3(-9 - 2) + 2(3 - 5) + 12(6 + 1) + 4(3 - 5)$$

$$= -33 - 4 + 84 - 8 = 39.$$

The *rank* of a matrix is the dimension of the largest submatrix whose determinant is nonzero.

Examples

1. The rank of the matrix in the example above is 4.
2. Find the rank of

$$\begin{pmatrix} 1 & 2 & 3 & 4 \\ 0 & 2 & 1 & -1 \\ 6 & 1 & 2 & 3 \end{pmatrix}.$$

Solution. The maximum possible value of the rank is 3 as that is the largest square submatrix it contains.

We compute

$$\det\begin{pmatrix} 1 & 2 & 3 \\ 0 & 2 & 1 \\ 6 & 1 & 2 \end{pmatrix} = 2\det\begin{pmatrix} 1 & 3 \\ 6 & 2 \end{pmatrix} - \det\begin{pmatrix} 1 & 2 \\ 6 & 1 \end{pmatrix}$$

$$= 2(2 - 18) - (1 - 12) = -21 \neq 0.$$

Thus the rank is 3.

3. Find the rank of

$$\begin{pmatrix} -1 & 2 & 3 \\ 0 & 1 & 1 \\ 1 & -2 & -3 \end{pmatrix}.$$

Solution. The determinant of the matrix is given by

$$\det\begin{pmatrix} -1 & 3 \\ 1 & -3 \end{pmatrix} - \det\begin{pmatrix} -1 & 2 \\ 1 & -2 \end{pmatrix} = 0.$$

The rank cannot be 3; however the determinant of

$$\begin{pmatrix} -1 & 2 \\ 0 & 1 \end{pmatrix}$$

is not zero, so the rank of the original matrix is 2.

A square matrix is said to be *triangular* if all the entries below the main diagonal are zero. Note that the determinant of a triangular matrix is just the product of the diagonal elements. We state without proof

Theorem 3. Any $n \times n$ matrix is row-equivalent to a triangular matrix.

Example. Find a triangular matrix row-equivalent to

$$\begin{pmatrix} 4 & 3 & 1 & 0 \\ 2 & 1 & 5 & 1 \\ 2 & 0 & 1 & 1 \\ -3 & -1 & -4 & 1 \end{pmatrix}.$$

Solution. The idea is to get zeros in the lower left-hand corner.

$$\begin{pmatrix} 4 & 3 & 1 & 0 \\ 2 & 1 & 5 & 1 \\ 2 & 0 & 1 & 1 \\ -3 & -1 & -4 & 1 \end{pmatrix} \xrightarrow[\text{from row 2}]{\text{subract row 3}} \begin{pmatrix} 4 & 3 & 1 & 0 \\ 0 & 1 & 4 & 0 \\ 2 & 0 & 1 & 1 \\ -3 & -1 & 4 & 1 \end{pmatrix} \xrightarrow[\text{to row 4}]{\text{add } \frac{3}{2} \text{ of row 3}}$$

$$\begin{pmatrix} 4 & 3 & 1 & 0 \\ 0 & 1 & 4 & 0 \\ 2 & 0 & 1 & 1 \\ 0 & -1 & \frac{11}{2} & \frac{5}{2} \end{pmatrix} \xrightarrow[\text{to row 4}]{\text{add row 2}} \begin{pmatrix} 4 & 3 & 1 & 0 \\ 0 & 1 & 4 & 0 \\ 2 & 0 & 1 & 1 \\ 0 & 0 & \frac{19}{2} & \frac{5}{2} \end{pmatrix} \xrightarrow[\text{of row 1 from row 3}]{\text{subtract } \frac{1}{2}}$$

$$\begin{pmatrix} 4 & 3 & 1 & 0 \\ 0 & 1 & 4 & 0 \\ 0 & -\frac{3}{2} & \frac{1}{2} & 1 \\ 0 & 0 & \frac{19}{2} & \frac{5}{2} \end{pmatrix} \xrightarrow[\text{to row 3}]{\text{add } \frac{3}{2} \text{ of row 2}} \begin{pmatrix} 4 & 3 & 1 & 0 \\ 0 & 1 & 4 & 0 \\ 0 & 0 & \frac{13}{2} & 1 \\ 0 & 0 & \frac{19}{2} & \frac{5}{2} \end{pmatrix} \xrightarrow[\text{of row 3 from row 4}]{\text{subtract 19/13}}$$

$$\begin{pmatrix} 4 & 3 & 1 & 0 \\ 0 & 1 & 4 & 0 \\ 0 & 0 & \frac{13}{2} & 1 \\ 0 & 0 & 0 & \frac{27}{26} \end{pmatrix}.$$

We now introduce a method of solving a system of linear equations by finding a triangular matrix which is row equivalent to the coefficient matrix and then solving the system whose coefficient matrix is the triangular matrix.

The *Gauss elimination process* for solving a system of linear equations is as follows:

1) Write down the augmented matrix of the system.

2) Transform this matrix by row operations into a matrix which has a triangular submatrix in the first n rows and n columns, n = rank of the coefficient matrix, and has zeros in the remaining rows of the coefficient matrix. To facilitate the transformation, we allow columns to be interchanged, provided the variables are also interchanged, i.e., if column i and column j are interchanged, the new column i still represents the coefficients of x_j.

3) If the rows which have all zero entries from the coefficient matrix have nonzero entries in the augmentation column, those rows represent equations of the form

$$0x_1 + \cdots + 0x_n = c \neq 0.$$

Such equations do not have a solution; hence the original system fails to have a solution and is said to be *inconsistent.*

4) If the rows which have all zero entries from the coefficient matrix have zero entries in the augmentation column, the system has a solution and is said to be *consistent.* We then proceed to solve the simplified system represented by the transformed matrix.

The proof of the following theorem is contained in Appendix C.

Theorem 4. The system

$$\begin{aligned} a_{11}x_1 + a_{12}x_2 + \cdots + a_{1n}x_n &= k_1 \\ a_{21}x_1 + a_{22}x_2 + \cdots + a_{2n}x_n &= k_2 \\ &\vdots \\ a_{m1}x_1 + a_{m2}x_2 + \cdots + a_{mn}x_n &= k_m \end{aligned}$$

has a solution if and only if the rank of the coefficient matrix $A = (a_{ij})$ is equal to the rank of the augmented matrix $(A \mid K)$.

Corollary. If $m = n$, there is a unique solution if and only if rank $A = n$. In particular, if the system is homogeneous, the system has a nontrivial solution if and only if rank $A < n$.

Examples

1. Solve

$$\begin{aligned} x + y &= 1 \\ 2x - y &= 0 \\ x - y &= 3. \end{aligned}$$

Solution. We have the coefficient matrix

$$\begin{pmatrix} 1 & 2 \\ 2 & -1 \\ 1 & -1 \end{pmatrix}$$

which is of rank 2. On the other hand, the rank of

$$\left(\begin{array}{cc|c} 1 & 2 & 1 \\ 2 & -1 & 0 \\ 1 & -1 & 3 \end{array}\right)$$

is 3, so the system is inconsistent.

2. Solve

$$\begin{aligned} x + y + 3z &= 1 \\ 2x - y + z &= 0 \\ x - y + 3z &= 3. \end{aligned}$$

Solution. We reduce to a row-equivalent triangular matrix

$$\left(\begin{array}{ccc|c} 1 & 1 & 3 & 1 \\ 2 & -1 & 1 & 0 \\ 1 & -1 & 3 & 3 \end{array}\right) \to \left(\begin{array}{ccc|c} 1 & 1 & 3 & 1 \\ 2 & -1 & 1 & 0 \\ 0 & -2 & 0 & 2 \end{array}\right)$$

$$\to \left(\begin{array}{ccc|c} 1 & 3 & 1 & 1 \\ 2 & 1 & -1 & 0 \\ 0 & 0 & -2 & 2 \end{array}\right) \to \left(\begin{array}{ccc|c} 1 & 3 & 1 & 1 \\ 0 & -5 & -3 & -2 \\ 0 & 0 & -2 & 2 \end{array}\right)$$

by the operations

1) replace row 3 by row 3 − row 1,
2) interchange columns 2 and 3, (this interchanges the roles of y and z),
3) replace row 2 by row 2 − twice row 1.

The system is consistent with solution $y = -1$ (obtained immediately from the equation represented by the third row), $x = -1$, and $z = 1$ (obtained by substituting the value -1 for y in the other equations).

3. Solve

$$\begin{aligned} x + y - z + w &= 2 \\ 3x - y + z &= 0 \\ 2x - 2y + 2z - w &= -2 \\ -x + 2y + 3w &= 1. \end{aligned}$$

Solution. We have

$$\left(\begin{array}{rrrr|r} 1 & 1 & -1 & 1 & 2 \\ 3 & -1 & 1 & 0 & 0 \\ 2 & -2 & 2 & -1 & -2 \\ -1 & 2 & 0 & 3 & 1 \end{array}\right) \to \left(\begin{array}{rrrr|r} 1 & 1 & -1 & 1 & 2 \\ 3 & -1 & 1 & 0 & 0 \\ 2 & -2 & 2 & -1 & -2 \\ 0 & 3 & -1 & 4 & 3 \end{array}\right)$$

$$\to \left(\begin{array}{rrrr|r} 1 & 1 & -1 & 1 & 2 \\ 3 & -1 & 1 & 0 & 0 \\ 0 & -4 & 4 & -3 & -6 \\ 0 & 3 & -1 & 4 & 3 \end{array}\right) \to \left(\begin{array}{rrrr|r} 1 & 1 & -1 & 1 & 2 \\ 3 & -1 & 1 & 0 & 0 \\ 0 & -4 & 4 & -3 & -6 \\ 0 & 0 & 2 & \frac{7}{4} & -\frac{3}{2} \end{array}\right)$$

$$\to \left(\begin{array}{rrrr|r} 1 & 1 & -1 & 1 & 2 \\ 0 & -4 & 4 & -3 & -6 \\ 0 & -4 & 4 & -3 & -6 \\ 0 & 0 & 2 & \frac{7}{4} & -\frac{3}{2} \end{array}\right) \to \left(\begin{array}{rrrr|r} 1 & 1 & -1 & 1 & 2 \\ 0 & -4 & 4 & -3 & -6 \\ 0 & 0 & 0 & 0 & 0 \\ 0 & 0 & 2 & \frac{7}{4} & -\frac{3}{2} \end{array}\right)$$

$$\to \left(\begin{array}{rrrr|r} 1 & 1 & -1 & 1 & 2 \\ 0 & -4 & 4 & -3 & -6 \\ 0 & 0 & 2 & \frac{7}{4} & -\frac{3}{2} \\ 0 & 0 & 0 & 0 & 0 \end{array}\right)$$

where the determination of the operations is left to the reader.

The rank of the matrix is 3. From the third row we have that the solution set of the system consists of 4-tuples (x, y, z, w) such that $2z + \frac{7}{4}w = -3/2$ and

$$z = -\tfrac{7}{8}w - \tfrac{3}{4}.$$

From the first and second rows we have that $-4y = -4z + 3w - 6$ so that

$$y = z - \tfrac{3}{4}w + \tfrac{3}{2}$$

and

$$x = -y + z - w + 2.$$

There are an infinite number of 4-tuples satisfying these conditions. We can write the solution set as

$$\left\{\left(-y + z - w + 2,\ z - \frac{3}{4}w + \frac{3}{2},\ \frac{-7}{8}w - \frac{3}{4},\ w\right)\right\} =$$

$$\left\{\left(\frac{-w}{4} + \frac{1}{2},\ \frac{-13}{8}w + \frac{3}{4},\ \frac{-7}{8}w - \frac{3}{4},\ w\right)\right\}.$$

Assignment of a value to w produces a particular solution, for example, $(\frac{1}{2}, \frac{3}{4}, -\frac{3}{4}, 0)$.

Hence the system has a solution, but it is not unique.

4. To produce a certain blend of coffee we need 6 units of variety A, 10 units of variety B, and 8 units of variety C. A can of blend 1 contains 1, 3, and 4 units

respectively of these varieties. A can of blend 2 contains 3, 3 and 3 units respectively and blend 3 contains 2 units of variety A and 5 units of variety B. How much of each blend should be used to produce the desired blend?

Solution. We set up the following system:

Let

$$
\begin{aligned}
x &= \text{units of blend 1}\\
y &= \text{units of blend 2}\\
z &= \text{units of blend 3.}
\end{aligned}
$$

Then

$$
\begin{aligned}
x + y + z &= 24 && \text{(total)}\\
\tfrac{1}{8}x + \tfrac{1}{3}y + \tfrac{2}{7}z &= 6 && \text{(variety } A\text{)}\\
\tfrac{3}{8}x + \tfrac{1}{3}y + \tfrac{5}{7}z &= 10 && \text{(variety } B\text{)}\\
\tfrac{1}{2}x + \tfrac{1}{3}y \phantom{+ \tfrac{5}{7}z} &= 8 && \text{(variety } C\text{).}
\end{aligned}
$$

The augmented matrix is

$$
\left(\begin{array}{ccc|c}
1 & 1 & 1 & 24\\
\frac{1}{8} & \frac{1}{3} & \frac{2}{7} & 6\\
\frac{3}{8} & \frac{1}{3} & \frac{5}{7} & 10\\
\frac{1}{2} & \frac{1}{3} & 0 & 8
\end{array}\right).
$$

We proceed to apply the Gauss elimination process:

$$
\left(\begin{array}{ccc|c}
1 & 1 & 1 & 24\\
\frac{1}{8} & \frac{1}{3} & \frac{2}{7} & 6\\
\frac{3}{8} & \frac{1}{3} & \frac{5}{7} & 10\\
\frac{1}{2} & \frac{1}{3} & 0 & 8
\end{array}\right)
\xrightarrow[\text{columns 1 and 3}]{\text{interchange}}
\left(\begin{array}{ccc|c}
1 & 1 & 1 & 24\\
\frac{2}{7} & \frac{1}{3} & \frac{1}{8} & 6\\
\frac{5}{7} & \frac{1}{3} & \frac{3}{8} & 10\\
0 & \frac{1}{3} & \frac{1}{2} & 8
\end{array}\right)
\xrightarrow[\times\text{ second row}]{\text{third row } - \frac{5}{2}}
$$

$$
\left(\begin{array}{ccc|c}
1 & 1 & 1 & 24\\
\frac{2}{7} & \frac{1}{3} & \frac{1}{8} & 6\\
0 & -\frac{1}{2} & \frac{1}{16} & -5\\
0 & \frac{1}{3} & \frac{1}{2} & 8
\end{array}\right)
\xrightarrow[\times\text{ third row}]{\text{fourth row } - \frac{2}{3}}
\left(\begin{array}{ccc|c}
1 & 1 & 1 & 24\\
\frac{2}{7} & \frac{1}{3} & \frac{1}{8} & 6\\
0 & -\frac{1}{2} & \frac{1}{16} & -5\\
0 & 0 & \frac{13}{24} & \frac{14}{3}
\end{array}\right)
\xrightarrow[\text{row}]{\text{second row } + 2/7 \times \text{ first}}
$$

$$
\left(\begin{array}{ccc|c}
1 & 1 & 1 & 24\\
0 & \frac{1}{21} & -\frac{9}{56} & -\frac{6}{7}\\
0 & -\frac{1}{2} & \frac{1}{16} & -5\\
0 & 0 & \frac{13}{24} & \frac{14}{3}
\end{array}\right)
\xrightarrow[\times\text{ second row}]{\text{third row } + 21/2}
\left(\begin{array}{ccc|c}
1 & 1 & 1 & 24\\
0 & \frac{1}{21} & -\frac{9}{56} & -\frac{6}{7}\\
0 & 0 & -\frac{13}{8} & -14\\
0 & 0 & \frac{13}{24} & \frac{14}{3}
\end{array}\right)
\xrightarrow[\times\text{ third row}]{\text{fourth row } + \frac{1}{3}}
$$

$$
\left(\begin{array}{ccc|c}
1 & 1 & 1 & 24\\
0 & \frac{1}{21} & -\frac{9}{56} & -\frac{6}{7}\\
0 & 0 & -\frac{13}{8} & -14\\
0 & 0 & 0 & 0
\end{array}\right)
$$

Thus the unique solution is given by $x = \frac{112}{13}$, $y = \frac{144}{13}$, $z = \frac{56}{13}$. (Don't forget that columns were interchanged.)

5. Suppose we vary the composition of the available blends in the last problem as follows: blend 1 has 1 unit of A and 3 of B, blend 2 has 3 units of A and 3 of B, and blend 3 contains 2 units of A and 5 of B. We want a blend containing 6 units of A and 10 of B.

Solution. Let x, y, and z be as in Example 4. We have the following system:

$$\begin{aligned} x + y + z &= 16 \\ \tfrac{1}{4}x + \tfrac{1}{2}y + \tfrac{2}{7}z &= 6 \\ \tfrac{3}{4}x + \tfrac{1}{2}y + \tfrac{5}{7}z &= 10. \end{aligned}$$

The augmented matrix is transformed as follows:

$$\left(\begin{array}{ccc|c} 1 & 1 & 1 & 16 \\ \frac{1}{4} & \frac{1}{2} & \frac{2}{7} & 6 \\ \frac{3}{4} & \frac{1}{2} & \frac{5}{7} & 10 \end{array}\right) \to \left(\begin{array}{ccc|c} 1 & 1 & 1 & 16 \\ \frac{1}{4} & \frac{1}{2} & \frac{2}{7} & 6 \\ 0 & -1 & -\frac{1}{7} & -8 \end{array}\right)$$

$$\left(\begin{array}{ccc|c} 1 & 1 & 1 & 16 \\ 0 & \frac{1}{4} & \frac{1}{28} & 2 \\ 0 & -1 & -\frac{1}{7} & -8 \end{array}\right) \to \left(\begin{array}{ccc|c} 1 & 1 & 1 & 16 \\ 0 & \frac{1}{4} & \frac{1}{28} & 2 \\ 0 & 0 & 0 & 0 \end{array}\right)$$

Thus the rank of the matrix is 2. We have as the solution set

$$\left\{\left(16 - y - z,\ 8 - \frac{z}{7},\ z\right)\right\} = \left\{\left(8 - \frac{6}{7}z,\ 8 - \frac{z}{7},\ z\right)\right\}.$$

Assigning a value to z produces corresponding values for x and y. However, we are restricted by the fact that we may add only nonnegative amounts, i.e., $x \geq 0$, $y \geq 0$, $z \geq 0$. Thus we must have

$$8 - \frac{z}{7} \geq 0, \qquad z \leq 56$$

$$8 - \frac{6}{7}z \geq 0, \qquad z \leq \frac{28}{3},$$

that is, we may assign any value between 0 and $\frac{28}{3}$ to z and compute x and y accordingly.

Warning. Check the solutions by substituting the values obtained into the equations of the original system.

PROBLEMS

1. Find the determinant of

a) $\begin{pmatrix} 1 & 2 \\ 0 & 3 \end{pmatrix}$.

b) $\begin{pmatrix} 1 & 2 & 1 \\ 3 & 1 & 0 \\ 4 & 1 & 1 \end{pmatrix}$.

c) $\begin{pmatrix} -1 & 2 & 3 \\ 1 & 0 & 2 \\ -3 & 2 & -1 \end{pmatrix}$.

d) $\begin{pmatrix} 1 & 0 & 5 & 4 \\ 3 & 1 & 5 & 1 \\ 2 & 8 & -1 & 2 \\ 0 & -3 & 5 & 1 \end{pmatrix}$.

e) $\begin{pmatrix} 1 & 0 & 1 & 0 \\ 1 & 1 & 0 & 1 \\ 0 & 1 & 0 & 1 \\ 0 & 1 & 1 & 0 \end{pmatrix}$.

f) $\begin{pmatrix} 1 & 0 & 5 & 6 & 7 \\ 0 & 0 & 4 & 6 & 9 \\ 0 & 0 & 0 & 0 & 0 \\ 7 & 1 & 2 & 1 & -5 \\ 2 & 1 & 5 & 4 & 1 \end{pmatrix}$.

g) $\begin{pmatrix} 2 & 1 & 3 & 4 \\ -2 & 1 & -1 & 0 \\ 3 & 4 & -5 & 1 \\ 0 & 5 & 5 & 1 \end{pmatrix}$.

h) $\begin{pmatrix} 3 & 1 & 2 \\ 0 & 1 & 1 \\ 3 & 5 & 9 \end{pmatrix}$.

i) $\begin{pmatrix} 2 & -1 & 0 \\ 1 & -2 & 1 \\ 3 & 0 & -2 \end{pmatrix}$.

j) $\begin{pmatrix} 4 & 1 & 0 & 1 \\ 2 & 0 & 1 & 3 \\ 4 & -1 & -1 & 3 \\ -1 & 2 & 3 & 1 \end{pmatrix}$.

k) $\begin{pmatrix} -1 & 3 & 2 & 1 \\ 0 & 2 & 1 & 0 \\ 3 & 4 & 1 & -1 \\ 2 & 2 & -1 & 0 \end{pmatrix}$.

2. Find the rank of each of the matrices of Problem 1.

3. Find a triangular matrix which is row-equivalent to each of the matrices of Problem 1.

4. Determine whether each of the following systems of linear equations is consistent or inconsistent. Solve those which are consistent.

a) $\begin{aligned} x - y + 5z &= 2 \\ 2x + 2y + 4z &= 10 \\ x + 4y + 3z &= -2. \end{aligned}$

b) $\begin{aligned} 2x - y + 3z &= 8 \\ 5x + 3y + 2z &= 6 \\ 2x + 2y + 6z &= 11. \end{aligned}$

c) $\begin{aligned} x + y + z &= 2 \\ x - y - z &= 0 \\ x + w &= -1 \\ y + w + z &= 1. \end{aligned}$

d) $\begin{aligned} x + 4y + z + 2w &= 4 \\ 3x + 2y + 2z + w &= 5 \\ 2x + y - 4z + 3w &= 1 \\ 4x - y + z - w &= -2. \end{aligned}$

e) $x - w = 0$
$y + z = 2.$

f) $2x + y - z = 0$
$x - y + 6z = 5.$

g) $x - y + z = 1$
$2x + y = 4$
$-x + 3y - z = 1.$

h) $2x - 2y + z = 2$
$x + y - 2z = 1$
$4x + - 3z = 0.$

i) $x + w = 1$
$x - y = 0$
$x + z = 4.$

j) $2x + 2y - w = 1$
$x - y - z = 0$
$3x + 4y + z = 5$
$2x - y - 2w = 4.$

5. Belch Baby Food wants to produce a cereal which has 600 units of protein, 20 units of carbohydrates, 45 units of iron and 195 units of calcium. It has available for blending the ingredients whose contents are described in the table below. How much of each should be used?

Ingredient	Protein	Carbohydrates	Iron	Calcium
A	100	4	6	20
B	50	5	9	40
C	150	0	4	20
D	0	10	35	175

10.3 PROPERTIES OF MATRICES

What can we do with matrices other than use them to solve systems of linear equations?

1. We can add them. The *sum* of two $m \times n$ matrices is defined by

$$\begin{pmatrix} a_{11} & a_{12} & \cdots & a_{1n} \\ a_{21} & & \cdots & a_{2n} \\ \vdots & & & \\ a_{m1} & & \cdots & a_{mn} \end{pmatrix} + \begin{pmatrix} b_{11} & b_{12} & \cdots & b_{1n} \\ b_{21} & & \cdots & b_{2n} \\ \vdots & & & \\ b_{m1} & & \cdots & b_{mn} \end{pmatrix}$$

$$= \begin{pmatrix} a_{11} + b_{11} & a_{12} + b_{12} & \cdots & a_{1n} + b_{1n} \\ a_{21} + b_{21} & & \cdots & a_{2n} + b_{2n} \\ \vdots & & & \\ a_{n1} + b_{n1} & & \cdots & a_{mn} + b_{mn} \end{pmatrix}$$

or $(a_{ij}) + (b_{ij}) = (a_{ij} + b_{ij})$ in condensed notation. Observe that only matrices with the same dimension can be added.

Example

$$\begin{pmatrix} 1 & 2 & 3 \\ 0 & -1 & 4 \end{pmatrix} + \begin{pmatrix} -1 & 0 & 5 \\ 1 & 2 & -3 \end{pmatrix} = \begin{pmatrix} 0 & 2 & 8 \\ 1 & 1 & 1 \end{pmatrix}.$$

This addition, being defined straightforwardly from addition of real numbers, retains the properties of commutativity and associativity, i.e., for $m \times n$ matrices A, B, C

$$A + B = B + A$$
$$A + (B + C) = (A + B) + C.$$

A matrix, all of whose entries are zero, is denoted by (0), regardless of dimension; it is an identity for this operation:

$$(a_{ij}) + (0) = (a_{ij} + 0) = (a_{ij}),$$

where (a_{ij}) and (0) have the same dimension. The additive inverse of the matrix (a_{ij}) is the matrix whose (i, j)th entry is $-a_{ij}$:

$$(a_{ij}) + (-a_{ij}) = (0).$$

2. We can multiply a matrix by a real number (called a *scalar*):

$$k\begin{pmatrix} a_{11} & \cdots & a_{1n} \\ \vdots & & \\ a_{m1} & \cdots & a_{mn} \end{pmatrix} = \begin{pmatrix} ka_{11} & \cdots & ka_{1n} \\ \vdots & & \\ ka_{m1} & \cdots & ka_{mn} \end{pmatrix}.$$

Example

$$5\begin{pmatrix} -1 & 2 \\ 3 & -4 \end{pmatrix} = \begin{pmatrix} -5 & 10 \\ 15 & -20 \end{pmatrix}.$$

3. We can find the *transpose* ${}^tA$ of a matrix A:

$${}^tA = (a'_{ij}) \qquad \text{where} \qquad a'_{ij} = a_{ji}.$$

That is, we reverse the position of the rows and columns. If ${}^tA = A$, we say that A is *symmetric*.

Examples

1. Find ${}^tA$ if

$$A = \begin{pmatrix} 1 & 0 & -4 & 5 \\ 2 & -3 & 1 & -1 \end{pmatrix}.$$

Solution

$${}^tA = \begin{pmatrix} 1 & 2 \\ 0 & -3 \\ -4 & 1 \\ 5 & -1 \end{pmatrix}.$$

2. The matrix A of Example 1 is not symmetric. A symmetric matrix must be square.

$$B = \begin{pmatrix} 1 & 0 & 3 \\ 0 & 3 & -1 \\ 3 & -1 & 2 \end{pmatrix}$$

is symmetric.

4. We can multiply two matrices if the first has the same number of columns as the second has rows, i.e., an $m \times n$ matrix times an $n \times p$ matrix. The *product* is defined by

$$(a_{ij})(b_{ij}) = (c_{ij}),$$

where

$$c_{ij} = \sum_{k=1}^{n} a_{ik}b_{kj},$$

(a_{ij}) an $m \times n$ matrix, (b_{ij}) an $n \times p$ matrix. Note that the product (c_{ij}) is an $m \times p$ matrix.

As a device for remembering how matrix multiplication is defined, think of running a finger of your left hand across the first row of (a_{ij}) and a finger of your right hand down the first column of (b_{ij}) in Fig. 10.1, taking the product of the matching entries and adding them up to get the first row, first column entry of the product matrix.

$$\begin{pmatrix} a_{11} & a_{12} & \cdots & a_{1n} \\ a_{12} & a_{22} & \cdots & a_{2n} \\ \vdots & & & \vdots \\ a_{m1} & & \cdots & a_{mn} \end{pmatrix} \begin{pmatrix} b_{11} & b_{12} & \cdots & b_{1p} \\ b_{21} & b_{22} & \cdots & b_{2p} \\ \vdots & & & \vdots \\ b_{n1} & & \cdots & b_{np} \end{pmatrix}$$

Fig. 10.1

Proceed to run your left finger across the first row again, but this time go down the second column with your right finger; the sum of the products of the matching entries gives the (1, 2) entry. Repeat the process with the first row as many times as there are columns in the second matrix. Then go through the same procedure with the second row and the successive columns to get the entries for the second row of the product matrix. Continue until the rows of the first matrix have all been used.

Examples

1. $$\begin{pmatrix} 1 & 2 & -4 \\ 0 & -1 & 2 \end{pmatrix} \begin{pmatrix} 0 & 3 \\ -1 & 2 \\ 1 & 0 \end{pmatrix}$$

$$= \begin{pmatrix} 1(0) + 2(-1) + (-4)1 & 1(3) + 2(2) + (-4)0 \\ 0(0) + (-1)(-1) + 2(1) & 0(3) + (-1)2 + 2(0) \end{pmatrix}$$

$$= \begin{pmatrix} -6 & 7 \\ 3 & -2 \end{pmatrix}.$$

2.
$$\begin{pmatrix} 5 & 4 & -1 \\ 2 & 0 & 2 \\ -3 & 0 & -1 \end{pmatrix}\begin{pmatrix} 2 & 3 & 1 \\ 1 & 4 & -5 \\ 2 & 1 & 0 \end{pmatrix} = \begin{pmatrix} 12 & 30 & -15 \\ 8 & 8 & 2 \\ -8 & -10 & -3 \end{pmatrix}.$$

Note that

$$\begin{pmatrix} 2 & 3 & 1 \\ 1 & 4 & -5 \\ 2 & 1 & .0 \end{pmatrix}\begin{pmatrix} 5 & 4 & -1 \\ 2 & 0 & 2 \\ -3 & 0 & -1 \end{pmatrix} = \begin{pmatrix} 13 & 8 & 3 \\ 28 & 4 & 12 \\ 12 & 8 & 0 \end{pmatrix}.$$

That is, multiplication of matrices is *not necessarily* commutative.

Warning. For two matrices A and B, AB and BA are both defined if and only if A has the same number of rows as B has columns *and* the same number of columns as B has rows.

2. A lawn furniture company produces three different articles and wants to compute the total manufacturing and transportation costs of the output from its plant to San Francisco. The input composition of each product is represented by the following matrix, which we denote by A:

	Plastic	Aluminum	Machine time	Labor time	Miscella-neous
Table	0.5	3.1	2	1	0.2
Chair	2.4	2.1	1.5	2	0.4
Lounger	3	2	1	2.5	0.1

The total and transportation costs are represented by the matrix B:

	Cost	Transportation
Plastic	0.75	0.02
Aluminum	1.90	0.03
Machine time	0.25	0
Labor time	3.25	0
Miscellaneous	0.20	0.04

The product yields

$$\begin{pmatrix} 0.5 & 3.1 & 2 & 1 & 0.2 \\ 2.4 & 2.1 & 1.5 & 2 & 0.4 \\ 3 & 2 & 1 & 2.5 & 0.1 \end{pmatrix}\begin{pmatrix} 0.75 & 0.02 \\ 1.90 & 0.03 \\ 0.25 & 0 \\ 3.25 & 0 \\ 0.20 & 0.04 \end{pmatrix} = \begin{matrix} \text{Cost} & \text{Transportation} & \\ \begin{pmatrix} 10.06 & 0.11 \\ 12.75 & 0.13 \\ 16.63 & 0.12 \end{pmatrix} & & \begin{matrix} \text{Table} \\ \text{Chair} \\ \text{Lounger} \end{matrix} \end{matrix}$$

4. Now we see why we chose to write

$$\begin{aligned} a_{11}x_1 + \cdots + a_{1n}x_n &= k_1 \\ \vdots \qquad\qquad & \quad\ \vdots \\ a_{m1}x_1 + \cdots + a_{mn}x_n &= k_m \end{aligned}$$

as

$$\begin{pmatrix} a_{11} & \cdots & a_{1n} \\ \vdots & & \vdots \\ a_{m1} & \cdots & a_{mn} \end{pmatrix} \begin{pmatrix} x_1 \\ \vdots \\ x_n \end{pmatrix} = \begin{pmatrix} k_1 \\ \vdots \\ k_m \end{pmatrix}.$$

Each equation is the expression for an entry of the product of the $m \times n$ matrix (a_{ij}) and the $n \times 1$ matrix (x_j); that is, (k_i) is the product of (a_{ij}) and (x_j).

The $n \times n$ matrix which has 1's on the main diagonal:

$$\begin{pmatrix} 1 & 0 & \cdots & 0 \\ 0 & 1 & \cdots & 0 \\ & \vdots & & \\ 0 & 0 & \cdots & 1 \end{pmatrix},$$

and zeros elsewhere is an identity for the operation of multiplication of $n \times n$ matrices. We denote such a matrix by $I = (\delta_{ij})$, where

$$\delta_{ij} = \begin{cases} 1, & \text{when } i = j \\ 0, & \text{otherwise.} \end{cases}$$

Then

$$(a_{ij})(\delta_{ij}) = (c_{ij}),$$

where

$$c_{ij} = \sum_{k=1}^{n} a_{ik}\,\delta_{kj} = a_{ij}(1) = a_{ij}.$$

Also

$$(\delta_{ij})(a_{ij}) = (a_{ij}).$$

For example,

$$\begin{pmatrix} 1 & 0 & 0 & 0 \\ 0 & 1 & 0 & 0 \\ 0 & 0 & 1 & 0 \\ 0 & 0 & 0 & 1 \end{pmatrix} \begin{pmatrix} 4 & 3 & 2 & 1 \\ 0 & 5 & -1 & 2 \\ 1 & -1 & 2 & 3 \\ 2 & -2 & 1 & 1 \end{pmatrix} = \begin{pmatrix} 4 & 3 & 2 & 1 \\ 0 & 5 & -1 & 2 \\ 1 & -1 & 2 & 3 \\ 2 & -2 & 1 & 1 \end{pmatrix}.$$

The problem of finding an inverse for matrix multiplication is more complicated. First of all a matrix must be square in order to have a multiplicative inverse, but not all square matrices have multiplicative inverses. There are several methods for determining the inverse of a matrix; we use a variation of the Gauss elimination process. (Another method is given in Appendix C.)

Let A be an $n \times n$ matrix whose determinant is not zero (such a matrix is said to be *nonsingular*). We augment A by the $n \times n$ identity matrix

$$(A \mid I) = \begin{pmatrix} a_{11} & \cdots & a_{1n} \\ \vdots & & \vdots \\ a_{m1} & \cdots & a_{mn} \end{pmatrix} \begin{pmatrix} 1 & 0 & \cdots & 0 \\ \vdots & 1 & \ddots & \vdots \\ 0 & & \cdots & 1 \end{pmatrix}.$$

By elementary row operations we transform this into a matrix of the form $(I \mid P)$. We claim that $AP = PA = I$; we denote the inverse of A by A^{-1}. The difficulty with this method is that it is not really clear why the matrix thus computed should turn out to be the inverse. We shall not prove that the method is valid; however, in any particular case it is easy to verify that the matrix obtained in this way is the inverse—we simply multiply it by the original matrix and obtain an identity matrix.

Example. Find the inverse of

$$\begin{pmatrix} 1 & 0 & 3 \\ 2 & 1 & -1 \\ 4 & -2 & 1 \end{pmatrix}.$$

This matrix is nonsingular so we augment it by the identity and proceed with the necessary elementary row operations:

$$\left(\begin{array}{ccc|ccc} 1 & 0 & 3 & 1 & 0 & 0 \\ 2 & 1 & -1 & 0 & 1 & 0 \\ 4 & -2 & 1 & 0 & 0 & 1 \end{array}\right) \xrightarrow[\times \text{ second row}]{\text{third row } -2}$$

$$\left(\begin{array}{ccc|ccc} 1 & 0 & 3 & 1 & 0 & 0 \\ 2 & 1 & -1 & 0 & 1 & 0 \\ 0 & -4 & 3 & 0 & -2 & 1 \end{array}\right)$$

$$\xrightarrow[\times \text{ first row}]{\text{second row } - 2} \left(\begin{array}{ccc|ccc} 1 & 0 & 3 & 1 & 0 & 0 \\ 0 & 1 & -7 & -2 & 1 & 0 \\ 0 & -4 & 3 & 0 & -2 & 1 \end{array}\right)$$

$$\xrightarrow[\times \text{ second row}]{\text{third row } + 4} \left(\begin{array}{ccc|ccc} 1 & 0 & 3 & 1 & 0 & 0 \\ 0 & 1 & -7 & -2 & 1 & 0 \\ 0 & 0 & -25 & -8 & 2 & 1 \end{array}\right)$$

$$\xrightarrow[\times(-1/25)]{\text{third row}} \left(\begin{array}{ccc|ccc} 1 & 0 & 3 & 1 & 0 & 0 \\ 0 & 1 & -7 & -2 & 1 & 0 \\ 0 & 0 & 1 & \frac{8}{25} & -\frac{2}{25} & -\frac{1}{25} \end{array}\right)$$

$$\xrightarrow[\times \text{ third row}]{\text{second row } + 7} \left(\begin{array}{ccc|ccc} 1 & 0 & 3 & 1 & 0 & 0 \\ 0 & 1 & 0 & \frac{6}{25} & \frac{11}{25} & -\frac{7}{25} \\ 0 & 0 & 1 & \frac{8}{25} & -\frac{2}{25} & -\frac{1}{25} \end{array}\right)$$

$$\xrightarrow[\times \text{ third row}]{\text{first row } - 3} \left(\begin{array}{ccc|ccc} 1 & 0 & 0 & \frac{1}{25} & \frac{6}{25} & \frac{3}{25} \\ 0 & 1 & 0 & \frac{6}{25} & \frac{11}{25} & -\frac{7}{25} \\ 0 & 0 & 1 & \frac{8}{25} & -\frac{2}{25} & -\frac{1}{25} \end{array}\right)$$

Thus the inverse should be

$$\frac{1}{25}\begin{pmatrix} 1 & 6 & 3 \\ 6 & 11 & -7 \\ 8 & -2 & -1 \end{pmatrix}.$$

We verify that it is the inverse:

$$\frac{1}{25}\begin{pmatrix} 1 & 6 & 3 \\ 6 & 11 & -7 \\ 8 & -2 & -1 \end{pmatrix}\begin{pmatrix} 1 & 0 & 3 \\ 2 & 1 & -1 \\ 4 & -2 & 1 \end{pmatrix} = \begin{pmatrix} 1 & 0 & 0 \\ 0 & 1 & 0 \\ 0 & 0 & 1 \end{pmatrix}.$$

The use of the inverse sometimes simplifies the solution of systems of linear equations in which the coefficient matrix is a nonsingular square matrix.

If $AX = B$, A a nonsingular square matrix, then $X = IX = A^{-1}AX = A^{-1}B$. We could simply go through the procedure for finding A^{-1} and then multiply by B. However, we can save some steps by simply carrying the Gauss process a little further. We augment A by B (instead of the identity) and carry out the necessary steps to convert A to the identity matrix. The matrix which results from the same operations being performed on B is the desired matrix, $A^{-1}B$.

Examples

1. Solve

$$\begin{aligned} x + 3y - z &= 0 \\ 2x - y \phantom{{}+ z} &= 1 \\ -x + 2y + z &= 2. \end{aligned}$$

Solution. We write down the augmented matrix and transform the left portion into the identity matrix:

$$\left(\begin{array}{rrr|r} 1 & 3 & -1 & 0 \\ 2 & -1 & 0 & 1 \\ -1 & 2 & 1 & 2 \end{array}\right) \xrightarrow{\text{third row + first row}} \left(\begin{array}{rrr|r} 1 & 3 & -1 & 0 \\ 2 & -1 & 0 & 1 \\ 0 & 5 & 0 & 2 \end{array}\right) \xrightarrow{\text{second row} - 2 \times \text{first row}}$$

$$\left(\begin{array}{rrr|r} 1 & 3 & -1 & 0 \\ 0 & -7 & 2 & 1 \\ 0 & 5 & 0 & 2 \end{array}\right) \xrightarrow{\text{second row} + 7/5 \times \text{third row}} \left(\begin{array}{rrr|r} 1 & 3 & -1 & 0 \\ 0 & 0 & 2 & \frac{19}{5} \\ 0 & 5 & 0 & 2 \end{array}\right)$$

$$\xrightarrow{\text{second row} \times \frac{1}{2} \text{ and third row} \times \frac{1}{5}} \left(\begin{array}{rrr|r} 1 & 3 & -1 & 0 \\ 0 & 0 & 1 & \frac{19}{10} \\ 0 & 1 & 0 & \frac{2}{5} \end{array}\right) \xrightarrow{\text{first row + second row}} \left(\begin{array}{rrr|r} 1 & 3 & 0 & \frac{19}{10} \\ 0 & 0 & 1 & \frac{19}{10} \\ 0 & 1 & 0 & \frac{2}{5} \end{array}\right)$$

$$\xrightarrow{\text{first row} - 3 \times \text{third row}} \left(\begin{array}{rrr|r} 1 & 0 & 0 & \frac{7}{10} \\ 0 & 0 & 1 & \frac{19}{10} \\ 0 & 1 & 0 & \frac{2}{5} \end{array}\right) \xrightarrow{\text{interchange rows two and three}} \left(\begin{array}{rrr|r} 1 & 0 & 0 & \frac{7}{10} \\ 0 & 1 & 0 & \frac{2}{5} \\ 0 & 0 & 1 & \frac{19}{10} \end{array}\right).$$

Thus the solution is

$$A^{-1}B = \begin{pmatrix} \frac{7}{10} \\ \frac{2}{5} \\ \frac{19}{10} \end{pmatrix},$$

which can be read off from the simplified system

$$\begin{aligned} x &= \tfrac{7}{10} \\ y &= \tfrac{2}{5} \\ z &= \tfrac{19}{10} \end{aligned}$$

represented by the final augmented matrix.

2. A laboratory has two pieces of equipment which process two different vaccines. Apparatus A can operate 12 hrs per day and apparatus B, 16 hrs. To process one unit of vaccine P requires that work be done for 3 hrs a day by apparatus A and 1 hr a day by apparatus B. Processing vaccine Q requires that apparatus A work 2 hrs a day and apparatus B, 3 hrs. Find the number of units of each vaccine that should be processed each day to keep the equipment working to capacity.

Solution. Let

$$\begin{aligned} x &= \text{units of vaccine } P \\ y &= \text{units of vaccine } Q. \end{aligned}$$

The equations are

$$\begin{aligned} 3x + 2y &= 12 \qquad \text{(for apparatus } A) \\ x + 3y &= 16 \qquad \text{(for apparatus } B). \end{aligned}$$

We write down the augmented matrix and proceed with the row operations:

$$\left(\begin{array}{cc|c} 3 & 2 & 12 \\ 1 & 3 & 16 \end{array}\right) \to \left(\begin{array}{cc|c} 3 & 2 & 12 \\ 0 & \frac{7}{3} & 12 \end{array}\right) \to \left(\begin{array}{cc|c} 3 & 2 & 12 \\ 0 & 1 & \frac{36}{7} \end{array}\right)$$

$$\to \left(\begin{array}{cc|c} 3 & 0 & \frac{12}{7} \\ 0 & 1 & \frac{36}{7} \end{array}\right) \to \left(\begin{array}{cc|c} 1 & 0 & \frac{4}{7} \\ 0 & 1 & \frac{36}{7} \end{array}\right).$$

Thus $x = \frac{4}{7}$ and $y = \frac{36}{7}$.

If it is necessary to solve several systems of equations in which the coefficient matrix remains unchanged but the right-hand side varies, then it is preferable to solve for the inverse directly. Suppose in Example 2 above we had found that the inverse of the coefficient matrix is

$$\frac{1}{7}\begin{pmatrix} 3 & -2 \\ -1 & 3 \end{pmatrix}.$$

Then we would have the solution as above:

$$\frac{1}{7}\begin{pmatrix} 3 & -2 \\ -1 & 3 \end{pmatrix}\begin{pmatrix} 12 \\ 16 \end{pmatrix} = \begin{pmatrix} \frac{4}{7} \\ \frac{36}{7} \end{pmatrix}.$$

Now suppose the operation time of A is increased to 16 hrs and that of B to 20 hrs. Then the solution is found by simple multiplication:

$$\frac{1}{7}\begin{pmatrix} 3 & -2 \\ -1 & 3 \end{pmatrix}\begin{pmatrix} 16 \\ 20 \end{pmatrix} = \begin{pmatrix} \frac{8}{7} \\ \frac{44}{7} \end{pmatrix}.$$

Warning. The method of solution by inverses works only if the system has a unique solution.

PROBLEMS

1. If $A = \begin{pmatrix} 1 & 0 & 3 \\ 2 & 1 & 4 \\ -1 & -2 & 5 \end{pmatrix}$ and $B = \begin{pmatrix} 2 & 1 & -1 \\ 0 & 4 & 5 \\ 1 & 3 & 0 \end{pmatrix}$, find

   a) $3A + B$. b) $2A - 3B$. c) AB. d) $A^2 - 5A$.
   e) ${}^t(BA)$. f) $3AB - 2BA + B^2$. g) ${}^tA{}^tB$.

2. If $A = \begin{pmatrix} 0 & 1 & 2 \\ 4 & -1 & 1 \\ 2 & 3 & -2 \end{pmatrix}$ and $B = \begin{pmatrix} 1 & 2 & -1 \\ 2 & 0 & 2 \\ -1 & 3 & 1 \end{pmatrix}$, find

   a) the determinant of A. b) $3A - 2B$.
   c) AB. d) $B{}^tA$.
   e) $2A^2 + {}^tB$. f) $AB - BA$.
   g) the determinant of B. h) the determinant of AB.

3. Find the transpose of

   a) $\begin{pmatrix} 1 & 2 \\ 3 & -4 \\ 2 & 1 \\ 6 & 7 \end{pmatrix}$. b) $\begin{pmatrix} 1 & 2 & 4 \\ 2 & 2 & 3 \\ 4 & 3 & -1 \end{pmatrix}$. c) $\begin{pmatrix} 1 & 0 & 1 & 0 \\ 1 & 0 & 1 & 0 \\ 0 & 1 & 0 & 1 \\ 1 & 0 & 0 & 1 \end{pmatrix}$.

4. Find AB and BA if possible:

   a) $A = \begin{pmatrix} 1 & 0 & 4 & 0 \\ 6 & 5 & 2 & 0 \\ 7 & 5 & 5 & 6 \\ 8 & 6 & 2 & 3 \end{pmatrix}$, $B = \begin{pmatrix} 4 & 5 & 4 & 2 \\ 8 & 3 & 4 & -5 \end{pmatrix}$.

   b) $A = \begin{pmatrix} 2 & 0 & -1 \\ 3 & 4 & 1 \\ -1 & 1 & -2 \end{pmatrix}$, $B = \begin{pmatrix} -1 & 0 & 3 \\ 4 & 0 & 2 \\ 2 & 1 & 3 \\ 4 & 1 & -1 \end{pmatrix}$.

   c) $A = \begin{pmatrix} 1 & 0 & 1 & 1 & 0 \\ 2 & 0 & -1 & 0 & 1 \\ 3 & -1 & 2 & 1 & 3 \\ 4 & 5 & 1 & 5 & 0 \end{pmatrix}$, $B = \begin{pmatrix} 6 & 1 & 6 & 1 \\ 2 & 8 & 4 & -5 \\ 1 & 0 & 2 & 3 \\ 4 & 5 & 1 & 0 \\ -5 & 1 & 3 & -2 \end{pmatrix}$.

d) $A = \begin{pmatrix} 2 & 0 & -2 & 0 \\ -2 & 0 & 2 & 0 \\ 3 & 1 & 0 & 1 \\ 7 & 0 & 2 & 1 \end{pmatrix}, \quad B = \begin{pmatrix} 1 & 0 & 0 & 2 \\ 6 & 0 & 0 & -6 \\ 1 & 0 & 0 & 2 \\ -9 & 0 & 0 & 0 \end{pmatrix}.$

e) $A = \begin{pmatrix} 1 & 2 & -3 \\ 0 & 1 & 2 \end{pmatrix}, \quad B = \begin{pmatrix} 1 & 0 \\ -1 & 2 \\ 3 & 1 \end{pmatrix}.$

f) $A = \begin{pmatrix} 1 & 2 & -1 \\ 0 & -1 & 2 \\ 3 & 4 & 1 \end{pmatrix}, \quad B = \begin{pmatrix} -1 & 3 & 1 \\ 2 & 4 & -2 \end{pmatrix}.$

g) $A = \begin{pmatrix} 1 & 0 & 2 & 3 \\ 2 & 1 & 4 & 0 \\ 3 & 1 & -1 & 1 \\ 4 & 0 & -2 & 2 \end{pmatrix}, \quad B = \begin{pmatrix} 1 & 3 & 4 & -1 \\ 2 & 0 & 3 & -2 \end{pmatrix}.$

h) $A = \begin{pmatrix} 1 & 0 & 1 \\ 1 & 2 & 6 \\ -1 & 3 & -1 \end{pmatrix}, \quad B = \begin{pmatrix} 2 & 3 & 1 \\ 2 & 0 & -1 \\ 0 & 1 & 4 \end{pmatrix}.$

5. The input and cost matrices for products of the Dashing Dachshund Co. are as follows:

	Input					
	Leather	Wool	Cedar chips	Machine time	Labor time	Misc.
Collar and leash	2.3	0.2	0	2	0.7	0.3
Coat	0.3	3.1	0	1.5	1	0.1
Cushion	0.4	5.6	3	1	0.5	0.1

	Cost	
	Cost	Transportation cost
Leather	2	0.02
Wool	1	0.03
Cedar chips	0.2	0.04
Machine time	0.3	0
Labor time	3	0
Miscellaneous	0.2	0.04

Find the cost and transportation cost for each of the products.

6. Find the inverse, if any, of:

a) $\begin{pmatrix} 1 & 0 & 2 \\ -2 & 1 & 3 \\ 0 & 1 & 1 \end{pmatrix}$. b) $\begin{pmatrix} 4 & -1 \\ 2 & 5 \end{pmatrix}$.

c) $\begin{pmatrix} 3 & 4 & 1 \\ -2 & -3 & -2 \\ 0 & 1 & 5 \end{pmatrix}$. d) $\begin{pmatrix} 1 & 0 & 2 & 1 \\ 2 & 0 & 1 & 2 \\ 3 & -1 & 0 & -1 \\ 4 & 0 & 0 & 5 \end{pmatrix}$.

e) $\begin{pmatrix} 3 & 2 \\ 1 & 2 \end{pmatrix}$. f) $\begin{pmatrix} 0 & 1 & 2 \\ 3 & -2 & 1 \\ 3 & 0 & 5 \end{pmatrix}$.

g) $\begin{pmatrix} 3 & -1 & 2 \\ 0 & 1 & 4 \\ 2 & 2 & 1 \end{pmatrix}$. h) $\begin{pmatrix} 4 & 3 & 2 \\ 0 & 2 & 1 \\ 2 & 0 & -3 \end{pmatrix}$.

i) $\begin{pmatrix} 0 & -1 & 2 \\ 4 & 5 & 1 \end{pmatrix}$.

7. Solve the following systems by finding the inverse of the coefficient matrix.

a) $$\begin{aligned} 2x - y + z &= 1 \\ 3x - y + 2z &= 5 \\ -2x + y - z &= 4. \end{aligned}$$

b) $$\begin{aligned} x + w &= 2 \\ x - y &= 0 \\ x + z &= 1 \\ y - z &= 3. \end{aligned}$$

c) $$\begin{aligned} 3x - y + 4z &= 0 \\ 2x - 2y - z &= 5 \\ -x + y - 4z &= 2. \end{aligned}$$

8. We may represent a communications system in which some stations may not speak to other stations by means of a matrix. If the ith station can speak to the jth station, define $a_{ij} = 1$, otherwise $a_{ij} = 0$. Suppose we have $A = (a_{ij})$ given by

	Receivers				
Speakers	1	2	3	4	5
1	0	1	0	1	0
2	0	0	1	1	0
3	1	0	0	0	1
4	1	1	0	0	1
5	0	0	1	1	0

Let $B = A^2$. Then $b_{ij} = \sum_{k=1}^{5} a_{ik}a_{kj}$. Any one of the products $a_{ik}a_{kj}$ is 1 if station i can talk to station k and station k can talk to station j. Thus if $a_{ik}a_{kj} = 1$, station i can talk to station j *through* station k, and the sum b_{ij} is equal to the number of ways i can talk to j through one relay. Similarly, the entry c_{ij} in $C = A^3$ is the number of two-stage relays between i and j. The total number of ways station i can communicate with station j through no more than n relays is given by the (i, j)th entry of the matrix

$$A + A^2 + A^3 + \cdots + A^{n+1}.$$

Find the number of ways station 3 can communicate with station 5 through no more than three relays.

9. Show that ${}^t(AB) = {}^tB\,{}^tA$ for any two matrices A and B such that AB is defined.

10. If

$$\begin{pmatrix} 0 & 1 & 1 & 0 \\ 1 & 0 & 0 & 0 \\ 0 & 1 & 0 & 1 \\ 1 & 0 & 1 & 0 \end{pmatrix}$$

represents a communications system, find the number of ways station 2 can communicate with station 3 through no more than three relays.

10.4 CHARACTERISTIC VALUES AND CHARACTERISTIC VECTORS

Given an $n \times n$ matrix A we sometimes want to find a real number k such that for

$$X = \begin{pmatrix} x_1 \\ \vdots \\ x_n \end{pmatrix}$$

the equation $AX = kX$ has a nontrivial solution, i.e., a solution $X \neq 0$. Let us restate the above equation as

$$(A - kI)X = 0,$$

I the $n \times n$ identity matrix. Since if $\det(A - kI) \neq 0$, by the corollary to Theorem 4 the system has no nontrivial solution, we are interested in finding out for what values of k, $\det(A - kI) = 0$.

Thus given the matrix A we form the matrix $A - kI$, whose entries will be the same as those of A except on the main diagonal

$$A - kI = \begin{pmatrix} a_{11} - k & a_{12} & \cdots & a_{1n} \\ a_{21} & a_{22} - k & \cdots & a_{2n} \\ \vdots & & & \\ a_{n1} & & \cdots & a_{nn} - k \end{pmatrix}.$$

The determinant of this matrix is a polynomial in k. Its roots are called the *characteristic* (or *eigen*) *values* of A.

Warning. Some or all of the roots may be complex numbers. If all of the roots are complex, there is no real number k such as described above.

Once the characteristic values $k_1, \ldots, k_n$, have been found, we can solve each of the resulting equations

$$(A - k_i I)X = 0,$$

for the *characteristic* (or *eigen*) vector X_i associated with the characteristic value k_i.

Example. Find the characteristic values and characteristic vectors of

$$A = \begin{pmatrix} 1 & 1 \\ 3 & -1 \end{pmatrix}.$$

Solution

$$A - kI = \begin{pmatrix} 1 - k & 1 \\ 3 & -1 - k \end{pmatrix}.$$

$\det(A - kI) = (1 - k)(-1 - k) - 3 = k^2 - 4$. The roots of this polynomial are ± 2, which are the characteristic values of A. To find the associated characteristic vectors we write

1.

$$(A - 2I)X = \begin{pmatrix} -1 & 1 \\ 3 & -3 \end{pmatrix} X = 0.$$

Transforming the augmented matrix, we have

$$\left(\begin{array}{cc|c} -1 & 1 & 0 \\ 3 & -3 & 0 \end{array}\right) \rightarrow \left(\begin{array}{cc|c} -1 & 1 & 0 \\ 0 & 0 & 0 \end{array}\right).$$

Thus

$$-x + y = 0, \qquad x = y.$$

Therefore

$$X = \begin{pmatrix} x \\ x \end{pmatrix},$$

for any real number x, is a characteristic vector associated with the characteristic value 2. In particular,

$$\begin{pmatrix} 1 \\ 1 \end{pmatrix}$$

is a characteristic vector.

2.

$$(A + 2I)X = \begin{pmatrix} 3 & 1 \\ 3 & 1 \end{pmatrix} X = 0.$$

We obtain

$$\left(\begin{array}{cc|c} 3 & 1 & 0 \\ 0 & 0 & 0 \end{array}\right)$$

so that

$$3x + y = 0, \qquad y = -3x.$$

Hence for any real number x,

$$X = \begin{pmatrix} x \\ -3x \end{pmatrix}$$

is a characteristic vector associated with -2. In particular,

$$\begin{pmatrix} 1 \\ -3 \end{pmatrix}$$

is such a vector.

PROBLEMS

1. Find the characteristic values and characteristic vectors of the following matrices:

a) $\begin{pmatrix} 1 & -1 \\ -4 & 1 \end{pmatrix}$.

b) $\begin{pmatrix} 4 & 3 \\ 1 & 2 \end{pmatrix}$.

c) $\begin{pmatrix} 0 & 2 \\ 1 & 0 \end{pmatrix}$.

d) $\begin{pmatrix} 0 & 3 \\ 1 & 2 \end{pmatrix}$.

e) $\begin{pmatrix} 5 & 1 \\ -2 & 8 \end{pmatrix}$.

f) $\begin{pmatrix} 2 & 1 & 1 \\ 1 & 1 & 0 \\ 1 & 0 & 1 \end{pmatrix}$.

g) $\begin{pmatrix} 0 & 0 & 3 \\ 0 & 2 & 0 \\ 3 & 0 & 0 \end{pmatrix}$.

h) $\begin{pmatrix} 0 & \frac{1}{2} & \frac{1}{2} \\ \frac{1}{2} & 0 & \frac{1}{2} \\ \frac{1}{2} & \frac{1}{2} & 0 \end{pmatrix}$.

i) $\begin{pmatrix} 7 & 0 & 0 & 0 \\ 0 & 2 & 0 & 0 \\ 0 & 0 & 3 & 0 \\ 0 & 0 & 0 & -1 \end{pmatrix}$.

j) $\begin{pmatrix} 2 & 0 \\ -1 & 1 \end{pmatrix}$.

k) $\begin{pmatrix} 2 & 3 \\ 1 & 4 \end{pmatrix}$.

l) $\begin{pmatrix} 0 & 1 \\ 3 & 2 \end{pmatrix}$.

m) $\begin{pmatrix} 1 & 2 \\ 1 & -1 \end{pmatrix}$.

n) $\begin{pmatrix} 3 & 2 & 2 \\ 1 & 4 & 1 \\ -2 & -4 & -1 \end{pmatrix}$.

o) $\begin{pmatrix} 2 & -2 & 3 \\ 1 & 1 & 1 \\ 1 & 3 & -1 \end{pmatrix}$.

p) $\begin{pmatrix} 4 & -2 & 1 \\ -2 & 1 & 2 \\ 1 & 2 & 4 \end{pmatrix}$.

q) $\begin{pmatrix} -1 & 0 & 0 \\ 0 & 1 & 2 \\ 0 & 3 & 1 \end{pmatrix}$.

r) $\begin{pmatrix} 1 & -4 & 3 \\ 0 & 1 & 6 \\ 0 & 0 & 2 \end{pmatrix}$.

2. A *diagonal matrix* is a matrix (a_{ij}) such that $a_{ij} = 0$ for $i \neq j$, e.g., (i) in Problem 1. Formulate a conjecture about the characteristic values of diagonal matrices.
3. In the Leontief input-output model, we assume that there are n industries, each of which produces a single commodity. Each commodity may reenter the system as an input to any or all of the industries. In particular, a unit output of the jth commodity requires a quantity a_{ij} of the ith commodity as input. We have a matrix of input coefficients:

$$A = \text{Input commodity} \overset{\text{Output commodity}}{\begin{pmatrix} a_{11} & \cdots & a_{1n} \\ \vdots & & \vdots \\ a_{n1} & \cdots & a_{nn} \end{pmatrix}}.$$

Note that the entries in the jth column specify the input requirements for production of one unit of the jth commodity. In a *closed* system, no outside inputs are required and each output serves only as input to the other elements of the system. If the matrix A has 1 as an eigenvalue, an associated eigenvector $X \neq 0$, i.e., a vector $X \neq 0$ such that $AX = X$, is called an *interior equilibrium* of the system. If

$$A = \begin{pmatrix} \frac{1}{2} & 0 & \frac{1}{2} \\ \frac{1}{3} & 0 & \frac{2}{3} \\ \frac{1}{4} & \frac{1}{4} & \frac{1}{2} \end{pmatrix},$$

find an interior equilibrium.

10.5 LINEAR PROGRAMMING

Since so many variables are usually involved in real-life situations, the solution of most systems of linear equations is efficiently accomplished by means of computers. The same is true of systems of linear inequalities, the subject of linear programming. First we look at a very much simplified problem and then we describe one method of dealing with more complicated situations.

To graph a linear inequality

$$ax + by + c \geq 0,$$

we graph the linear equation (Fig. 10.2)

$$ax + by + c = 0.$$

Its graph divides the plane into two parts: all the points in one half-plane and all the points on the line satisfy the inequality and all the points in the other half do not.

Examples

1. Graph

$$3x - y + 4 \geq 0.$$

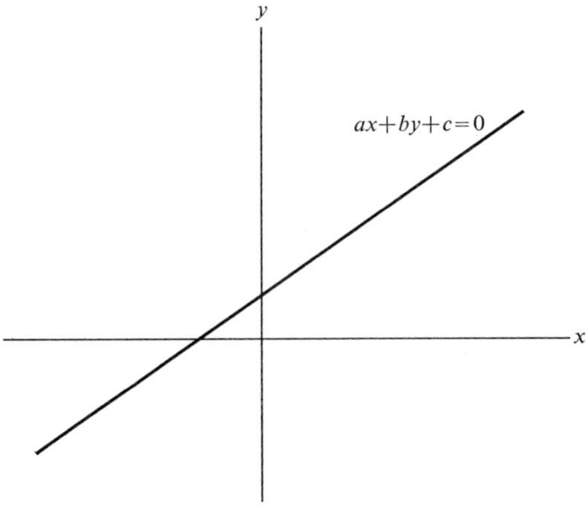

Fig. 10.2

Solution. If we write the equation $3x - y + 4 = 0$ (see Fig. 10.3) as

$$y = 3x + 4,$$

we see that it is a straight line of slope 3, y-intercept 4. If we write the original inequality as

$$y \leq 3x + 4,$$

it is clear that all points on or below the line satisfy the inequality.

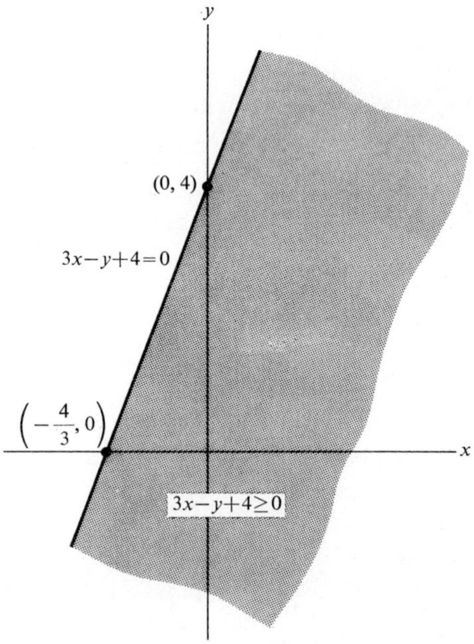

Fig. 10.3

2. Graph

$$2x + 3y - 2 \geq 0.$$

Solution. We rewrite this as $y \geq -\frac{2}{3}x + \frac{2}{3}$. We graph the straight line $y = -\frac{2}{3}x + \frac{2}{3}$ (see Fig. 10.4). The solution set of the inequality consists of all points on or above this line.

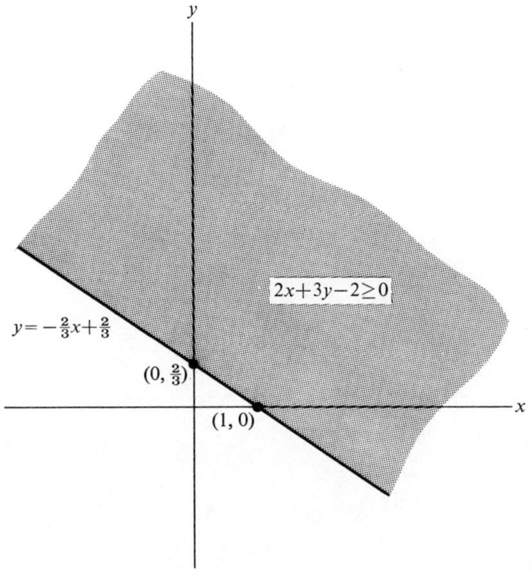

Fig. 10.4

Just as the solution set of a system of linear equations in two variables is the intersection of their graphs, the solution set of a system of linear inequalities is the intersection of the graphs of the inequalities.

Example. Graph the solution set of the system

$$2x + y - 8 \leq 0 \tag{10.3}$$

$$-x - y + 4 \leq 0. \tag{10.4}$$

Solution. The graph of (10.3) is the set of all points on or below the line $y = -2x + 8$; the graph of (10.4) is the set of all points on or above the line $y = -x + 4$. The solution set of the system is seen in Fig. 10.5.

In linear programming we have a system of linear inequalities, which represent constraints or limitations on certain variables. The problem is to maximize (or minimize) some quantity subject to these restraints.

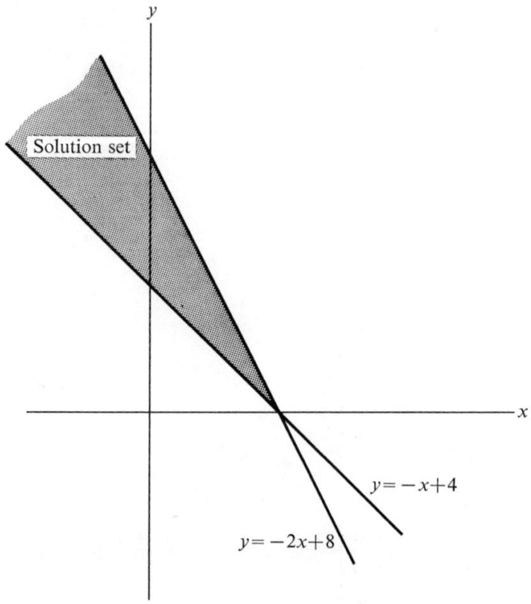

Fig. 10.5

Example. Let us return to the laboratory which processes two types of vaccine. We summarize the available data:

	Vaccine P	Vaccine Q	Apparatus limitation
Apparatus A	3 hrs/unit	1 hr/unit	12 hrs/day
Apparatus B	2 hrs/unit	3 hrs/unit	16 hrs/day
Profit per unit	400	300	

We want to maximize the profit, given these conditions.

Solution. Letting

$$x = \text{units of } P \text{ per day}$$
$$y = \text{units of } Q \text{ per day,}$$

we have as the function we want to maximize

$$P = 400x + 300y.$$

The constraints are

$$\begin{aligned} 3x + y &\le 12 \\ 2x + 3y &\le 16 \\ x &\ge 0 \\ y &\ge 0. \end{aligned}$$

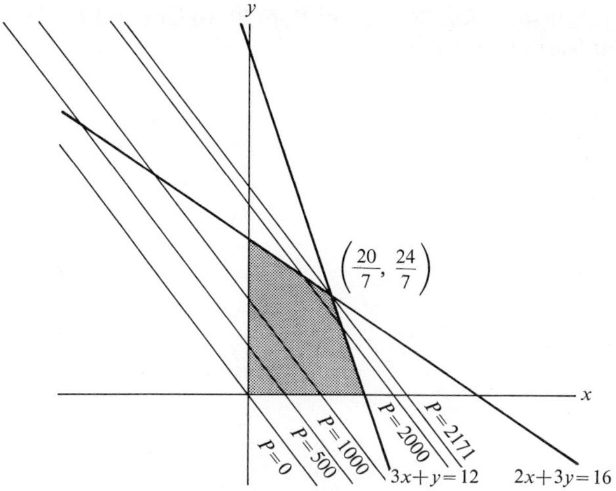

Fig. 10.6

We graph each inequality; then the intersection of the graphs represents the set of all points which satisfy the system of inequalities; such a set is called a *feasible set*. In Fig. 10.6 it is the shaded area.

To get a graphical solution, we observe that different values of P yield parallel lines of slope $-4/3$. The highest value for P for which any point of the line falls within the shaded area is that value for P which yields the line through the intersection of

$$y = -3x + 12,$$

and

$$y = \frac{-2}{3}x + \frac{16}{3},$$

namely $(\frac{20}{7}, \frac{24}{7})$. The corresponding maximum value for P is

$$P = 400(\tfrac{20}{7}) + 300(\tfrac{24}{7}) = \tfrac{15200}{7}.$$

Note that the minimum value for P subject to the given constraints is at the origin, $P = 0$.

Although we shall not prove it, whenever the feasible set is a polygon (or a polyhedron in 3-space), the extreme values (maximum and minimum) always occur at vertices of the figure. This is a central result in linear programming.

PROBLEMS

1. Graph the inequalities
   a) $3x + 4y - 5 \geq 0.$
   b) $4y - 2x \geq 2.$
   c) $-3x - 6y \leq 2.$
   d) $x/2 - y/2 \leq 3.$
   e) $2x - 3y + 1 \leq 0.$
   f) $x + y - 2 \geq 0.$
   g) $4y + x - 3 \geq 0.$
   h) $x/2 - y/3 \leq 2.$

2. A grocer has shelf space for 25 boxes of Pure Phosphate and Dingy Duds detergents. If he wants at least two boxes of each on the shelf, graph the set of possible numbers of boxes of the two brands.
3. Graph the solution set of the following systems of linear inequalities:

   a) $2x + y - 5 \geq 0$, $x - y + 4 \geq 0.$
   b) $3x - y + 4 \leq 2$, $2x - 2y - 3 \leq 1.$
   c) $4x + y - 3 \leq 0$, $3x - y + 2 \geq 0.$
   d) $3x - 2y - 4 \leq 0$, $2x + 3y + 4 \leq 0.$
   e) $x - y + 4 \geq 0$, $2x + y - 3 \geq 0.$
   f) $2x - 3y - 3 \leq 0$, $-x - y + 4 \geq 0.$
   g) $3x + y - 3 \geq 0$, $x - y + 4 \leq 0.$
   h) $4x + 2y - 3 \leq 0$, $3x + y + 4 \leq 0.$

4. Maximize $10x + 4y$ subject to
$$\begin{aligned} 4x + 3y &\leq 8 \\ 5x + 2y &\leq 6 \\ x, y &\geq 0. \end{aligned}$$
5. Maximize $2x - 3y$ subject to
$$\begin{aligned} 7x + 4y &\leq 20 \\ 2x - y &\geq 0 \\ x, y &\geq 0. \end{aligned}$$
6. Maximize $100x + 80y$ subject to the restraints
$$\begin{aligned} 40x + 60y &\leq 200 \\ 4x + 5y &\leq 20 \\ x, y &\geq 0. \end{aligned}$$
7. Smog Associates produces two types of autos—Monster and Super Monster. The production plant needs two man-days on each model; the assembly plant needs one man-day on the Monster and eight man-days on the Super Monster. The production plant has 5000 man-days per week available and the assembly plant 6000. If the profit on each Monster is \$400 and on each Super Monster \$500, how many of each should be produced in order to maximize the profit?
8. Fish and Chips Unlimited, Ltd., produces two sizes of frozen fish-and-chip dinners. The snack size takes one man-minute for production and eight minutes for fast-freezing; the regular size takes $1\frac{1}{2}$ minutes for production and ten minutes for fast-freezing; 11,500 man-minutes are available for production and 90,000 minutes for fast-freezing. If the profit on the snack size is \$0.04 and on the regular size \$0.07, how many of each should be produced to maximize the profit?

10.6 THE SIMPLEX METHOD

The general form of a linear programming problem is as follows:

Maximize
$$z = c_1x_1 + c_2x_2 + \cdots + c_nx_n$$
(called the *objective function*) subject to
$$x_i \geq 0, \qquad i = 1, \ldots, n, \tag{10.5}$$

and

$$\begin{aligned} a_{11}x_1 + \cdots + a_{1n}x_n &\geq b_1 \\ &\vdots \\ a_{m1}x_1 + \cdots + a_{mn}x_n &\geq b_m. \end{aligned} \tag{10.6}$$

A solution which satisfies (1) and (2) is a *feasible solution.* A feasible solution which has m positive variables is called a *nondegenerate basic feasible solution.* An *optimum nondegenerate feasible solution* is one which also maximizes (or minimizes) the objective function. Obtaining such a solution is our goal. To apply the simplex method, it is necessary that the rank of the coefficient matrix $A = (a_{ij})$ be m; in this case the problem is said to be nondegenerate.

Since we want to deal with equations rather than with inequalities, we add so-called "slack variables" to transform (2) into a system of linear equations:

$$\begin{aligned} a_{11}x_1 + \cdots + a_{1n}x_n + x_{n+1} &= b_1 \\ a_{21}x_1 + \cdots + a_{2n}x_n + x_{n+2} &= b_2 \\ &\vdots \\ a_{m1}x_1 + \cdots + a_{mn}x_n + x_{n+m} &= b_m. \end{aligned}$$

Now the coefficient matrix is the original matrix A augmented by the $m \times m$ identity matrix:

$$\begin{pmatrix} a_{11} & \cdots & a_{1n} & 1 & 0 & & \cdots & 0 \\ a_{21} & \cdots & a_{2n} & 0 & 1 & 0 & \cdots & 0 \\ \vdots & & & & & & & \\ a_{m1} & \cdots & a_{mn} & 0 & 0 & & \cdots & 1 \end{pmatrix}.$$

The simplex method may be characterized as a sophisticated trial and error process.

1. We take an initial nondegenerate basic feasible solution, i.e., a solution with m nonzero variables, setting the other variables equal to zero. Those variables included as nonzero variables are called *included variables* and those set equal to zero the *excluded variables.*

2. Then we include other variables one-by-one and, with the included variables, evaluate the objective function. If there is any improvement (increase if we want to maximize, decrease if we want to minimize) by including a new variable while excluding one of the others, we are approaching the optimum solution. If there is no improvement, we proceed with the other variables.

3. We repeat (1) and (2) until no further improvement is possible, i.e., we have reached the optimum solution.

In particular, suppose we have an objective function

$$z = c_1x_1 + \cdots + c_nx_n,$$

and constraints

$$x_i \geq 0, \qquad i = 1, \ldots, n,$$

and

$$\begin{aligned} a_{11}x_1 + \cdots + a_{1n}x_n &= b_1 \\ &\vdots \\ a_{m1}x_1 + \cdots + a_{mn}x_n &= b_n. \end{aligned} \tag{10.7}$$

(If the constraints originally involved inequalities other than $x_i \geq 0$, we assume that they have already been transformed by the addition of slack variables.) We consider the maximization problem. The minimization problem is treated analogously.

Now we choose an initial nondegenerate basic solution. How to do this is in general a complicated problem which we shall not go into. In the case of the addition of m slack variables, we choose the slack variables as the included variables and exclude (set equal to zero) the others. At any rate, we make the selection such that the rank of the coefficient matrix for the included variables is m. For example, we can choose the first m variables if and only if the rank of

$$\begin{pmatrix} a_{11} & \cdots & a_{1m} \\ \vdots & & \vdots \\ a_{m1} & \cdots & a_{mm} \end{pmatrix}$$

is m.

Since the $m \times m$ coefficient matrix has rank m, the system represented by the included variables has a unique solution, say $(b_1, \ldots, b_m)$. We substitute these values for the m included variables and zero for the remaining variables in the objective function and obtain a value for z.

Suppose, for notational simplicity, the first m variables were included. Now we select an excluded variable, say x_k, and form the system

$$\begin{aligned} a_{11}x_{1k} + \cdots + a_{1m}x_{mk} &= a_{1k} \\ &\vdots \\ a_{m1}x_{1k} + \cdots + a_{mm}x_{mk} &= a_{mk}. \end{aligned} \tag{10.8}$$

The coefficient matrix of this system has rank m so it also has a unique solution, $(x_{1k}, \ldots, x_{mk})$. Let

$$e_k = \min_j \left\{ \left| \frac{x_j}{x_{jk}} \right| \right\}.$$

Multiplying (10.8) by e_k and subtracting from (10.7) we get

$$\begin{aligned} a_{1k}e_k + a_{11}(x_1 - e_kx_{1k}) + \cdots + a_{1m}(x_m - e_kx_{mk}) &= b_1 \\ &\vdots \\ a_{mk}e_k + a_{m1}(x_1 - e_kx_{1k}) + \cdots + a_{mm}(x_m - e_kx_{mk}) &= b_k. \end{aligned}$$

Then the value of the objective function for e_k (including x_k) is

$$z(e_k) = c_1(x_1 - e_kx_{1k}) + \cdots + c_m(x_m - e_kx_{mk}) + c_ke_k.$$

We are interested in the improvement (or lack of it) in the value of the objective function

$$z(e_k) - z_0 = e_k(c_k - (c_1 x_{1k} + \cdots + c_m x_{mk})).$$

Letting

$$z_k = c_1 x_{1k} + \cdots + c_m x_{mk},$$

we can write this difference as

$$z(e_k) - z_0 = e_k(c_k - z_k).$$

The quantity $c_k - z_k$ is called the *simplex criterion* for x_k. If $c_k - z_k > 0$, we exclude x_l, where x_l is the numerator of the smallest positive fraction among the x_j/x_{jk}, and include x_k. If $c_k - z_k \leq 0$, there is no improvement; so the initial solution is the best we have so far.

Now we use the new basic solution (or the same one if $c_k - z_k \leq 0$) and try including a different variable. This iteration process continues until further alteration in the set of variables included fails to result in improvement.

A problem which we have ignored is that of "ties" in the selection of e_k, i.e., which variable is to be excluded if

$$\frac{x_l}{x_{lk}} = \frac{x_i}{x_{ik}}, \qquad i \neq l?$$

Methods of selection range from choosing the variable with the smallest subscript through picking the one whose subscript appears first in a random number table to more complicated procedures; any one will do.

Example. Let us look at a company which produces five articles—$x_1, \ldots, x_5$, where commodities x_3 and x_4 do not enter into the cost picture. We want to minimize the cost

$$z = x_1 - 4x_2 + 2x_5,$$

subject to the constraints

$$x_i \geq 0, \qquad i = 1, \ldots, 5,$$

and

$$\begin{aligned} 2x_1 + x_2 + x_3 \qquad\qquad &= 7 \quad \text{(raw material)} \\ -x_1 - 2x_2 \qquad + x_4 \qquad &= 8 \quad \text{(capacity)} \\ 4x_1 + 3x_2 \qquad\qquad + x_5 &= 4 \quad \text{(miscellaneous)} \end{aligned}$$

Since there are no inequalities we do not add slack variables. However, the coefficient matrix for x_3, x_4, x_5 has the same form as if they were slack variables, namely it is the 3×3 identity matrix. Thus its rank is 3 and we might as well choose x_3, x_4, x_5 as the included variables in our initial nondegenerate basic feasible solution. We have

$$\begin{aligned} x_3 &= 7 \\ x_4 &= 8 \\ x_5 &= 4. \end{aligned}$$

This yields $z_0 = 2x_5 = 8$.

We test the excluded variable x_1 for inclusion. We form the system

$$\begin{aligned} a_{13}x_{31} + a_{14}x_{41} + a_{15}x_{51} &= a_{11} \\ a_{23}x_{31} + a_{24}x_{41} + a_{25}x_{51} &= a_{21} \\ a_{33}x_{31} + a_{34}x_{41} + a_{35}x_{51} &= a_{31} \end{aligned}$$

which is

$$\begin{aligned} x_{31} \qquad\qquad &= 2 \\ x_{41} \qquad &= -1 \\ x_{51} &= 4. \end{aligned}$$

The simplex criterion is found as follows:

$$\begin{aligned} z_1 &= c_3x_{31} + c_4x_{41} + c_5x_{51} \\ &= 2(4) = 8 \end{aligned}$$

so that

$$c_1 - z_1 = 1 - 8 = -7 < 0.$$

This represents improvement since we want to minimize the function.

To see which variable is to be excluded in order to include x_1, we look at

$$\frac{x_3}{x_{31}} = \frac{7}{2}, \qquad \frac{x_4}{x_{41}} = \frac{8}{-1}, \qquad \frac{x_5}{x_{51}} = \frac{4}{4}.$$

Thus x_5 is to be excluded.

To get our new basic solution we form the system

$$\begin{aligned} 2x_1 + x_3 \qquad &= 7 \\ -x_1 \qquad + x_4 &= 8 \\ 4x_1 \qquad\qquad &= 4. \end{aligned}$$

We solve this system, find the corresponding value for z, and test an excluded variable as above.

The solution of this system is

$$x_1 = 1, \qquad x_4 = 9, \qquad x_3 = 5.$$

Substituting these values in the expression for z gives a new value for z_0 of 1. Testing x_2 for inclusion, we have:

$$\begin{aligned} a_{11}x_{12} + a_{13}x_{32} + a_{14}x_{42} &= a_{12} \\ a_{21}x_{12} + a_{23}x_{32} + a_{24}x_{42} &= a_{22} \\ a_{31}x_{12} + a_{33}x_{32} + a_{34}x_{42} &= a_{32}. \end{aligned}$$

which is

$$\begin{aligned} 2x_{12} + x_{32} \qquad &= 1 \\ - x_{12} \qquad + x_{42} &= -2 \\ 4x_{12} \qquad\qquad &= 3 \end{aligned}$$

Solving this system gives us

$$x_{12} = \tfrac{3}{4}$$
$$x_{42} = \frac{-5}{4}$$
$$x_{32} = \frac{-1}{2}.$$

Now

$$z_2 = \tfrac{3}{4} \qquad \text{and} \qquad c_2 - z_2 = -4 - \tfrac{3}{4} < 0.$$

Hence x_2 is to be included. A check of the ratios

$$\frac{x_1}{x_{12}} = \frac{4}{3}, \qquad \frac{x_3}{x_{32}} = -10, \qquad \frac{x_4}{x_{42}} = \frac{-36}{5}$$

tells us that x_1 is to be excluded.

Our new basic solution is the solution to the system

$$\begin{aligned} x_2 + x_3 \qquad &= 7 \\ -2x_2 \qquad + x_4 &= 8 \\ 3x_2 \qquad\qquad &= 4, \end{aligned}$$

which is

$$x_2 = \tfrac{4}{3}, \qquad x_3 = \tfrac{17}{3}, \qquad x_4 = \tfrac{32}{3}, \qquad z = \frac{-16}{3}.$$

Testing x_5 for inclusion, we have:

$$\begin{aligned} a_{12}x_{25} + a_{13}x_{35} + a_{14}x_{45} &= a_{15} \\ a_{22}x_{25} + a_{23}x_{35} + a_{24}x_{45} &= a_{25} \\ a_{32}x_{25} + a_{33}x_{35} + a_{34}x_{45} &= a_{35}. \end{aligned}$$

which is

$$\begin{aligned} x_{25} + x_{35} \qquad &= 0 \\ -2x_{25} \qquad + x_{45} &= 0 \\ 3x_{25} \qquad\qquad &= 1 \end{aligned}$$

Solving this system gives us

$$x_{25} = \tfrac{1}{3}$$
$$x_{45} = \tfrac{2}{3}$$
$$x_{35} = \frac{-1}{3}.$$

Now

$$z_5 = (-4)(\tfrac{1}{3}) = \frac{-4}{3} \qquad \text{and} \qquad c_5 - z_5 = 2 + \tfrac{4}{3} > 0.$$

Hence we do not include x_5.

It remains to test x_1. We have

$$\begin{aligned} a_{12}x_{21} + a_{13}x_{31} + a_{14}x_{41} &= a_{11} \\ a_{22}x_{21} + a_{23}x_{31} + a_{24}x_{41} &= a_{21} \\ a_{32}x_{21} + a_{33}x_{31} + a_{34}x_{41} &= a_{31}, \end{aligned}$$

which is

$$\begin{aligned} x_{21} + x_{31} \phantom{+ x_{41}} &= 2 \\ -2x_{21} \phantom{+ x_{31}} + x_{41} &= -1 \\ 3x_{21} \phantom{+ x_{31} + x_{41}} &= 4 \end{aligned}$$

Solving this system gives us

$$\begin{aligned} x_{21} &= \tfrac{4}{3} \\ x_{31} &= \tfrac{2}{3} \\ x_{41} &= \tfrac{5}{3}. \end{aligned}$$

Now we have

$$z_1 = \frac{-16}{3}, \qquad c_1 - z_1 = 1 + \tfrac{16}{3} > 0.$$

Thus no further improvement in the value of z is possible, and $z = -16/3$ is the minimum.

PROBLEMS

1. Minimize

$$z = 6x_1 + 3x_2 + 5x_3$$

subject to the constraints

$$\begin{aligned} 2x_1 + 3x_2 + x_3 &\geq 10 \\ x_1 + x_2 + 3x_3 &\geq 12 \\ 3x_1 + x_2 + x_3 &\geq 10. \end{aligned}$$

2. Maximize

$$z = 2x_1 - 4x_2 + 5x_3$$

subject to the constraints

$$\begin{aligned} 2x_1 + x_2 - x_3 + x_4 &= 15 \\ -x_1 - x_2 + 3x_3 - x_5 &= 8 \\ 3x_1 + 2x_2 - x_3 + x_6 &= 17. \end{aligned}$$

3. Minimize

$$z = 2x_1 + 3x_2 + 4x_3$$

subject to the constraints

$$\begin{aligned} x_i \geq 0, \qquad i &= 1, 2, 3 \\ 2x_1 - x_2 + x_3 &\geq 15 \\ x_1 + x_2 + x_3 &\geq 40 \\ 4x_1 + 5x_2 - x_3 &\geq 16. \end{aligned}$$

4. Maximize

$$z = 3x_1 - 4x_2 + 7x_6$$

subject to the constraints

$$\begin{aligned} x_1 + x_2 - x_3 + x_4 &= 18 \\ 2x_1 + 3x_2 + x_3 + x_5 &= 25 \\ x_1 - x_2 + 3x_3 + x_6 &= 40. \end{aligned}$$

5. The mathematics department of Perish University has x full, y associate, and z assistant professors. By administrative decree, $x + y \leq z$. Each full professor makes \$24,000 a year, each associate \$18,000 and each assistant \$12,000. The department's salary budget is at most \$2,200,000 a year. Each full professor teaches three hrs a week and writes four papers a year; each associate teaches six hrs a week and writes two papers a year; each assistant teaches eight hrs a week and writes one paper a year. If the department must schedule at least 300 hrs of teaching a week, find the number of professors in each rank to be hired to maximize the number of publications in the department.

11 / VECTOR SPACES AND LINEAR TRANSFORMATIONS

11.1 R^n

In Chapter 2 we discussed the representation of the points on a line by real numbers and of points in the plane by pairs of real numbers. It is possible to generalize these notions in a natural way. We define *n-dimensional Euclidean space* to be the set of all ordered n-tuples of real numbers, that is,

$$R^n = \{(x_1, \ldots, x_n) \mid x_i \text{ a real number, } i = 1, \ldots, n\}.$$

To the uninitiated the notion of spaces of dimension higher than three may seem strange. Conditioned by science fiction he may think of time as the fourth dimension, but what about the fifth, sixth, and higher dimensions? The mind boggles. But it need not, for the definition above tells us precisely what is meant by n-dimensional Euclidean space. We call an n-tuple of real numbers a *vector* or *point* of *n-space.*

We know that $(x_1, x_2) = (y_1, y_2)$ if and only if $x_i = y_i$, $i = 1, 2$. We define equality of points of n-space analogously:

$$(x_1, \ldots, x_n) = (y_1, \ldots, y_n),$$

if and only if $x_i = y_i$, $i = 1, \ldots, n$.

Next we observe that it is easy to develop the operation of addition of points in n-space:

$$(x_1, \ldots, x_n) + (y_1, \ldots, y_n) = (x_1 + y_1, \ldots, x_n + y_n).$$

The sum of two points in n-space is another point in n-space. Note that the operation is commutative:

$$(x_1, \ldots, x_n) + (y_1, \ldots, y_n) = (y_1, \ldots, y_n) + (x_1, \ldots, x_n).$$

For $n = 2$, the addition of vectors is illustrated by Fig. 11.1. We indicate the vectors (x_1, x_2) and (y_1, y_2) by arrows from the origin to the points (x_1, x_2) and (y_1, y_2) respectively. Then the vector $(x_1 + y_1, x_2 + y_2)$ is their sum and can be represented as the diagonal of the parallelogram formed by the two original vectors.

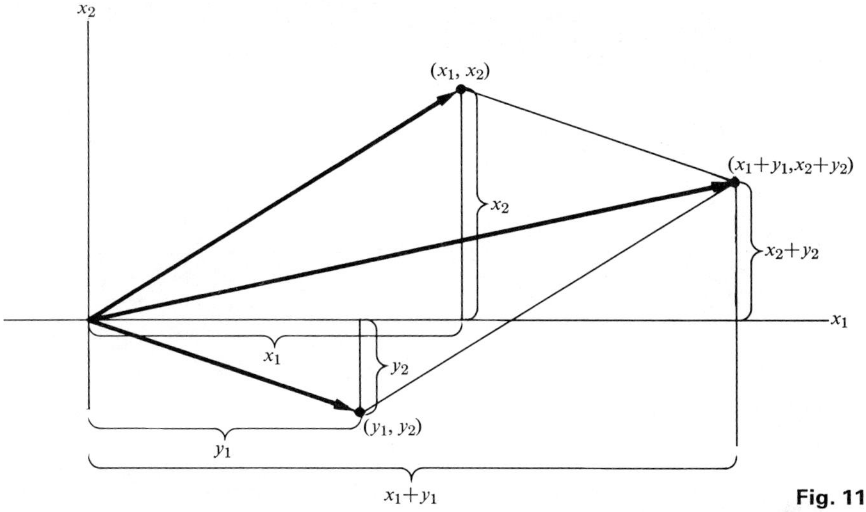

Fig. 11.1

Another operation can be defined on points of n-space. If k is a real number, let

$$k(x_1, \ldots, x_n) = (kx_1, \ldots, kx_n).$$

This operation is called *scalar multiplication*; it is illustrated in 2-space by Fig. 11.2.

Let $k > 0$; then

$$k(x_1, x_2) = (kx_1, kx_2),$$

is a vector k times as long as the original vector (an elongation if $k > 1$, a shortening if $k < 1$). If $k < 0$, then the arrow representing the point is reversed in direction. See Fig. 11.3.

If all of this looks very familiar, it is, of course, because vectors are just $1 \times n$ matrices and hence subject to the same operations as any matrices.

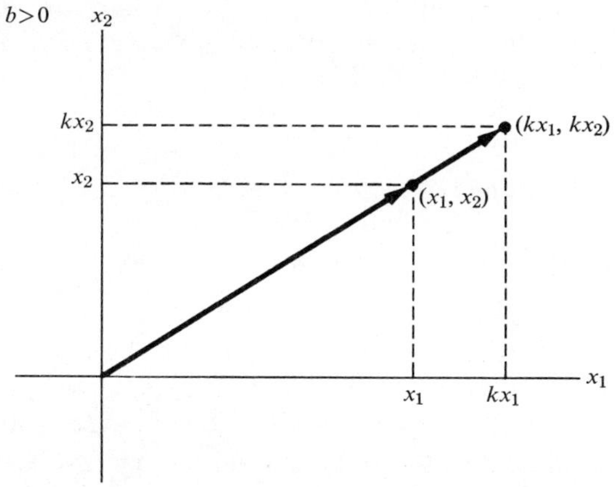

Fig. 11.2

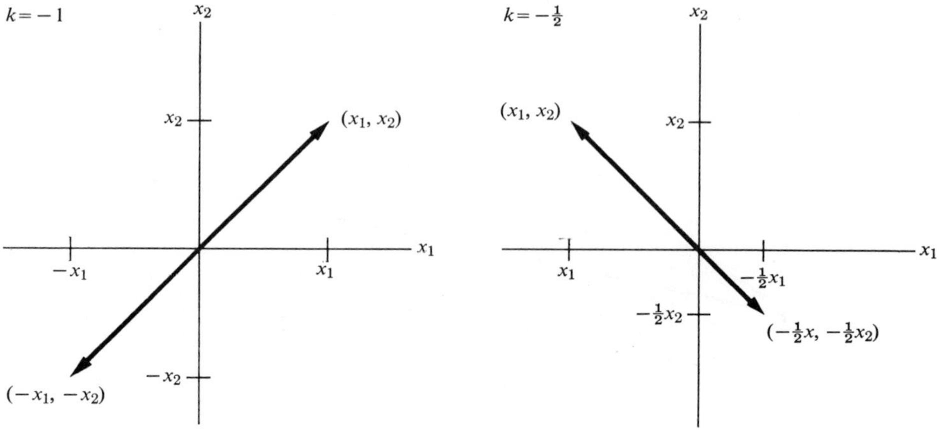

Fig. 11.3

Examples

1. Find the sum of (3, 4, 2) and (−1, 0, −2).

Solution

$$(3, 4, 2) + (-1, 0, -2) = (2, 4, 0).$$

2. For any n, we call the point of n-space $(x_1, \ldots, x_n)$ such that $x_i = 0$ for all i the *zero vector*. We note that for any n-vector $(y_1, \ldots, y_n)$,

$$(y_1, \ldots, y_n) + (0, \ldots, 0) = (y_1, \ldots, y_n).$$

That is, the zero vector is an identity for the operation of vector addition.

3. If $(x_1, \ldots, x_n)$ is a point of n-space so is $(-1)(x_1, \ldots, x_n) = (-x_1, \ldots, -x_n)$. The vector $(-x_1, \ldots, -x_n)$ is the additive inverse of $(x_1, \ldots, x_n)$; that is,

$$(x_1, \ldots, x_n) + (-x_1, \ldots, -x_n) = (0, \ldots, 0).$$

4. Find the additive inverse of (4, 0, −2, 1).

Solution. The additive inverse is given by

$$(-1)(4, 0, -2, 1) = (-4, 0, 2, -1).$$

5. Compute $5(4, 2, 1, 0, -1) + (-3)(0, 2, -1, 5, 1)$.

Solution. We have

$$\begin{aligned} 5(4, 2, 1, 0, -1) &+ (-3)(0, 2, -1, 5, 1) \\ &= (20, 10, 5, 0, -5) + (0, -6, 3, -15, -3) \\ &= (20, 4, 8, -15, -8). \end{aligned}$$

Warning. Addition of $(x_1, \ldots, x_n)$ and $(y_1, \ldots, y_m)$ is defined if and only if $n = m$. For example,

$$(3, 4, 1, 0) + (-2, 1, 5)$$

has no meaning.

Sometimes we denote the vector $(x_1, \ldots, x_n)$ by X. Then the zero vector is written as 0 and the additive inverse of X as $-X$.

6. Suppose we have four bits of information about each of the refineries of Brand X oil company: capacity in thousands of barrels per day, number of workers employed, number of gallons of gasoline produced daily, and production cost per gallon. Since each bit of information is in the form of a real number, we are really dealing with points in 4-space and the information on each refinery can be represented as a vector

$$(x_1, x_2, x_3, x_4),$$

where

x_1 = capacity in thousands of barrels per day.

x_2 = number of workers employed.

x_3 = number of gallons of gasoline produced daily.

x_4 = production cost per gallon.

If we want to pool the information on two refineries, we can represent this as the vector sum

$$(x_1, \ldots, x_4) + (y_1, \ldots, y_4) = (x_1 + y_1, \ldots, x_4 + y_4),$$

where $x_1 + y_1$ is the combined capacity, $x_2 + y_2$ the total work force, etc.

We summarize the properties of vector addition and scalar multiplication, the verification of which is left to the reader: for vectors $A, B, C \in R^n$, k_1, k_2 real numbers,

1. $(A + B) + C = A + (B + C)$.
2. $A + B = B + A$.
3. $k_1(A + B) = k_1A + k_1B$.
4. $(k_1 + k_2)A = k_1A + k_2A$.
5. $k_1(k_2A) = (k_1k_2)A$.
6. $A + 0 = A$.
7. $1 \cdot A = A$.
8. $A + (-1)A = 0$.

There is one more operation on vectors of R^n which we want to discuss: the *inner product*, which we denote by $\langle , \rangle$:

$$\langle (x_1, \ldots, x_n), (y_1, \ldots, y_n) \rangle = x_1y_1 + \cdots + x_ny_n.$$

Example

$$\langle(1, 2, -3, 0), (2, 4, -1, -1)\rangle = 2 + 8 + 3 + 0 = 13.$$

We note that scalar multiplication of a vector by a constant yields a vector, vector addition of vectors yields a vector, but the inner product of two vectors is a real number. We summarize this information in functional notation:

$$\begin{aligned} \cdot: &\quad R \times R^n \to R^n \\ +: &\quad R^n \times R^n \to R^n \\ \langle , \rangle: &\quad R^n \times R^n \to R, \end{aligned}$$

where $A \times B$ as the domain indicates that the function is a function of two variables, the first from A and the second from B.

We observe that the inner product is a special case of matrix multiplication.

$$(x_1 \cdots x_n)\begin{pmatrix} y_1 \\ \vdots \\ y_n \end{pmatrix} = x_1 y_1 + \cdots + x_n y_n = \langle(x_1, \ldots, x_n), (y_1, \ldots, y_n)\rangle.$$

Indeed the product of an $m \times n$ matrix (a_{ij}) and an $n \times p$ matrix (b_{ij}) is given by

$$(a_{ij})(b_{ij}) = (c_{ij}),$$

where

$$c_{ij} = \sum_{k=1}^{n} a_{ik} b_{kj},$$

the inner product of the ith row of A, considered as an n-vector, and the jth row of B, considered as an n-vector.

We list, again leaving the verification to the reader, the properties of the inner product. Let $A, B, C \in R^n$, k a real number. Then

1. $\langle A, B\rangle = \langle B, A\rangle$.
2. $\langle A, B + C\rangle = \langle A, B\rangle + \langle A, C\rangle$.
3. $\langle kA, B\rangle = k\langle A, B\rangle = \langle A, kB\rangle$.
4. $\langle A, A\rangle \geq 0$ with equality if and only if A is a zero vector.

Sometimes the inner product is called the *scalar product*, or the *dot product*. Note that scalar product is *not* the same as scalar multiplication. We use the inner product to generalize the notion of the length of a vector. In R^2 the length of a vector (x, y) (Fig. 11.4) is, by the Pythagorean theorem,

$$\sqrt{x^2 + y^2}.$$

But

$$\sqrt{x^2 + y^2} = \sqrt{\langle(x, y), (x, y)\rangle}.$$

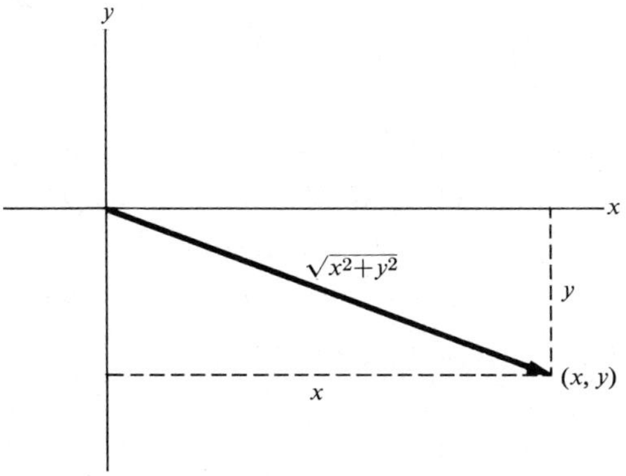

Fig. 11.4

Thus we define the *norm* (or *length*) of a vector $X \in R^n$ by

$$\|X\| = \sqrt{\langle X, X \rangle}.$$

The norm of a vector is a nonnegative real number; $\|X\| = 0$ if and only if X is a zero vector. If $\|X\| = 1$, X is said to be a *unit vector*. If $A \neq 0$, the vector

$$\frac{1}{\|A\|} A$$

is a *unit vector in direction A*.

Examples

1. Find the norm of $(1, 4, -3, 2, 0)$.

Solution

$$\|(1, 4, -3, 2, 0)\| = \sqrt{1 + 16 + 9 + 4 + 0}$$
$$= \sqrt{30}.$$

2. Find a unit vector in direction $(1, 4, -3, 2, 0)$.

Solution. The vector we want is

$$\frac{1}{\|(1, 4, -3, 2, 0)\|}(1, 4, -3, 2, 0) = \frac{1}{\sqrt{30}}(1, 4, -3, 2, 0)$$
$$= \left(\frac{1}{\sqrt{30}}, \frac{4}{\sqrt{30}}, \frac{-3}{\sqrt{30}}, \frac{2}{\sqrt{30}}, 0\right).$$

It can be shown that for vectors X and Y, and k a real number,

$$\|kX\| = |k|\|X\| \qquad \text{and} \qquad \|X + Y\| \leq \|X\| + \|Y\|.$$

PROBLEMS

1. If $A = (2, -1, 3, 4)$, $B = (0, 2, -3, 1)$, find

   a) $3A + B$. b) $2A - 5B$. c) $\langle A, B\rangle$.
   d) $\langle 4A, -2B\rangle$. e) $\langle A + B, 3B\rangle$. f) $\|A\|$.
   g) $\|B\|$. h) $\|A + 2B\|$. i) $\langle A, 2B\rangle\|A - 3B\|$.

2. If A and B are as in Problem 1, find a unit vector in

   a) direction A. b) direction $-B$. c) direction $A - 2B$.

3. If $A = (-1, 0, 2, 3)$ and $B = (2, 1, -1, 1)$, find

   a) $2A - B$. b) $\langle A, B\rangle$. c) $\langle 3A, -B\rangle$.
   d) $\|A + B\|$. e) $\langle A, 3B\rangle\|A - 2B\|$.

4. If A and B are as in Problem 3, find a unit vector in

   a) direction $-A$. b) direction $2B$. c) direction $-A + 3B$.

5. Find the product of the following pairs of matrices:

   a) $(1, 2, -3)\begin{pmatrix}0\\1\\5\end{pmatrix}$.

   b) $(4, 0, 1, -2)\begin{pmatrix}6\\9\\4\\1\end{pmatrix}$.

   c) $(2, 1, 3)\begin{pmatrix}-3\\3\\1\end{pmatrix}$.

   d) $\begin{pmatrix}2 & 1 & 5 & 6\\0 & 1 & -2 & 5\end{pmatrix}\begin{pmatrix}2 & 1\\0 & 3\\-5 & 4\\-1 & 2\end{pmatrix}$.

   e) $\begin{pmatrix}6 & -1 & 3\\2 & 5 & 9\\4 & -5 & 4\end{pmatrix}\begin{pmatrix}2 & 8 & 3\\-4 & 6 & 5\\0 & 2 & 0\end{pmatrix}$.

   f) $(-1, 0, 3, 4)\begin{pmatrix}3\\2\\-1\\0\end{pmatrix}$.

   g) $(2, 3, 1)\begin{pmatrix}-1\\-1\\2\end{pmatrix}$.

   h) $\begin{pmatrix}2 & 3 & -1 & 0\\1 & 4 & 5 & 9\\2 & 1 & 0 & 1\end{pmatrix}\begin{pmatrix}2\\0\\-1\\5\end{pmatrix}$.

   i) $\begin{pmatrix}0 & -1 & 2\\-1 & 0 & -2\\1 & 0 & 2\end{pmatrix}\begin{pmatrix}2 & 3\\1 & -3\\5 & 1\end{pmatrix}$.

6. Two vectors A and B are said to be *orthogonal* if $\langle A, B\rangle = 0$. Find a nonzero vector orthogonal to

   a) $(2, 1, 3)$. b) $(-1, 0, 1, 4)$. c) $(-2, 0, 0)$.
   d) $(1, 1, 1, 1)$. e) $(4, -1, 3, 5, 1)$.

7. Suppose the consumption of three industries is as follows:

	A	B	C
Gas	100	25	40
Coal	5	40	50
Electricity	200	100	50

Find the total consumption by taking the sum of vectors.

8. Suppose 100 units of quantity A, 200 units of quantity B, 150 units of quantity C, 50 units of quantity D, and 250 units of quantity E are purchased at prices of \$10, \$15, \$25, \$12, and \$18 per unit, respectively. Find the total expenditure by taking the inner product of the quantity vector and the price vector.

11.2 VECTOR SPACES

Mathematics is "done" in large part by looking at some familiar system, asking what makes the system "tick," and then abstracting the essential properties to make a general definition. Mathematicians noticed long ago that the properties which enable us to operate algebraically with the real numbers as we do were the existence of the operations of addition and multiplication subject to the following axioms: For all real numbers a, b, c,

1. $a + b = b + a$.
2. $a + (b + c) = (a + b) + c$.
3. $ab = ba$.
4. $a(bc) = (ab)c$.
5. $a(b + c) = ab + ac$.
6. there is a real number 0 such that for every real number a, $a + 0 = a$.
7. there is a real number 1 such that for every real number a, $a \cdot 1 = a$.
8. for each real number a, there is a real number $-a$ such that $a + (-a) = 0$.
9. for each real number $a \neq 0$, there is a real number $1/a$ such that $a(1/a) = 1$. (see Section 1.3.)

Therefore, they made the definition.

A *field* F is a nonempty set of elements together with two operations on F

$$+ : F \times F \to F$$
$$\cdot : F \times F \to F$$

such that for $a, b, c \in F$ the nine axioms above hold. The examples of fields with which one works in elementary mathematics are the field of real numbers and the field of complex numbers. Another example with which the reader may be familiar is the integers modulo 2 denoted by Z_2. Z_2 has 2 elements, 0 and 1, and addition and multiplication are defined by the tables

$+$	0	1
0	0	1
1	1	0

$\cdot$	0	1
0	0	0
1	0	1

The verification that Z_2 together with these operations is a field is left to the reader.

The set of all vectors in R^n is not a field since we do *not* have a multiplication of vectors

$$\cdot : R^n \times R^n \to R^n,$$

which produces another vector (scalar multiplication involves a real number and the inner product of vectors produces a real number.) However, we can look at R^n and the operations on it and abstract the properties which we have found that they have. We make the following definition:

A (real) *vector space* V is a nonempty set of elements together with two operations

$$+ : V \times V \to V$$
$$\cdot : R \times V \to V$$

subject to

1. For $A, B, C \in V$, $A + (B + C) = (A + B) + C$.
2. For $A, B \in V$, $A + B = B + A$.
3. For $k \in R$, $A, B \in V$, $k(A + B) = kA + kB$.
4. For $k_1, k_2 \in R$, $A \in V$, $(k_1 + k_2)A = k_1A + k_2A$.
5. For $k_1, k_2 \in R$, $A \in V$, $k_1(k_2A) = (k_1k_2)A$.
6. There is an element $0 \in V$ such that for each $A \in V$, $A + 0 = A$.
7. For each $A \in V$, $A + (-1)A = 0$.
8. $1 \cdot A = A$.

Note: it is possible to define a vector space over any field, but we confine our attention to vector spaces over the real numbers.

Examples

1. To be a good generalization a definition should include the starting point as a special case. Indeed, R^n is a vector space. In particular R^1 = the real numbers is a vector space.
2. The set of all $m \times n$ matrices (for any n and m) with entries from the real numbers is a real vector space.

3. Let $\mathscr{C}$ be the set of all continuous real-valued functions defined on the real numbers. Then for $f, g \in \mathscr{C}$, k a real number, we define addition and scalar multiplication by

$$(f + g)(x) = f(x) + g(x)$$

and

$$kf(x) = k(f(x)).$$

The sum and scalar multiplication are also continuous real-valued functions, i.e., elements of $\mathscr{C}$. It is not hard to verify that the axioms for a vector space are satisfied by $\mathscr{C}$ together with these operations. This is an important example to keep in mind to counteract the tendency to think of all vector spaces as collections of n-tuples.

4. The set

$$\{a_0 + a_1x + a_2x^2 \mid a_0, a_1, a_2 \quad \text{real numbers}\}$$

of all polynomials of degree ≤ 2, together with the operations of addition of polynomials and the multiplication of a polynomial by a constant, is a vector space.

Suppose W is a nonempty subset of a vector space V such that for $v, w \in W$ and k a real number, $v + w \in W$ and $kv \in W$. Then W is a *subspace* of V. It is not difficult to check that W is itself a vector space with the same operations as V.

Examples

1.

$$W = \{(x_1, x_2, 0) \mid x_1, x_2 \quad \text{real numbers}\}$$

is a subspace of R^3. However, $R^2 = \{(x_1, x_2) \mid x_1, x_2 \text{ real numbers}\}$ is not a subspace of R^3 since it is not even a subset.

2. The set of all differentiable functions defined on the real numbers is a subspace of the set of all continuous functions on the real numbers.

3. Consider the following subset of R^3:

$$\{(a, b, 3a)\},$$

that is, all vectors whose third coordinate is three times the first. We have

$$(a, b, 3a) + (a', b', 3a') = (a + a', b + b', 3(a + a')),$$

so that the sum of elements of the subset is in the subset. Similarly,

$$k(a, b, 3a) = (ka, kb, k(3a)) = (ka, kb, 3ka).$$

Also, the subset is nonempty; for example, (0, 0, 0) is in it. Therefore it is a subspace.

4. Consider the subset of R^3 consisting of all vectors of the form $(a, b, a + 3)$. We have

$$(a, b, a + 3) + (a', b', a' + 3) = (a + a', b + b', a + a' + 6).$$

Thus the sum of elements of the subset is not in the subset and the subset is not a subspace.

PROBLEMS

1. Check whether each of the following is a vector space:
   a) the complex numbers.
   b) the integers.
   c) the set of real-valued functions defined on $[-\pi, \pi]$.
   d) $\{a \sin x + b \cos x \mid a, b \text{ real numbers}\}$.
   e) $\{\sin ax + \cos bx \mid a, b \text{ real numbers}\}$.
   f) the set of all polynomials with real coefficients.
   g) the set of all polynomials with integer coefficients.
   h) the set of all real-valued functions defined on $[-1, 1]$.
   i) the rational numbers.
   j) the set of all differentiable functions whose second derivative is zero.
   k) the set of all triangular 2×2 matrices.
2. Is each of the following a subspace of R^3?
   a) $\{(x_1, x_2, 2x_1 + x_2) \mid x_1, x_2 \in R\}$.
   b) $\{(x_1, x_2, x_2^2) \mid x_1, x_2 \in R\}$.
   c) $\{(x_1, x_2, x_3) \mid x_1, x_2, x_3 \in R, x_1 + x_3 = 0\}$.
   d) $\{(x_1, x_2, x_1^2 - x_2^2) \mid x_1, x_2 \in R\}$.
   e) $\{(x_1, x_2, x_1 - x_2) \mid x_1, x_2 \in R\}$.
   f) $\{(3, x_1, x_2) \mid x_1, x_2 \in R\}$.
   g) $\{(x, 3x, 0) \mid x \in R\}$.
   h) the intersection of (e) and (g).
3. Is each of the following a subspace of R^4?
   a) $\{(x_1, x_2, 0) \mid x_1, x_2 \in R\}$.
   b) $\{(x_1, 2x_1, x_2, x_3) \mid x_1, x_2, x_3 \in R\}$.
   c) $\{(0, x_1, x_2, x_3) \mid x_1, x_2, x_3 \in R\}$.
   d) The intersection of the sets of (b) and (c).
   e) The union of the sets of (b) and (c).
   f) $\{a_1v_1 + a_2v_2 \mid a_1, a_2 \in R, v_1, v_2 \text{ fixed elements of } R^4\}$.
4. Is the set of all polynomials of degree ≤ 2 a subspace of the vector space of all polynomials of degree ≤ 3?

11.3 INNER PRODUCTS AND QUADRATIC FORMS

The notion of inner product also generalizes nicely:

$$\langle , \rangle : V \times V \to R$$

is an *inner product* on a vector space V if for all $A, B \in V$, k a real number,

1. $\langle A, B \rangle = \langle B, A \rangle$.
2. $\langle A, B + C \rangle = \langle A, B \rangle + \langle A, C \rangle$.

3. $\langle kA, B\rangle = k\langle A, B\rangle = \langle A, kB\rangle$.
4. $\langle A, A\rangle \geq 0$ with equality if and only if $A = 0$.

Examples

1. The ordinary inner product on R^n is an inner product under this definition.
2. For

$$X, Y \in R^2, \qquad X = (x_1, x_2), \qquad Y = (y_1, y_2),$$

   let

$$\langle X, Y\rangle = x_1y_1 + 3x_1y_2 + 3x_2y_1 + 10x_2y_2.$$

   This defines an inner product on R^2 (not the ordinary one).
3. Consider the set $\mathscr{C}[-1, 1]$ of all continuous functions on $[-1, 1]$. This is a vector space. For $f, g \in \mathscr{C}[-1, 1]$, define

$$\langle f, g\rangle = \int_{-1}^{1} f(x)g(x)\, dx.$$

   Using the properties of integrals, we see that

$$\langle f, g\rangle = \int_{-1}^{1} f(x)g(x)\, dx = \int_{-1}^{1} g(x)f(x)\, dx = \langle g, f\rangle. \tag{11.1}$$

$$\langle f, g + h\rangle = \int_{-1}^{1} f(x)(g(x) + h(x))\, dx = \int_{-1}^{1} f(x)g(x)\, dx + \int_{-1}^{1} f(x)h(x)\, dx = \langle f, g\rangle + \langle f, h\rangle. \tag{11.2}$$

$$\langle kf, g\rangle = \int_{-1}^{1} kf(x)g(x)\, dx = k\int_{-1}^{1} f(x)g(x)\, dx = k\langle f, g\rangle. \tag{11.3}$$

$$\langle f, f\rangle = \int_{-1}^{1} f(x)f(x)\, dx \geq 0, \tag{11.4}$$

with equality if and only if $f(x) = 0$ for all $x \in [-1, 1]$, i.e., f is the zero function on $[-1, 1]$.

Any polynomial in n variables all of whose terms are of degree two is a *quadratic form* on R^n. In particular, an inner product $\langle , \rangle$ defines a quadratic form by

$$q(X) = \langle X, X\rangle.$$

The quadratic form associated with Example 1 above is

$$q(X) = \langle X, X\rangle = x_1^2 + \cdots + x_n^2,$$

for $X = (x_1, \ldots, x_n)$. That associated with Example 2 is

$$q(X) = x_1^2 + 3x_1x_2 + 3x_2x_1 + 10x_2^2 = x_1^2 + 10x_2^2 + 6x_1x_2.$$

Other examples of quadratic forms are

$$3x^2 + 5xy - 5xz + 4z^2,$$

$$6x_1^2 - 9x_1x_4 + 15x_1x_6 - 3x_3^2 + x_4x_5,$$

$$x_1x_2 - x_1x_3 + x_1x_4 - x_2x_3 + x_2x_4.$$

We can represent any quadratic form by means of a symmetric matrix. If $q(X)$ is a polynomial in n variables each of whose terms is of degree two, the coefficient of the x_i^2 term is entered in the (i, i) position in an $n \times n$ matrix S and if $c_{ij}x_ix_j$, $i \neq j$, is a term of $q(X)$, $c_{ij}/2$ is entered in the (i, j) position and in the (j, i) position. Then

$$q(X) = (x_1 \cdots x_n)S \begin{pmatrix} x_1 \\ \vdots \\ x_n \end{pmatrix}.$$

For example, let

$$X = (x_1, \ldots, x_4)$$

and

$$q(X) = 3x_1^2 + 2x_1x_2 - 3x_2^2 + x_1x_3 - 3x_2x_4 + x_3^2.$$

Then this quadratic form is represented by the matrix

$$S = \begin{pmatrix} 3 & 1 & \frac{1}{2} & 0 \\ 1 & -3 & 0 & -\frac{3}{2} \\ \frac{1}{2} & 0 & 1 & 0 \\ 0 & -\frac{3}{2} & 0 & 0 \end{pmatrix},$$

since

$$(x_1 \cdots x_4)S \begin{pmatrix} x_1 \\ \vdots \\ x_4 \end{pmatrix} = q(X).$$

PROBLEMS

1. Does each of the following define an inner product?
   a) $\langle (x_1, y_1), (x_2, y_2) \rangle = x_1^2 + y_1^2 + 2x_2^2 + y_2^2$.
   b) $\langle (x_1, x_2, x_3), (y_1, y_2, y_3) \rangle = 2x_2y_2 + x_1y_1 + 3x_3y_3$.
   c) $\langle (x_1, y_1), (x_2, y_2) \rangle = x_1x_2 + 3x_2y_1 - 2x_1y_2 - 3y_1y_2$.
   d) $\langle X, Y \rangle = Y \begin{pmatrix} 2 & 1 \\ 1 & 1 \end{pmatrix} X$; X, Y 2×2 matrices.
   e) $\langle (x_1, y_1), (x_2, y_2) \rangle = 2x_1x_2 - 3y_1y_2$.
   f) $\langle (x_1, x_2), (y_1, y_2) \rangle = 3x_1 + 2y_1 - 4y_2$.
   g) $\langle X, Y \rangle = Y \begin{pmatrix} -1 & 2 \\ 1 & 0 \end{pmatrix} X$; X, Y 2×2 matrices.

h) $\langle (x_1, y_1), (x_2, y_2) \rangle = 4x_1x_2 + x_1y_2 + x_2y_1.$
i) $\langle (x_1, y_1), (x_2, y_2) \rangle = 3x_1x_2 - x_1y_2 + y_1y_2.$

j) $\langle X, Y \rangle = Y \begin{pmatrix} 0 & 1 \\ 1 & 3 \end{pmatrix} X;$ X, Y 2×2 matrices.

k) $\langle X, Y \rangle = Y \begin{pmatrix} 1 & 2 \\ 0 & 1 \end{pmatrix} {}^t X;$ X, Y 2×2 matrices.

2. For each of the inner products of Problem 1, find the associated quadratic form.

3. Find the symmetric matrix which represents each of the following quadratic forms:
a) $x^2 + 3xy + y^2.$
b) $3x^2 - 4xy + 6yz - z^2.$
c) $2x_1^2 - 4x_2^2 - 5x_3^2 + x_4^2.$
d) $x_1x_2 + x_1x_3 + x_1x_4 - x_2x_3 - x_3x_4.$
e) $\sum_{i=1}^{8} (i + 2)x_i^2.$
f) $x^2 - 2xy - 3y^2.$
g) $2x_1^2 - 3x_1x_2 + 2x_1x_3 - x_3^2 + x_4^2.$
h) $4x_1x_2 - x_2^2 - x_2x_3 + 2x_3^2.$
i) $\sum_{\substack{i \neq j \\ i,j=1}}^{4} ij(x_ix_j).$

4. Find the quadratic form which each symmetric matrix represents:

a) $\begin{pmatrix} 1 & 0 & -1 \\ 0 & 2 & 3 \\ -1 & 3 & 4 \end{pmatrix}.$
b) $\begin{pmatrix} 2 & -1 & 0 \\ -1 & 0 & 1 \\ 0 & 1 & 0 \end{pmatrix}.$

c) $\begin{pmatrix} 2 & 0 & 2 \\ 0 & 1 & 4 \\ 2 & 4 & 3 \end{pmatrix}.$
d) $\begin{pmatrix} 1 & 2 & 3 \\ 2 & -1 & 1 \\ 3 & 1 & 0 \end{pmatrix}.$

11.4 LINEAR INDEPENDENCE

We have introduced the important concept of limit. Another major idea of modern mathematics is the notion of linearity. We have already observed its manifestations —linear equations, linear combinations of equations, linear programming, etc. In some sense, linearity complements continuity and enables us to formulate and solve problems, for example, the maximizations of the last chapter, not amenable to attack by the methods of calculus. A thorough understanding of the definition of linear independence is essential.

Let $v_1, \ldots, v_n$ be elements of the vector space V, $a_1, \ldots, a_n$ real numbers; $a_1v_1 + \cdots + a_nv_n$ is said to be a *linear combination* of the v_i. If

$$a_1v_1 + \cdots + a_nv_n = 0$$

implies that $a_i = 0, i = 1, \ldots, n$, then $v_1, \ldots, v_n$ are said to be *linearly independent*.

If $v_1, \ldots, v_n$ are not linearly independent, i.e., if there exist real numbers $a_1, \ldots, a_n$, not *all* zero, such that

$$a_1 v_1 + \cdots + a_n v_n = 0,$$

then $v_1, \ldots, v_n$ are *linearly dependent.* In other words, $v_1, \ldots, v_n$ are linearly dependent if and only if one of them can be written as a linear combination of the others: to see this, let $a_j \neq 0$ and

$$a_1 v_1 + \cdots + a_n v_n = 0$$

that is, let $v_1, \ldots, v_n$ be linearly dependent. Then

$$v_j = \frac{1}{a_j}\,(a_1 v_1 + \cdots + \widehat{a_j v_j} + \cdots + a_n v_n)$$

(where the caret indicates omission of the term). Conversely, if the vector v_j is a linear combination of the others, then we need only subtract v_j from the linear combination to produce the required equation of linear dependence.

Warning. It is important to observe the quantifiers in the definition of linear independence. That is, $v_1, \ldots, v_n$ are linear independent if *whenever*

$$a_1 v_1 + \cdots + a_n v_n = 0,$$

then the a_i are all necessarily zero. It is always possible for any vectors $w_1, \ldots, w_n$ to write

$$a_1 w_1 + \cdots + a_n w_n = 0,$$

for $a_i = 0, i = 1, \ldots, n$. Also, it is possible to produce *some* linear combination (with nonzero coefficients) of linearly dependent vectors which is not equal to zero; however, if the vectors are linearly independent, *every* linear combination with nonzero coefficients must be nonzero.

Examples

1. Determine whether $(1, 2, 3)$, $(2, 0, -1)$, and $(0, -1, 2)$ are linearly independent elements of R^3.

Solution. Suppose for real numbers a_1, a_2, a_3

$$a_1(1, 2, 3) + a_2(2, 0, -1) + a_3(0, -1, 2) = 0. \tag{11.5}$$

Since this is a sum of elements of R^3, it is an element of R^3. Therefore the 0 is the zero vector of R^3, namely $(0, 0, 0)$. But two vectors of R^3 are equal if and only if each pair of coordinates is equal. Thus

$$\begin{aligned} a_1 + 2a_2 \phantom{{}- a_3} &= 0 \\ 2a_1 \phantom{{}+ 2a_2} - a_3 &= 0 \\ 3a_1 - a_2 + 2a_3 &= 0. \end{aligned} \tag{11.6}$$

The coefficient matrix

$$\begin{pmatrix} 1 & 2 & 0 \\ 2 & 0 & -1 \\ 3 & -1 & 2 \end{pmatrix}$$

is of rank 3. Therefore the system (11.6) has the unique solution $a_1 = a_2 = a_3 = 0$. But (11.5) was an arbitrary linear combination of the vectors equal to zero. Therefore $v_1, \ldots, v_n$ are linearly independent.

2. Determine whether (0, 1, 1), (2, 0, 1), and (2, −1, 0) are linearly independent elements of R^3.

Solution. Suppose

$$a_1(0, 1, 1) + a_2(2, 0, 1) + a_3(2, -1, 0) = 0,$$

for real numbers a_1, a_2, a_3. Then

$$\begin{aligned} 2a_2 + 2a_3 &= 0 \\ a_1 \qquad - a_3 &= 0 \\ a_1 + a_2 \qquad &= 0. \end{aligned}$$

The coefficient matrix

$$\begin{pmatrix} 0 & 2 & 2 \\ 1 & 0 & -1 \\ 1 & 1 & 0 \end{pmatrix}$$

is of rank 2 and therefore the system has a nontrivial solution. It does not matter *what* the solution is so long as we know that it is nontrivial, i.e., at least one of a_1, a_2, a_3 is not zero. Then we have a linear combination with nonzero coefficients equal to zero and the vectors are linearly dependent.

3. Determine whether x and x^2 are linearly independent as elements of the vector space of differentiable functions on the real numbers.

Solution. Suppose for real numbers a_1, a_2

$$a_1x + a_2x^2 = 0.$$

Taking the derivative of both sides of this equation yields

$$a_1 + 2a_2x = 0;$$

repeating the process gives

$$2a_2 = 0.$$

Substituting $a_2 = 0$, we obtain $a_1 = 0$. That is, the only way to have a linear combination of x and x^2 equal to zero is to have both coefficients zero. Thus x and x^2 are linearly independent.

Note that $a_1x + a_2x^2 = 0$ is not an equation for which we are seeking a solution, i.e., a value of x which makes it a true statement. Rather we are trying to find values for a_1 and a_2 such that the equation holds for all values of x.

Suppose S is a set of elements of a vector space V such that every $v \in V$ can be written as

$$v = a_1 v_1 + \cdots + a_n v_n, \qquad v_i \in S, \quad a_i \text{ real numbers.}$$

Then the set S is said to *generate* (or *span*) V. If we let $S = V$, then this definition is satisfied, that is, the elements of V generate V. However, the notion of generating is more interesting in the case in which S is finite.

Example. (1, 1) and (2, 0) generate R^2. For, let $(x, y) \in R^2$. We claim that there are real numbers a and b such that

$$a(1, 1) + b(2, 0) = (x, y).$$

We have

$$\begin{aligned} a + 2b &= x \\ a &= y \end{aligned}$$

for which we obtain $a = y$, $b = (x - y)/2$. Thus we have $(x, y) = y(1, 1) + (x - y)/2\ (2, 0)$. For example, (3, 4) can be written as

$$4(1, 1) - \tfrac{1}{2}(2, 0) = (3, 4).$$

Now we are ready to make the following definition: if S is a set of linearly independent vectors which generate a vector space V, S is called a *basis* for V.

It turns out, although we shall not prove it, that every vector space has a basis. Again it is the case in which S has only a finite number of elements in which we are most interested. If V has a basis with a finite number of elements, it is said to be a *finite-dimensional vector space.*

Example. We claim that {(1, 0, 0), (0, 1, 0), and (0, 0, 1)} is a basis for R^3.

We must show

1. that these vectors are linearly independent;
2. that they generate R^3.

For the first assertion, suppose

$$a_1(1, 0, 0) + a_2(0, 1, 0) + a_3(0, 0, 1) = 0.$$

Then

$$\begin{aligned} a_1 &= 0 \\ a_2 &= 0 \\ a_3 &= 0, \end{aligned}$$

so the vectors are linearly independent. Now let (x, y, z) be an arbitrary element of R^3. Then

$$x(1, 0, 0) + y(0, 1, 0) + z(0, 0, 1) = (x, y, z).$$

Thus

$$\{(1, 0, 0), (0, 1, 0), (0, 0,1)\}$$

is a basis. In fact, we can show that

$$\{(1, 0, \ldots, 0), \ldots, (0, \ldots, 1, \ldots, 0), \ldots, (0, \ldots, 1)\}$$

is a basis for R^n, usually called the *canonical* basis. Its elements are called *the unit vectors* for R^n.

We prove a few theorems, the purpose of which is to enable us to define the dimension of a vector space.

Theorem 1. Let $\{v_1, \ldots, v_m\}$ be a basis for a vector space V. If $w_1, \ldots, w_n$ are elements of V, $n > m$, then $w_1, \ldots, w_n$ are linearly dependent.

Proof. Since $\{v_1, \ldots, v_m\}$ is a basis, we can write

$$\begin{aligned} w_1 &= a_{11}v_1 + \cdots + a_{1m}v_m \\ &\vdots \\ w_n &= a_{n1}v_1 + \cdots + a_{nm}v_m, \end{aligned}$$

for some real numbers a_{ij}.

Now suppose $x_1, \ldots, x_n$ are real numbers and form the linear combination

$$\begin{aligned} &x_1w_1 + \cdots + x_nw_n \\ &\qquad = (x_1a_{11} + \cdots + x_na_{n1})v_1 + \cdots + (x_1a_{1m} + x_1a_{m1} + \cdots + x_na_{nm})v_m. \end{aligned}$$

Now

$$\begin{aligned} x_1a_{11} + \cdots + x_na_{n1} &= 0 \\ \vdots \qquad\qquad &\vdots \\ x_1a_{m1} + \cdots + x_na_{mn} &= 0 \end{aligned}$$

has a nontrivial solution $(x_1, \ldots, x_n)$ since $n > m$. But this gives us real numbers $x_1, \ldots, x_n$, not all zero, such that

$$x_1w_1 + \cdots + x_nw_n = 0,$$

so that the w_i are linearly dependent.

Corollary. Let V be a vector space with a basis of n elements. Then any other basis for V has n elements.

Proof. Suppose

$$\{v_1, \ldots, v_n\} \qquad \text{and} \qquad \{w_1, \ldots, w_m\}$$

are bases. Then two applications of the theorem show that $m \not> n$ and $n \not> m$. Hence $n = m$.

Since we now know that all bases have the same number of elements it makes sense to say: If V is a vector space with a basis of n elements, we say that it is of *dimension n*.

Theorem 2. Let V be a vector space and $\{v_1, \ldots, v_n\}$ a maximal set of linearly independent elements of V, i.e., any subset of V of more than n elements is linearly dependent. Then $\{v_1, \ldots, v_n\}$ is a basis for V.

Proof. We need only show that $\{v_1, \ldots, v_n\}$ generates V. Let $v \in V$. Then by the maximality of $\{v_1, \ldots, v_n\}$ there are real numbers $a_1, \ldots, a_n, a$, not all zero, such that

$$a_1v_1 + \cdots + a_nv_n + av = 0.$$

Moreover, $a \neq 0$; for if $a = 0$ then we have a relation of linear dependence among $v_1, \ldots, v_n$. Hence

$$v = \frac{-1}{a}(a_1v_1 + \cdots + a_nv_n);$$

that is, an arbitrary element of V can be written as a linear combination of $v_1, \ldots, v_n$. Thus $\{v_1, \ldots, v_n\}$ generates V.

Corollary. If V is a vector space of dimension n, any subset of n linearly independent elements of V is a basis.

Proof. By Theorem 1, the subset is a maximal linearly independent subset, so by Theorem 2 it is a basis.

Example. We have observed that the n unit vectors in R^n form a basis. Hence, as we would expect, R^n is of dimension n. Thus, for example, any three linearly independent elements of R^3 constitute a basis, so that given three elements as a potential basis we need only check their linear independence.

Example. Is $\{(0, -1, 3), (5, 2, 1), (-1, -1, -1)\}$ a basis for R^3?

Solution. Let

$$a_1(0, -1, 3) + a_2(5, 2, 1) + a_3(-1, -1, -1) = 0.$$

Then

$$\begin{aligned} 5a_2 - a_3 &= 0 \\ -a_1 + 2a_2 - a_3 &= 0 \\ 3a_1 + a_2 - a_3 &= 0. \end{aligned}$$

The rank of

$$\begin{pmatrix} 0 & 5 & -1 \\ -1 & 2 & -1 \\ 3 & 1 & -1 \end{pmatrix}$$

is 3 so the only solution is the trivial one, $a_1 = a_2 = a_3 = 0$. Therefore the vectors are linearly independent and

$$\{(0, -1, 3), (5, 2, 1), (-1, -1, -1)\}$$

is a basis for R^3. Note that any subset of 4 or more elements of R^3 is linearly dependent.

Warning. Although in the rest of this chapter we shall be mainly concerned with finite-dimensional vector spaces, many important vector spaces are infinite-dimensional. In these cases we do not have the luxury of writing down a finite set of elements as a basis and expressing any element of the vector space as a linear combination of them.

Example. The vector space of differentiable functions defined on the real numbers is infinite-dimensional. To see this, consider the set of functions

$$\{x^i \mid i \text{ a positive integer}\}.$$

By extending the method we used for x and x^2, we can show that any finite subset of elements of this set must be linearly independent. Hence no finite set of elements of the vector space can be a basis, for if the supposed basis has n elements, then

$$\{x, x^2, \ldots, x^n, x^{n+1}\}$$

is a set of more than n linearly independent elements, contradicting Theorem 1.

Warning. Although the number of elements in a basis of a finite-dimensional vector space is uniquely determined, the basis itself is not. For example, we have already looked at several different bases for R^3.

If $\{v_1, \ldots, v_n\}$ is a basis of a vector space V, we can write $v \in V$ as

$$v = a_1v_1 + \cdots + a_nv_n.$$

Then (see Problem 6), $(a_1, \ldots, a_n)$ is a uniquely determined element of R^n, called the *coordinate vector of v with respect to the basis* $\{v_1, \ldots, v_n\}$. If $\{w_1, \ldots, w_n\}$ is another basis and

$$v = b_1w_1 + \cdots + b_nw_n,$$

the coordinate vectors $(a_1, \ldots, a_n)$ and $(b_1, \ldots, b_n)$ are equal if and only if $v_i = w_i$, $i = 1, \ldots, n$.

Example. Find the coordinate vector of $(2, -1, 0)$ with respect to each of the following bases of R^3:

a) $\{(1, 0, 0), (0, 1, 0), (0, 0, 1)\}$,

b) $\{(2, 3, 1), (0, 2, 0), (1, 1, 1)\}$ and,

c) $\{(0, -1, 3), (5, 2, 1), (-1, -1, -1)\}$.

Solution

a) $(2, -1, 0) = 2(1, 0, 0) + (-1)(0, 1, 0) + 0(0, 0, 1)$, so the coordinate vector is $(2, -1, 0)$.

b) $(2, -1, 0) = a(2, 3, 1) + b(0, 2, 0) + c(1, 1, 1)$. The solution to

$$\begin{aligned} 2a \quad\quad + c &= 2 \\ 3a + 2b + c &= -1 \\ a \quad\quad + c &= 0, \end{aligned}$$

is $a = 2$, $b = -5/2$, $c = -2$, so the coordinate vector is $(2, -5/2, -2)$.

c) $(2, -1, 0) = a(0, -1, 3) + b(5, 2, 1) + c(-1, -1, -1)$. The solution to

$$\begin{aligned} 5b - c &= 2 \\ -a + 2b - c &= -1 \\ 3a + b - c &= 0 \end{aligned}$$

is the coordinate vector $(\frac{6}{13}, \frac{11}{13}, \frac{29}{13})$.

PROBLEMS

1. Determine whether each of the following sets of vectors is linearly independent:
   a) $(0, 1, 2), (3, 4, 5), (3, 3, 0)$.
   b) $(2, 5, 4), (-1, 0, 4), (4, 5, -4)$.
   c) $(1, 2, 0, 1), (0, 1, 2, 0), (1, 0, 2, 1)$.
   d) $\sin x$, $\cos x$ (as elements of the vector space of differentiable real-valued functions).
   e) $\begin{pmatrix} 0 & 1 \\ 2 & 0 \end{pmatrix}, \begin{pmatrix} 2 & 3 \\ 1 & 1 \end{pmatrix}, \begin{pmatrix} 1 & -1 \\ -1 & 1 \end{pmatrix}, \begin{pmatrix} 4 & 0 \\ 0 & 0 \end{pmatrix}.$
   f) $x^2 - 2x$, $3x - 4$, $x^3 + 1$ (as elements of the vector space of all polynomials with real coefficients).
   g) $(0, 1, 3), (2, 1, -1), (-1, 6, 2), (0, 1, 0)$.
   h) $(1, -1, 1), (2, 0, 1), (0, 3, 2)$.
   i) $\begin{pmatrix} 1 & 2 \\ -1 & 0 \end{pmatrix}, \begin{pmatrix} 3 & 4 \\ 2 & 1 \end{pmatrix}, \begin{pmatrix} -1 & 0 \\ 4 & 0 \end{pmatrix}, \begin{pmatrix} 0 & 2 \\ 3 & 0 \end{pmatrix}.$
   j) $\sin 2x$, $\sin x$.
   k) $x^2 + 2$, $x - 5$, x.
2. Show that any subset S of a vector space V spans a subspace of V. (*Hint:* consider the set of linear combinations of elements of S.)
3. Find a basis for R^3 which contains $(-1, -1, -1)$.
4. Find a basis for R^3 which contains $(2, 1, 3)$.
5. Find the dimension of the vector space spanned by
   a) $(2, 1, 1), (0, -1, 3), (2, 3, -5)$.
   b) $(2, 1), (-1, 3)$.
   c) $1, x, x^2, x^3$.
   d) $1, 2x, 3x$.
   e) $(1, 0, -1), (2, 1, 1), (3, 4, 0), (0, -1, 2)$.
   f) $(2, 1, 1), (3, -1, 0), (1, 3, 2)$.
   g) $3x^3 - 1$, $x^2 + 1$, x, $x + 1$.
   h) $(-1, 0, 2), (0, 1, 3), (2, 1, 1)$.

6. If $\{v_1, \ldots, v_n\}$ is a basis of V and $v \in V$, show that

$$a_1v_1 + \cdots + a_nv_n = v = b_1v_1 + \cdots + b_nv_n$$

implies that $a_i = b_i$, $i = 1, \ldots, n$.

7. Find the coordinate vector of $(-1, 2)$ with respect to each of the following bases of R^2:

a) $\{(1, 0), (0, 1)\}$. b) $\{(0, -1), (7, 1)\}$. c) $\{(2, 3), (-1, 2)\}$.

8. Find the coordinate vector of $3x^2 - 4x + 5$ with respect to each of the following bases of the vector space of all polynomials of degree ≤ 2:

a) $\{1, x, x^2\}$ b) $\{2x^2 + 1, 3x - 4, -x + 1\}$.

9. Find the coordinate vector of $(2, 3)$ with respect to

a) $\{(1, 0), (1, 1)\}$. b) $\{(-1, 2), (2, 3)\}$. c) $\{(5, 1), (-4, 7)\}$.

11.5 LINEAR TRANSFORMATIONS AND MATRICES

In calculus the only functions of real interest are those which are at least continuous. The definition below characterizes the kind of function which is important in the theory of vector spaces.

Suppose V and W are vector spaces and $F: V \to W$ is a function. If for $v, v' \in V$, k a real number,

$$F(v + v') = F(v) + F(v'), \tag{11.7}$$

and

$$F(kv) = kF(v), \tag{11.8}$$

then F is a *linear transformation*. Note that the addition and multiplication on the left-hand sides of (11.7) and (11.8) are those in the vector space V and on the right-hand sides those in the vector space W.

Examples

1. It is always possible to define the *identity* function on a vector space V, $I: V \to V$, by

$$I(v) = v \qquad \text{for} \qquad v \in V.$$

The verification that I is a linear transformation is straightforward. Given any two vector spaces V and W, we can define a linear transformation $0: V \to W$ by

$$0(v) = 0 \qquad \text{for all} \qquad v \in V.$$

The 0 on the right-hand side is the zero vector of the vector space W. We call such a function a *zero transformation*.

2. Define $F: R^3 \to R^2$ by

$$F(x, y, z) = (x, y + z).$$

Then

$$\begin{aligned} F[(x, y, z) + (x', y', z')] \\ &= F(x + x', y + y', z + z') = (x + x', y + y' + z + z') \\ &= (x + x', y + z + y' + z) = (x, y + z) + (x', y' + z') \\ &= F(x, y, z) + F(x', y', z'). \end{aligned}$$

Also

$$\begin{aligned} F[k(x, y, z)] &= F(kx, ky, kz) = (kx, ky + kz) \\ &= \big(kx, k(y + z)\big) = k(x, y + z) = kF(x, y, z). \end{aligned}$$

Thus F is a linear transformation.

3. Define $P\colon R^n \to R^{n-k}$ by

$$P(x_1, \ldots, x_n) = (x_1, \ldots, x_{n-k}) \qquad (n > k).$$

It is easy to verify that P is a linear transformation. P is called a *projection.*

4. Let V be the vector space of C^∞ functions defined on the real numbers. Define $D\colon V \to V$ by

$$D(f) = f',$$

that is, D is the function which assigns to each element of V its derivative. From the properties of derivatives,

$$D(f + g)d = D(f) + D(g)$$

and

$$D(kf) = kD(f).$$

Therefore D is a linear transformation.

5. Let F and G be two linear transformations with the same domain and range, $F, G\colon V \to W$. Define $F + G$ and kF, k a real number, by

$$(F + G)(v) = F(v) + G(v) \qquad \text{and} \qquad (kF)(v) = kF(v).$$

Then

$$\begin{aligned} (F + G)(v + v') &= F(v + v') + G(v + v') = F(v) + F(v') + G(v) + G(v') \\ &= F(v) + G(v) + F(v') + G(v') \\ &= (F + G)(v) + (F + G)(v'). \end{aligned}$$

Similarly, $(kF)(cv) = ckF(v)$ for real numbers k and c. Thus we have an operation of addition and an operation of multiplication by a scalar. Under these operations the set of all linear transformations $V \to W$ is itself a vector space, denoted by $\mathscr{L}(V, W)$.

6. Let $G\colon V \to W$, $F\colon W \to Z$ be linear transformations of vector spaces. Then their composition as functions is given by

$$F \circ G(t) = F\big(G(t)\big).$$

A straightforward verification shows that the composite function is a linear transformation $F \circ G: V \to Z$.

7. Consider the function

$$F: R^2 \to R^2,$$

given by

$$F(x, y) = (x^2, y).$$

Is it a linear transformation?

Solution.

$$\begin{aligned} F\big((x, y) + (x', y')\big) &= F(x + x', y + y') = \big((x + x')^2, y + y'\big) \\ &= (x^2 + 2xx + (x')^2, y + y'), \end{aligned}$$

which is not in general equal to

$$F(x, y) + F(x', y') = (x^2 + (x')^2, y + y').$$

Therefore, it is not a linear transformation.

A linear transformation $F: V \to W$ which is one-to-one, i.e., $F(v) = F(v')$ implies that $v = v'$, is said to be *injective*. If it is onto, that is, if for each $w \in W$ there is a $v \in V$ such that $F(v) = w$, it is said to be *surjective*. A linear transformation which is both surjective and injective is said to be an *isomorphism*. If $F: V \to W$ is an isomorphism, V and W are *isomorphic*.

Examples

1. An identity linear transformation is an isomorphism.

2. The linear transformation

$$D(f) = f'$$

is not injective since two functions which differ by a constant have the same derivative.

3. The projections

$$P: R^n \to R^{n-k} \qquad (n > k)$$

are surjective.

4. If $F: V \to W$ is an isomorphism we can define an inverse

$$F^{-1}(w) = v,$$

where v is the unique element of V such that $F(v) = w$. Then $F^{-1} \circ F =$ identity transformation on V and $F \circ F^{-1} =$ identity transformation on W. F^{-1} is also a linear transformation and indeed it is also an isomorphism.

5. Let $F: R^3 \to R^3$ be given by

$$F(x, y, z) = (3x, -y, 2z).$$

Then if

$$F(x, y, z) = F(x', y', z'), \qquad 3x = 3x', \qquad -y = -y', \qquad 2z = 2z',$$

so that

$$x = x', \qquad y = y', \qquad z = z',$$

and F is injective.

To show that F is surjective, let (u, v, w) be an element of R^3. Then

$$F\left(\frac{u}{3}, -v, \frac{w}{2}\right) = \left(\frac{3u}{3}, -(-v), \frac{2w}{2}\right) = (u, v, w),$$

so that F is surjective.

6. Let $F: R^3 \to R^4$ be given by

$$F(x, y, z) = (2x, y + z, -z, x + z).$$

Then if

$$F(x, y, z) = F(x', y', z'),$$

we have

$$2x = 2x', \qquad y + z = y' + z'$$
$$-z = -z', \qquad x + z = x' + z'$$

from which $x = x'$, $y = y'$, $z = z'$. Therefore F is injective. On the other hand, F is not surjective. In particular, $(1, 1, 1, 1)$ is not the image under F of any point in R^3; for it to be the image we would need to be able to find x and z such that $2x = 1$, $-z = 1$ and $x + z = 1$, which is not possible.

7. R^4 is isomorphic to the vector space of 2×2 matrices with entries from the real numbers. We can define an isomorphism

$$F: R^4 \to M_{2 \times 2}$$

by

$$F(x_1, x_2, x_3, x_4) = \begin{pmatrix} x_1 & x_2 \\ x_3 & x_4 \end{pmatrix}.$$

The verification that this is a linear transformation, that it is one-to-one and that it is onto is left to the reader.

8. R^3 is isomorphic to the subspace of the 2×2 matrices consisting of all symmetric matrices (verify that this *is* a subspace) with an isomorphism given by

$$F(x_1, x_2, x_3) = \begin{pmatrix} x_1 & x_2 \\ x_2 & x_3 \end{pmatrix}.$$

Observe that an isomorphism "identifies" the elements of one vector space with those of another. Isomorphic vector spaces are, except for the names by which their elements are known, the same space.

Now suppose V is a finite-dimensional vector space with basis $v_1, \ldots, v_n$ and $F\colon V \to W$ a linear transformation. Then if we write

$$\begin{aligned} v &= a_1v_1 + \cdots + a_nv_n, \\ F(v) &= F(a_1v_1 + \cdots + a_nv_n) = a_1F(v_1) + \cdots + a_nF(v_n). \end{aligned}$$

Thus in order to specify the value of F on each element of V, it suffices to specify its value on each basis element. In fact we have

Theorem 3. If $\{v_1, \ldots, v_n\}$ is a basis for a vector space V and $w_1, \ldots, w_n$ are arbitrary elements of a vector space W, then there is a unique linear transformation $F\colon V \to W$ such that $F(v_i) = w_i$, $i = 1, \ldots, n$.

Proof. For $v = a_1v_1 + \cdots + a_nv_n$ in V, we define

$$F\colon V \to W \qquad \text{by} \qquad F(v) = a_1w_1 + \cdots + a_nw_n.$$

Then

$$F(v_i) = w_i, \qquad i = 1, \ldots, n.$$

Moreover, if

$$v' = a_1'v_1 + \cdots + a_n'v_n,$$

then

$$\begin{aligned} F(v + v') &= F[(a_1 + a_1')v_1 + \cdots + (a_n + a_n')v_n] \\ &= (a_1 + a_1')w_1 + \cdots + (a_n + a_n')w_n \\ &= F(v) + F(v'). \end{aligned}$$

For k a real number,

$$\begin{aligned} F(kv) &= F[ka_1v_1 + \cdots + ka_nv_n] = ka_1w_1 + \cdots + ka_nw_n \\ &= k(a_1w_1 + \cdots + a_nw_n) = kF(v). \end{aligned}$$

Therefore F is a linear transformation. To show uniqueness we suppose that G is a linear transformation such that

$$G(v_i) = w_i, \qquad i = 1, \ldots, n.$$

Then by the linearity of G we have

$$\begin{aligned} G(v) &= G(a_1v_1 + \cdots + a_nv_n) = a_1G(v_1) + \cdots + a_nG(v_n) \\ &= a_1w_1 + \cdots + a_nw_n = F(v). \end{aligned}$$

Thus $F = G$.

The importance of this theorem needs to be emphasized. It means that in order to define completely a linear transformation on an n-dimensional vector space V we need only say what it does to the n elements of some basis of V; moreover it is the *only* linear transformation which has these values on the basis elements.

Example. Define a linear transformation $F: R^3 \to R^2$ by

$$\begin{aligned} F(1, 0, 0) &= (2, 3) \\ F(0, 1, 0) &= (0, 1) \\ F(0, 0, 1) &= (-1, 1). \end{aligned}$$

Find $F(2, 3, -1)$.

Solution. Since we have specified that F be a linear transformation, we have

$$\begin{aligned} F(x, y, z) &= F\big(x(1, 0, 0) + y(0, 1, 0) + z(0, 0, 1)\big) \\ &= x(2, 3) + y(0, 1) + z(-1, 1). \end{aligned}$$

In particular

$$\begin{aligned} F(2, 3, -1) &= 2(2, 3) + 3(0, 1) - 1(-1, 1) \\ &= (4, 6) + (0, 3) + (1, -1) \\ &= (5, 8). \end{aligned}$$

Theorem 3 enables us to represent a linear transformation by means of a matrix. Let $\{v_1, \ldots, v_n\}$ be a basis for V, $\{w_1, \ldots, w_m\}$ a basis for W, $F: V \to W$ a linear transformation. Then for $v \in V$,

$$\begin{aligned} F(v) &= F(x_1v_1 + \cdots + x_nv_n) \\ &= x_1F(v_1) + \cdots + x_nF(v_n). \end{aligned}$$

On the other hand, each $F(v_i)$ is in W so we can write

$$F(v_i) = b_{1i}w_1 + \cdots + b_{mi}w_m, \qquad i = 1, \ldots, n.$$

Combining these we have

$$\begin{aligned} F(v) &= x_1(b_{11}w_1 + \cdots + b_{m1}w_m) + \cdots + x_n(b_{1n}w_1 + \cdots + b_{mn}w_m) \\ &= (x_1b_{11} + \cdots + x_nb_{1n})w_1 + \cdots + (x_1b_{m1} + \cdots + x_nb_{mn})w_m. \end{aligned}$$

Now

$$(x_1b_{11} + \cdots + x_nb_{1n}, \ldots, x_1b_{m1} + \cdots + x_nb_{mn})$$

is the coordinate vector of $F(v)$ with respect to the basis $\{w_1, \ldots, w_m\}$. Moreover, $x_1b_{i1} + \cdots + x_nb_{in}$ is the scalar product of the coordinate vector $(x_1, \ldots, x_n)$ and the coordinate vector $(b_{i1}, \ldots, b_{in})$ of $F(v_i)$.

For convenience's sake we sometimes write the coordinate vector $(x_1, \ldots, x_n)$ as a column vector

$$X = \begin{pmatrix} x_1 \\ \vdots \\ x_n \end{pmatrix}.$$

Using this notation, the coordinate vector of $F(v)$ is $(b_{ij})X$. We have just proved the following theorem.

Theorem 4. Let $F: V \to W$ be a linear transformation, $\{v_1, \ldots, v_n\}$ a basis of V, $\{w_1, \ldots, w_m\}$ a basis of W. Let X be the column coordinate vector of v with respect to $\{v_1, \ldots, v_n\}$ and B the $m \times n$ matrix whose jth column is the coordinate vector of $F(v_i)$ with respect to $\{w_1, \ldots, w_m\}$. Then the coordinate vector of $F(v)$ with respect to $\{w_1, \ldots, w_m\}$ is given by

$$F(v) = BX.$$

(Note that the matrix representation depends upon the choice of basis for the domain and range spaces.)

Examples

1. Find the matrix associated with the projection map $R^4 \to R^2$ (with respect to the canonical bases).

Solution. We look at the action of the transformation on the basis

$$\begin{aligned} P(1, 0, 0, 0) &= 1(1, 0) + 0(0, 1) \\ P(0, 1, 0, 0) &= 0(1, 0) + 1(0, 1) \\ P(0, 0, 1, 0) &= 0(0, 0) + 0(0, 0) \\ P(0, 0, 0, 1) &= 0(0, 0) + 0(0, 0). \end{aligned}$$

The required matrix is

$$\begin{pmatrix} 1 & 0 & 0 & 0 \\ 0 & 1 & 0 & 0 \end{pmatrix}.$$

Note that in general the matrix representing a linear transformation $F: V \to W$ is obtained by taking the transpose of the coefficient matrix of the system of linear equations in the basis vectors of W obtained by evaluating the linear transformation on the basis of V.

2. Find the matrix representation of the projection map of Problem 1 with respect to the basis $\{(1, 2, 0, 1), (-1, 2, -1, 0), (2, 1, 1, 1), (0, 1, 4, 3)\}$ and $\{(2, 1), (-3, 0)\}$.

Solution

$$\begin{aligned} P(1, 2, 0, 1) &= (1, 2) = a_{11}(2, 1) + a_{21}(-3, 0) \\ P(-1, 2, -1, 0) &= (-1, 2) = a_{12}(2, 1) + a_{22}(-3, 0) \\ P(2, 1, 1, 1) &= (2, 1) = a_{13}(2, 1) + a_{23}(-3, 0) \\ P(0, 1, 4, 3) &= (0, 1) = a_{14}(2, 1) + a_{24}(-3, 0). \end{aligned}$$

$$\begin{aligned} a_{11} &= 2, & a_{21} &= 1, & a_{12} &= 2, & a_{22} &= \tfrac{5}{3}, \\ a_{13} &= 1, & a_{23} &= 0, & a_{14} &= 1, & a_{24} &= \tfrac{2}{3}. \end{aligned}$$

The required matrix is

$$\begin{pmatrix} 2 & 2 & 1 & 1 \\ 1 & \frac{5}{3} & 0 & \frac{2}{3} \end{pmatrix}.$$

3. Suppose we want to rotate the axes in R^2. For example, to eliminate an xy term in an equation, we will replace the canonical basis vectors $E_1 = (1, 0)$, $E_2 = (0, 1)$ by $T(E_1)$, $T(E_2)$ (see Fig. 11.5), T a linear transformation such that

$$T(E_1) = \cos\theta E_1 + \sin\theta E_2,$$
$$T(E_2) = -\sin\theta E_1 + \cos\theta E_2.$$

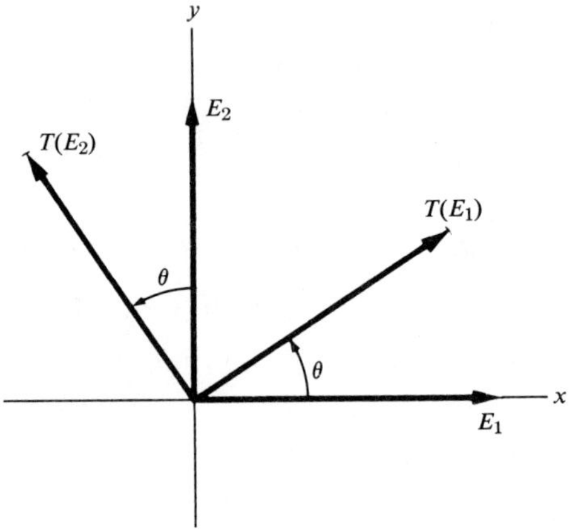

Fig. 11.5

Then the matrix representation of T is

$$\begin{pmatrix} \cos\theta & -\sin\theta \\ \sin\theta & \cos\theta \end{pmatrix}.$$

Suppose we want to find the coordinate vector of $(2, -1)$ with respect to a new coordinate system obtained by a 45° rotation.

Solution. The matrix is

$$\begin{pmatrix} \frac{1}{\sqrt{2}} & \frac{-1}{\sqrt{2}} \\ \frac{1}{\sqrt{2}} & \frac{1}{\sqrt{2}} \end{pmatrix},$$

and the image is

$$\begin{pmatrix} \frac{1}{\sqrt{2}} & \frac{-1}{\sqrt{2}} \\ \frac{1}{\sqrt{2}} & \frac{1}{\sqrt{2}} \end{pmatrix} \begin{pmatrix} 2 \\ -1 \end{pmatrix} = \begin{pmatrix} \frac{3}{\sqrt{2}} \\ \frac{1}{\sqrt{2}} \end{pmatrix}.$$

4. Let $T\colon V \to W$ be a linear transformation whose matrix representation is A and $S\colon W \to Z$ a linear transformation whose matrix representation is B. Then for $v \in V$,

$$(S \circ T)(v) = S(T(v)) = S(AX) = BAX,$$

where X is the coordinate vector of v with respect to a given basis. Thus composition of linear transformations can be represented by the multiplication of matrices.

5. If $F\colon R^3 \to R^3$ is represented by the matrix

$$\begin{pmatrix} 1 & 2 & 3 \\ -2 & 0 & 1 \\ 4 & 3 & 0 \end{pmatrix}$$

and $G\colon R^3 \to R^4$ is represented by the matrix

$$\begin{pmatrix} 1 & 0 & 0 \\ -1 & 3 & 2 \\ 0 & 2 & -2 \\ 2 & 0 & 1 \end{pmatrix},$$

both with respect to the canonical bases. Find $(G \circ F)(-1, 2, 3)$.

Solution.

$$(G \circ F)(-1, 2, 3) = \begin{pmatrix} 1 & 0 & 0 \\ -1 & 3 & 2 \\ 0 & 2 & -2 \\ 2 & 0 & 1 \end{pmatrix} \begin{pmatrix} 1 & 2 & 3 \\ -2 & 0 & 1 \\ 4 & 3 & 0 \end{pmatrix} \begin{pmatrix} -1 \\ 2 \\ 3 \end{pmatrix}$$

$$= \begin{pmatrix} 12 \\ 7 \\ 6 \\ 26 \end{pmatrix}.$$

If

$$A = \begin{pmatrix} a_{11} & \cdots & a_{1n} \\ \vdots & & \\ a_{m1} & \cdots & a_{mn} \end{pmatrix}$$

is an $m \times n$ matrix, we associate with it a linear transformation

$$F\colon V \to W \qquad (\text{dimension } V = n, \text{ dimension } W = m),$$

such that the coordinate vector of $F(v)$ with respect to a given basis of W is given by

$$F(v) = (a_{ij}) \begin{pmatrix} x_1 \\ \vdots \\ x_n \end{pmatrix},$$

where $(x_1, \ldots, x_n)$ is the coordinate vector of v with respect to a given basis of V.

Example. If

$$\begin{pmatrix} 1 & 0 & 2 \\ -1 & 2 & -1 \\ 3 & 4 & 5 \\ 2 & 1 & 4 \end{pmatrix}$$

represents a linear transformation $F: R^3 \to R^4$ with respect to the canonical bases of R^3 and R^4, find $F(1, 2, -3)$.

Solution

$$F(1, 2, -3) = \begin{pmatrix} 1 & 0 & 2 \\ -1 & 2 & -1 \\ 3 & 4 & 5 \\ 2 & 1 & 4 \end{pmatrix} \begin{pmatrix} 1 \\ 2 \\ -3 \end{pmatrix} = \begin{pmatrix} -5 \\ 6 \\ -4 \\ -8 \end{pmatrix}.$$

Because each linear transformation on a finite-dimensional vector space defines a matrix and conversely, there is a surjective and injective function from $\mathscr{L}(V, W)$ (space of linear transformations from V to W) to the vector space of m, n matrices, where dimension $V = n$ and dimension $W = m$. This function is also a linear transformation, so the spaces are isomorphic.

PROBLEMS

1. Determine whether each of the following is a linear transformation.
   a) $F(x, y, z) = (3x, y + z, 2z)$.
   b) $F(x, y, z) = (x, 2y, -z, 0)$.
   c) $F(x, y, z) = x + y + z$.
   d) $F(x, y, z) = (x^2 + y^2, z)$.
   e) $F(x, y, z) = (3, y, z)$.
   f) $F(x, y, z) = (x + 3, 2y)$.
   g) D^2 (second derivative, as a function on the space of C^∞ functions).
   h) $F(x, y, z) = (-x, y - z, 2z)$.
   i) $F(x, y) = x + y + 1$.
   j) $F(x, y, z) = (0, y, x)$
   k) $D^2 + D$.
   l) $F(x, y, z, w) = x^2 + y^2 + z^2 + w^2$.
2. Determine which of the linear transformations of Problem 1 are injective and which are surjective. Which are isomorphisms?
3. Define an isomorphism from $\{(x_1, x_2, x_1 + x_2) \mid x_i \in R, i = 1, 2\}$ to R^2.
4. Find the matrix which represents the following linear transformations with respect to the canonical bases:
   a) $F(x, y, z) = (3x - y, z)$.
   b) $F(x, y, z) = (2x, -y, z, y + z)$.
   c) $F(x, y, z) = (x, 0, y)$.
   d) $F(x, y, z) = (3x, y + z)$.
   e) $F(x, y) = (x, 2y, 2x)$.
   f) $F(x, y, z) = (2x, y - z, -z)$.
5. Find the matrix representation of $F(x, y, z) = (2x + y, -z)$ with respect to the bases $\{(2, 1, 3), (0, -1, 2), (-1, 1, 0)\}$ and $\{(3, 1), (2, 1)\}$.
6. If

$$\begin{pmatrix} 2 & 0 & 1 \\ 6 & 5 & 6 \\ -1 & 2 & 3 \end{pmatrix}$$

is the matrix representation of $F\colon R^3 \to R^3$ with respect to the canonical basis, find $F(1, -1, 2)$.

7. If

$$\begin{pmatrix} 2 & 1 \\ 3 & 5 \\ -1 & 0 \end{pmatrix}$$

represents $F\colon R^2 \to R^3$ with respect to $\{(1, 1), (2, -1)\}$ as a basis for R^2 and $\{(0, 1, -1), (2, 1, 3), (4, 1, 0)\}$ as a basis for R^3, find $F(1, -1)$.

8. If

$$\begin{pmatrix} 0 & -1 & 2 \\ 3 & 2 & -1 \\ 0 & 5 & 1 \end{pmatrix}$$

is the matrix representation of $F\colon R^3 \to R^3$ with respect to the canonical basis, find $F(0, -4, 2)$.

9. Find the matrix representation of $F(x, y, z) = (x, y + z)$ with respect to the bases $\{(1, 2, 1), (-1, 0, 2), (3, 1, 3)\}$ and $\{(2, 0), (-1, 2)\}$.

10. Find the matrix representation of the derivative as a function from
    a) the space of polynomials of degree ≤ 2 to itself with respect to the basis $\{1, x, x^2\}$.
    b) same as (a) with respect to the basis $\{3x - 1, 2x^2 + 1, 5x\}$.
    c) the space $V = \{a \sin x + b \cos x \mid a, b \in R\}$ to itself with respect to the basis $\{\sin x, \cos x\}$.
    d) the space of polynomials of degree ≤ 3 to itself with respect to the basis $\{2 - x, x^2 - 2, x^3, x\}$.

11.6 APPLICATIONS

1. Markov Chains. Suppose that we have a sequence of experiments such that the outcome of each experiment is one of a finite number of possible outcomes $a_1, \ldots, a_n$. We assume that the probability of outcome a_j in any given experiment is not necessarily independent of the outcomes of previous experiments but depends at most upon the outcome of the immediately preceding experiment. We denote by p_{ij} the probability of outcome a_j given that a_i occurred on the preceding experiment. The outcomes $a_1, \ldots, a_n$ are called *states* and the p_{ij} *transition probabilities*. If we assume that the process begins in some particular state we can determine the probability measure for the process and calculate the probabilities of statements relating to the whole sequence of experiments. Such a process is called a *Markov chain process*.

We can represent the probabilities by a matrix

$$\begin{pmatrix} p_{11} & \cdots & p_{1n} \\ \vdots & & \\ p_{n1} & \cdots & p_{nn} \end{pmatrix}.$$

The sum $\sum_{j=1}^{n} p_{ij}$ of entries in the ith row is the sum of all probabilities when the process is in state a_i and therefore must equal 1. In general, a vector $(x_1, \ldots, x_n)$ such that $\sum_{i=1}^{n} x_i = 1$ and $x_i \geq 0$ for all i is called a *probability vector.* A square matrix with each entry nonnegative and the sum of the entries in each row equal to one is called a *transition matrix.*

To obtain a Markov chain it is necessary to specify an initial state. We suppose that the initial state is chosen by a random device that selects state a_j with probability $p_j^{(0)}$. Then the initial probability is represented by the vector

$$p^{(0)} = (p_1^{(0)}, \ldots, p_n^{(0)}).$$

A type of problem of particular interest is to determine the probability p_j^n of the process being in state a_j after n steps. Let us look at a three state situation. We have

$$\begin{aligned} p_1^{(n)} &= p_1^{(n-1)}p_{11} + p_2^{(n-1)}p_{21} + p_3^{(n-1)}p_{31}, \\ p_2^{(n)} &= p_1^{(n-1)}p_{12} + p_2^{(n-1)}p_{22} + p_3^{(n-1)}p_{32}, \\ p_3^{(n)} &= p_1^{(n-1)}p_{13} + p_2^{(n-1)}p_{23} + p_3^{(n-1)}p_{33}. \end{aligned} \tag{11.9}$$

That is, the probability of being in state a_j after n steps is the sum of the probabilities of being at each of the three possible states after $n - 1$ steps and then moving to state a_j on the nth step. This holds for any finite number m of states so that

$$p^{(n)} = p^{(n-1)}P,$$

where P is the transition matrix (p_{ij}), and $p^{(n)} = (p_1^{(n)}, \ldots, p_m^{(n)})$. Thus

$$p^{(n)} = p^{(n-1)}P = (p^{(n-2)}P)P = \cdots = p^{(0)}P^n.$$

Example. If the transition matrix for a Markov chain process is given by

$$\begin{pmatrix} \frac{1}{2} & \frac{1}{2} & 0 \\ \frac{1}{3} & 0 & \frac{2}{3} \\ 0 & 0 & 1 \end{pmatrix},$$

find the probability of the process being in state a_2 after four steps, given that the initial probability is $p^{(0)} = (\frac{1}{4}, \frac{1}{4}, \frac{1}{2})$.

Solution

$$\begin{aligned} p^{(4)} &= (\tfrac{1}{4}, \tfrac{1}{4}, \tfrac{1}{2}) \begin{pmatrix} \frac{1}{2} & \frac{1}{2} & 0 \\ \frac{1}{3} & 0 & \frac{2}{3} \\ 0 & 0 & 1 \end{pmatrix}^4 \\ &= (\tfrac{1}{4}, \tfrac{1}{4}, \tfrac{1}{2}) \begin{pmatrix} \frac{31}{144} & \frac{7}{48} & \frac{23}{36} \\ \frac{7}{72} & \frac{5}{72} & \frac{5}{6} \\ 0 & 0 & 1 \end{pmatrix} \\ &= (\tfrac{45}{576}, \tfrac{31}{576}, \tfrac{125}{144}). \end{aligned}$$

Thus $p_2^{(4)} = \frac{31}{576}$, the probability of being in state a_2 after four steps.

If v is a probability vector such that

$$v = vP$$

for a transition matrix P, v is called a *fixed point* of P. If $p^{(0)}$, the initial probability vector of a Markov chain process is a fixed point of P, then

$$p^{(n)} = p^{(0)}P^n = p^{(0)};$$

that is, the probability of being at any particular state is the same at all steps of the process. In such a case the process is said to be in *equilibrium*. We state two theorems without proof.

A transition matrix P is *regular* if there is a positive integer n such that every element of P^n is positive.

Theorem 5. If $P = (p_{ij})$ is a $k \times k$ regular transition matrix, then

1) there is a $k \times k$ matrix W such that given $\varepsilon > 0$, there exists a positive integer N such that for $n > N$

$$|p_{ij}^{(n)} - w_{ij}| < \varepsilon \qquad \text{for all} \quad i, j,$$

where $p_{ij}^{(n)}$ is the i, jth entry of P^n;

2) each row of W is the same probability vector

$$w = (w_1, \ldots, w_k);$$

3) $w_i > 0$ for all i.

Theorem 6. If P, W, and w are as in Theorem 5, then

1) if p is a probability vector, let pP^n be denoted by $(x_1^{(n)}, \ldots, x_k^{(n)})$; then given $\varepsilon > 0$ there is a positive integer N such that for $n > N$

$$|x_i^{(n)} - w_i| < \varepsilon,$$

$i = 1, \ldots, k$;

2) w is the unique probability vector which is a fixed point of P.

We see that since a regular transition matrix approaches a fixed matrix, any Markov chain process represented by such a matrix approaches equilibrium.

Example. The probability that someone whose father was a registered Democrat will be a registered Democrat is 0.7; that he will be a Republican 0.2, and an Independent 0.1. The children of registered Republicans split 0.6 Republican, 0.2 Democrat, and 0.2 Independent; while the children of Independents go 0.5 Independent, 0.3 Democrat, and 0.2 Republican. Set up the process as a Markov chain, compute the probability of a grandchild of a registered Democrat being a Democrat and find the long-range fraction to be expected in each party.

Solution. The transition matrix is given by

$$P = \begin{cases} & \text{Democrat} & \text{Republican} & \text{Independent} \\ \text{Democrat} & 0.7 & 0.2 & 0.1 \\ \text{Republican} & 0.2 & 0.6 & 0.2 \\ \text{Independent} & 0.3 & 0.2 & 0.5 \end{cases}$$

If we are interested in the grandchild of a registered Democrat, the initial state probability vector is given by

$$p^{(0)} = (1, 0, 0).$$

We want to find

$$\begin{aligned} p^{(2)} &= p^{(0)}P^2 \\ &= (1, 0, 0)\begin{pmatrix} 0.7 & 0.2 & 0.1 \\ 0.2 & 0.6 & 0.2 \\ 0.3 & 0.2 & 0.5 \end{pmatrix}\begin{pmatrix} 0.7 & 0.2 & 0.1 \\ 0.2 & 0.6 & 0.2 \\ 0.3 & 0.2 & 0.5 \end{pmatrix} \\ &= (0.56, 0.28, 0.16). \end{aligned}$$

Thus the probability that the grandchild of a registered Democrat will be a registered Democrat is 0.56.

For the second part of the question we want to find a fixed point of P, that is, a vector w such that $w = wP$. Let $w = (w_1, w_2, w_3)$. Then we must have

$$\begin{aligned} 0.7w_1 + 0.2w_2 + 0.3w_3 &= w_1 \\ 0.2w_1 + 0.6w_2 + 0.2w_3 &= w_2 \\ 0.1w_1 + 0.2w_2 + 0.5w_3 &= w_3 \end{aligned}$$

This converts to the homogeneous system

1) $$-0.3w_1 + 0.2w_2 + 0.3w_3 = 0$$

2) $$0.2w_1 - 0.4w_2 + 0.2w_3 = 0$$

3) $$0.1w_1 + 0.2w_2 - 0.5w_3 = 0.$$

(w_1, w_2, w_3) must be a probability vector all of whose entries are positive, so we have the further conditions

4) $$w_1 + w_2 + w_3 = 1, \qquad w_1, w_2, w_3 \geq 0.$$

Solving the system (1)–(4), yields the probability vector $(\frac{4}{9}, \frac{1}{3}, \frac{2}{9})$. Thus the long-range fraction of Democrats is $\frac{4}{9}$, of Republicans $\frac{1}{3}$, and of Independents $\frac{2}{9}$.

Observe that the requirement that w be a fixed point row vector

$$w = wP,$$

is equivalent to saying that the column vector ${}^t w$ must be a characteristic vector associated with the characteristic value 1 of ${}^t P$:

$${}^t w = {}^t P \, {}^t w.$$

*2. **Matrix Equations.*** a) Recall the example of the lawn furniture company in Chapter 10 whose input matrix was

	Plastic	Aluminum	Machine time	Labor time	Miscellaneous
Table	0.5	3.1	2	1	0.2
Chair	2.4	2.1	1.5	2	0.4
Lounger	3	2	1	2.5	0.1

Suppose an order is received for x tables, y chairs, and z loungers. What raw materials must the production department order and what time must be scheduled?

Solution. Let A be the input matrix, $X = (x, y, z)$ the order vector, and P the production vector. We have

$$XA = P.$$

In particular, suppose 100 tables, 400 chairs, and 150 loungers are ordered. Then

$$(100, 400, 150)\begin{pmatrix} 0.5 & 3.1 & 2 & 1 & 0.2 \\ 2.4 & 2.1 & 1.5 & 2 & 0.4 \\ 3 & 2 & 1 & 2.5 & 0.1 \end{pmatrix} = (1460, 1450, 950, 1275, 195),$$

so that 1460 units of plastic, 1450 units of aluminum, and 195 units of miscellaneous must be available for production, together with 950 hours of machine time and 1275 hours of labor.

b) Also in Chapter 10 we considered a laboratory which processed vaccine. The processing matrix P was given by

$$\begin{pmatrix} 3 & 2 \\ 1 & 3 \end{pmatrix}$$

and the capacity vector by

$$C = \begin{pmatrix} 12 \\ 16 \end{pmatrix}.$$

Then we solved the matrix equation

$$PX = C,$$

by finding P^{-1} and then

$$X = P^{-1}B.$$

Suppose the capacity vector is altered; for example, another piece of apparatus of type A is purchased. Then the capacity vector is

$$\begin{pmatrix} 24 \\ 16 \end{pmatrix}.$$

The hard part of the earlier solution was finding

$$P^{-1} = \frac{1}{7}\begin{pmatrix} 3 & -2 \\ -1 & 3 \end{pmatrix},$$

so a change in C does not present much difficulty. The new solution is

$$X = P^{-1}C = \frac{1}{7}\begin{pmatrix} 3 & -2 \\ -1 & 3 \end{pmatrix}\begin{pmatrix} 24 \\ 16 \end{pmatrix}$$

$$= \frac{1}{7}\begin{pmatrix} 40 \\ 24 \end{pmatrix} = \begin{pmatrix} \frac{40}{7} \\ \frac{24}{7} \end{pmatrix}.$$

c) A certain company generates electricity, purifies water and manufactures a chemical. In order to produce one unit of chemical it uses two units of electricity and one unit of water. To produce one unit of water it uses $\frac{1}{2}$ unit of electricity and $\frac{1}{10}$ unit of chemical; to produce one unit of electricity requires $\frac{1}{5}$ unit of electricity, $\frac{1}{10}$ unit of water and $\frac{1}{10}$ unit of chemical. We have what is called a consumption matrix:

	Production of			Consumed
	electricity	water	chemical	
$C =$	$\frac{1}{5}$	$\frac{1}{2}$	2	electricity
	$\frac{1}{10}$	0	1	water
	$\frac{1}{10}$	$\frac{1}{10}$	0	chemical

Let

$$X = \begin{pmatrix} x_1 \\ x_2 \\ x_3 \end{pmatrix}$$

be the production vector, x_1 = number of units of electricity produced, x_2 = number of units of water produced, and x_3 = number of units of chemical produced. The product CX is the internal consumption necessary for this level of production.

Now suppose that the company has orders for 1000 units of electricity, 500 units of water, and 100 units of chemical. How much must be produced to fill the orders?

Solution. We seek a solution to

$$X - CX = D,$$

where D is the demand vector:

$$\begin{pmatrix} 1000 \\ 500 \\ 100 \end{pmatrix}.$$

We rewrite the equation as

$$(I - C)X = D,$$

where I is the 3×3 identity matrix. Then

$$X = (I - C)^{-1}D,$$

if this inverse exists. Note that once $(I - C)^{-1}$ is calculated, finding the production vector corresponding to a new demand vector is simply a matter of finding the product of a matrix and a vector. We have

$$X = \left[\begin{pmatrix} 1 & 0 & 0 \\ 0 & 1 & 0 \\ 0 & 0 & 1 \end{pmatrix} - \begin{pmatrix} \frac{1}{5} & \frac{1}{2} & 2 \\ \frac{1}{10} & 0 & 1 \\ \frac{1}{10} & \frac{1}{10} & 0 \end{pmatrix}\right]^{-1} \begin{pmatrix} 1000 \\ 500 \\ 100 \end{pmatrix}$$

$$= \begin{pmatrix} \frac{4}{5} & -\frac{1}{2} & -2 \\ -\frac{1}{10} & 1 & -1 \\ -\frac{1}{10} & -\frac{1}{10} & 1 \end{pmatrix}^{-1} \begin{pmatrix} 1000 \\ 500 \\ 100 \end{pmatrix}$$

$$= \frac{5}{2}\begin{pmatrix} \frac{9}{10} & \frac{7}{10} & \frac{5}{2} \\ \frac{1}{5} & \frac{3}{5} & 1 \\ \frac{11}{100} & \frac{13}{100} & \frac{3}{4} \end{pmatrix} \begin{pmatrix} 1000 \\ 500 \\ 100 \end{pmatrix}$$

$$= \begin{pmatrix} 3750 \\ 1500 \\ 625 \end{pmatrix}.$$

Thus to meet the demand it is necessary to produce 3750 units of electricity, 1500 units of water, and 625 units of chemical.

3. *Diagonalization*

$$f(x, y, z) = 5x^2 + 3y^2 + 3z^2 - 2xy + 2yz - 2xz = 1,$$

is the equation of a surface in R^3. Suppose we want to transform the equation into

$$g(u, v, w) = c_1u^2 + c_2v^2 + c_3w^2 = 1.$$

That is, we want to introduce new variables in order to eliminate the cross-products. The geometric interpretation is that of the rotation of axes. We see a simple case in R^2 (Fig. 11.6):

$$f(x, y) = ax^2 + by^2 + cxy = 1 \quad (a, b > 0)$$
$$g(u, v) = a_1u^2 + b_1v^2 = 1,$$

where rotation eliminates the xy term in the equation of an ellipse. (See also Section 4.) But to return to the original problem, we represent $f(x, y, z)$ as

$$(x, y, z)\begin{pmatrix} 5 & -1 & -1 \\ -1 & 3 & 1 \\ -1 & 1 & 3 \end{pmatrix}\begin{pmatrix} x \\ y \\ z \end{pmatrix} = 1,$$

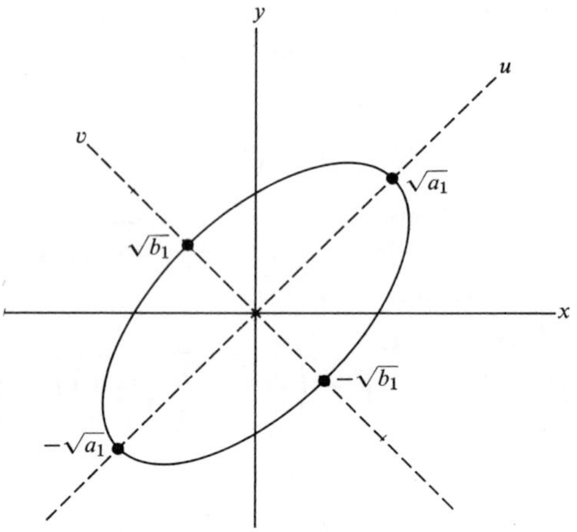

Fig. 11.6

or ${}^tXAX = 1$, and $g(u, v, w)$ as

$$(u, v, w)\begin{pmatrix} c_1 & 0 & 0 \\ 0 & c_2 & 0 \\ 0 & 0 & c_3 \end{pmatrix}\begin{pmatrix} u \\ v \\ w \end{pmatrix} = 1,$$

or ${}^tUDU$. A square matrix (a_{ij}) with $a_{ij} = 0, i \neq j$, is a *diagonal* matrix. Since the introduction of new variables really is just a change of basis, we should be able to represent the change of variables as a linear transformation and hence as a matrix T. Thus we want a transformation represented by a matrix T such that

$$T(U) = X.$$

If such a T exists, then

$${}^tXAX = {}^t(TU)ATU = {}^tU({}^tTAT)U$$

(by Problem 9 of Sec. 10.3). That is, the desired diagonal matrix is ${}^tTAT$. The following theorem tells us that under certain conditions such a T exists and how we may find it.

Theorem 7. Let A be a symmetric $n \times n$ matrix with distinct characteristic values $k_1, \ldots, k_n$ and corresponding characteristic unit vectors $X_1, \ldots, X_n$. If T is the matrix whose column vectors are the X_i, $i = 1, \ldots, n$, then

$$T^{-1}AT = \begin{pmatrix} k_1 & & 0 \\ & \ddots & \\ 0 & & k_n \end{pmatrix}$$

and ${}^tT = T^{-1}$.

Proof. By matrix multiplication we have that the column vectors of AT are $AX_1, \ldots, AX_n$. But the X_i are characteristic vectors so the columns of AT are

$k_1X_1, \ldots, k_nX_n$. But the product of the matrix whose columns are $X_1, \ldots, X_n$, namely T, and

$$\begin{pmatrix} k_1 & & 0 \\ & \ddots & \\ 0 & & k_n \end{pmatrix}$$

also yields a matrix whose column vectors are $k_1X_1, \ldots, k_nX_n$:

$$AT = T\begin{pmatrix} k_1 & & 0 \\ & \ddots & \\ 0 & & k_n \end{pmatrix}.$$

Now we multiply by T^{-1} to get

$$T^{-1}AT = T^{-1}T\begin{pmatrix} k_1 & & 0 \\ & \ddots & \\ 0 & & k_n \end{pmatrix} = \begin{pmatrix} k_1 & & 0 \\ & \ddots & \\ 0 & & k_n \end{pmatrix}.$$

For the proof that $T^{-1} = {}^tT$ see, e.g., Serge Lang, *Linear Algebra*; 2nd. ed., Addison-Wesley, Reading, Mass., 1971.

To apply this theorem to our example we need to find the characteristic values of the given matrix A:

$$A - kI = \begin{pmatrix} 5-k & -1 & -1 \\ -1 & 3-k & 1 \\ -1 & 1 & 3-k \end{pmatrix}.$$

Thus the characteristic values are the roots of

$$-k^3 + 11k^2 - 36k + 36,$$

namely 2, 3, 6.

The corresponding characteristic vectors are found by solving the equations

$$\begin{pmatrix} 5 & -1 & -1 \\ -1 & 3 & 1 \\ -1 & 1 & 3 \end{pmatrix}\begin{pmatrix} x \\ y \\ z \end{pmatrix} = 2\begin{pmatrix} x \\ y \\ z \end{pmatrix},$$

$$\begin{pmatrix} 5 & -1 & -1 \\ -1 & 3 & 1 \\ -1 & 1 & 3 \end{pmatrix}\begin{pmatrix} x \\ y \\ z \end{pmatrix} = 3\begin{pmatrix} x \\ y \\ z \end{pmatrix},$$

and

$$\begin{pmatrix} 5 & -1 & -1 \\ -1 & 3 & 1 \\ -1 & 1 & 3 \end{pmatrix}\begin{pmatrix} x \\ y \\ z \end{pmatrix} = 6\begin{pmatrix} x \\ y \\ z \end{pmatrix}.$$

The characteristic vectors are

$$\begin{pmatrix} \frac{1}{2} \\ 0 \\ -\frac{1}{2} \end{pmatrix}, \quad \begin{pmatrix} \frac{1}{3} \\ \frac{1}{3} \\ \frac{1}{3} \end{pmatrix}, \quad \begin{pmatrix} \frac{1}{6} \\ -\frac{2}{6} \\ \frac{1}{6} \end{pmatrix}.$$

Thus

$$T = \begin{pmatrix} \frac{1}{2} & \frac{1}{3} & \frac{1}{6} \\ 0 & \frac{1}{3} & -\frac{2}{6} \\ -\frac{1}{2} & \frac{1}{3} & \frac{1}{6} \end{pmatrix},$$

and

$${}^{t}TAT = T^{-1}AT = \begin{pmatrix} 2 & 0 & 0 \\ 0 & 3 & 0 \\ 0 & 0 & 6 \end{pmatrix}.$$

The transformed quadratic form is $g(u, v, w) = 2u^2 + 3v^2 + 6w^2 = 1$.

Note that T is not unique as the characteristic vectors assembled in a different order produce a matrix which also diagonalizes A. Observe also that in order to find $T^{-1}AT$ it suffices to find the characteristic values; we need not find the characteristic vectors nor T itself. The matrix $T^{-1}AT$ is said to be *similar* to A.

This solves a special case of the general problem of whether for a given matrix A there exists a matrix T such that $T^{-1}AT$ is diagonal. The answer is no for the general case. Applications of diagonalization are found in many areas, e.g., factor analysis in statistics. We close the chapter with one more example.

Suppose x_n and y_n represent changes, say in certain economic indices, after n intervals of time and we can write

$$\begin{aligned} x_n &= a_{11}x_{n-1} + a_{12}y_{n-1} \\ y_n &= a_{21}x_{n-1} + a_{22}y_{n-1}. \end{aligned} \tag{11.10}$$

We are interested in the behavior of x_n and y_n as n becomes arbitrarily large—in particular we want to know whether they approach zero. Such systems are called *difference equations* (see Chapter 14), and the problem is a *stability* problem.

Writing (11.10) in matrix notation

$$X_n = AX_{n-1},$$

we see that

$$X_n = A^{n-1}X_1$$

(compare with the Markov chain process discussed earlier).

If there is a matrix T such that $T^{-1}AT = D$ is a diagonal matrix, then

$$\begin{aligned} A &= TDT^{-1} \\ A^2 &= (TDT^{-1})(TDT^{-1}) = TD^2T^{-1} \\ &\vdots \\ A^{n-1} &= TD^{n-1}T^{-1}. \end{aligned}$$

However, raising a diagonal matrix to the nth power produces a diagonal matrix whose diagonal elements are the original diagonal elements raised to the nth power. Thus we have

$$X_n = A^{n-1}X_1 = TD^{n-1}T^{-1}X_1$$
$$= T\begin{pmatrix} k_1^{n-1} & 0 \\ 0 & k_2^{n-1} \end{pmatrix} T^{-1}X_1,$$

where k_1 and k_2 are the characteristic values of A and X_1 represents the initial conditions. From this equation we see that the solution must be of the form

$$x_n = a'_{11}k_1^{n-1} + a'_{12}k_2^{n-1}$$
$$y_n = a'_{21}k_1^{n-1} + a'_{22}k_2^{n-1}.$$

Then the necessary and sufficient condition for x_n and y_n to approach zero as $n \to \infty$ is that

$$|k_i| < 1, \qquad i = 1, 2.$$

For example, let

$$x_n = -3x_{n-1} + 4y_{n-1}$$
$$y_n = 4x_{n-1} + 3y_{n-1}.$$

Then let

$$A = \begin{pmatrix} -3 & 4 \\ 4 & 3 \end{pmatrix}.$$

The characteristic values are $\pm$ 5. The associated unit characteristic vectors are

$$\begin{pmatrix} 1/\sqrt{5} \\ -2/\sqrt{5} \end{pmatrix} \quad \text{and} \quad \begin{pmatrix} 2/\sqrt{5} \\ 1/\sqrt{5} \end{pmatrix}.$$

Thus let

$$T = \begin{pmatrix} \dfrac{1}{\sqrt{5}} & \dfrac{-2}{\sqrt{5}} \\ \dfrac{2}{\sqrt{5}} & \dfrac{1}{\sqrt{5}} \end{pmatrix},$$

$$T^{-1} = \begin{pmatrix} \dfrac{1}{\sqrt{5}} & \dfrac{2}{\sqrt{5}} \\ \dfrac{2}{\sqrt{5}} & \dfrac{1}{\sqrt{5}} \end{pmatrix}.$$

As a check we compute that

$$T^{-1}AT = \begin{pmatrix} 5 & 0 \\ 0 & -5 \end{pmatrix}.$$

Now

$$X_n = \begin{pmatrix} \frac{1}{\sqrt{5}} & \frac{-2}{\sqrt{5}} \\ \frac{2}{\sqrt{5}} & \frac{1}{\sqrt{5}} \end{pmatrix} \begin{pmatrix} 5^{n-1} & 0 \\ 0 & (-5)^{n-1} \end{pmatrix} \begin{pmatrix} \frac{1}{\sqrt{5}} & \frac{2}{\sqrt{5}} \\ \frac{-2}{\sqrt{5}} & \frac{1}{\sqrt{5}} \end{pmatrix} X_1$$

and

$$x_n = \frac{(x_1 + 2y_1)5^{n-1}}{5} + \frac{(4x_1 - 2x_2)(-5)^{n-1}}{5}$$

$$y_n = \frac{(2x_1 + 4y_1)5^{n-1}}{5} + \frac{(-2x_1 + y_1)(-5)^{n-1}}{5},$$

where

$$X_1 = \begin{pmatrix} x_1 \\ y_1 \end{pmatrix}.$$

Note that the values of x_n and y_n do not approach zero since $|\pm 5| > 1$.

Some matrices which are not symmetric may be diagonalized in the same manner. The crucial step is to find linearly independent characteristic vectors to form the columns of the matrix T. Even if a matrix T is found such that if A is the given matrix, then $T^{-1}AT$ is diagonal, it is not necessarily the case that $T^{-1} = {}^tT$ as in the symmetric case.

Example. Diagonalize

$$\begin{pmatrix} 1 & 2 \\ 3 & 2 \end{pmatrix}.$$

Solution. The characteristic values are 4, -1. Although we shall not prove the fact, it is the case that distinct characteristic values produce linearly independent characteristic vectors; for this matrix the unit characteristic vectors are

$$\begin{pmatrix} \frac{2}{\sqrt{13}} \\ \frac{3}{\sqrt{13}} \end{pmatrix}, \quad \begin{pmatrix} \frac{1}{\sqrt{2}} \\ \frac{-1}{\sqrt{2}} \end{pmatrix}.$$

Let

$$T = \begin{pmatrix} \frac{2}{\sqrt{13}} & \frac{1}{\sqrt{2}} \\ \frac{3}{\sqrt{13}} & \frac{-1}{\sqrt{2}} \end{pmatrix}.$$

Then

$$T^{-1} = \frac{-\sqrt{26}}{5}\begin{pmatrix} \dfrac{-1}{\sqrt{2}} & \dfrac{-1}{\sqrt{2}} \\ \dfrac{-3}{\sqrt{13}} & \dfrac{2}{\sqrt{13}} \end{pmatrix} \quad \text{and} \quad T^{-1}AT = \begin{pmatrix} 4 & 0 \\ 0 & -1 \end{pmatrix}.$$

PROBLEMS

1. If the transition matrix for a Markov chain process is given by

$$\begin{pmatrix} \frac{1}{2} & \frac{1}{4} & \frac{1}{4} \\ \frac{1}{2} & 0 & \frac{1}{2} \\ 1 & 0 & 0 \end{pmatrix},$$

find the probability of the process being in state a_3 after three steps, given that the initial probability is $(\frac{1}{3}, \frac{1}{3}, \frac{1}{3})$.

2. If the transition matrix for a Markov chain process is given by

$$\begin{pmatrix} 0 & 0 & 1 \\ \frac{1}{2} & 0 & \frac{1}{2} \\ \frac{1}{3} & \frac{1}{3} & \frac{1}{3} \end{pmatrix},$$

find the probability of the process being in state a_2 after three steps, given that the initial probability is $(\frac{1}{2}, \frac{1}{4}, \frac{1}{4})$.

3. If the transition matrix for a Markov chain process is given by

$$\begin{pmatrix} \frac{1}{3} & \frac{2}{3} & 0 \\ \frac{1}{4} & \frac{1}{4} & \frac{1}{2} \\ 1 & 0 & 0 \end{pmatrix},$$

find the probability of the process being in state a_1 after three steps, given that the initial probability is $(\frac{1}{3}, \frac{1}{3}, \frac{1}{3})$.

4. In Fun City the Republicans, Democrats, Conservatives, and Liberals nominate candidates for mayor. The probabilities of election depend on the party in control and are given by

	R	*D*	*C*	*L*
R	0.4	0.1	0.25	0.25
D	0.2	0.5	0.1	0.2
C	0.25	0.15	0.45	0.15
L	0.30	0.35	0	0.35

If a Democrat is mayor, find the probability that after two elections there will still be a Democrat as mayor.

5. Eros, Porno and Lib Publishers each have $\frac{1}{3}$ of the market. Each introduces a new magazine. During the year, Eros retains 50 percent of its customers and loses 30 percent to Porno and 20 percent to Lib. Porno retains 25 percent of its customers and

loses 60 percent to Eros and 15 percent to Lib. Lib retains 30 percent of its market and loses 40 percent to Eros and 30 percent to Porno. If the trend continues, what percentage of the market will each have at the end of three years?

6. The probability that someone whose mother has red hair will have red hair is 0.6; that he will be blond is 0.1 and that he will have brown hair is 0.3. The children of blonds split 0.7 blond, 0.2 brown, 0.1 red, and the children of brown-haired mothers are 0.5 brown, 0.3 blond and 0.2 red. Set up the process as a Markov chain, compute the probability of a great grandchild of a red-haired woman having red hair and find the long term percentage of the population with red hair.

7. For the input matrix

	Leather	Wool	Cedar chips	Machine time	Labor time	Misc.
Collar and leash	2.3	0.2	0	2	0.7	0.3
Coat	0.3	3.1	0	1.5	1	0.1
Cushion	0.4	5.6	3	1	1	0.1

what raw materials must the production department order and what time must be scheduled for an order of 5450 collar and leash sets, 2160 coats, and 4120 cushions?

8. If the consumption matrix for a certain firm is given by

$$\begin{pmatrix} \frac{1}{10} & 0 & 0 \\ 0 & 0 & \frac{1}{2} \\ \frac{1}{5} & \frac{1}{3} & 0 \end{pmatrix} \quad \text{and the order vector is} \quad \begin{pmatrix} 1400 \\ 2100 \\ 5000 \end{pmatrix},$$

find the production vector.

9. Diagonalize the quadratic forms:
   a) $f(x, y) = 2x^2 + 4xy + 5y^2$.
   b) $f(x, y) = 2x^2 - 3xy + 5y^2$.
   c) $f(x, y) = -x^2 - 6xy - 6y^2$.
   d) $f(x, y, z) = x^2 - xy - 6xz + \frac{5}{2}y^2 + 12yz + 5z^2$.
   e) $f(x, y, z) = xy + yz + zx$.
   f) $f(x, y) = x^2 + 3xy + 4y^2$.
   g) $f(x, y) = x^2 + 8y^2$.
   h) $f(x, y) = 3x^2 + 4xy$.
   i) $f(x, y) = -6xy + y^2$.

10. Find a diagonal matrix similar to

a) $\begin{pmatrix} 1 & \sqrt{2} \\ \sqrt{2} & 2 \end{pmatrix}$.
b) $\begin{pmatrix} 1 & 1 \\ 2 & 3 \end{pmatrix}$.
c) $\begin{pmatrix} 0 & 2 \\ 2 & 3 \end{pmatrix}$.

d) $\begin{pmatrix} 1 & 4 \\ 3 & 5 \end{pmatrix}$.
e) $\begin{pmatrix} 5 & -6 & -6 \\ -1 & 4 & 2 \\ 3 & -6 & -4 \end{pmatrix}$.
f) $\begin{pmatrix} -1 & 3 \\ 3 & 0 \end{pmatrix}$.

g) $\begin{pmatrix} 2 & 1 \\ 2 & 3 \end{pmatrix}$.
h) $\begin{pmatrix} 1 & -1 \\ -1 & 2 \end{pmatrix}$.
i) $\begin{pmatrix} 1 & 3 \\ 2 & 5 \end{pmatrix}$.

j) $\begin{pmatrix} 2 & 4 & 1 \\ 1 & -2 & -1 \\ 0 & 0 & 0 \end{pmatrix}$.

11.7 THE DIFFERENTIAL

We remarked in Chapter 9 that the problem of differentiating a function of several variables was a difficult one; we now have the machinery to tackle it.

Suppose $F\colon R^n \to R^m$ is a function; that is, to each element of R^n, F assigns a unique element of R^m. If $n > 1$, we say that F is a *function of several variables.* If $m = 1$, F is a *real-valued function* of several (real) variables, whereas if $m > 1$, we say that F is a *vector-valued function.* If $F\colon R^n \to R^m$ is given by

$$F(X) = (f_1(X), f_2(X), \ldots, f_m(X)),$$

the f_i, $i = 1, \ldots, m$, are called *coordinate functions.* Each coordinate function is a real-valued function of several variables.

Example. Let $F\colon R^3 \to R^4$ be given by

$$F(x, y, z) = (xy, x^2, y - z, xz^2).$$

Then the coordinate functions are

$$f_1(x, y, z) = xy, \qquad f_2(x, y, z) = x^2,$$
$$f_3(x, y, z) = y - z, \qquad f_4(x, y, z) = xz^2.$$

Let $F\colon R^n \to R^m$. We define

$$\lim_{X \to X_0} F(X) = A,$$

if for every neighborhood U of A there is a neighborhood V of X_0 such that for $X \in V$, $X \neq X_0$, $F(X) \in U$. Remember that $A = (a_1, \ldots, a_m)$ is an element of R^m. If

$$\lim_{X \to X_0} F(X) = F(X_0),$$

we say that F is *continuous* at X_0.

So far everything has proceeded as in the case of functions of one variable. But we are in trouble when we try to define the derivative of $F\colon R^n \to R^m$ as

$$\lim_{X \to X_0} \frac{F(X) - F(X_0)}{X - X_0},$$

because we are required to divide by a *vector* $X - X_0$, rather than by a real number. Such an operation is not defined.

Since it is not possible to divide by a vector, let us form the quotient whose denominator is $\|X - X_0\|$, a real number, and consider its limit:

$$\lim_{X \to X_0} \frac{F(X) - F(X_0)}{\|X - X_0\|}.$$

Note that the norm is a continuous function so that as $X \to X_0$, $\|X - X_0\| \to 0$.

For a function $f(x)$ of a single variable,

$$\lim_{x \to x_0} \left(\frac{f(x) - f(x_0)}{x - x_0} - f'(x) \right) = 0 \tag{*}$$

if $f'(x)$ is continuous at x_0, i.e., if $\lim_{x \to x_0} f'(x) = f'(x_0)$.

We want to find a function which acts for several variables as $f'(x)$ does for a single variable. In particular, we want a linear transformation $L: R^n \to R^m$, such that

$$\lim_{X \to X_0} \frac{F(X) - F(X_0) - L(X - X_0)}{\|X - X_0\|} = 0.$$

If such a function exists and if $F(X)$ is defined on a neighborhood of X_0, we say that F is *differentiable* at X_0, and the linear function L is the *differential* of F at X_0.

Note that this coincides with the case of one variable if we take $df = f'(x)(x - x_0)$, a linear function of $x = x - x_0$ and write (*) as

$$\lim_{x \to x_0} \left(\frac{f(x) - f(x_0) - df}{x - x_0} \right) = 0.$$

To get down to practical matters, how can we find such a function? We have the following.

Theorem 1. If $F: R^n \to R^m$ is defined on an open set $D \subset R^n$ on which the partial derivatives of its coordinate functions, $\partial f_i / \partial x_j$ are continuous, then F is differentiable at every point of D.

Proof. We shall define the required linear function and indicate the proof, which depends on repeated applications of the Mean Value Theorem for real-valued functions of a single variable.

We recall that an $m \times n$ matrix defines a linear transformation $R^n \to R^m$. Let $J(X_0)$ be the matrix of partial derivatives of the coordinate functions at a point $X_0 \in D$

$$J(X_0) = \begin{pmatrix} \dfrac{\partial f_1}{\partial x_1}(X_0) & \cdots & \dfrac{\partial f_1}{\partial x_n}(X_0) \\ \vdots & & \vdots \\ \dfrac{\partial f_m}{\partial x_1}(X_0) & \cdots & \dfrac{\partial f_m}{\partial x_n}(X_0) \end{pmatrix}. \tag{1}$$

This is called the *Jacobian matrix* of F at X_0. Let L be the linear transformation defined by the matrix $J(X_0)$ (with respect to the canonical bases).

Using without proof the fact that a function of several variables approaches a limit if and only if the coordinate functions approach the coordinates of the limit, we see that it suffices to prove the theorem for the coordinate functions.

Let

$$X_0 = (a_1, \ldots, a_n), \qquad X = (x_1, \ldots, x_n)$$

and

$$Y_k = (x_1, \ldots, x_k, a_{k+1}, \ldots, a_n), \qquad k = 0, \ldots, n.$$

Then

$$f_i(X) - f_i(X_0) = \sum_{k=1}^{n} \big(f_i(Y_k) - f_i(Y_{k-1})\big).$$

Now $f_i(Y_k)$ and $f_i(Y_{k-1})$ differ only in the kth coordinate, so we apply the Mean Value Theorem (Chapter 5) to obtain

$$f_i(Y_k) - f_i(Y_{k-1}) = (x_k - a_k)\,\frac{\partial f_i}{\partial x_k}(Z_k),$$

where Z_k is on the line segment joining Y_k and Y_{k-1}. Thus

$$f_i(X) - f_i(X_0) = \sum_{k=1}^{n} (x_k - a_k)\,\frac{\partial f_i}{\partial x_k}(Z_k).$$

Let L_i be the linear transformation defined by the Jacobian matrix for the coordinate function f_i, i.e., L_i is the function defined by the ith row of the matrix (1). Then

$$L_i(X - X_0) = \left(\frac{\partial f_i}{\partial x_1}(X_0) \quad \cdots \quad \frac{\partial f_i}{\partial x_n}(X_0)\right)\begin{pmatrix} x_1 - a_1 \\ \vdots \\ x_n - a_n \end{pmatrix}$$

$$= \sum_{k=1}^{n} (x_k - a_k)\,\frac{\partial f_i}{\partial x_k}(X_0).$$

Hence

$$|f_i(X) - f_i(X_0) - L_i(X - X_0)| = \left|\sum_{k=1}^{n} \left|\frac{\partial f_i}{\partial x_k}(Z_k) - \frac{\partial f_i}{\partial x_k}(X_0)\right|(x_k - a_k)\right|$$

$$\leq \sum_{k=1}^{n} \left|\frac{\partial f_i}{\partial x_k}(Z_k) - \frac{\partial f_i}{\partial x_k}(X_0)\right| \|X - X_0\|$$

by the triangle inequality (Chapter 1, Section 4) and the fact that

$$|x_k - a_k| \leq \sqrt{(x_1 - a_1)^2 + \cdots + (x_n - a_n)^2} = \|X - X_0\|, \qquad k = 1, \ldots, n.$$

But the partial derivatives are continuous and $Z_k \to X_0$ as $X \to X_0$ so that as $X \to X_0$,

$$\sum_{k=1}^{n} \left|\frac{\partial f_i}{\partial x_k}(Z_k) - \frac{\partial f_i}{\partial x_k}(X_0)\right| \to 0.$$

Therefore

$$\lim_{X \to X_0} \frac{|f_i(X) - f_i(X_0) - L_i(X - X_0)|}{\|X - X_0\|} = 0.$$

The linear transformation L which is the differential of F at X_0 is denoted by $dF(X_0)$.

Examples

1. If $f(x, y) = x^2 + y^2$, find $df(1, 0)$.

Solution. The partial derivatives are

$$\frac{\partial f}{\partial x} = 2x \qquad \text{and} \qquad \frac{\partial f}{\partial y} = 2y.$$

Note that they are continuous in the whole plane and therefore in any open set containing (1, 0); therefore the function is differentiable at (1, 0).

$$df(x, y) = (2x, 2y), \qquad df(1, 0) = (2, 0).$$

2. Find dF at $(2, -1)$ if $F(x, y) = (x + y, x^2 - y, 1)$.

Solution. The coordinate functions are

$$\begin{aligned} f_1(x, y) &= x + y \\ f_2(x, y) &= x^2 - y \\ f_3(x, y) &= 1. \end{aligned}$$

The partial derivatives are

$$\frac{\partial f_1}{\partial x} = 1, \qquad \frac{\partial f_1}{\partial y} = 1, \qquad \frac{\partial f_2}{\partial x} = 2x,$$

$$\frac{\partial f_2}{\partial y} = -1, \qquad \frac{\partial f_3}{\partial x} = 0, \qquad \frac{\partial f_3}{\partial y} = 0.$$

They are continuous, so the function is differentiable and

$$dF = \begin{pmatrix} 1 & 1 \\ 2x & -1 \\ 0 & 0 \end{pmatrix}$$

$$dF(2, -1) = \begin{pmatrix} 1 & 1 \\ 4 & -1 \\ 0 & 0 \end{pmatrix}.$$

Warning. $dF(X_0)$ is the *function* defined by the matrix. In the example above,

$$\begin{aligned} dF(2, -1)(x, y) &= \begin{pmatrix} 1 & 1 \\ 4 & -1 \\ 0 & 0 \end{pmatrix} \begin{pmatrix} x \\ y \end{pmatrix} \\ &= (x + y, 4x - y, 0). \end{aligned}$$

Note that for any function $F: R^n \to R^m$ differentiable at X_0, $dF(X_0)$ is a function

$$dF(X_0): R^n \to R^m.$$

Moreover $dF(X_0)$ is a linear transformation, but dF is *not* a linear function of X_0.

Recall that in Chapter 5 in introducing the notion of the differential of a function of a single variable, we used the variable dx. If $F: R^n \to R$ is a real-valued function of several variables, we sometimes use the variables $dx_1, \ldots, dx_n$ to define the *total differential* of F to be

$$dF(x_1, \ldots, x_n)(dx_1, \ldots, dx_n) = \left(\frac{\partial F}{\partial x_1}, \ldots, \frac{\partial F}{\partial x_n}\right)\begin{pmatrix} dx_1 \\ \vdots \\ dx_n \end{pmatrix}$$

$$= \frac{\partial F}{\partial x_1}\, dx_1 + \cdots + \frac{\partial F}{\partial x_n}\, dx_n.$$

Some authors call $dF(x_1, \ldots, x_n)$ the *derivative* of F and reserve the term differential for $dF(x_1, \ldots, x_n)(dx_1, \ldots, dx_n)$.

The chain rule, although we do not prove it, operates for functions of several variables as it does for functions of a single variable. Namely, if $F: R^n \to R^m$ is differentiable at X_0 and $G: R^m \to R^k$ is differentiable at $F(X_0)$, then

$$d(G \circ F)(X_0) = dG(F(X_0))\, dF(X_0).$$

Here the notation means that the matrix which defines the differential of the composite function is the product of the matrix which defines $dG(F(X_0))$ and the matrix which defines $dF(X_0)$.

Example. If $F(x, y) = (x^2 + xy + 1, y^2 + 2)$ and $G(u, v) = (u + v, 2u, v^2)$, find $d(G \circ F)(1, 1)$.

Solution. $d(G \circ F)(x, y) = dG(F(x, y))\, dF(x, y)$. But

$$dG(u, v) = \begin{pmatrix} 1 & 1 \\ 2 & 0 \\ 0 & 2v \end{pmatrix},$$

so

$$dG(x^2 + xy + 1, y^2 + 2) = \begin{pmatrix} 1 & 1 \\ 2 & 0 \\ 0 & 2y^2 + 4 \end{pmatrix}.$$

On the other hand

$$dF(x, y) = \begin{pmatrix} 2x + y & x \\ 0 & 2y \end{pmatrix}.$$

Thus

$$d(G \circ F)(x, y) = \begin{pmatrix} 1 & 1 \\ 2 & 0 \\ 0 & 2y^2 + 4 \end{pmatrix} \begin{pmatrix} 2x + y & x \\ 0 & 2y \end{pmatrix}$$

$$= \begin{pmatrix} 2x + y & x + 2y \\ 4x + 2y & 2x \\ 0 & 4y^3 + 8y \end{pmatrix}$$

and

$$d(G \circ F)(1, 1) = \begin{pmatrix} 3 & 3 \\ 6 & 2 \\ 0 & 12 \end{pmatrix}.$$

PROBLEMS

1. If $F(x, y) = 2xy - x^2y$, find $dF(x, y)$.
2. If $F(x_1, \ldots, x_n) = \sum_{i=1}^{n} x_i^2$, find $dF(x_1, \ldots, x_n)$.
3. If $F(x, y, z) = 4xz - 3xy + 2yz$, find $dF(x, y, z)$.
4. If $F(x, y, z) = (x^2, y, z^2)$, find $dF(x, y, z)$.
5. If $F(x, y, z) = x \log yz - e^z$, find $dF(x, y, z)$.
6. If $F(x, y, z) = 3x^2 - 3xyz + 4y^2z$, find $dF(0, 1, -2)$.
7. If $F(x, y) = (x^2, 3y^3, x + y)$, find $dF(1, -1)$.
8. If $F(x, y, z) = (3x^2z, ye^x, ze^{xy})$, find $dF(2, 0, 1)$.
9. If $F: R^2 \to R^3$ is given by $F(x, y) = (x^2, y^2, x^2 + y^2)$, find $(dF(0, 1))(5, 4)$.
10. If $F: R^3 \to R^3$ is given by $F(x, y, z) = (xe^z, ye^z, z)$, find $(dF(0, 1, 1))(-1, 2, -1)$.
11. If $F: R^3 \to R^2$ is given by $F(x, y, z) = (x \sin z, \cos y)$, find $(dF(\pi, -\pi, \pi))(0, 1, 1)$.
12. Find the total differential of:
    a) $f(x, y) = x^2y - 3xy^2$.
    b) $f(x, y) = 2x^3 - xe^y$.
    c) $f(x, y, z) = x^2y^2 - z^2x + 3xyz$.
    d) $f(x, y, z) = x \log y/z$.
13. If $f(x, y, z) = 3x + y^2 + z$ and $F(t) = (\sin t, \cos t, t)$, find $d(f \circ F)(\pi)$.
14. If $F(x, y) = (x^2, y + x)$, $G(u, v) = (3u - 3, v - 2)$, find $d(G \circ F)(2, 3)$.
15. If $F(x, y, z) = (x + z, y + z)$, $G(u, v) = (2u^2, v^2, u + v)$, find $d(G \circ F)(x, y, z)$.
16. If $F(x, y, z) = x^2y^2z^2 - 4xy$ and $x = u^2$, $y = -3v$, $z = w - u$, find $\partial F/\partial u$, $\partial F/\partial v$, $\partial F/\partial w$.
17. If $F(x, y, z) = 2xyz - x^2y^2 + y^2z^2$, find $dF(x, y, z)$.
18. If $F(x, y) = x \log y$, find $dF(x, y)$.
19. If $F(x, y, z) = z \sin xy$, find $dF(x, y, z)$.
20. If $F(x, y, z) = x/(y + z)$, find $dF(x, y, z)$.
21. If $F(x, y) = xe^y$, find $dF(1, 1)$.

22. If $F(x, y, z) = x^2y + z + y^2z$, find $dF(-1, 0, 2)$.
23. If $F(x, y, z) = (x^2, yz)$, find $dF(2, -1, 1)$.
24. If $F(x, y) = (3x, y + x, y^2)$, find $dF(3, 1)$.
25. If $F(x, y) = (x^2, y^2)$, find $(dF(1, 1))(2, 3)$.
26. If $F(x, y, z) = (x \sin y, z)$, find $(dF(1, \pi, 1))(1, 2, -1)$.
27. If $F(x, y, z) = (x + y, z^2, xy)$, find $(dF(1, 1, 0))(2, 1, 3)$.
28. If $F(x, y) = (x^2 + y^2, xy, 1)$, find $(dF(2, -2))(-1, 1)$.
29. Find the total differential of:
    a) $F(x, y) = 3x^2 - xy$.
    b) $F(x, y) = \sin xy$.
    c) $F(x, y, z) = xe^y - ze^{x^2}$.
    d) $F(x, y, z) = x^2y^2 - z^3 + 4yz$.
30. If $f(x, y, z) = x^2y^2z^2$ and $F(t) = (e^t, t, t^2)$, find $d(f \circ F)(1)$.
31. If $F(x, y) = (x^2, y^2)$, $G(u, v) = (2u, u + v)$, find $d(G \circ F)(1, 2)$.
32. If $F(x, y, z) = 2xy - x^2y^2 - xyz$ and $x = 2u$, $y = u + v$, $z = w^2$, find $\partial F/\partial u$, $\partial F/\partial v$, $\partial F/\partial w$.
33. If $F(x, y, z) = x \sin yz$ and $x = e^u$, $y = v^2 - w$, $z = ve^u$, find $\partial F/\partial u$, $\partial F/\partial v$.
34. Suppose all the partial derivatives of the real-valued function $F: R^n \to R$ exist at X. We form the vector

$$\left(\frac{\partial F}{\partial x_1}, \cdots, \frac{\partial F}{\partial x_n}\right) = (D_1F(X), \cdots, D_nF(X)),$$

called the *gradient* of F and denoted by

$$\text{grad } F(X) \qquad \text{or} \qquad \nabla F(X).$$

If $F(x, y, z) = 2x^2y - 3xyz^2$, find grad $F(X)$.
(Note that the differential is a transformation, represented by a matrix or vector, whereas the gradient is a vector.)

35. If $f(x, y) = |xy|$, find $\nabla f(2, -1)$.
36. If $f(x, y, z) = x^2y + xyz - 3z^3$, find $\nabla f(0, 1, 2)$.
37. If $f(x, y, z) = e^{x+y+z}$, find $\nabla f(1, 0, 1)$.
38. If $f(x, y) = e^{x+y}$, find $\nabla f(1, 1)$.
39. If $f(x, y, z) = e^{x/z}$, find $\nabla f(1, 1, -2)$.
40. If $f(x, y, z) = xz^2 - x^2y + z$, find $\nabla f(1, -1, 1)$.

11.8 LAGRANGE MULTIPLIERS

Sometimes we want to find maxima or minima for a function subject to some constraints. We discussed this problem for linear functions in Chapter 10. If the function is not linear, sometimes it is possible to use the constraints to reduce the problem to one involving only one variable and to apply the techniques of Chapter 6.

Example. Minimize $x^2 + y$ subject to the constraint $x - y - 1 = 0$.

Solution. From the constraint

$$y = x - 1,$$

so we are to minimize $x^2 + x - 1$. The derivative of this function is zero at $x = -1/2$. Its second derivative is positive, so it is a minimum.

However, a more general method is that of Lagrange multipliers.

Suppose $G: R^n \to R^m$, $n > m$, is differentiable with continuous partial derivatives and coordinate functions $G_1, \ldots, G_m$. Suppose

$$\begin{aligned} G_1(x_1, \ldots, x_n) &= 0 \\ &\vdots \\ G_m(x_1, \ldots, x_n) &= 0 \end{aligned}$$

implicitly define a surface in R^n and that the matrix $dG(X_0)$ has some m columns linearly independent (as elements of R^m). We state without proof: If X_0 is a maximum or minimum of a differentiable function $F: R^n \to R$, when restricted to S, then X_0 is a critical point of the function

$$F + \lambda_1 G_1 + \cdots + \lambda_m G_m,$$

for some constants $\lambda_1, \ldots, \lambda_m$.

Examples

1. Find the maxima and minima of $f(x, y, z) = x + y + z$ on the intersection of $x^2 + y^2 = 1$ and $z = 2$.

Solution. We write

$$\begin{aligned} x^2 + y^2 - 1 &= 0 \\ z - 2 &= 0. \end{aligned}$$

We want to find the critical points of

$$(x + y + z) + \lambda_1(x^2 + y^2 - 1) + \lambda_2(z - 2).$$

The critical points occur when

$$1 + 2\lambda_1 x = 0, \qquad 1 + 2\lambda_1 y = 0, \qquad 1 + \lambda_2 = 0.$$

But the points we seek must also satisfy $x^2 + y^2 = 1$ and $z = 2$. Therefore

$$\lambda_2 = -1, \qquad \lambda_1 = \frac{\pm 1}{\sqrt{2}},$$

and we have

$$x = y = \frac{\mp 1}{\sqrt{2}}, \qquad z = 2.$$

Thus the critical points are

$$\left(\frac{1}{\sqrt{2}}, \frac{1}{\sqrt{2}}, 2\right) \quad \text{and} \quad \left(\frac{-1}{\sqrt{2}}, \frac{-1}{\sqrt{2}}, 2\right).$$

It is easy to see that the first yields a maximum value of $\sqrt{2} + 2$ for f and the second a minimum of $-\sqrt{2} + 2$.

2. Let a utility function be defined by

$$U = x + y + z,$$

subject to the constraint $x^2 + y^2 + z^2 = 100$. Find its maximum and minimum values.

Solution. Forming the function

$$f(x, y, z) = x + y + z + \lambda(x^2 + y^2 + z^2 - 100),$$

we compute its partial derivatives and set them equal to zero.

$$\frac{\partial f}{\partial x} = 1 + 2\lambda x = 0$$

$$\frac{\partial f}{\partial y} = 1 + 2\lambda y = 0$$

$$\frac{\partial f}{\partial z} = 1 + 2\lambda z = 0.$$

We must find a solution to this system which also satisfies

$$x^2 + y^2 + z^2 - 100 = 0.$$

We find that

$$\lambda = \frac{\pm\sqrt{3}}{20}$$

and $x = y = z = \mp 10/\sqrt{3}$.

There is a maximum at $(10/\sqrt{3}, 10/\sqrt{3}, 10/\sqrt{3})$ of $10\sqrt{3}$ and a minimum of $-10\sqrt{3}$ at $(-10/\sqrt{3}, -10/\sqrt{3}, -10/\sqrt{3})$.

PROBLEMS

1. Maximize $f(x, y) = x^2 - y^2 - 2y$ subject to the restriction $x^2 + y^2 = 1$.
2. Maximize $f(x, y, z) = x + y + z$ subject to the restriction $x^2 + y^2 + z^2 = 1$.
3. Find the distance between the point $(-2, 1, 3)$ and the plane $2x + y - 2z = 3$.

4. Find the distance between the point $(2, -1, 0)$ and the plane $x - y + 2z = 1$.
5. Maximize $f(x, y) = x + y^2$ subject to $2x^2 + y^2 = 1$.
6. Maximize $f(x, y, z) = xyz$ subject to $x^2 + y^2 + z^2 = 9$.
7. Maximize $f(x, y, z) = 3x - y + 2z^2$ subject to $x + y + z = 1$ and $x - y + 2z = 2$.
8. Maximize $f(x, y) = x^2 - y + 4$ subject to $x^2 + y^2 = 1$.
9. Maximize $f(x, y) = 3x - y^2 + 8x^2$ subject to $x - y = 1$ and $x^2 + y^2 = 4$.
10. If $U = 3x^{1/3}y^{2/3}$ is a utility function and x and y are subject to the budget equation $x + 4y = 16$, find the values of x and y which maximize U.

12 / TECHNIQUES OF INTEGRATION

This chapter is a very brief look at some of the methods used to evaluate integrals. There are certain rational patterns in applying the various techniques, but sometimes a high degree of mathematical ingenuity is required. We shall not try to reproduce, for example, the thought pattern which first led someone to try the common trigonometric substitutions. The sad fact is that there are integrals which defy the application of any of the many available formulas. Included in this chapter is a short section on numerical methods, that is, ways to get a good enough approximation of the value of an integral to use in applications.

12.1 INTEGRATION BY PARTS

We recall that the derivative of a product is given by

$$\frac{d(uv)}{dx} = u\frac{dv}{dx} + v\frac{du}{dx},$$

where u and v are differentiable functions of x. From this we get

$$u\frac{dv}{dx} = \frac{d(uv)}{dx} - u\frac{dv}{dx}.$$

Writing this equation in differential form, we integrate both sides

$$\boxed{\begin{aligned} \int u\,dv &= \int d(uv) - \int v\,du \\ &= uv - \int v\,du. \end{aligned}}$$

This is called the formula for *integration by parts.*

Examples

1. Find

$$\int x \sin x\,dx.$$

Solution. Let $u = x$, $dv = \sin x\, dx$. Then $du = dx$ and $v = -\cos x$. Integration by parts gives

$$\int x \sin x\, dx = -x \cos x - \int -\cos x\, dx$$

$$= -x \cos x + \sin x + C.$$

2. Find

$$\int e^x \cos x\, dx.$$

Solution. Let $u = e^x$, $dv = \cos x\, dx$. Then $du = e^x\, dx$ and $v = \sin x$.

$$\int e^x \cos x\, dx = e^x \sin x - \int e^x \sin x\, dx. \tag{12.1}$$

Apparently we are not making much progress, but we repeat the process to evaluate

$$\int e^x \sin x\, dx.$$

Let $u = e^x$, $dv = \sin x\, dx$. Then $du = e^x\, dx$ and $v = -\cos x$.

$$\int e^x \sin x\, dx = -e^x \cos x - \int e^x(-\cos x)\, dx.$$

Substituting this in (12.1) we obtain

$$2 \int e^x \cos x\, dx = e^x \sin x + e^x \cos x + C$$

so that

$$\int e^x \cos x\, dx = \frac{e^x \sin x + e^x \cos x + C}{2}.$$

Warning. It is recommended that as a check one differentiate the result to obtain the integrand.

We are now equipped to integrate some of the elementary functions which we learned to differentiate in Chapter 8.

3. Find

$$\int \log x\, dx.$$

Solution. Let $u = \log x$, $dv = dx$. Then $du = dx/x$ and $v = x$.

$$\int \log x\, dx = x \log x - \int x \frac{dx}{x}$$

$$= x \log x - x + C.$$

4. Find

$$\int \arctan x \, dx.$$

Solution. Let $u = \arctan x$, $dv = dx$. Then

$$du = \frac{1}{1 + x^2} dx, \qquad v = x.$$

$$\int \arctan x = x \arctan x - \int \frac{x}{1 + x^2} dx.$$

To evaluate this new integral we let $u = 1 + x^2$, $du = 2x\,dx$:

$$\int \frac{x}{1 + x^2} dx = \frac{1}{2} \int \frac{2x\,dx}{1 + x^2} = \frac{1}{2} \int \frac{du}{u} = \frac{1}{2} \log u = \frac{1}{2} \log (1 + x^2) + C.$$

Finally

$$\int \arctan x = x \arctan x - \tfrac{1}{2} \log (1 + x^2) + C.$$

5. Find

$$\int \sin^3 x \, dx.$$

Solution. Let $u = \sin^2 x$, $dv = \sin x \, dx$. Then $du = 2 \sin x \cos x \, dx$ and $v = -\cos x$.

$$\int \sin^3 x \, dx = -\sin^2 x \cos x + 2 \int \cos^2 x \sin x \, dx.$$

To evaluate the new integral, let $u = \cos x$, $du = -\sin x \, dx$.

$$\int \cos^2 x \sin x \, dx = -\int u^2 \, du = \frac{-u^3}{3} = \frac{-\cos^3 x}{3} + C.$$

Thus

$$\int \sin^3 x \, dx = -\sin^2 x \cos x - \frac{2 \cos^3 x}{3} + C.$$

PROBLEMS

Integrate

1. $\int x \log x \, dx.$
2. $\int x^2 \sin x \, dx.$
3. $\int \sin^2 x \cos^2 x \, dx.$
4. $\int xe^{-x} \, dx.$

5. $\displaystyle\int (\log x)^2\,dx.$

6. $\displaystyle\int \frac{x}{\sqrt{x+1}}\,dx.$

7. $\displaystyle\int x^2e^x\,dx.$

8. $\displaystyle\int x\arctan x\,dx.$

9. $\displaystyle\int x(2x+1)^3\,dx.$

10. $\displaystyle\int \sin^2 x\,dx.$

11. $\displaystyle\int x^2\cos x\,dx.$

12. $\displaystyle\int \arcsin x\,dx.$

13. $\displaystyle\int e^x\sin x\,dx.$

14. $\displaystyle\int \cos^3 x\,dx.$

15. $\displaystyle\int \frac{x}{1+3x}\,dx.$

12.2 PARTIAL FRACTIONS

We know how to integrate some rational functions, for example:

$$\int\frac{dx}{ax^2+c}\quad\left(\frac{c}{a}\geq 0\right),\qquad \int\frac{x\,dx}{ax^2+c},\qquad \int\frac{(2ax+b)\,dx}{ax^2+bx+c}.$$

Also, if the degree of the numerator is greater than the degree of the denominator we can try performing the indicated division. Suppose, however, we have

$$\int\frac{x}{3x^2+2x-1}\,dx.$$

This doesn't fit into any of our molds. We observe that

$$3x^2+2x-1=(3x-1)(x+1).$$

So we write

$$\frac{c_1}{3x-1}+\frac{c_2}{x+1}=\frac{x}{3x^2+2x-1}.$$

What must the constants c_1 and c_2 be?

Since

$$\frac{c_1(x+1)+c_2(3x-1)}{(3x-1)(x+1)}=\frac{x}{3x^2+2x-1},$$

the numerators must be equal; in particular the coefficients of each power of x must be equal so

$$c_1+3c_2=1$$

and

$$c_1-c_2=0.$$

Solving these simultaneously, we obtain $c_1 = \frac{1}{4}$ and $c_2 = \frac{1}{4}$. Thus we write

$$\int \frac{x\,dx}{3x^2 + 2x - 1} = \int \frac{dx}{4(3x - 1)} + \int \frac{dx}{4(x + 1)}$$

$$= \tfrac{1}{4} \cdot \tfrac{1}{3} \log |3x - 1| + \tfrac{1}{4} \log |x + 1| + C.$$

We can apply a similar procedure for any rational function of the form

$$\frac{ax + b}{cx^2 + dx + e} \qquad (a \text{ or } b \text{ may be zero})$$

if we can factor the denominator into linear factors with rational coefficients. Let us look at other examples.

Examples

1. Find

$$\int \frac{(x + 1)\,dx}{6x^2 - 7x + 1}.$$

Solution. Write

$$\frac{x + 1}{6x^2 - 7x + 1} = \frac{c_1}{6x - 1} + \frac{c_2}{x - 1}.$$

Then

$$c_1(x - 1) + c_2(6x - 1) = x + 1$$
$$c_1 + 6c_2 = 1$$
$$-c_1 - c_2 = 1.$$

From these we obtain $c_1 = -7/5$, $c_2 = \frac{2}{5}$. Hence

$$\int \frac{(x + 1)\,dx}{6x^2 - 7x + 1} = \int \frac{-7\,dx}{5(6x - 1)} + \int \frac{2\,dx}{5(x - 1)}$$

$$= \frac{-7}{30} \log |6x - 1| + \frac{2}{5} \log |x - 1| + C.$$

2. Find

$$\int \frac{6\,dx}{2x^2 - 5x - 3}.$$

Solution. Write

$$\frac{6}{2x^2 - 5x - 3} = \frac{c_1}{2x + 1} + \frac{c_2}{x - 3}.$$

Then

$$c_1(x - 3) + c_2(2x + 1) = 6$$
$$c_1 + 2c_2 = 0$$
$$-3c_1 + c_2 = 6$$

$$c_1 = \frac{-12}{7}, \qquad c_2 = \frac{6}{7}.$$

Thus

$$\int \frac{6\,dx}{2x^2 - 5x - 3} = \int \frac{-12\,dx}{7(2x + 1)} + \int \frac{6\,dx}{7(x - 3)}$$

$$= \frac{-6}{7} \log |2x + 1| + \frac{6}{7} \log |x - 3| + C.$$

If we have a rational function $f(x)/g(x)$ with a quadratic polynomial as denominator and a polynomial of second or higher degree as numerator, we can perform the indicated division to obtain

$$s(x) + \frac{r(x)}{g(x)},$$

where $r(x)$ is a linear polynomial or a constant. Then we can treat $r(x)/g(x)$ as above. For example, suppose we want to evaluate

$$\int \frac{2x^4 + x^3}{x^2 + 4x + 3}\,dx.$$

Dividing the numerator by the denominator yields

$$\int \frac{2x^4 + x^3}{x^2 + 4x + 3}\,dx = \int (2x^2 - 7x + 22)\,dx + \int \frac{-67x - 66}{x^2 + 4x + 3}\,dx.$$

Now we have

$$\frac{-67x - 66}{x^2 + 4x + 3} = \frac{c_1}{x + 1} + \frac{c_2}{x + 3}$$

$$c_1(x + 3) + c_2(x + 1) = -67x - 66,$$

$$c_1 + c_2 = -67,$$

$$3c_1 + c_2 = -66,$$

$$c_1 = \tfrac{1}{2}, \qquad c_2 = \frac{-135}{2}.$$

Finally we have

$$\int \frac{2x^4 + x^3}{x^2 + 4x + 3}\,dx = \int (2x^2 - 7x + 22)\,dx + \frac{1}{2}\int \frac{dx}{x + 1} - \frac{135}{2}\int \frac{dx}{x + 3}$$

$$= \tfrac{2}{3}x^3 - \tfrac{7}{2}x^2 + 22x + \tfrac{1}{2} \log |x + 1| - \tfrac{135}{2} \log |x + 3| + C.$$

What happens if the denominator cannot be factored into linear factors with rational coefficients? If the roots are real, we go ahead and find them by means of the quadratic formula and write

$$ax^2 + bx + c = a\left(x + \frac{b - \sqrt{b^2 - 4ac}}{2a}\right)\left(x + \frac{b + \sqrt{b^2 - 4ac}}{2a}\right),$$

and proceed as above. Suppose the roots are complex. There are two cases to consider: a linear numerator or a constant numerator.

Consider

$$\int \frac{x + 1}{x^2 + x + 1}\, dx.$$

We can write this as

$$\int \frac{x + \frac{1}{2}}{x^2 + x + 1}\, dx + \int \frac{\frac{1}{2}}{x^2 + x + 1}\, dx.$$

The first integral is of the form

$$\frac{1}{2}\int \frac{du}{u},$$

where

$$u = x^2 + x + 1 \qquad \text{and} \qquad du = 2x + 1.$$

It is easily evaluated. Thus we can reduce the linear numerator case to the constant numerator case. Hence let us consider

$$\int \frac{k\, dx}{ax^2 + bx + c}.$$

If we could write the denominator as the sum of squares, we might be able to use the arctangent integration formula

$$\int \frac{1}{1 + u^2}\, du = \arctan u + C.$$

Therefore we *complete the square*:

$$ax^2 + bx + c = a\left(x + \frac{b}{2a}\right)^2 + c - \frac{b^2}{4a}.$$

The roots of the polynomial are given by the quadratic formula

$$\frac{-b \pm \sqrt{b^2 - 4ac}}{2a},$$

(which is obtained in the first place by completing the square). The roots are not real if and only if $c - b^2/4a > 0$. Since we are currently interested in the case where the roots are not real, let

$$c - \frac{b^2}{4a} = c_2^2 \qquad (c_2 \textit{ a real number}).$$

Also we let $b/2a = c_1$. Then we write

$$\int \frac{k\, dx}{ax^2 + bx + c} = \int \frac{k\, dx}{a(x + c_1)^2 + c_2^2}.$$

To evaluate this integral we let

$$u = \frac{\sqrt{a}(x + c_1)}{c_2}, \qquad du = \sqrt{a}dx/c_2.$$

Then

$$\begin{aligned}\int \frac{k\,dx}{a(x + c_1)^2 + c_2^2} &= \frac{c_2}{\sqrt{a}} \int \frac{k(\sqrt{a}dx/c_2)}{c_2^2((\sqrt{a}(x + c_1)/c_2)^2 + 1)} \\ &= \frac{k}{c_2\sqrt{a}} \int \frac{du}{u^2 + 1} \\ &= \frac{k}{c_2\sqrt{a}} \arctan u \\ &= \frac{k}{c_2\sqrt{a}} \arctan \frac{\sqrt{a}(x + c_1)}{c_2} + C.\end{aligned}$$

Thus partial fractions handle the quadratic denominators with real roots, the arctangent those with nonreal roots.

The method of partial fractions is also applicable to rational functions with denominator of degree greater than two. Unless it is clear how to factor the denominator into linear or quadratic factors, however, the procedure is quite complicated. We do some simple ones.

1. Find

$$\int \frac{dx}{x^3 - 8}.$$

Solution. We write

$$\frac{1}{x^3 - 8} = \frac{c_1}{x - 2} + \frac{c_2 + c_3x}{x^2 + 2x + 4}.$$

Note the use of a linear polynomial instead of a constant as numerator when the quadratic cannot be reduced to linear factors with real coefficients. Now we have

$$\begin{aligned}c_1(x^2 + 2x + 4) + (c_2 + c_3x)(x - 2) &= 1 \\ c_1 + c_3 &= 0 \\ 2c_1 + c_2 - 2c_3 &= 0 \\ 4c_1 - 2c_2 &= 1\end{aligned}$$

$$c_1 = \frac{1}{12}, \qquad c_2 = \frac{-1}{3}, \qquad c_3 = \frac{-1}{12}.$$

Therefore

$$\int \frac{1}{x^3 - 8}\,dx = \frac{1}{12}\int \frac{dx}{x - 2} - \frac{1}{12}\int \frac{4 + x}{x^2 + 2x + 4}\,dx$$

$$= \frac{1}{12}\log|x - 2| - \frac{1}{12}\int \frac{3\,dx}{x^2 + 2x + 4} - \frac{1}{24}\int \frac{(2x + 2)\,dx}{x^2 + 2x + 4}.$$

Letting $u = x^2 + 2x + 4$, $du = (2x + 2)\,dx$, this gives us

$$\frac{1}{12}\log|x - 2| - \frac{1}{4}\int \frac{dx}{(x + 1)^2 + 3} - \frac{1}{24}\int \frac{du}{u}$$

$$= \frac{1}{12}\log|x - 2| - \frac{1}{4\sqrt{3}}\int \frac{(1/\sqrt{3})\,dx}{((x + 1)/\sqrt{3})^2 + 1} - \frac{1}{24}\log|x^2 + 2x + 4|$$

$$= \frac{1}{12}\log|x - 2| - \frac{1}{4\sqrt{3}}\arctan\left(\frac{x + 1}{\sqrt{3}}\right) - \frac{1}{24}\log|x^2 + 2x + 4| + C.$$

2. Find

$$\int \frac{5\,dx}{(x - 2)(x + 1)^2}.$$

Solution. This time we write

$$\frac{5}{(x - 2)(x + 1)^2} = \frac{c_1}{x - 2} + \frac{c_2}{x + 1} + \frac{c_3}{(x + 1)^2}.$$

Note that the linear factor $x + 1$ appears to both the first and second powers. We have

$$c_1(x + 1)^2 + c_2(x - 2)(x + 1) + c_3(x - 2) = 5,$$

$$c_1 + c_2 = 0,$$

$$2c_1 - c_2 + c_3 = 0,$$

$$c_1 - 2c_2 - 2c_3 = 5,$$

$$c_1 = \frac{5}{9}, \qquad c_2 = \frac{-5}{9}, \qquad c_3 = \frac{-5}{3}.$$

Finally

$$\int \frac{5\,dx}{(x - 2)(x + 1)^2} = \frac{5}{9}\int \frac{dx}{x - 2} - \frac{5}{9}\int \frac{dx}{x + 1} - \frac{5}{3}\int \frac{dx}{(x + 1)^2}$$

$$= \tfrac{5}{9}\log|x - 2| - \tfrac{5}{9}\log|x + 1| + \tfrac{5}{3}(x + 1)^{-1} + C.$$

PROBLEMS

Integrate

1. $\displaystyle\int \frac{2}{x^2 + 2x - 15}\, dx.$

2. $\displaystyle\int \frac{x^2 + 3x + 4}{x + 2}\, dx.$

3. $\displaystyle\int \frac{dx}{(2x - 5)^3}.$

4. $\displaystyle\int \frac{x^3 + 1}{x^2 + 4}\, dx.$

5. $\displaystyle\int \frac{x + 3}{x^2 + 2x + 2}\, dx.$

6. $\displaystyle\int \frac{x - 3}{(x + 1)^2(x - 2)}\, dx.$

7. $\displaystyle\int \frac{3x + 1}{x^2 - 4x - 5}\, dx.$

8. $\displaystyle\int \frac{3x}{x^3 - 27}\, dx.$

9. $\displaystyle\int \frac{x^2}{x^2 - 5x + 6}\, dx.$

10. $\displaystyle\int \frac{x}{2x + 1}\, dx.$

11. $\displaystyle\int \frac{x}{2x^2 + 1}\, dx.$

12. $\displaystyle\int \frac{1}{2x^2 + 1}\, dx.$

13. $\displaystyle\int \frac{3}{x^2 - 2x - 8}\, dx.$

14. $\displaystyle\int \frac{3x + 1}{x^2 - 5x - 14}\, dx.$

15. $\displaystyle\int \frac{x + 1}{x^2 + x + 1}\, dx.$

16. $\displaystyle\int \frac{x^2}{x^2 - 5x + 6}\, dx.$

17. $\displaystyle\int \frac{x + 1}{(x - 1)^2(x - 3)}\, dx.$

18. $\displaystyle\int \frac{2}{x^2 + 6x + 9}\, dx.$

19. $\displaystyle\int \frac{x}{x^2 + 1}\, dx.$

20. $\displaystyle\int \frac{1}{9x^2 + 1}\, dx.$

21. Assuming that the birth rate is proportional to population size and that the death rate is proportional to the square of the population size, we can write

$$\frac{dP}{dt} = aP - bP^2.$$

Find an expression for P. What happens to P as $t \to \infty$.

12.3 SUBSTITUTION

We have already employed the method of substitution, as in the example of the last section where we let $u = \cos x$ in evaluating $\int \cos^2 x \sin x\, dx$. The success of the method depends simply on the change-of-variables theorem of Chapter 7.

In this section we introduce somewhat more esoteric substitutions. A very useful device is to substitute $u = \sin t$ in an integrand involving the expression $\sqrt{a^2 - u^2}$.

Examples

1. Evaluate $\int \sqrt{1 - x^2}\, dx$.

Solution. Let $x = \sin t$, $dx = \cos t\, dt$. We have

$$\sqrt{1 - x^2}\, dx = \sqrt{1 - \sin^2 t} \cos t\, dt.$$

But $1 - \sin^2 t = \cos^2 t$ so our integral becomes $\int \cos^2 t\, dt$.

Now we use the half-angle formula

$$\cos t = \sqrt{\frac{1 + \cos 2t}{2}},$$

so that we get

$$\tfrac{1}{2} \int (1 + \cos 2t)\, dt = \frac{t}{2} + \frac{1}{2} \int \cos 2t\, dt$$

$$= \frac{t}{2} + \frac{1}{4} \sin 2t + C.$$

Using that $\sin 2t = 2 \sin t \cos t$ and $\cos t = \sqrt{1 - x^2}$, we get

$$\int \sqrt{1 - x^2}\, dx = \tfrac{1}{2} \arcsin x + \frac{x}{2} \sqrt{1 - x^2} + C.$$

2. Find

$$\frac{x^2\, dx}{\sqrt{9 - x^2}}.$$

Solution. Let $x = 3 \sin t$ and $dx = 3 \cos t\, dt$ so that

$$\sqrt{9 - x^2} = \sqrt{9 - 9 \sin^2 t} = 3 \cos t.$$

Then

$$\int \frac{x^2\, dx}{\sqrt{9 - x^2}} = \int \frac{(9 \sin^2 t)(3 \cos t)}{3 \cos t}\, dt = 9 \int \sin^2 t\, dt$$

$$= 9 \int \frac{1 - \cos 2t}{2}\, dt$$

$$= \tfrac{9}{2} t - \tfrac{9}{4} \sin 2t + C$$

$$= \frac{9}{2} \arcsin \frac{x}{3} - \frac{9}{4} (2 \sin t \cos t) + C$$

$$= \frac{9}{2} \arcsin \frac{x}{3} - \frac{9}{4} \cdot 2 \cdot \frac{x}{3} \sqrt{\frac{1 - x^2}{9}} + C$$

$$= \frac{9}{2} \arcsin \frac{x}{3} - \frac{x}{2} \sqrt{9 - x^2} + C.$$

Warning. Do not get carried away. Integrals of the form

$$\int \sqrt{a^2 - x^2}\, x\, dx, \qquad \int \frac{x\, dx}{\sqrt{a^2 - x^2}} \qquad \text{or} \qquad \int \frac{dx}{\sqrt{a^2 - x^2}},$$

do not require trigonometric substitutions, the first two being amenable to solution by use of the power formula

$$\int u^r\, du = \frac{u^{r+1}}{r+1} + C \qquad (r \neq 1),$$

and the third by means of the arcsin

$$\int \frac{dx}{\sqrt{a^2 - x^2}} = \int \frac{(1/a)\, dx}{\sqrt{1 - (x/a)^2}} = \arcsin \frac{x}{a} + C.$$

3. We can finally "find" the area of a circle, the formula for which we have known since early childhood. Suppose we want to find the area of the circle bounded by the graph of $x^2 + y^2 = 25$ as shown in Fig. 12.1.

Solution. $y = \sqrt{25 - x^2}$ is a function which we can integrate and, by symmetry, the area of the circle, according to our intuitive notion of area, is

$$4 \int_0^5 \sqrt{25 - x^2}\, dx.$$

Letting $x = 5 \sin t$ and $dx = 5 \cos t\, dt$, we have

$$\begin{aligned} \int_0^5 \sqrt{25 - x^2}\, dx &= \int_0^{\pi/2} \sqrt{25 - 25 \sin^2 t}\, 5 \cos t\, dt \\ &= 25 \int_0^{\pi/2} \cos^2 t\, dt \\ &= 25 \int_0^{\pi/2} \frac{1 + \cos 2t}{2}\, dt \\ &= \tfrac{25}{2} t \Big|_0^{\pi/2} + \tfrac{25}{4} \int_0^{\pi/2} 2 \cos 2t\, dt \\ &= [\tfrac{25}{2} t + \tfrac{25}{4} \sin 2t]_0^{\pi/2} \\ &= \tfrac{25}{4}\pi. \end{aligned}$$

Thus the desired area is $4(25\pi/4) = 25\pi$.

Any standard integral table contains pages and pages of integrals evaluated by various means. If an integral is not immediately recognizable as being known or being adaptable to integration by parts, by a familiar substitution, or by partial fractions, one can turn to a table. Most are organized in an orderly fashion,

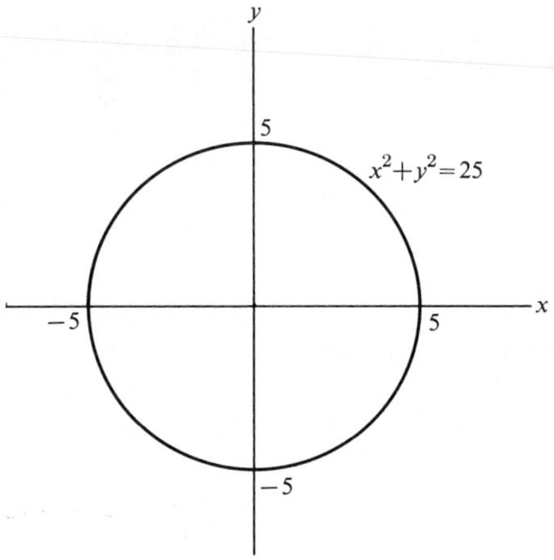

Fig. 12.1

e.g., all integrals involving $\sqrt{x^2 - a^2}$ are together, all integrals involving log x are together, etc. The only skill required is in the algebraic manipulation which may be necessary to put the given integral in a standard form.

Example. Find

$$\int \frac{dx}{\sqrt{3x^2 - 2}}.$$

Solution. Suppose we have found in a table of integrals:

$$\int \frac{du}{\sqrt{u^2 - a^2}} = \log |u + \sqrt{u^2 - a^2}| + C.$$

We let $\sqrt{3}\, x = u$ and $\sqrt{2} = a$. Then

$$\frac{1}{\sqrt{3}} \int \frac{\sqrt{3}\, dx}{\sqrt{3x^2 - 2}} = \frac{1}{\sqrt{3}} \log |\sqrt{3}\, x + \sqrt{3x^2 - 2}| + C.$$

PROBLEMS

1. Integrate by means of an appropriate substitution:

a) $\int 7\sqrt{4 - x^2}\, dx.$ b) $\int \frac{dx}{4 + 3x^2}.$

c) $\int \frac{4 - 3x}{2 - 3x}\, dx.$ d) $\int \sqrt{1 - e^{2x}}\, dx.$

e) $\int \frac{dx}{\sqrt{2x - x^2}}.$ f) $\int \frac{dx}{\sqrt{25 - x^2}}.$

g) $\displaystyle\int \frac{dx}{x\sqrt{\log x}}$.

h) $\displaystyle\int \frac{2}{\sqrt{9 - x^2}}\, dx$.

i) $\displaystyle\int \frac{1}{1 + 3x^2}\, dx$.

j) $\displaystyle\int \frac{1}{\sqrt{x - x^2}}\, dx$.

k) $\displaystyle\int \frac{1}{x^2\sqrt{16 - x^2}}\, dx$.

l) $\displaystyle\int \frac{1}{\sqrt{e^{2x} - 1}}\, dx$.

2. Integrate by whatever method works:

a) $\displaystyle\int \sin(\log x)\, dx$.

b) $\displaystyle\int \frac{x}{\sqrt{x - x^2}}\, dx$.

c) $\displaystyle\int x \sec^2 x\, dx$.

d) $\displaystyle\int \sin\sqrt{x}\, dx$.

e) $\displaystyle\int \frac{5}{x^2 + 4x + 5}\, dx$.

f) $\displaystyle\int \sqrt{1 - 9x^2}\, dx$.

g) $\displaystyle\int \frac{x}{\sqrt{9 - x^2}}\, dx$.

h) $\displaystyle\int \frac{x}{9 - x^2}\, dx$.

i) $\displaystyle\int \frac{dx}{\sqrt{9 - x^2}}$.

j) $\displaystyle\int \frac{dx}{4 + x^2}$.

k) $\displaystyle\int \frac{dx}{\sqrt{1 - 9x^2}}$.

l) $\displaystyle\int \frac{x^2}{\sqrt{4 - x^2}}\, dx$.

m) $\displaystyle\int \frac{dx}{\sqrt{2 - 5x^2}}$.

n) $\displaystyle\int \frac{|\sin x|}{\sqrt{1 - \cos^2 x}}\, dx$.

o) $\displaystyle\int \frac{x^2}{\sqrt{1 - x^2}}\, dx$.

p) $\displaystyle\int \frac{\sin x}{\cos^3 x}\, dx$.

q) $\displaystyle\int \frac{1}{\sqrt{4 - x^2}}\, dx$.

r) $\displaystyle\int \frac{1}{\sqrt{4 - x^2}}\, dx$.

s) $\displaystyle\int \frac{x}{\sqrt{4 - x^2}}\, dx$.

t) $\displaystyle\int \frac{3}{x^2 - 5x - 6}\, dx$.

u) $\displaystyle\int \frac{1}{x^2 + 3x + 4}\, dx$.

12.4 NUMERICAL METHODS

Numerical solutions used to be the last resort—what to do if all else fails. If the computations must be made by hand, achieving any degree of accuracy can be very tedious. However, the advent of computers has changed the situation; sometimes we use numerical methods of approximation even though an exact solution is available.

Differentiation of algebraic combinations of the elementary functions we have studied can be worked out by following the explicit formulas we have developed. In the absence of such a set of rules for integration, we should at least like to have a procedure for approximating the numerical value of a definite integral, together with a means of checking the accuracy of the approximation.

A finite set of mathematical formulas, together with instructions for their application in order to obtain a solution to a problem is called an *algorithm*; algorithms have the advantage of being adaptable to programming for the machine solution of problems. What we seek then are formulas which can be routinely applied to the integrand in order to evaluate

$$\int_a^b f(x)\,dx$$

approximately and to estimate the error in the approximation.

We recall that in computing the area under a parabola we used rectangles as an approximation (Fig. 12.2). Consider the area under $y = x^2$ from 0 to 1 and use four exterior rectangles as an approximation. Their areas are:

$$A_1 = (\tfrac{1}{4})(\tfrac{1}{4})^2 \qquad A_2 = (\tfrac{1}{4})(\tfrac{1}{2})^2$$
$$A_3 = (\tfrac{1}{4})(\tfrac{3}{4})^2 \qquad A_4 = (\tfrac{1}{4})1^2$$

and the sum of the areas is $\frac{15}{32}$. The actual area is

$$A = \int_0^1 x^2\,dx = \left.\frac{x^3}{3}\right|_0^1 = \frac{1}{3}.$$

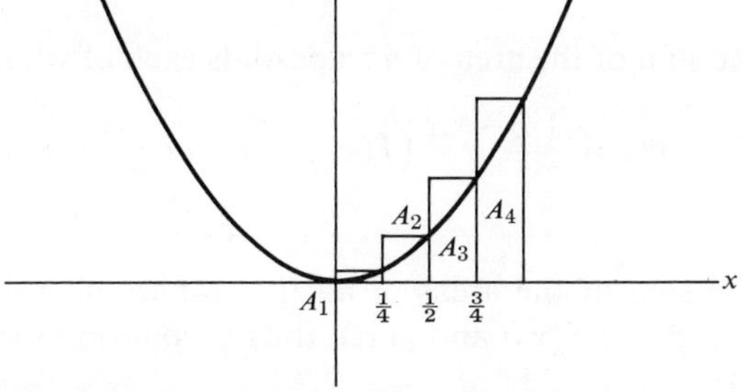

Fig. 12.2

Using eight exterior rectangles, we get an area of $\frac{51}{128}$, and finally the area of 100 exterior rectangles is $\frac{6767}{20000}$, a fairly close approximation to $\frac{1}{3}$.

However, we can obtain a good approximation much quicker if we replace the rectangles by trapezoids (Fig. 12.3). For four trapezoids we obtain

$$\begin{aligned} A_1 &= (\tfrac{1}{2})(\tfrac{1}{4})(0 + (\tfrac{1}{4})^2) \\ A_2 &= (\tfrac{1}{2})(\tfrac{1}{4})((\tfrac{1}{4})^2 + (\tfrac{1}{2})^2) \\ A_3 &= (\tfrac{1}{2})(\tfrac{1}{4})((\tfrac{1}{2})^2 + (\tfrac{3}{4})^2) \\ A_4 &= (\tfrac{1}{2})(\tfrac{1}{4})((\tfrac{3}{4})^2 + 1^2) \end{aligned}$$

for a total of $\frac{11}{32}$, already a fair approximation to the actual area.

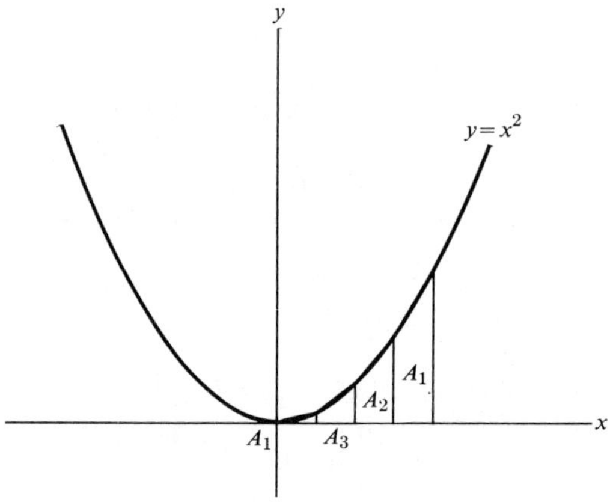

Fig. 12.3

The above is an example of the trapezoid rule of approximation:

> If $y = f(x)$ is continuous on an interval containing a and b, $\int_a^b f(x)\,dx$ is approximated by
>
> $$T_n = \frac{b-a}{n}\left(\tfrac{1}{2}f(x_0) + f(x_1) + \cdots + f(x_{n-1}) + \tfrac{1}{2}f(x_n)\right), \qquad (12.2)$$
>
> where T_n is the sum of the area of n trapezoids each of whose areas is given by
>
> $$\tfrac{1}{2}h(b + b') = \frac{1}{2}\frac{b-a}{n}\left(f(x_{i-1}) + f(x_i)\right) \quad \text{(see Fig. 12.4.)}$$

(Note that in the sum of the areas of all the trapezoids, each of the values $f(x_i)$ appears twice except for $f(x_0)$ and $f(x_n)$, thus accounting for the form of the sum given by (12.2).)

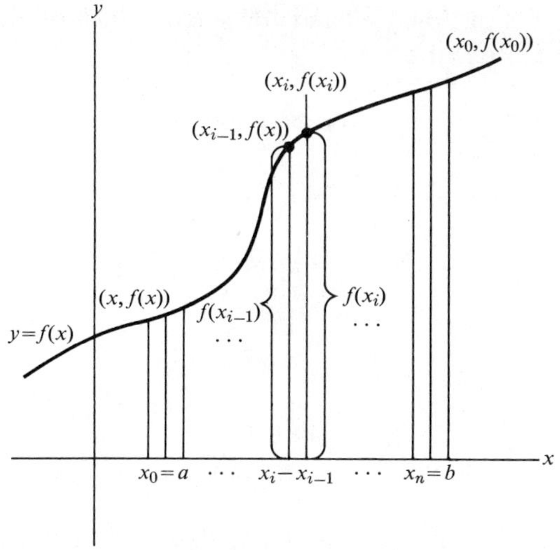

Fig. 12.4

We saw that in our simple example our trapezoidal approximation was "good." In fact, although we shall not prove it,

$$\lim_{n \to \infty} T_n = \int_a^b f(x)\, dx.$$

Even more useful to know is that not only do we know that the error can be made as small as we would like, but it is easy to estimate the error in the approximation. We state without proof:

Theorem 1. If f is a C^2 function on an interval containing a and b, then there exists a number c such that $a < c < b$ and

$$\int_a^b f(x)\, dx = T_n + \frac{b-a}{12} f''(c) \left(\frac{b-a}{n}\right)^2.$$

Let us see why this is a good thing to know.

Suppose we use the trapezoid rule to approximate $\int_1^3 1/x\, dx$.

$$T_n = \frac{2}{n}\left(\frac{1}{2}(1) + \frac{1}{1 + (2/n)} + \frac{1}{1 + (2 \cdot 2/n)} + \cdots + \frac{1}{1 + (2i/n)} + \cdots + \frac{1}{2} \cdot \frac{1}{3}\right)$$

$$= \frac{2}{n}\left[\sum_{i=0}^{n} \frac{1}{1 + (2i/n)} - \frac{1}{2} - \frac{1}{6}\right]$$

$$= \frac{2}{n}\left[\sum_{i=0}^{n} \frac{n}{n + 2i} - \frac{2}{3}\right].$$

This sum is easily computed—by hand for small values of n or by machine for large values. How close is it to

$$\int_1^3 \frac{1}{x}\,dx\ (=\log 3)?$$

By Theorem 1 there is a c such that $1 < c < 3$ and

$$\int_1^3 \frac{1}{x}\,dx - T_n = \frac{2}{12}\frac{1}{c^3}\left(\frac{2}{n}\right)^2 = \frac{2}{3n^2c^3}.$$

But for all $c \in (1, 3)$, $1/c^3 < 1$. Thus the error is less than $2/3n^2$. To make the error as small as we like, say less than ε, we simply make n sufficiently large so that n^2 is greater than $\frac{2}{3\varepsilon}$.

Are there other efficient methods for approximating integrals? Suppose we pick a different trapezoid. Instead of connecting $f(x_i)$ and $f(x_{i-1})$ in each subinterval, suppose we draw a line tangent to the curve at the midpoint of the subinterval (see Fig. 12.5). (This method requires that $f(x)$ be differentiable on an interval containing a and b.) Then the area of each trapezoid is

$$\frac{1}{2}\frac{b-a}{n}(y_2 + y_1). \tag{12.3}$$

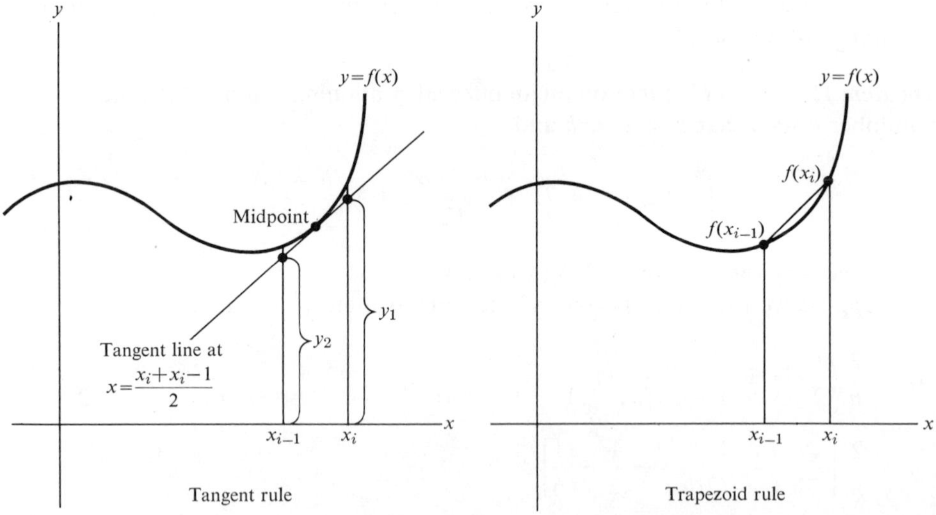

Fig. 12.5

The equation of the tangent is

$$y - f(m) = f'(m)(x - m),$$

where

$$m = a + \frac{i(b - a)/n + (i - 1)(b - a)/n}{2},$$

the midpoint of the interval $[x_{i-1}, x_i]$. Therefore

$$y_1 = f'(m)\left(a + \frac{i(b - a)}{n} - m\right) + f(m),$$

and

$$y_2 = f'(m)\left(a + \frac{(i - 1)(b - a)}{n} - m\right) + f(m).$$

So the area of each trapezoid is (from (3))

$$\frac{1}{2}\frac{b - a}{n}\left[f'(m)\left(\frac{i(b - a)}{2n} - \frac{(i - 1)(b - a)}{2n}\right) + f(m)\right.$$

$$\left. + f'(m)\left(\frac{(i - 1)(b - a)}{2n} - \frac{i(b - a)}{2n}\right) + f(m)\right]$$

$$= \frac{b - a}{n} f(m).$$

But this is just the area of a rectangle of base the width of the subinterval, namely $(b - a)/n$, and height the value of the function $f(x)$ at the midpoint of the subinterval.

This method of approximation is called the *tangent* or *midpoint rule:*

> $\int_a^b f(x)\, dx$ is approximated by
>
> $$M_n = \frac{b - a}{n} \sum_{i=1}^{n} f\left(a + \frac{i(b - a)/n + (i - 1)(b - a)/n}{2}\right).$$

Let us see how well the midpoint rule works with $y = x^2$, from 0 to 1. With four rectangles, the midpoints are $\frac{1}{8}$, $\frac{3}{8}$, $\frac{5}{8}$, and $\frac{7}{8}$.

$$M_4 = (\tfrac{1}{4})(\tfrac{1}{8})^2 + (\tfrac{1}{4})(\tfrac{3}{8})^2 + (\tfrac{1}{4})(\tfrac{5}{8})^2 + (\tfrac{1}{4})(\tfrac{7}{8})^2$$

$$= \tfrac{21}{64}.$$

Comparing this with

$$T_4 = \tfrac{11}{32},$$

and the actual value of $\frac{1}{3}$, we see that

$$\int_0^1 x^2\,dx - T_4 = \tfrac{1}{96},$$

whereas

$$\int_0^1 x^2\,dx - M_4 = \tfrac{1}{192}.$$

That is, the error in this method is only half as large.

It can be shown that

$$\lim_{n\to\infty} M_n = \int_a^b f(x)\,dx,$$

and in fact we have

Theorem 2. If $f(x)$ is C^2 on an interval containing a and b then there is a number c such that $a < c < b$ and

$$\int_a^b f(x)\,dx = M_n + \frac{b-a}{24} f''(c) \left(\frac{b-a}{n}\right)^2.$$

There is one more approximation which we shall discuss briefly. Both the trapezoid and the midpoint rule depend on approximations to the curve $y = f(x)$ by a straight line. That is, we replace the original function by one which is *piecewise linear*—made up of straight line segments—and compute the area under the new function. We now introduce *Simpson's rule*, which uses a parabola rather than a straight line as an approximating function on each subinterval.

Suppose $f(x)$ is continuous on an interval containing a and b and that we want to approximate

$$\int_a^b f(x)\,dx.$$

We divide the interval $[a, b]$ into an *even* number of equal subintervals

$$\left[a + (i-1)(a-b),\ a + \frac{i(b-a)}{n}\right], \qquad i = 1, \ldots, n.$$

Then there are an integral number of double intervals

$$\left[a + \frac{(2i-2)(b-a)}{n},\ a + \frac{2i(b-a)}{n}\right], \qquad i = 1, \ldots, n/2.$$

Let $x_i = a + i(b-a)/n$. There exists precisely one polynomial of degree less than 3 passing through the three points shown in Fig. 12.6:

$$(x_{2i-2}, f(x_{2i-2})), \qquad (x_{2i-1}, f(x_{2i-1})), \qquad (x_{2i}, f(x_{2i})). \tag{12.4}$$

Let us call this polynomial $p_i(x)$, $i = 1, \ldots, n/2$.

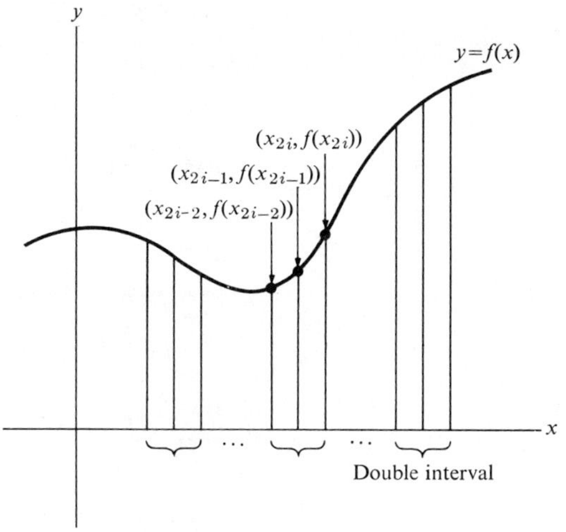

Fig. 12.6

If $f(x)$ is a straight line on the double interval $[x_{2i-2}, x_2]$, the polynomial $p_i(x)$ will be linear. If $f(x)$ is itself linear on $[a, b]$, we shall have $p_i(x) = f(x)$, $i = 1, \ldots, n/2$. Also, if $f(x)$ is a quadratic polynomial, $p_i(x) = f(x)$, $i = 1, \ldots, n/2$. Thus Simpson's rule works, but it is not of much interest in these cases.

It turns out to be easier to write p_i as a polynomial in $x - x_{2i-1}$.

$$p_i(x) = \alpha_i(x - x_{2i-1})^2 + \beta_i(x - x_{2i-1}) + \gamma_i. \tag{12.5}$$

Then let $u = x - x_{2i-1}$, and since $x_{2i-2} - x_{2i-1} = -(b - a)/n$ and $x_{2i} - x_{2i-1} = (b - a)/n$, we may write

$$\begin{aligned}
\int_{x_{2i-2}}^{x_{2i}} p_i(x)\,dx &= \int_{-(b-a)/n}^{(b-a)/n} (\alpha_i(x - x_{2i-1})^2 + \beta_i(x - x_{2i-1}) + \gamma_i)\,dx \\
&= \int_{-(b-a)/n}^{(b-a)/n} (\alpha_i u^2 + \beta_i u + \gamma_i)\,du \\
&= \frac{\alpha_i u^3}{3} + \frac{\beta_i u^2}{2} + \gamma_i u \bigg|_{-(b-a)/n}^{(b-a)/n} \\
&= \tfrac{2}{3}\alpha_i \frac{(b - a)^3}{n^3} + 2\gamma_i \frac{b - a}{n} \\
&= \frac{b - a}{3n}\left[2\alpha_i \left(\frac{b - a}{n}\right)^2 + 6\gamma_i\right].
\end{aligned}$$

Since $p_i(x)$ is to pass through the three points given by (12.4), when each of these values of x is substituted in (12.5) we should obtain the corresponding value of y.

First let $x = x_{2i-1}$; then

$$f(x_{2i-1}) = p(x_{2i-1}) = \alpha_i 0 + \beta_i 0 + \gamma_i = \gamma_i.$$

Now let $x = x_{2i-2}$; then

$$f(x_{2i-2}) = p(x_{2i-2}) = \frac{\alpha_i(b-a)^2}{n^2} - \frac{\beta_i(b-a)}{n} + \gamma_i.$$

Finally, let $x = x_{2i}$; then

$$f(x_{2i}) = p(x_{2i}) = \frac{\alpha_i(b-a)^2}{n^2} + \frac{\beta_i(b-a)}{n} + \gamma_i.$$

We see that

$$f(x_{2i-2}) + f(x_{2i}) = \frac{2\alpha_i(b-a)^2}{n^2} + 2\gamma_i,$$

or

$$f(x_{2i-2}) + f(x_{2i}) + 4f(x_{2i-1}) = \frac{2\alpha_i(b-a)^2}{n^2} + 6\gamma_i.$$

Therefore

$$\int_{x_{2i-2}}^{x_{2i}} p_i(x)\,dx = \frac{b-a}{3n}(f(x_{2i-2}) + 4f(x_{2i-1}) + f(x_{2i})).$$

We are now ready to state *Simpson's rule*:

$\int_a^b f(x)\,dx$ is approximated by

$$S_n = \sum_{i=1}^{n/2} \int_{x_{2i-2}}^{x_{2i}} p_i(x)\,dx$$

$$= \frac{b-a}{3n} \sum_{i=1}^{n/2} (f(x_{2i-2}) + 4f(x_{2i-1}) + f(x_{2i})).$$

It can be shown that

$$\lim_{n\to\infty} S_n = \int_a^b f(x)\,dx.$$

Moreover,

Theorem 3. If $f(x) \in C^4$ on an interval containing a and b, then there exists a number c such that $a < c < b$ and

$$\int_a^b f(x)\,dx = S_n + \frac{b-a}{180} f^{(4)}(c) \left(\frac{b-a}{n}\right)^4.$$

If $f(x)$ is a polynomial of degree less than 4, $f^{(4)}(x)$ is identically zero, so Simpson's rule gives an exact value for the integral of cubic as well as quadratic and linear functions.

Let us work out an example: Find S_4 for $\int_0^\pi \sin x\, dx$.

Solution.

$$S_4 = \frac{b-a}{12} \sum_{i=1}^{2} (f(x_{2i-2}) + 4f(x_{2i-1}) + f(x_{2i}))$$

$$= \frac{\pi}{12}\left(f(0) + 4f\left(\frac{\pi}{4}\right) + f\left(\frac{\pi}{2}\right)\right) + \left(f\left(\frac{\pi}{2}\right) + 4f\left(\frac{3\pi}{4}\right) + f(\pi)\right)$$

$$= \frac{\pi}{12}(0 + 2\sqrt{2} + 1 + 1 + 2\sqrt{2} + 0)$$

$$= \frac{\pi}{6}(1 + 2\sqrt{2})$$

$$= 2.03 \text{ (correct to two decimal places).}$$

This compares with

$$\int_0^\pi \sin x\, dx = -\cos x\Big|_0^\pi = -\cos \pi + \cos 0 = 2.$$

The trapezoid rule for four subintervals gives

$$T_4 = \frac{\pi}{4}\left(\frac{1}{2}\sin 0 + \sin\frac{\pi}{4} + \sin\frac{\pi}{2} + \sin\frac{3\pi}{4} + \frac{1}{2}\sin \pi\right)$$

$$= \frac{\pi}{4}\left(\frac{\sqrt{2}}{2} + 1 + \frac{\sqrt{2}}{2}\right)$$

$$= 1.91.$$

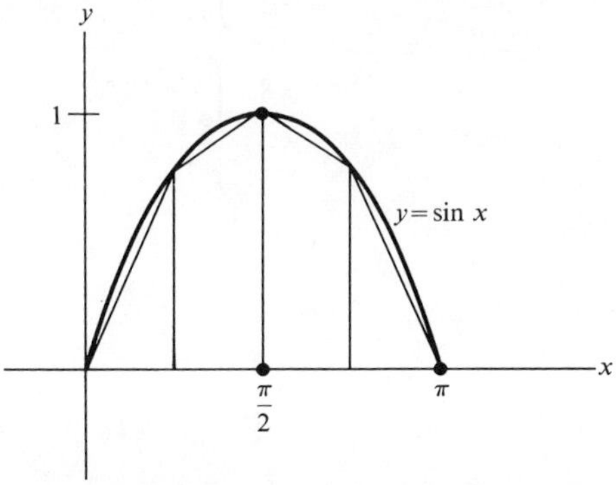

Fig. 12.7

The midpoint rule approximation for four subintervals gives

$$\begin{aligned} M_4 &= \frac{\pi}{4}\left(\sin\frac{\pi}{8} + \sin\frac{3\pi}{8} + \sin\frac{5\pi}{8} + \sin\frac{7\pi}{8}\right) \\ &= 1.95. \end{aligned}$$

We close the chapter by using Simpson's rule on an integral for which we have no method of obtaining an exact answer, namely

$$\int_0^1 e^{x^2}\,dx.$$

We have

$$\int_0^1 e^{x^2}\,dx \sim S_n = \frac{1}{3n}\sum_{i=1}^{n/2}\left(e^{x_{2i-2}^2} + 4e^{x_{2i-1}^2} + e^{x_{2i}^2}\right).$$

Let us choose $n = 10$. Then

$$\begin{aligned} S_{10} &= \tfrac{1}{30}[(e^0 + 4e^{1/100} + e^{1/25}) + (e^{1/25} + 4e^{9/100} + e^{4/25}) \\ &\quad + (e^{4/25} + 4e^{1/4} + e^{9/25}) + (e^{9/25} + 4e^{49/100} + e^{16/25}) \\ &\quad + (e^{16/25} + 4e^{81/100} + e)] \\ &\sim 1.46102. \end{aligned}$$

PROBLEMS

Use the trapezoid method, the midpoint rule and Simpson's rule, each with $n = 4$, to approximate:

1. $\int_1^2 x^3\,dx.$
2. $\int_0^2 (x^2 - 2)\,dx.$
3. $\int_0^\pi \sin x\,dx.$
4. $\int_1^3 \log x\,dx.$
5. $\int_1^5 \frac{1}{x}\,dx.$
6. $\int_0^{2\pi} \sin^4 x\,dx.$
7. $\int_0^2 \frac{1}{1+x^2}\,dx.$
8. $\int_0^1 e^{x^2}\,dx.$

13 / SERIES

13.1 SEQUENCES

A favorite form of question on intelligence tests is: What is the next term in the sequence

$$2, 4, 6, 8, \ldots ?$$

The examinee is expected to observe that for the terms listed the nth term is equal to $2n$ and that the next term should be $2 \cdot 5 = 10$. However, one with a more devious mind might notice that for

$$\begin{aligned} n = 1, \quad & n^4 - 10n^3 + 35n^2 - 48n + 24 = 2 \\ n = 2, \quad & n^4 - 10n^3 + 35n^2 - 48n + 24 = 4 \\ n = 3, \quad & n^4 - 10n^3 + 35n^2 - 48n + 24 = 6 \\ n = 4, \quad & n^4 - 10n^3 + 35n^2 - 48n + 24 = 8, \end{aligned}$$

and thus the next term should be

$$5^4 - 10 \cdot 5^3 + 35 \cdot 5^2 - 48 \cdot 5 + 24 = 34.$$

It is apparent from this example that a sequence cannot be specified by writing down its first few terms. Let us formalize the notion of sequence. A *sequence* (of real numbers) is a function from the positive integers to the real numbers,

$$s\colon Z^+ \to R.$$

The image $s(n)$ of $n \in Z^+$ under s is denoted by s_n. We usually identify a sequence with its image and write a sequence as

$$\{s_n\}.$$

In this notation we write the two sequences described above as

$$\{2n\}$$

and

$$\{n^4 - 10n^3 + 35n^2 - 48n + 24\}.$$

We concern ourselves first of all with two especially useful kinds of sequences. An *arithmetic sequence* is a sequence $\{s_n\}$ such that $s_{n+1} - s_n = d$, a constant. The constant d is called the common *difference*. If s_1 is the initial term of an arithmetic sequence, then $s_2 = s_1 + d$, $s_3 = s_1 + 2d$, and in general

$$s_n = s_1 + (n - 1)\, d.$$

Examples

$$2, 4, 6, 8, \ldots, 2 + 2(n - 1), \ldots$$
$$1, 5, 9, 13, \ldots, 1 + 4(n - 1), \ldots$$
$$4, 2, -1, -4, \ldots, 4 - 3(n - 1), \ldots$$

are arithmetic sequences.

Suppose we are interested in the sum S_n of the first n terms of an arithmetic sequence. We have

$$S_n = s_1 + (s_1 + d) + \cdots + s_n$$

and

$$S_n = s_n + (s_n - d) + \cdots + s_1.$$

Adding these two expressions for S_n, we obtain

$$2S_n = (s_1 + s_n) + (s_1 + s_n) + \cdots + (s_n + s_1) = n(s_1 + s_n).$$

Thus

$$S_n = \frac{n}{2}(s_1 + s_n) = \frac{n}{2}(2s_1 + (n - 1)d).$$

Examples

1. Associated Eggshells produces 100 units of eggshells in its first month. If its production increases at the rate of 15 units per month, when will the firm have reached a monthly production of over 2000 units? How much will it have produced by then?

Solution

$$s_1 = 100, \qquad d = 15, \qquad s_n \geq 2000.$$
$$s_n = s_1 + (n - 1)d$$
$$= 100 + (n - 1)15.$$

Therefore,

$$100 + (n - 1)15 \geq 2000$$
$$15n \geq 1915$$
$$n \geq 127\tfrac{2}{3}.$$

Thus in the 128th month the production will be greater than 2000. In fact,

$$s_{128} = 100 + 127(15) = 2005.$$

Its total production at the end of 128 months is given by

$$S_{128} = \tfrac{128}{2}(100 + 2005) = 134{,}720 \text{ units.}$$

2. Suburbia View has a population of 5000 now. If its population increases at the rate of 3000 a year, what will its population be after 20 years?

Solution. Again we use

$$\begin{aligned} s_n &= s_1 + (n - 1)d \\ s_{21} &= 5000 + 20(3000) \\ &= 65{,}000. \end{aligned}$$

We choose $n = 21$ instead of 20 since 5000 represents year one and 20 years later represents year 21.

Consider the sequence

$$1, 3, 9, 27, 81, \ldots, 3^{n-1}, \ldots$$

There is no common difference: the first and second terms differ by 2, the second and third by 6, etc. However, we observe that there is a common ratio:

$$\frac{1}{3} = \frac{3}{9} = \frac{9}{27} = \frac{27}{81} = \cdots = \frac{3^{n-1}}{3^n} = \cdots.$$

This is a special case of another common kind of sequence.

A *geometric sequence* is a sequence $\{s_n\}$ such that

$$s_{n+1} = rs_n$$

where r is a constant. The nth term of such a sequence is given by

$$s_n = s_1 r^{n-1}.$$

The constant r is called the common *ratio*.

Examples

$$2, 1, \tfrac{1}{2}, 1, \ldots, 2(\tfrac{1}{2})^{n-1}, \ldots$$

$$3, 6, 12, 24, \ldots, 3(2)^{n-1}, \ldots$$

$$1, -3, 9, -27, \ldots, (-3)^{n-1}, \ldots$$

are geometric sequences. To find the sum of the first n terms of a geometric sequence we write

$$S_n = s_1 + s_1 r + \cdots + s_1 r^{n-1}, \tag{13.1}$$

and multiply by r

$$S_n r = s_1 r + s_1 r^2 + \cdots + s_1 r^n. \tag{13.2}$$

Subtracting (13.2) from (13.1) gives

$$S_n - S_n r = s_1 - s_1 r^n$$

so that

$$S_n = \frac{s_1(1 - r^n)}{1 - r}, \qquad \text{if } r \neq 1$$

and

$$S_n = ns_1, \qquad \text{if } r = 1.$$

Examples

1. The number of bacteria in a culture is 600,000. If the population increases at the rate of six percent per minute, how many are there at the end of five minutes?

Solution. We use

$$s_n = s_1 r^{n-1}.$$

$$s_6 = 600{,}000(1.06)^5 \sim 802{,}800.$$

2. Greg starts out by practicing his piano lesson for two hrs on Monday. Each day he practices half as long as he has the day before. How much has Greg practiced in one week?

Solution. This time we want the sum of the first seven terms of a squence. The formula is

$$S_n = \frac{s_1(1 - r^n)}{1 - r}.$$

We have

$$S_7 = \frac{2(1 - (\frac{1}{2})^7)}{1 - \frac{1}{2}}$$

$$= \frac{2 \cdot \frac{127}{128}}{\frac{1}{2}}$$

$$= \tfrac{127}{32} \text{ hrs.}$$

Recall that in Chapter 7 the notion of integration was introduced by the use of a sequence of sums of areas of rectangles $\{S_n\}$ and that we were concerned with

$$\lim_{n \to \infty} S_n.$$

Sequences which have such a limit are particularly useful. We formalize this notion of limit. We need a definition which tells us that for large values of n, the nth term and all subsequent terms of the sequence are as close to the limit as we want.

If $\{s_n\}$ is a sequence such that given a neighborhood U of L there is a positive integer N such that

$$n > N \quad \text{implies that} \quad s_n \in U,$$

then we say that the sequence *converges* to L and write

$$\lim_{n\to\infty} s_n = L.$$

Let us see what this means geometrically (see Fig. 13.1). The function

$$y = s_n$$

has as its domain the positive integers. If $\{s_n\}$ converges to L, then if we move sufficiently far to the right, the values of y must lie arbitrarily close to the line $y = L$. If a sequence fails to converge, it is said to *diverge*.

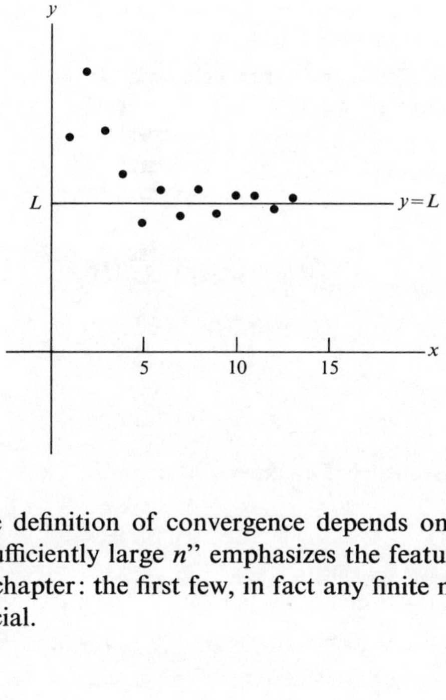

Fig. 13.1

The fact that the definition of convergence depends on the behavior of the sequence only for "sufficiently large n" emphasizes the feature which we noted at the beginning of the chapter: the first few, in fact any finite number, of terms of a sequence are not crucial.

Examples

1. Test for convergence:

$$\{(-1)^n\}.$$

Solution. No matter how large we choose N there will be terms s_n, $n > N$, of the sequence equal to 1 and terms s_n, $n > N$, equal to -1. Therefore, if L is any real number, no neighborhood of L of diameter less than 2 will contain all these terms. Thus the sequence does not converge.

2. Test for convergence:

$$\left\{-1, 1, \frac{3}{7}, \frac{2}{7}, \frac{5}{23}, \ldots, \frac{n}{n^2 - 2}, \ldots\right\}.$$

Solution. We have

$$\lim_{n\to\infty}\left[\frac{n}{n^2 - 2}\right] = \lim_{n\to\infty}\frac{(1/n)}{1 - 2/n^2} = \frac{\lim_{n\to\infty}(1/n)}{\lim_{n\to\infty}(1 - 2/n^2)} = 0.$$

So the sequence converges to 0.

PROBLEMS

1. Belch Baby Food produces 10,000 units the first month of operation. If production increases by 2000 each month, how much will be produced in the twentieth month? What will be the total production for five years? By what month will monthly production exceed one million units?
2. Smog Associates build 400 cars in their first week. If production increases 20 percent each week, how many cars will be built in six weeks?
3. Find the indicated sums:

a) $\sum_{n=1}^{49} (3 + 7n)$.
b) $\sum_{n=0}^{100} (5 + n/2)$.
c) $\sum_{n=1}^{75} (\frac{1}{3})^n$.
d) $\sum_{n=2}^{40} (\frac{3}{4})^n$.
e) $\sum_{n=1}^{100} (2 + 4n)$.
f) $\sum_{n=1}^{50} (1 - 5n)$.
g) $\sum_{n=0}^{14} (\frac{2}{3})^n$.
h) $\sum_{n=1}^{100} (-1)^n$.

4. Test the following sequences for convergence:

a) $\{3/(4 + 2n)\}$.
b) $\{\sqrt{n}/n\}$.
c) $\{(2n + 1)/\sqrt{n}\}$.
d) $\{-2/\sqrt{n}\}$.
e) $\{(n^2 + 3n - 1)/(n^3 + 2n)\}$.
f) $\{\sqrt{2n}/(\sqrt{2n} + 1)\}$.
g) $\{(n^2 + 5n - 1)/(2n + 5)\}$.

5. Show that a geometric sequence converges if $0 < r < 1$.
6. Show that an arithmetic sequence with $d \neq 0$ diverges.
7. A wise man asked a king whom he had assisted for his reward to be paid in grain as follows: one grain on the first square of a chessboard, two grains on the second, four grains on the third and 2^{n-1} grains on the nth. How much grain would he receive?
8. If a sum of money P is deposited at compound interest, the amounts on deposit at successive time periods form a geometric sequence. For example, if the rate is six percent and the compounding period one year, we have

$$P,\ 1.06P,\quad (1.06)^2P, \ldots,\quad (1.06)^nP.$$

Find the amount to which \$1000 has grown at the end of ten years. When will the original amount triple? (see also Chapter 2).

9. If the population of a certain city increases by 5000 a year for 20 years and at the end of that time has a population of 250,000, how big was it at the beginning of the period?
10. If the number of bacteria in a culture increases by 20 percent every hour and there are 1000 bacteria originally, how many will there be at the end of 4 hours?
11. If a first week's production of 100 chairs is increased by two percent per week, how many chairs will have been produced at the end of a year?

13.2 SERIES

One of the paradoxes of Zeno concerns the story of Achilles and the tortoise. Achilles travels 100 times as fast as the tortoise. The tortoise is given a head start in a race and their progress is as follows.

	Tortoise	Achilles
1st time period (1 hour)	1 mile	0
2nd time period ($\frac{1}{100}$ hour)	$\frac{1}{100}$ mile	1 mile
3rd time period ($\frac{1}{10000}$ hour)	$\frac{1}{10000}$ mile	$\frac{1}{100}$ mile
⋮		
nth time period ($(\frac{1}{100})^{n-1}$ hour)	$(\frac{1}{100})^{n-1}$ mile	$(\frac{1}{100})^{n-2}$ mile
⋮		

Thus at the end of the nth time period the tortoise has traveled

$$S_n = \frac{(1 - (\frac{1}{100})^n)}{1 - \frac{1}{100}} \text{ miles}$$

(sum of first n terms of sequence beginning with 1, $r = \frac{1}{100}$) and Achilles has traveled only

$$S'_n = \frac{(1 - (\frac{1}{100})^{n-1})}{1 - \frac{1}{100}} \text{ miles,}$$

$(\frac{1}{100})^n$ miles less. Hence it seems the fleet Achilles will never catch up.

However, let us consider the sequences of sums

$$\{S_n\}, \qquad \{S'_n\}.$$

We have

$$\lim_{n\to\infty} S_n = \lim_{n\to\infty} S'_n = \frac{1}{1 - \frac{1}{10}} = \frac{100}{99}.$$

Thus if the race were to continue for an infinite number of time periods—whatever that might be—Achilles and the tortoise would both cover the same distance.

Just how long would the "infinite number of time periods" last? The tortoise must cover $\frac{100}{99}$ miles and his speed is a constant 1 mile per hour. Thus the elapsed time is $\frac{100}{99}$ hours. Therefore, since he started one hour after the tortoise, it takes Achilles $\frac{1}{99}$ hour to catch him.

The process described above is basically an infinite analog of the summation process, which we indicate by

$$\sum_{k=1}^{\infty} \left(\frac{1}{100}\right)^k.$$

In general we call an expression

$$\sum_{k=1}^{\infty} s_k, \qquad \text{where } \{s_k\} \text{ is a sequence,}$$

an *infinite series.* Note that the letter used as an index is not significant:

$$\sum_{k=1}^{\infty} s_k = \sum_{i=1}^{\infty} s_i, \qquad \text{etc.}$$

Also we may want to consider a series with an index beginning at some integer other than one, for example,

$$\sum_{k=0}^{\infty} s_k,$$

but these variations present no real difficulties. Let

$$S_n = \sum_{k=1}^{n} s_i$$

be the *nth partial sum* of the sequence $\{s_i\}$, that is, the sum of the first n terms. Then let us consider the sequence of partial sums $\{S_n\}$. If the sequence of partial sums converges, i.e., if

$$\lim_{n\to\infty} S_n = L,$$

then we write

$$\sum_{k=1}^{\infty} s_k = L,$$

and say that the *series converges.*

If $\{s_k\}$ is a geometric sequence, then $\sum_{k=1}^{\infty} s_k$ is a *geometric series.*

Examples

1. In a geometric series if $0 < r < 1$, $r^n \to 0$ as $n \to \infty$ so that

$$\lim_{n\to\infty} S_n = \lim_{n\to\infty} \frac{s_1(1 - r^n)}{1 - r} = \frac{s_1}{1 - r}.$$

Thus the series converges to $s_1/(1 - r)$. (If the index starts at 0, we replace s_1 by s_0.) On the other hand, if $r > 1$, $r^n \to \infty$ and the series does not converge. Nor does it converge for $r = 1$ since then $S_n = ns_1 \to \infty$ as $n \to \infty$.

2. Suppose in a simple Keynesian model of the economy the marginal propensity to consume is denoted by p and that an increased consumption expenditure of ΔC is injected into the system. We want to find the new equilibrium level of income after the consumption expenditure increment takes effect. We have a sequence

$$\Delta C, \Delta Cp, \Delta Cp^2, \ldots$$

and its sum is the equilibrium increment in income

$$Y = \sum_{k=0}^{\infty} \Delta C p^k.$$

This is a geometric series and $0 < p < 1$ (by definition of marginal propensity) so the sum is given by

$$Y = \frac{\Delta C}{1 - p}.$$

Suppose the marginal propensity to consume is 0.9 and a new consumption expenditure of one billion dollars is injected into the economy. Then the increase in national income at the new equilibrium is given by

$$Y = \frac{\Delta C}{1 - p} = \frac{1{,}000{,}000{,}000}{1 - 0.9} = 10{,}000{,}000{,}000.$$

3. A ball is dropped from a height of 6 ft; at each bounce the ball rises to one third of its previous height. Find the total distance traveled by the ball. The distance is given by

$$6 + 2 \cdot \tfrac{1}{3} \cdot 6 + 2 \cdot (\tfrac{1}{3})^2 \cdot 6 + \cdots = 6 + \sum_{k=1}^{\infty} 12(\tfrac{1}{3})^k.$$

This is a geometric series with ratio $\frac{1}{3}$ and initial term $s_0 = 6$. Hence the sum is

$$6 + \frac{4}{1 - \frac{1}{3}} = 12.$$

Since a series really only has meaning if it converges, the central problem is to determine convergence. We have solved the problem for geometric series and we summarize briefly some more general methods.

The first result is a negative one:

Theorem 1. If $\lim_{k\to\infty} s_k \neq 0$, then the series $\sum_{k=1}^{\infty} s_k$ does not converge.

Proof. Suppose the series converges, say $\lim_{n\to\infty} S_n = L$, where $S_n = \sum_{k=1}^{n} s_k$. Then given a neighborhood U of L, there is an N such that for $n > N$, $S_n \in U$; in particular, for sufficiently large n, S_n and S_{n+1} must both be in U. But this is true for arbitrarily small neighborhoods of L, so we must have

$$\lim_{n\to\infty} (S_{n+1} - S_n) = \lim_{n\to\infty} s_n = 0.$$

The converse does not hold. There are series

$$\sum_{k=1}^{\infty} s_k,$$

such that $\lim_{k\to\infty} s_k = 0$, but the series does not converge. $\sum_{k=1}^{\infty} 1/k$ is an example of such a series.

Warning. Do not confuse convergence of the sequence of terms of the series $\{s_k\}$ with convergence of the sequence of partial sums $\{S_n\}$, which is by definition equivalent to the convergence of the series. $\{s_k\}$ may converge without $\{S_n\}$ converging, but if $\{S_n\}$ converges, then $\{s_k\}$ not only must converge, but must converge to 0.

The next few theorems, which we state without proof, concern series all of whose terms are nonnegative.

Theorem 2 (*Comparison Test*). If $\sum_{k=1}^{\infty} u_k$, is a series of nonnegative terms which converges and $\sum_{k=1}^{\infty} v_k$ is a series of nonnegative terms such that $v_k \le u_k$ for each k, then $\sum_{k=1}^{\infty} v_k$ also converges. On the other hand, if $\sum_{k=1}^{\infty} u_k$ diverges and $\sum_{k=1}^{\infty} v_k$ is such that $v_k \ge u_k$ for each k, then $\sum_{k=1}^{\infty} v_k$ diverges.

We need some handy series for comparison purposes. At the moment we have only the geometric series, convergent for values of r between 0 and 1, divergent for $r > 1$, but any new series which is shown to be convergent or divergent either by comparison or by more direct means can, of course, subsequently be used for comparison.

Example. $\sum_{k=1}^{\infty} 1/k!$ converges since $\sum_{k=0}^{\infty} 1/2^{k-1}$ converges and $1/k! \le 1/2^{k-1}$. Note that in making comparisons one may neglect the first n terms, for any n, if necessary, e.g.,

$$15 + 2 + 3 + \sum_{k=1}^{\infty} \frac{1}{k!}$$

also converges by the comparison test since eventually each term is less than the corresponding term of

$$1 + 1 + 1 + \sum_{k=0}^{\infty} \frac{1}{2^{k-1}}.$$

Theorem 3 (*Integral Test*). If $f(x)$ is a decreasing continuous function such that $f(k) = u(k) \geq 0$, then the series

$$\sum_{k=1}^{\infty} u_k \qquad \text{and the integral} \qquad \int_1^{\infty} f(x)\,dx,$$

both converge or both diverge.

Examples

1. $\sum_{k=1}^{\infty} 1/k$ diverges since

$$\int_1^{\infty} \frac{dx}{x} = \lim_{b\to\infty} \int_1^b \frac{dx}{x} = \lim_{b\to\infty} \log b = \infty.$$

   This series is usually called the *harmonic series*. As we noted, $\lim_{k\to\infty} 1/k = 0$, even though the series diverges.

2. The series $\sum_{k=1}^{\infty} 1/k^p$ is called a p-series. If $p > 1$, it converges since

$$\int_1^{\infty} x^{-p}\,dx = \lim_{b\to\infty} \int_1^b x^{-p}\,dx = \lim_{b\to\infty} \left. \frac{x^{-p+1}}{-p+1} \right|_1^b = \frac{1}{p-1}.$$

   If $p = 1$, we have the harmonic series of the example above. If $p < 1$, the series diverges by comparison with the harmonic series.

Theorem 4 (*Ratio Test*). If $\sum_{k=1}^{\infty} u_k$ is a series of nonnegative terms such that $\lim_{k\to\infty} u_{k+1}/u_k$ exists, then

1) if $\lim_{k\to\infty} u_{k+1}/u_k < 1$, the series converges.
2) if $\lim_{k\to\infty} u_{k+1}/u_k > 1$, the series diverges.
3) if $\lim_{k\to\infty} u_{k+1}/u_k = 1$, the series may either converge or diverge.

Examples

1. $\sum_{k=1}^{\infty} k^2/2^k$ converges since

$$\lim_{k\to\infty} \frac{u_{k+1}}{u_k} = \lim_{k\to\infty} \frac{(k+1)^2}{2^{k+1}} \cdot \frac{2^k}{k^2} = \frac{1}{2}.$$

2. For both the divergent harmonic series $\sum_{k=1}^{\infty} 1/k$ and the convergent series $\sum_{k=1}^{\infty} (1/k^2)$,

$$\lim_{k\to\infty} \frac{u_{k+1}}{u_k} = 1.$$

What about series all of whose terms are not nonnegative? We make the following definition.

A series $\sum_{k=1}^{\infty} u_k$ is *absolutely convergent* if $\sum_{k=1}^{\infty} |u_k|$ is convergent. This is a useful concept since we have

Theorem 5. If a series converges absolutely, then it converges. Thus if $\sum_{k=1}^{\infty} u_k$ is an arbitrary series, we can apply the various tests for convergence for series of nonnegative terms to the series $\sum_{k=1}^{\infty} |u_k|$. If it converges so does the original series.

Example. Test for convergence:

$$\sum_{k=1}^{\infty} \frac{(-1)^k}{2^k}.$$

Solution

$$\sum_{k=1}^{\infty} \frac{(-1)^k}{2^k},$$

is a p-series, $p > 1$, so it converges. Hence by Theorem 5, $\sum_{k=1}^{\infty} (-1)^k/2^k$ converges.

The converse of Theorem 5 is not true. For example, $\sum_{k=1}^{\infty} (-1)^k/k$ does not converge absolutely since

$$\sum_{k=1}^{\infty} \frac{(-1)^k}{k},$$

is just the harmonic series, which we know diverges. However, it can be shown, by methods we shall not describe, that $\sum_{k=1}^{\infty} (-1)^k/k$ does converge.

PROBLEMS

1. Suppose the marginal propensity to consume is 0.99 and a new consumption expenditure of one million dollars is injected into the economy. What is the resulting increase in national income?
2. If a ball is dropped from a height of ten feet and rises on each bounce to half its previous height, how far does the ball travel?
3. Determine whether each of the given series converges or diverges:

a) $\sum_{k=1}^{\infty} 3^k/k!.$

b) $\sum_{n=1}^{\infty} \frac{n}{2n^2+1}.$

c) $\sum_{k=1}^{\infty} \frac{\sin^2 k}{3^k}.$

d) $\sum_{n=1}^{\infty} \frac{n^2}{2^n}.$

e) $\sum_{n=1}^{\infty} 1/(2n-1)!.$

f) $\sum_{n=1}^{\infty} \frac{(n+2)!}{2!\, n!\, 2^n}.$

g) $\sum_{k=1}^{\infty} \frac{k}{k^2+1}.$

h) $\sum_{n=1}^{\infty} \frac{2}{2n+1}.$

i) $\sum_{k=1}^{\infty} \frac{1}{k \log k}.$

j) $\sum_{n=1}^{\infty} \frac{2}{n^2-5}.$

k) $\sum_{n=1}^{\infty} \frac{\log n}{n^2}.$

l) $\sum_{k=1}^{\infty} \frac{\log k}{k}.$

m) $\sum_{k=1}^{\infty} \frac{\log k}{2k}.$

n) $\sum_{k=1}^{\infty} \frac{2}{k^2+1}.$

o) $\sum_{k=1}^{\infty} \frac{2^k}{(2k)!}.$

p) $\sum_{k=1}^{\infty} \frac{3}{2k^{3/2}}.$

4. Determine whether each of the following series is absolutely convergent.

a) $\sum_{n=1}^{\infty} \frac{(-1)^n n^2}{2^n}$.

b) $\sum_{n=1}^{\infty} \frac{(-1)^n 3n}{n!}$.

c) $\sum_{n=1}^{\infty} \frac{(-1)^{n-1} n!}{10^n}$.

d) $\sum_{n=1}^{\infty} \frac{(-1)^n}{n\sqrt{n}}$.

e) $\sum_{n=1}^{\infty} \frac{(-1)^n}{2n^2}$.

f) $\sum_{n=1}^{\infty} \frac{(-1)^n 2^n}{n!}$.

g) $\sum_{n=1}^{\infty} \frac{(-1)^{n+1}(\sqrt{n-1})}{n^2}$.

h) $\sum_{n=1}^{\infty} \frac{(-1)^n n!}{10^n}$.

5. A political action group chooses to spread its message by leafleting. Assuming that a constant proportion k of the handbills survive a given time period t and that each surviving leaflet reaches p people, find the total number of people reached by handing out N leaflets. Compute the number reached if $N = 10{,}000$, $k = \frac{1}{10}$, and $p = 3$.

13.3 POWER SERIES

We want to approximate a C^{∞} function

$$y = f(x)$$

by a sequence of polynomials $f_n(x)$ of the form

$$f_n(x) = a_0 + a_1 x + a_2 x^2 + \cdots + a_n x^n. \tag{13.3}$$

If we expect the approximations to improve as n increases, it seems reasonable to take small values of x so that as $n \to \infty$, $a_n x^n \to 0$. Thus we concern ourselves with that part of the graph of $y = f(x)$ near the point $(0, f(0))$ (see Fig. 13.2).

$$y = f(0)$$

is a 0-degree approximation of $f(x)$ which agrees with the function at the origin. To get a first-degree approximation we take the tangent line at the origin:

$$y = f(0) + f'(0)x.$$

In general we want a polynomial of degree n which has its first n-derivatives equal to the first n-derivatives of

$$y = f(x);$$

agreement of the derivatives should give us a good "fit" to the given curve. Thus if we write the polynomial as in (13.3), we have

$$\begin{aligned} f_n'(x) &= a_1 + 2a_2 x + 3a_3 x^2 + \cdots + n a_n x^{n-1} \\ &\vdots \\ f_n^{(n)}(x) &= n!\, a_n \end{aligned}$$

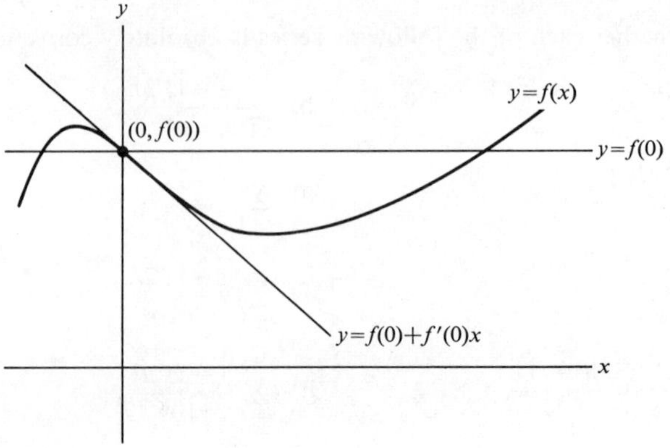

Fig. 13.2

Substituting $x = 0$ and setting each of these derivatives equal to $f(0), f'(0), \ldots, f^{(n)}(0)$, we solve for the $n + 1$ coefficients:

$$\begin{aligned} a_0 &= f(0) \\ a_1 &= f'(0) \\ a_2 &= \frac{f''(0)}{2!} \\ &\vdots \\ a_n &= \frac{f^{(n)}(0)}{n!}. \end{aligned}$$

Thus

$$f_n(x) = f(0) + f'(0)x + \frac{f''(0)}{2!}x^2 + \cdots + \frac{f^{(n)}(0)}{n!}x^n.$$

Note that $f_n(x)$ is the nth partial sum of the infinite series

$$\sum_{n=0}^{\infty} \frac{f^{(n)}(0)}{n!} x^n.$$

Thus the convergence of the sequence of approximating polynomials is equivalent to the convergence of this series, called the *Taylor series expansion of the function* $f(x)$ *at zero*:

$$\sum_{n=0}^{\infty} \frac{f^{(n)}(0)}{n!} x^n.$$

For all reasonable C^∞ functions, if the Taylor series converges, it converges to $f(x)$ and we write

$$f(x) = \sum_{k=0}^{\infty} \frac{f^{(n)}(0)x^n}{n!},$$

for values of x near 0.

If we are interested in approximating the function

$$y = f(x)$$

near $x = a \neq 0$, $x - a$ will be small and $(x - a)^n \to \infty$ as $n \to \infty$, so we expand in terms of $x - a$:

$$f(x) = \sum_{n=0}^{\infty} \frac{f^{(n)}(a)}{n!}(x - a)^n.$$

The Taylor series of a function is a special case of a *power series*

$$\sum_{k=0}^{\infty} a_k x^k,$$

or more generally

$$\sum_{k=0}^{\infty} a_k(x - a)^k \qquad (a_k \text{ real numbers}).$$

The usual tests for convergence may be applied to power series; however the series may converge for certain values of x and diverge for others.

Examples

1.
$$\sum_{k=1}^{\infty} x^k.$$

Consider the ratio

$$\lim_{k\to\infty} \frac{|x^{k+1}|}{|x^k|} = |x|.$$

By the ratio test, the series converges for $|x| < 1$ and diverges for $|x| > 1$. We must test $x = \pm 1$ separately, but we see at once that neither

$$\sum_{k=1}^{\infty} 1 \qquad \text{nor} \qquad \sum_{k=1}^{\infty} (-1)^k$$

converges.

2.
$$\sum_{k=1}^{\infty} \frac{x^k}{k^2}.$$

We have

$$\lim_{k\to\infty} \frac{|x^{k+1}|}{|x^k|} \cdot \frac{k^2}{(k + 1)^2} = |x|.$$

Therefore this series also converges for $|x| < 1$ and diverges for $|x| < 1$. Testing ± 1 yields

$$\sum_{k=1}^{\infty} \frac{1}{k^2} \qquad \text{and} \qquad \sum_{k=1}^{\infty} \frac{(-1)^k}{k^2}.$$

The first is convergent; the second is absolutely convergent and hence convergent.

3.
$$\sum_{k=1}^{\infty} \frac{x^k}{k!}.$$

Again we use the ratio test:

$$\lim_{k\to\infty} \frac{|x^{k+1}|}{(k+1)!} \cdot \frac{k!}{|x^k|} = \lim_{k\to\infty} \frac{|x|}{k+1} = 0.$$

Thus the series converges for all values of x.

4. Suppose that p_k is the probability that a variable x assumes the positive integer value k. We require that $p_k \geq 0$ for all k and that $\sum_{k=1}^{\infty} p_k = 1$. The expected value of x is defined as

$$E(x) = \sum_{k=1}^{\infty} kp_k ,$$

provided that the series converges. If

$$p_k = \frac{2^{k-1}}{3^k},$$

find $E(x)$.

Solution. First of all we note that

$$\sum_{k=1}^{\infty} p_k,$$

is a geometric series whose first term is $\frac{1}{3}$ and whose ratio is $\frac{2}{3}$; thus its sum is

$$\frac{\frac{1}{3}}{1 - \frac{2}{3}} = 1.$$

Now

$$E(x) = \sum_{k=1}^{\infty} k \frac{2^{k-1}}{3^k}$$

$$= \frac{1}{3} \sum_{k=1}^{\infty} k \frac{2^{k-1}}{3^{k-1}}.$$

Note that

$$\frac{1}{1-x} = 1 + x + x^2 + \cdots + x^n + \frac{x^{n+1}}{1-x},$$

and differentiating yields

$$\frac{1}{(1-x)^2} = 1 + 2x + 3x^2 + \cdots + nx^{n-1} + R,$$

R a remainder term. Moreover, if $|x| < 1$, since R involves powers of x, $R \to 0$. Therefore, we feel justified in writing

$$\frac{1}{(1-x)^2} = \sum_{k=1}^{\infty} kx^{k-1}.$$

Thus we have

$$\frac{1}{3}\sum_{k=1}^{\infty} k\left(\frac{2}{3}\right)^{k-1} = \frac{1}{3}\left(\frac{1}{(1-\frac{2}{3})^2}\right) = 3.$$

5. $$\sum_{k=1}^{\infty} \frac{(x-1)^k}{k}.$$

Solution. We have

$$\lim_{k\to\infty} \frac{|x-1|^{k+1}}{|x-1|^k} \cdot \frac{k}{k+1} = |x-1|.$$

Therefore the series converges for $|x - 1| < 1$ and diverges for $|x - 1| > 1$. Testing $x - 1 = 1$, $x = 2$, or $x - 1 = -1$, $x = 0$, yields $\sum_{k=1}^{\infty} 1/k$ which diverges and $\sum_{k=1}^{\infty} (-1)^k/k$, which converges.

If a power series

$$\sum_{k=1}^{\infty} a_k(x-a)^k$$

converges for $|x - a| < R$, R is called the *radius of convergence* of the series. We can speak of a *function defined by a power series* for values of x within the interval $(a - R, a + R)$ of convergence and write

$$f(x) = \sum_{k=1}^{\infty} a_k(x-a)^k \qquad \text{for} \quad |x-a| < R.$$

We turn our attention again to Taylor series, this time for the approximation of values of trigonometric, logarithmic, and exponential functions.

Example. Express sin 1^0 in the form of a series.

Solution. The Taylor series for the sine function at 0 is calculated as follows:

$$\begin{aligned}
f(0) &= 0 \\
f'(0) &= \cos 0 = 1 \\
f''(0) &= -\sin 0 = 0 \\
f'''(0) &= -\cos 0 = -1 \\
&\vdots \\
f^{(4n)}(0) &= 0 \\
f^{(4n+1)}(0) &= 1 \\
f^{(4n+2)}(0) &= 0 \\
f^{(4n+3)}(0) &= -1.
\end{aligned}$$

Thus

$$\sin x = x - \frac{x^3}{3!} + \frac{x^5}{5!} + \cdots$$

$$= \sum_{n=0}^{\infty} \frac{(-1)x^{2n+1}}{(2n+1)!}.$$

This series converges for all values of x. In particular

$$\sin 1^0 = \sin \frac{\pi}{180} = \sum_{n=0}^{\infty} \frac{(-1)^n(\pi/180)^{2n+1}}{(2n+1)!}.$$

We should like to know how many terms in the series must be calculated in order to obtain an adequate approximation of the value of the function. We emphasize that if the series converges it gives the precise value of the function, but for any finite number of terms we are guaranteed only an approximation. To judge the possible error in approximation we write the Taylor series for a function as a polynomial of degree n plus a remainder term. The remainder term then tells us the size of the error resulting from using the nth degree polynomial as an approximation. We state Taylor's theorem without proof:

Theorem 6 (*Taylor's Theorem*). If f is a C^{n+1} function on an interval containing x and a, then

$$f(x) = f(a) + f'(a)(x - a) + \frac{f''(a)}{2!}(x - a)^2 + \cdots$$

$$+ \frac{f^{(n)}(a)}{n!}(x - a)^n + \int_a^x \frac{(x - t)^n}{n!} f^{(n+1)}(t)\, dt.$$

The Taylor series for $f(x)$ converges if and only if the remainder term

$$\int_a^x \frac{(x - t)^n}{n!} f^{(n+1)}(t)\, dt,$$

goes to zero as $n \to \infty$. Another form of the remainder term, called the *Lagrange form*, is given by

$$f^{(n+1)}(c)\frac{(x - a)^{n+1}}{(n+1)!},$$

for some c between a and x.

Example. Calculate the value of e correct to three decimal places.

Solution. The Taylor series for e^x at 0 is given by

$$e^x = \sum_{n=0}^{\infty} \frac{x^n}{n!}$$

(note that $0! = 1$). The ratio test establishes that the series converges for all values of x, in particular, for $x = 1$. By Taylor's theorem

$$e^x = 1 + x + \frac{x^2}{2!} + \cdots + \frac{x^n}{n!} + \int_0^x \frac{(x-t)^n}{n!} e^t \, dt.$$

For $0 < t < x$, $e^t < e^x < 3^x$, so that

$$\int_0^x \frac{(x-t)^n}{n!} e^t \, dt \le \int_0^x \frac{(x-t)^n}{n!} 3^x \, dt = \frac{3^x x^{n+1}}{(n+1)!}.$$

Thus for $x = 1$ the remainder term is less than

$$\frac{3}{(n+1)!},$$

and to insure three-place accuracy this error term must satisfy the inequality

$$\frac{3}{(n+1)!} < 0.0005,$$

$$(n+1)! > 6000.$$

Since $7! = 5040$ and $8! = 40{,}320$, we need eight terms in the series ($n = 0, \ldots, 7$) to get the desired accuracy:

$$\begin{aligned} e &= 1 + 1 + \frac{1}{2!} + \frac{1}{3!} + \frac{1}{4!} + \frac{1}{5!} + \frac{1}{6!} + \frac{1}{7!} \\ &= 2 + 0.5 + 0.1667 + .0416 + 0.0083 + 0.0014 + 0.0002 \\ &= 2.718. \end{aligned}$$

Note that errors in approximation can result either from truncation—dropping terms in the series—or from rounding off individual terms. In particular, in the example above, premature rounding off of the individual terms results in an approximation not accurate to three decimal places even though eight terms of the series are used.

PROBLEMS

1. Determine the interval of convergence of the following series:

a) $\sum_{n=1}^{\infty} (x/2)^n$. b) $\sum_{n=1}^{\infty} \frac{n!\, x^n}{10^n}$. c) $\sum_{n=1}^{\infty} n^2(x+1)^n$.

d) $\sum_{n=1}^{\infty} x^n/\log(n+1)$. e) $\sum_{k=1}^{\infty} \frac{(x-2)^k}{k^3}$. f) $\sum_{k=1}^{\infty} \frac{(2x+a)^k}{2^k}$.

g) $\sum_{n=1}^{\infty} \frac{(2x-1)^n}{\sqrt{n}}$. h) $\sum_{k=1}^{\infty} \frac{(2k)!\, x^k}{(10k)!}$. i) $\sum_{n=1}^{\infty} (x/3)^n$.

j) $\sum_{n=1}^{\infty} \frac{x^n}{2^n}$. k) $\sum_{n=1}^{\infty} \frac{x^n}{\log n}$. l) $\sum_{n=1}^{\infty} \frac{(2x-1)^n}{n^2}$.

2. Items are inspected at random from an assembly line. If the probability of an item's being defective is p and being accepted is $1 - p$, the probability that the first defective item found is the kth item inspected in $p(1-p)^{k-1}$. The average number inspected up to and including the first defective item found is

$$\sum_{k=1}^{\infty} kp(1-p)^{k-1}.$$

Using the fact that

$$\frac{1}{(1-x)^2} = 1 + 2x + \cdots nx^{n-1} + R,$$

where

$$R \to 0 \quad \text{as} \quad n \to \infty \quad \text{if} \quad |x| < 1,$$

find the average number of items inspected up to and including the first defective one if $p = \frac{1}{100}$.

3. In one throw of two dice the probability of getting an 11 is $p = \frac{1}{18}$, and the probability of getting an 11 for the first time on the kth throw is $p(1-p)^{k-1}$. The expected number of throws until an 11 first appears is

$$\sum_{k=1}^{\infty} kp(1-p)^{k-1}.$$

Using the technique of Problem 2, find this number.

4. Complete the computation of sin 1°, accurate to three decimal places.

5. Find the Taylor series for:
   a) $f(x) = e^{2x}$ at $x = 0$.
   b) $f(x) = \sqrt{x}$ at 4.
   c) $f(x) = e^x$ at 1.
   d) $f(x) = 1/\sqrt{1-x}$ at 0.
   e) $f(x) = \log x$ at $x = 1$.
   f) $f(x) = \cos x$ at $x = 0$.
   g) $f(x) = \sqrt{2x}$ at $x = 2$.

6. Find the first four terms of the Taylor series expansion of:
   a) $f(x) = e^{-x}$ at 0.
   b) $f(x) = \tan x$ at $\pi/4$.
   c) $f(x) = 1/(1 + x^2)$ at 0.
   d) $f(x) = \log \cos x$ at $\pi/3$.
   e) $f(x) = e^{x^2}$ at $x = 0$.

7. Find log 0.5 accurate to three decimal places.

8. Compute $e^{0.05}$, accurate to five decimal places. Why is Problem 8 easier than Problem 7?

9. If $f(x, y)$ has continuous nth-order partial derivatives in a neighborhood of (x_0, y_0), the Taylor expansion of degree n about (x_0, y_0) is the polynomial

$$f(x_0, y_0) + \frac{1}{1!}\left[(x - x_0)\frac{\partial f}{\partial x}(x_0, y_0) + (y - y_0)\frac{\partial f}{\partial y}(x_0, y_0)\right]$$

$$+ \frac{1}{2!}\left[(x - x_0)^2 \frac{\partial^2 f}{\partial x^2}(x_0, y_0) + 2(x - x_0)(y - y_0)\frac{\partial^2 f}{\partial x\,\partial y}(x_0, y_0) + (y - y_0)^2 \frac{\partial^2 f}{\partial y^2}(x_0, y_0)\right] + \cdots$$

$$+ \frac{1}{n!}\sum_{k=0}^{n}\binom{n}{k}(x - x_0)^k (y - y_0)^{n-k}\frac{\partial^n f}{\partial x^k\,\partial y^{n-k}}(x_0, y_0) + \cdots.$$

The term

$$\frac{1}{2!}\left[(x - x_0)^2 \frac{\partial^2 f}{\partial x^2}(x_0, y_0) + 2(x - x_0)(y - y_0)\frac{\partial^2 f}{\partial x\,\partial y}(x_0, y_0) + (y - y_0)^2 \frac{\partial^2 f}{\partial y^2}(x_0, y_0)\right],$$

which we denote by

$$d^2_{X_0} f(X - X_0) \qquad (X = (x, y))$$

is a quadratic polynomial in $(x - x_0)$ and $(y - y_0)$. Its associated matrix is

$$\frac{1}{2!}\begin{pmatrix} \dfrac{\partial^2 f}{\partial x^2}(x_0, y_0) & \dfrac{\partial f}{\partial x\,\partial y}(x_0, y_0) \\ \dfrac{\partial f}{\partial x\,\partial y}(x_0, y_0) & \dfrac{\partial^2 f}{\partial y^2}(x_0, y_0) \end{pmatrix}.$$

If all the eigenvalues of a matrix are positive, the matrix is said to be *positive definite*; if they are all negative, the matrix is *negative definite*. We have the following result:

If $f\colon R^2 \to R$ has all its second partials continuous in a neighborhood of X_0 and X_0 is a critical point, then

i) if $d^2_{X_0} f(X - X_0)$ is positive definite, then X_0 is a local minimum.
ii) if $d^2_{X_0} f(X - X_0)$ is negative definite, then X_0 is a local maximum.
iii) if the eigenvalues of $d^2_{X_0} f(X - X_0)$ are both positive and negative, then X_0 is a saddle point, i.e., in any neighborhood of X_0 there is a point X such that $f(X) \geq f(X_0)$ and a point Y such that $f(Y) \leq f(X_0)$.

Using this result, classify the critical points of the following functions:

a) $2x - 5x^2 + 2y - 3y^2 - 2$.
b) $x^2 + y^2$
c) $xy + x^2$.
d) $x^2 + y^2 - yx$.
e) $x \sin y$.
f) $x^2 + 2y^2 - y$.

13.4 FOURIER SERIES

Recall that the Taylor series of a function $f(x)$ may converge (if at all) for very restricted values of x and that the expansion depends on the choice of a, the point of expansion. Thus a polynomial approximation based on the Taylor series may be very accurate locally, but not much good globally.

To remedy this deficiency we introduce the notion of Fourier series, the application of which is very useful in, e.g., physics and economics.

We want to approximate a function $f(x)$ on the interval $[0, 2\pi)$ by a trigonometric polynomial

$$\begin{aligned}\varphi_n(x) &= a_0 + a_1 \cos x + b_1 \sin x + a_2 \cos x + b_2 \sin 2x + \cdots \\ &\quad + a_n \cos nx + b_n \sin nx \\ &= \sum_{k=0}^{n} (a_k \cos kx + b_k \sin kx). \end{aligned} \tag{13.6}$$

In what sense do we want this to be a good approximation? Agreement with the derivatives of $f(x)$ was the criterion for choosing the Taylor series; here we require agreement of integrals: we want

$$\int_0^{2\pi} \varphi_n(x)\, dx = \int_0^{2\pi} f(x)\, dx.$$

However, the expression for $\varphi_n(x)$ involves $2n + 1$ constants so we need to impose additional conditions, namely

$$\int_0^{2\pi} \varphi_n(x) \cos kx\, dx = \int_0^{2\pi} f(x) \cos kx\, dx,$$

and

$$\int_0^{2\pi} \varphi_n(x) \sin kx\, dx = \int_0^{2\pi} f(x) \sin kx\, dx, \qquad k = 1, \ldots, n.$$

To see how these conditions determine the values of $a_0, a_1, \ldots, a_n, b_0, \ldots, b_n$ let us replace $\varphi_n(x)$ in the integrand by the sum in (13.6). We have

$$\begin{aligned}\sum_{i=0}^{n} \left[\int_0^{2\pi} a_i \cos ix \cos kx\, dx + \int_0^{2\pi} b_i \sin ix \cos kx\, dx\right] \\ = \int_0^{2\pi} \varphi_n(x) \cos kx\, dx = \int_0^{2\pi} f(x) \cos kx\, dx,\end{aligned} \tag{13.7}$$

and

$$\begin{aligned}\sum_{i=0}^{n} \left[\int_0^{2\pi} a_i \cos ix \sin kx\, dx + \int_0^{2\pi} b_i \sin ix \sin kx\, dx\right] \\ = \int_0^{2\pi} \varphi_n(x) \sin kx\, dx = \int_0^{2\pi} f(x) \sin kx\, dx.\end{aligned} \tag{13.8}$$

The only nonzero term in (13.7) is

$$\int_0^{2\pi} a_k \cos^2 kx\, dx = \begin{cases} \pi a_k, & k = 1, \dots, n \\ 2\pi a_k, & k = 0 \end{cases}$$

and in (13.8) the only nonzero term is

$$\int_0^{2\pi} b_k \sin^2 kx\, dx = \pi b_k, \qquad k = 1, \dots, n.$$

Hence the conditions imposed on the φ_n reduce to

$$a_0 = \frac{1}{2\pi} \int_0^{2\pi} f(x)\, dx$$

$$a_k = \frac{1}{\pi} \int_0^{2\pi} f(x) \cos kx\, dx, \qquad k = 1, \dots, n$$

$$b_k = \frac{1}{\pi} \int_0^{2\pi} f(x) \sin kx\, dx, \qquad k = 1, \dots, n.$$

If $f(x)$ is sufficiently well-behaved, then these integrals exist for all n and the sequence of trigonometric polynomials converges to $f(x)$. In this case we write the *Fourier series*:

$$f(x) = \sum_{n=0}^{\infty} (a_n \cos nx + b_n \sin nx).$$

In addition to giving a global approximation, Fourier series have other useful properties. Since a Fourier series depends on the integrability of functions rather than on differentiability as does the Taylor series, it can be used for functions which are less regular in their behavior. The classic examples are step functions such as

$$f(x) = \begin{cases} 1, & 0 < x < \pi \\ 0, & \pi < x < 2\pi. \end{cases}$$

(see Fig. 13.3).

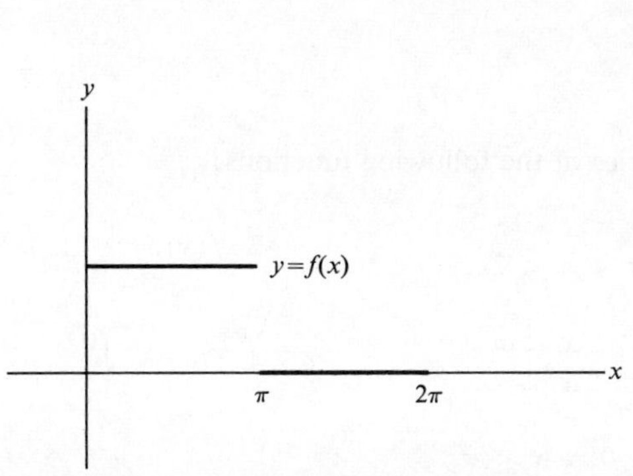

Fig. 13.3

We observe that $f(x)$ is not even continuous (much less differentiable) but it is integrable; and the integrals needed for the Fourier series can be found.

$$a_0 = \frac{1}{2\pi}\int_0^{2\pi} f(x)\,dx = \frac{1}{2\pi}\int_0^{\pi} dx + \int_{\pi}^{2\pi} 0\,dx = \frac{1}{2}.$$

For $k \geq 1$,

$$\begin{aligned}
a_k &= \frac{1}{\pi}\int_0^{2\pi} f(x)\cos kx\,dx \\
&= \frac{1}{\pi}\left[\int_0^{\pi} \cos kx\,dx + \int_{\pi}^{2\pi} 0 \cos kx\,dx\right] \\
&= \frac{1}{\pi}\,\frac{\sin kx}{k}\bigg|_0^{\pi} = 0 \\
b_k &= \frac{1}{\pi}\int_0^{2\pi} f(x)\sin kx\,dx \\
&= \frac{1}{\pi}\left[\int_0^{\pi} \sin kx\,dx + \int_{\pi}^{2\pi} 0 \sin kx\,dx\right] \\
&= \frac{1}{\pi}\,\frac{(-\cos kx)}{k}\bigg|_0^{\pi} = \frac{(-1)^{k+1}+1}{k\pi}.
\end{aligned}$$

Thus we have

$$\begin{aligned}
f(x) &= \frac{1}{2} + \frac{2}{\pi}\left(\sin x + \frac{\sin 3x}{3} + \frac{\sin 5x}{5} + \cdots\right) \\
&= \frac{1}{2} + \frac{2}{\pi}\sum_{k=1}^{\infty}\frac{\sin(2k-1)x}{2k-1}.
\end{aligned}$$

PROBLEMS

Find the Fourier series of the following functions:

1. $f(x) = \begin{cases} 1 & 0 < x < \pi \\ 2 & \pi < x < 2\pi. \end{cases}$

2. $f(x) = x \qquad 0 < x < 2\pi.$

3. $f(x) = \begin{cases} x & 0 < x < \pi \\ 0 & \pi < x < 2\pi. \end{cases}$

4. $f(x) = \begin{cases} 0 & 0 < x < \pi \\ x & \pi < x < 2\pi. \end{cases}$

5. $f(x) = \begin{cases} -1 & 0 < x < \pi \\ 1 & \pi < x < 2\pi. \end{cases}$

13.5 OPERATIONS WITH SERIES

Addition and multiplication of power series are defined as for polynomials, of which power series are the natural generalization. The *sum* is given by

$$\sum_{k=0}^{\infty} a_k x^k + \sum_{k=0}^{\infty} b_k x^k = \sum_{k=0}^{\infty} (a_k + b_k) x^k,$$

and the (*Cauchy*) *product* by

$$\left(\sum_{k=0}^{\infty} a_k x^k\right)\left(\sum_{k=0}^{\infty} b_k x^k\right) = \sum_{k=0}^{\infty} c_k x^k,$$

where

$$c_k = \sum_{i=0}^{k} a_i b_{k-i}.$$

We shall omit the proof of the following theorem.

Theorem 7. If

$$\sum_{k=0}^{\infty} a_k x^k \quad \text{and} \quad \sum_{k=0}^{\infty} b_k x^k$$

converge for $|x| < R$, then their sum and Cauchy product both converge for $|x| < R$.

Example. Find the first five terms of the Taylor series for $e^x \sin x$ at $x = 0$.

We know the series for

$$e^x = \sum_{k=0}^{\infty} \frac{x^k}{k!} = 1 + x + \frac{x^2}{2!} + \frac{x^3}{3!} + \frac{x^4}{4!} + \cdots$$

and for

$$\sin x = \sum_{k=1}^{\infty} \frac{(-1)^{k+1} x^{2k-1}}{(2k-1)!} = x - \frac{x^3}{3!} + \frac{x^5}{5!} - \frac{x^7}{7!} + \cdots.$$

Let

$$e^x \sin x = \sum_{k=1}^{\infty} c_k x^k$$

be the Cauchy product.

Then

$$c_0 = 0$$
$$c_1 = 1$$
$$c_2 = 1$$
$$c_3 = \frac{1}{2!} - \frac{1}{3!}$$
$$c_4 = \frac{-1}{3!} + \frac{1}{3!}$$
$$c_5 = \frac{1}{5!} - \frac{1}{3!\,2!} + \frac{1}{4!}$$

and

$$e^x \sin x = x + x^2 + \frac{x^3}{3} - \frac{x^5}{30} + \cdots.$$

We give without proof two theorems which enable us to differentiate and integrate series term-by-term.

Theorem 8. If $R > 0$ and

$$f(x) = \sum_{n=0}^{\infty} a_n x^n \qquad \text{for} \quad |x| < R,$$

then $f(x)$ is a C^∞ function for all $|x| < R$ and the derivatives of $f(x)$ are obtained by successive term-by-term differentiation.

Theorem 9. If $R > 0$ and

$$f(x) = \sum_{n=0}^{\infty} a_n x^n \qquad \text{for} \quad |x| < R,$$

then if

$$F(x) = \int_0^x f(t)\, dt,$$

we have

$$F(x) = \sum_{n=0}^{\infty} \frac{a_n x^{n+1}}{n+1}$$

for $|x| < R$.

Examples

1. Find the Taylor series for $\cos x$.

Solution. We have

$$\sin x = x - \frac{x^3}{3} + \frac{x^5}{5} + \cdots = \sum_{k=1}^{\infty} \frac{(-1)^{k+1} x^{2k-1}}{2k-1}$$

for all x. But $(d \sin x)/dx = \cos x$. Therefore, for all x,

$$\cos x = 1 - \frac{x^2}{2} + \frac{x^4}{4} + \cdots = \sum_{k=0}^{\infty} \frac{(-1)^k x^{2k}}{2k}.$$

2. Find a series expansion for $\log(1 + x)$ valid for $|x| < 1$.

Solution. We can write

$$\frac{1}{1+x} = \sum_{n=0}^{\infty} (-1)^n x^n$$

(by performing the indicated division). This series converges for $|x| < 1$. Let

$$F(x) = \log(1 + x)$$

so that

$$F'(x) = \frac{1}{1 + x} = \sum_{n=0}^{\infty} (-1)^n x^n, \qquad |x| < 1.$$

Term-by-term integration yields the series

$$F(x) = \int_0^x \frac{dt}{1 + t} = \sum_{n=0}^{\infty} \frac{(-1)^n x^{n-1}}{n + 1}, \qquad |x| < 1.$$

PROBLEMS

1. Find the Taylor series at $x = 0$ for each of the following functions:

   a) $f(x) = \dfrac{1}{(1 + x)^2}$.

   ($Hint$: differentiate the series for $1/(1 + x)$)

   b) $f(x) = \begin{cases} \dfrac{\sin x}{x} & x \neq 0 \\ 1 & x = 0 \end{cases}$

   c) $f(x) = \log \dfrac{1 + x}{1 - x}$.

   d) $f(x) = \arctan x$.

   ($Hint$: find the series for $(1 + x^2)^{-1}$ first)

   e) $f(x) = \cos x + \sin x$.

2. Find the first five terms of the Taylor series at $x = 0$ for each of the following functions:

   a) $e^x \cos x$.
   b) $\sin^2 x$.
   c) $\sqrt{1 + x} \log(1 + x)$.
   d) $f(x) = \sin 2x$ (using series for $\sin x$ and $\cos x$ and directly).
   e) $f(x) = \cos^2 x$.

14 / DIFFERENTIAL AND DIFFERENCE EQUATIONS

The elementary study of differential equations emphasizes the manipulation of equations into special forms for which solutions are known. The fact that the vast majority of differential equations cannot be forced into these forms or solved by these methods is frequently underemphasized. Indeed, the question of whether they can be solved by *any* methods is a difficult one. This problem—the existence of solutions to general differential equations—is treated in advanced texts on the subject. However, even when the existence question is settled we may be no further towards actually solving a particular equation, since knowing the existence of a solution does not produce the solution itself.

We treat the question of solutions to differential equations much as we did that of evaluating

$$\int f(x)\,dx,$$

for after all producing a function $F(x)$ whose derivative is $f(x)$ is solving the differential equation

$$\frac{dy}{dx} = f(x).$$

Our system there was to look at some elementary methods, keeping in mind their limited usefulness, and then to look at numerical methods of solution.

An (*ordinary*) *differential equation* is a function G of $n + 2$ variables:

$$G(x, y, y', \ldots, y^{(n)}) = 0, \tag{14.1}$$

where $y^{(i)}$ is the ith derivative of y with respect to x. A function $f(x)$ is a *solution* of the differential equation (14.1) if

$$G\big(x, f(x), f'(x), \ldots, f^{(n)}(x)\big) = 0,$$

for every x in the domain of $f(x)$. If the nth derivative is the highest order derivative occuring with nonzero coefficient in G, the differential equation is said to be of *order n*. An equation involving partial derivatives is said to be a *partial differential equation*.

14.1 SEPARABLE DIFFERENTIAL EQUATIONS

If M is a function of x and N is a function of y, then

$$M(x) + N(y)y' = 0 \tag{14.2}$$

is a first order (ordinary) differential equation. Its solution is given by the following procedure:

$$M(x) + N(y)\frac{dy}{dx} = 0$$

$$N(y)\frac{dy}{dx} = -M(x)$$

$$N(y)\,dy = -M(x)\,dx$$

$$\int N(y)\,dy + \int M(x)\,dx + C = 0. \tag{14.3}$$

That is, any function $f(x)$ which is a solution of (14.2) satisfies (14.3) (for some constant C), and conversely. Equation (14.3) is called the *general solution* of (14.2). The type of equation represented by (14.2) is called a *separable* differential equation, for the obvious reason that the functions of x and y are separated and then integrated to give the solution.

Whether any particular separable equation has a solution depends on whether the integrals of $M(x)$ and $N(y)$ exist and whether we can find a solution depends on whether we are able to evaluate the integrals.

Examples

1. Solve

$$3x^2 + 2x - 3y\frac{dy}{dx} = 0.$$

Solution. Here $M(x) = 3x^2 + 2x$ is a function of x and $N(y) = -3y$ is a function of y. Thus the general solution of this differential equation is given by

$$\int (3x^2 + 2x)\,dx - \int 3y\,dy + C = 0$$

$$x^3 + x^2 - \tfrac{3}{2}y^2 + C = 0.$$

This solution is an implicit solution, but we may solve it for y in terms of x if we desire:

$$y = \pm\sqrt{\frac{2(x^3 + x^2)}{3}} + C.$$

2. Solve

$$\frac{dy}{dx} = \frac{x + 2}{y - 3}.$$

Separating the variables, we have

$$(y - 3)\,dy - (x + 2)\,dx = 0.$$

The solution is given by:

$$\int (y - 3)\,dy - \int (x + 2)\,dx = 0$$

$$\frac{y^2}{2} - 3y - \frac{x^2}{2} - 2x + C = 0.$$

3. The exponential differential equations we solved in Chapter 8 are separable.

$$\frac{dy}{dx} = 2y - 10$$

$$\int \frac{dy}{2y - 10} = \int dx$$

$$\log \sqrt{2y - 10} - x + C = 0$$

$$\sqrt{2y - 10} = C_1 e^x.$$

We have a general solution for the separable form of differential equation; what about a particular solution?

A particular solution is obtained by assigning a value to the constant which appears in the general solution. An arbitrary assignment produces an arbitrary particular solution.

In Example 1 above let $C = 0$. Then

$$x^3 + x^2 - \tfrac{3}{2}y^2 = 0$$

is a particular solution to

$$3x^2 + 2x - 3y\frac{dy}{dx} = 0.$$

On the other hand,

$$x^3 + x^2 - \tfrac{3}{2}y^2 + 10 = 0,$$

is also a particular solution to the same differential equation.

In order to decide which particular solution is appropriate in applications, we need additional information.

Examples

1. The velocity of a particle is given by the differential equation

$$\frac{dy}{dt} = t^2 + 2t + 4.$$

Find the motion equation if when $t = 0$, then $y = 40$.

Solution. We have

$$\int dy = \int (t^2 + 2t + 4)\, dt$$

$$y = \frac{t^3}{3} + t^2 + 4t + C,$$

the general solution of the differential equation. When $t = 0$, $y = 40$, so $C = 40$. Therefore

$$y = \frac{t^3}{3} + t^2 + 4t + 40$$

is the particular solution we seek.

2. Suppose that capital is invested whenever total capital is below a certain equilibrium value C and is withdrawn if total capital is above C. Letting x represent the total capital invested at time t, we assume that the rate of investment is proportional to $x - C$:

$$\frac{dx}{dt} = k(x - C).$$

We rewrite this as

$$\frac{dx}{k(x - C)} = dt.$$

Then we integrate to obtain

$$(1/k) \log (x - C) = t + C_1$$

$$(x - C) = Ae^{kt}$$

$$x = Ae^{kt} + C,$$

where A is a constant. Now observe that k and C are given constants, and A is a constant arising from the integration. Thus passing from a general to a particular solution of the original differential equation involves evaluating A.

If the investment at time $t = 0$ is $C/2$, we have

$$\frac{C}{2} = Ae^0 + C$$

$$\frac{C}{2} = A + C$$

$$A = \frac{-C}{2}$$

and

$$x = \frac{-C}{2} e^{kt} + C.$$

PROBLEMS

1. Solve the following differential equations:

a) $\dfrac{dy}{dx} = e^{x+y}$.

b) $\sqrt{xy}\,\dfrac{dy}{dx} = 1$.

c) $\log x \,\dfrac{dx}{dy} = \dfrac{x}{y}$.

d) $y^2\sqrt{2x^2 - 3}\,dy + x\sqrt{6 - y^3}\,dx = 0$.

e) $\dfrac{dy}{dx} = \dfrac{-y}{x}$.

f) $y\,\dfrac{dy}{dx} = \sin x + x$.

g) $\dfrac{dy}{dx} = \dfrac{xy + y}{x + xy}$.

h) $\cos y + x \sin y\,\dfrac{dy}{dx} = 0$.

i) $y - (1 + x^2)\,\dfrac{dy}{dx} = 0$.

j) $x\,\dfrac{dx}{dy} + y^2 = y$.

k) $\dfrac{dy}{dx} = \cos x + x^2$.

l) $\sqrt{1 - x}\,dx = \sqrt{1 - y}\,dy$.

m) $\dfrac{dx}{\sqrt{1 - x^2}} = y\,dy$.

n) $e^x\,dy = e^y\,dx$.

2. Find the particular solution of the differential equation which satisfies the given initial condition.

a) $2x^2\,\dfrac{dy}{dx} + y = 0, \quad y(1) = 2$.

b) $v^2 - 4v + 4 = (v - 2)\,\dfrac{dv}{dt}, \quad v(0) = 1$.

c) $3y \cos x + 2 \sin x\,\dfrac{dy}{dx} = 0, \quad y(\pi/2) = 1$.

d) $\dfrac{ds}{dt} = \dfrac{1}{s} + s, \quad s(1) = 4$.

e) $3 \sin x + y\,\dfrac{dy}{dx} = 0, \quad y(0) = 4$.

f) $\dfrac{dx}{x} + y\,e^{y^2}\,dy = 0, \quad y(1) = 0$.

g) $\dfrac{dy}{dx} = x^2y^2, \quad y(0) = 7$.

h) $\sqrt{\dfrac{1 - x^2}{1 - y^2}}\,\dfrac{dy}{dx} = 2, \quad y(0) = 0$.

3. If P is the population at time t, and I the number of inventors in the population, a possible measure of the technological development of the society is

$$r(t) = \frac{I(t)}{P(t)}.$$

Using the assumption of Rashevsky that

$$\frac{dr}{dt} = [A + BP(t)][kr(t) - Lr^2(t)], \qquad (*)$$

$A, B, k, L \geq 0, L \geq k$, and that the population growth is exponential:

$$P(t) = P_0 e^{\lambda t},$$

solve (*) for $r(t)$.

14.2 EXACT DIFFERENTIAL EQUATIONS

In the differential equation

$$3x^2 + 2xy + (2y + x^2)\frac{dy}{dx} = 0,$$

it is not possible to separate the variables. However, we observe that

$$\frac{\partial}{\partial y}(3x^2 + 2xy) = 2x = \frac{\partial}{\partial x}(2y + x^2).$$

In general, an equation of the form

$$M(x, y) + N(x, y)y' = 0,$$

with

$$\frac{\partial M}{\partial y} = \frac{\partial N}{\partial x}$$

is called an *exact differential equation*. Frequently such an equation is written in differential form:

$$M(x, y)\,dx + N(x, y)\,dy = 0.$$

As the terminology indicates, the solution to such an equation involves a function $F(x, y)$ whose total differential is exactly

$$M(x, y)\,dx + N(x, y)\,dy. \qquad (14.4)$$

In fact, the function we seek has (14.4) as its differential and satisfies $F(x, y) = 0$. We obtain the solution by successive integrations.

We want to find a function $F(x, y)$ such that

$$M(x, y) = \frac{\partial}{\partial x} F(x, y),$$

and

$$N(x, y) = \frac{\partial}{\partial y} F(x, y).$$

Considering y as a constant, we integrate

$$F(x, y) = \int M(x, y)\, dx + g(y).$$

We need to add the function of y instead of just a constant since the derivative with respect to x of a function of y is also zero. Since

$$\frac{\partial}{\partial y} F(x, y) = N(x, y),$$

$$g'(y) = N(x, y) - \frac{\partial}{\partial y} \int M(x, y)\, dx.$$

Thus to find g we integrate with respect to y:

$$g(y) = \int N(x, y)\, dy - \int \left(\frac{\partial}{\partial y} \int M(x, y)\, dx \right) dy.$$

Finally

$$\boxed{F(x, y) = \int M(x, y)\, dx + \int N(x, y)\, dy - \int \left(\frac{\partial}{\partial y} \int M(x, y)\, dx \right) dy.}$$

Let us return to the example at the beginning of this section.

$$3x^2 + 2xy + (2y + x^2) \frac{dy}{dx} = 0.$$

The solution is given by

$$\begin{aligned}
F(x, y) &= \int (3x^2 + 2xy)\, dx + \int (2y + x^2)\, dy - \int \left(\frac{\partial}{\partial y} \int (3x^2 + 2xy)\, dx \right) dy \\
&= x^3 + x^2y + y^2 + x^2y - \int \frac{\partial}{\partial y} (x^3 + x^2y)\, dy \\
&= x^3 + x^2y + y^2 + x^2y - \int x^2\, dy \\
&= x^3 + x^2y + y^2 + x^2y - x^2y + C \\
&= x^3 + x^2y + y^2 + C.
\end{aligned}$$

Thus

$$x^3 + x^2y + y^2 + C = 0$$

is a solution to the given exact differential equation.

As a check we compute the total differential of this function:

$$dF(x, y) = (3x^2 + 2xy)\,dx + (x^2 + 2y)\,dy = 0.$$

We work out another example: find the general solution to

$$e^x + \sin y + \left(\frac{1}{y} + x \cos y\right)\frac{dy}{dx} = 0. \tag{14.5}$$

We note that

$$\frac{\partial}{\partial y}(e^x + \sin y) = \cos y = \frac{\partial}{\partial x}\left(\frac{1}{y} + x \cos y\right),$$

so that the differential equation is exact. Thus

$$\begin{aligned}
F(x, y) &= \int (e^x + \sin y)\,dx + \int\left(\frac{1}{y} + x\cos y\right)dy - \int\left(\frac{\partial}{\partial y}\int (e^x + \sin y)\,dx\right)dy \\
&= e^x + x \sin y + \log y + x \sin y - \int \frac{\partial}{\partial y}(e^x + x \sin y)\,dy \\
&= e^x + x \sin y + \log y + x \sin y - \int x \cos y\,dy \\
&= e^x + x \sin y + \log y + x \sin y - x \sin y + C \\
&= e^x + x \sin y + \log y + C.
\end{aligned}$$

The general solution to the differential equation (14.5) is

$$e^x + x \sin y + \log y + C = 0.$$

PROBLEMS

Solve the following differential equations:

1. $(3x + 8y - 3)\,dx + (8x - 5y + 6)\,dy = 0.$
2. $(x + y)\,dx + (x + y^2)\,dy = 0.$
3. $(xy^2 + y)\,dx = (x^2y + x)\,dy.$
4. $(e^y + e^x)\,dx + (xe^y + y^2)\,dy = 0.$
5. $(x + 3y) + 3x\dfrac{dy}{dx} = 0.$
6. $(y^2 + 2xy)\,dx + (x^2 + 2xy)\,dy = 0.$
7. $(e^x - 2x \cos y)\,dx + x^2 \sin y\,dy = 0.$

8. $\sin y\, dx + x \cos y\, dy = 0.$

9. $\arctan x\, dy + \dfrac{y}{1 + x^2}\, dx = 0.$

10. $\log y\, dx = \dfrac{-x}{y}\, dy.$

11. $(3x^2 - 8y)\, dx + (3y^2 - 8x)\, dy = 0.$

12. $ye^{xy}\, dx + xe^{xy}\, dy = 0.$

13. $(\sin y + y)\, dx + (x \cos y + x)\, dy = 0.$

14. $\dfrac{dx}{x} - \dfrac{dy}{y} = 0.$

15. $\left(y^2 e^x - \dfrac{1}{1 + x^2}\right) dx + 2y\, e^x\, dy = 0.$

16. $(6xy^3 - 8)\, dx + (9x^2y^2 + 5)\, dy = 0.$

14.3 HOMOGENEOUS EQUATIONS

Notice that in the equation

$$(x - 2y)\, dx + x\, dy = 0,$$

each term is of first degree in x or y, while in

$$(y^2 + 2xy)\, dx + (y^2 - x^2)\, dy = 0,$$

each term is of degree 2. More generally, a function $F(x, y)$ is *homogeneous of degree n* if

$$F(tx, ty) = t^n F(x, y),$$

for all real numbers t and all (x, y) in the domain of F. Then a differential equation

$$M(x, y) + N(x, y)\frac{dy}{dx} = 0,$$

is a *homogeneous differential equation* of degree n if M and N are homogeneous functions of degree n.

Examples

1.

$$x \sin \frac{x}{y} + y\frac{dy}{dx} = 0$$

is homogeneous of degree 1.

2.

$$\frac{x^3 + 3y^3}{x}\,dx + y^2\,dy = 0$$

is homogeneous of degree 2.

To solve a homogeneous equation we make an appropriate substitution to reduce the equation to a separable differential equation. Let

$$y = xv.$$

Now

$$dy = x\,dv + v\,dv,$$

so that

$$M(x, y)\,dx + N(x, y)\,dy = 0$$

is transformed as follows.

$$M(x, xv)\,dx + N(x, xv)[x\,dv + v\,dx] = 0$$

$$\frac{1}{x}\left(M(1, v) + vN(1, v)\right)dx + N(1, v)\,dv = 0$$

$$\frac{1}{x}\,dx + \frac{N(1, v)}{M(1, v) + vN(1, v)}\,dv + 0.$$

We now integrate with respect to each variable separately in order to obtain a solution.

Examples

1. Solve

$$(y + x)\,dx + (x + 2y)\,dy = 0.$$

Solution. Let $y = xv$. Then the transformed equation is

$$\frac{1}{x}\,dx + \frac{(1 + 2v)\,dv}{(1 + v) + v(1 + 2v)} = 0.$$

The solution is

$$\log x + \tfrac{1}{2}\log(1 + 2v + 2v^2) = \tfrac{1}{2}\log C$$

$$\log x + \tfrac{1}{2}\log\left(1 + \frac{2y}{x} + \frac{2y^2}{x^2}\right) = \tfrac{1}{2}\log C$$

$$x^2\left(1 + \frac{2y}{x} + \frac{2y^2}{x^2}\right) = C$$

$$x^2 + 2xy + 2y^2 = C.$$

2. Solve

$$2xy\,dx + (x^2 + y^2)\,dy = 0.$$

Solution. Proceeding as above we have

$$\frac{dx}{x} + \frac{(1 + v^2)\,dv}{2v + v(1 + v^2)} = 0$$

$$\log x + \tfrac{1}{3}\log(v^3 + 3v) = \tfrac{1}{3}\log C$$

$$x^3(v^3 + 3v) = C$$

$$x^3\left(\frac{y^3}{x^3} + \frac{3y}{x}\right) = C$$

$$y^3 + 3x^2y = C.$$

PROBLEMS

Solve the following differential equations:

1. $(x^2 + y^2)\,dx + 3xy\,dy = 0.$
2. $x^2\,dy + y^2\,dx = 0.$
3. $\dfrac{dy}{dx} = \dfrac{x + y}{x}.$
4. $\dfrac{dy}{dx} = \dfrac{y}{x} + \sin\dfrac{y}{x}.$
5. $(2x^2 - y^2)\,dx + x^2\,dy = 0.$
6. $(x^2 + xy)\,dy + y^2\,dx = 0.$
7. $\dfrac{dy}{dx} = \dfrac{x^2 + y^2}{x^2}.$
8. $2xy\,dx + x^2\,dy = 0.$
9. $(3x - y)\,dx + (x + 2y)\,dy = 0.$
10. $(x^2 + y^2)\,dx + y^2\,dy = 0.$
11. $\dfrac{2x}{y} + \cos\dfrac{x}{y} = \dfrac{dy}{dx}.$
12. $(x^2 - xy)\,dx + y^2\,dy = 0.$

14.4 LINEAR DIFFERENTIAL EQUATIONS

The equation

$$y' + P(x)y = Q(x), \tag{14.6}$$

P and Q continuous functions of x, is a *first order linear differential equation.* If $Q(x) = 0$, the equation is separable with the solution

$$y = Ce^{-\int P(x)\,dx},$$

usually written as

$$ye^{\int P(x)\,dx} = C.$$

Now suppose we multiply each side of (14.6) by $e^{\int P(x)dx}$; we have

$$y'e^{\int P(x)\,dx} + yP(x)e^{\int P(x)\,dx} = Q(x)e^{\int P(x)\,dx},$$

which can be written as

$$\frac{\partial}{\partial x}\, ye^{\int P(x)\,dx} = Q(x)e^{\int P(x)\,dx}.$$

Integrating with respect to x yields

$$ye^{\int P(x)\,dx} = \int Q(x)e^{\int P(x)\,dx} + C$$

$$y = e^{-\int P(x)\,dx}\left(\int Q(x)e^{\int P(x)\,dx} + C\right)$$

as the general solution of (14.6). $e^{\int P(x)dx}$ is called an *integrating factor.*

Examples

1. $y' + xy = x.$

Solution

$$ye^{\int x\,dx} = \int xe^{\int x\,dx} + C$$

$$ye^{x2/2} = e^{x^2/2} + C$$

$$y = 1 + Ce^{-x^2/2}.$$

2. $y' - y/x = x \sin x.$

Solution

$$ye^{-\int dx/x} = \int x \sin xe^{-\int dx/x} + C$$

$$ye^{-\log x} = \int x \sin xe^{-\log x} + C$$

$$\frac{y}{x} = \int \sin x + C$$

$$y = -x \cos x + Cx.$$

A differential equation of the form

$$y^{(n)} + A_1(x)y^{(n-1)} + \cdots + A_{n-1}(x)y' + A_n(x)y = Q(x)$$

is a *linear differential equation of order n.* We look at second-order differential equations with constant coefficients:

$$y'' + by' + cy = Q(x).$$

If $Q(x) = 0$, the equation is said to be a *homogeneous* linear differential equation. We use the solution of the homogeneous case to get a solution in the general setting.

We need the following:

Theorem 1. The set of solutions of the homogeneous linear differential equation

$$y'' + by' + cy = 0 \tag{14.7}$$

forms a vector space.

Proof. Observe that the zero-function

$$f(x) = 0 \qquad \text{for all real } x,$$

is a solution to (14.7). Let $u(x)$ and $v(x)$ be any two solutions to (14.7). We need only show that

$$f(x) = ku(x) + v(x) \qquad (k \text{ a real number}),$$

is also a solution. (We use the fact that the set of solutions is a subset of the vector space of all differentiable functions; thus, to show that it is a vector space, it is sufficient to show that it is a subspace.)

We use the notation $D^n f$ for the nth derivative of a function $f(x)$. In this notation (14.7) becomes

$$D^2 y + b\, Dy + y = 0.$$

Then

$$D(ku + v) = k\, Du + Dv$$

and

$$D^2(ku + v) = D(k\, Du + Dv) = kD^2u + D^2v.$$

Thus

$$\begin{aligned} &D^2(ku + v) + bD(ku + v) + c(ku + v) \\ &\quad = kD^2u + D^2v + bkDu + bDv + ku + v \\ &\quad = k(D^2u + bDu + cu) + (D^2v + bDv + cv). \end{aligned}$$

But since u and v are solutions, both of the quantities in parentheses are zero so $ku + v$ is a solution also. This vector space is called the *solution space* of the differential equation (14.7).

Since we have had good luck with exponential functions in the solution of differential equations, let us ask whether

$$y = e^{mx} \qquad (m \text{ a real number})$$

is a solution of (14.7). We have

$$D^2e^{mx} + bDe^{mx} + ce^{mx} = 0,$$

if and only if

$$m^2e^{mx} + bme^{mx} + ce^{mx} = 0,$$

if and only if

$$m^2 + bm + c = 0.$$

This last equation is called the *characteristic equation* of (14.7), and from it we see that e^{mx} is a solution to (14.7) if and only if m is a root of its characteristic equation.

Three possibilities arise.

CASE 1. $b^2 - 4c > 0$. Then

$$m^2 + bm + c = 0 \tag{14.8}$$

has distinct real roots m_1 and m_2. Since $y = e^{m_1x}$ and $y = e^{m_2x}$ are solutions and the solutions form a vector space,

$$y = C_1e^{m_1x} + C_2e^{m_2x},$$

for arbitrary constants C_1 and C_2, is also a solution.

CASE 2. $b^2 - 4c = 0$. Then (14.8) has a double root $m = -b/2$ and $y = e^{mx}$ is a solution of (14.7). Moreover, $y = xe^{mx}$ is also a solution, for

$$\begin{aligned}(D^2 + bD + 1)xe^{mx} &= m^2xe^{mx} + 2me^{mx} + bmxe^{mx} + be^{mx} + Cxe^{mx}\\ &= (m^2 + bm + c)xe^{mx} + (2m + b)e^{mx} = 0.\end{aligned}$$

Therefore, by Theorem 1,

$$y = C_1e^{mx} + C_2xe^{mx}$$

is a solution for arbitrary constants C_1 and C_2.

CASE 3. $b^2 - 4c < 0$. Then (14.8) has distinct imaginary roots. We have not treated the calculus of functions of complex variables. However, it is the case that

$$y = C_1e^{(\alpha+\beta i)x} + C_2e^{(\alpha-\beta i)x},$$

is a solution to (14.7), where

$$\alpha \pm \beta = \frac{-b}{2} \pm \frac{(\sqrt{b^2 - 4c})}{2},$$

and C_1, C_2 are complex numbers. Note that in this case the constants come from the complex numbers rather than the real numbers and in fact the solution space is a vector space over the complex numbers.

It is customary to write

$$C_1 e^{(\alpha+\beta i)x} + C_2 e^{(\alpha-\beta i)x}$$

as

$$e^{\alpha x}[C_1 \cos \beta x + C_2 \sin \beta x],$$

since

$$e^{i\beta x} = \cos \beta x + i \sin \beta x,$$

and

$$e^{-i\beta x} = \cos \beta x - i \sin \beta x.$$

It would be nice to know that

$$y = C_1 e^{m_1 x} + C_2 e^{m_2 x},$$

$$y = C_1 e^{mx} + C_2 x e^{mx},$$

and

$$y = C_1 e^{(\alpha+\beta i)x} + C_2 e^{(\alpha-\beta i)x}$$

are the general solutions for each of the cases, that is, that *any* solution is of one of these forms for an appropriate choice of constants. We state without proof

Theorem 2. The solution space of the differential equation (14.7) is a vector space of dimension 2.

We can show that the two solutions we have found in each case are linearly independent and therefore by Theorem 2 they form a basis, giving us the desired general solution.

Examples

1.

$$(D^2 - 2D - 8)y = 0.$$

Solution. The characteristic equation

$$m^2 - 2m - 8 = 0$$

has 4 and -2 as its roots. Thus the general solution is

$$y = C_1 e^{4x} + C_2 e^{-2x}.$$

2.

$$y'' + 2y' + y = 0.$$

Solution. The characteristic equation

$$m^2 + 2m + 1 = 0,$$

has -1 as a double root. Thus the general solution is

$$y = C_1 e^{-x} + C_2 x e^{-x}.$$

3.

$$\frac{d^2y}{dx^2} - 6\frac{dy}{dx} + 13y = 0.$$

Solution. The characteristic equation

$$m^2 - 6m + 13 = 0$$

has complex roots

$$\frac{6 \pm \sqrt{-16}}{2} = 3 \pm 2i.$$

Thus, the general solution is

$$y = C_1e^{3+2i} + C_2e^{3-2i} = e^{3x}[C_1 \cos 2x + C_2 \sin 2x].$$

Now let us look at the general second-order linear differential equation

$$y'' + by' + cy = Q(x), \tag{14.9}$$

where $Q(x)$ is a continuous function. If $Q(x) \neq 0$, we say that the equation is *non-homogeneous.*

We know that the derivative is a linear transformation, i.e.,

$$D(f + g) = Df + Dg,$$

and

$$D(kf) = kDf.$$

We call a linear transformation whose domain is a set of functions a *linear operator*. We have noted before that the sums and constant multiples of linear transformations are linear transformations; in the case of the derivative, its powers (iterated derivatives) are also linear operators, so that in particular $D^2 + bD + c$ is a linear operator.

Now let g_p be a particular solution of (14.9) and g any other solution and denote their difference by $f(x)$. Then

$$\begin{aligned}(D^2 + bD + c)f(x) & \\ &= (D^2 + bD + c)(g(x) - g_p(x)) \\ &= (D^2 + bD + c)(g(x)) - (D^2 + bD + c)(g_p(x)) \qquad \text{(by linearity)} \\ &= Q(x) - Q(x) \\ &= 0.\end{aligned}$$

But this means that $f(x)$ is a solution of the homogeneous equation

$$(D^2 + bD + c)y = 0. \tag{14.10}$$

On the other hand, if $f(x)$ is a solution of this homogeneous equation and g_p is a particular solution of (14.9), then

$$g(x) = g_p(x) + f(x),$$

is a solution of (14.9). Thus the general solution of (14.9) is

$$y = g_p(x) + f_G(x),$$

where g_p is a particular solution of (14.9) and $f_G(x)$ the general solution of (14.10).

Now we need only develop a method for finding a particular solution of (14.9). Sometimes it is possible to do this by inspection. For example,

$$(D^2 + 2)y = 3x$$

has $y = 3x/2$ as a particular solution. We give the general method known as *variation of parameters*. Other techniques may also be used.

Let

$$y = C_1u(x) + C_2v(x)$$

be the general solution of the homogeneous equation (14.10). We want to find functions $u_1(x)$ and $v_1(x)$ such that

$$y_p = u_1(x)u(x) + v_1(x)v(x)$$

is a particular solution of (14.9). We observe that if

$$y = u_1u + v_1v,$$

then

$$y' = u_1u' + u_1'u + v_1v' + v_1'v.$$

The computation of y'' will get out of hand without some restriction. Hence we require that the functions u_1 and v_1 satisfy

$$u_1'u + v_1'v = 0. \tag{14.11}$$

Then

$$y' = u_1u' + v_1v',$$

so that

$$y'' = u_1u'' + v_1v'' + u_1'u' + v_1'v'.$$

But now

$$\begin{aligned} Q(x) &= (D^2 + bD + c)(u_1u + v_1v) \\ &= u_1(D^2 + bD + c)u + v_1(D^2 + bD + c)v + u_1'u' + v_1'v' \\ &= u_1'u' + v_1'v'. \end{aligned}$$

Hence if we impose the condition (14.11), then

$$u_1u + v_1v$$

is a solution if and only if

$$u_1'u' + v_1'v' = Q(x).$$

Recall that the functions $u(x)$ and $v(x)$ are of one of the forms

$$e^{m_1 x}, \quad e^{m_2 x}, \quad e^{mx}, \quad xe^{mx},$$

(m_1, m_2 distinct roots (real or complex), m a double root, of the characteristic equation of (14.10)).

Hence $uv' - vu' \neq 0$ and the system of two linear equations in two variables,

$$u_1' u + v_1' v = 0$$

and

$$u_1' u' + v_1' v' = Q(x),$$

has the solution

$$u_1' = \frac{-vQ(x)}{uv' - vu'}, \qquad v_1' = \frac{uQ(x)}{uv' - vu'}.$$

We integrate to obtain the values of u_1 and v_1.

Example. Find the solution of

$$(D^2 + 2D + 1)y = e^x.$$

Solution. The general solution of

$$(D^2 + 2D + 1)y = 0$$

is given by

$$y = C_1 e^{-x} + C_2 x e^{-x},$$

since -1 is a double root of the characteristic equation.

Let

$$y = u_1 e^{-x} + v_1 x e^{-x}$$

be a particular solution of the original equation where

$$u_1' e^{-x} + v_1' x e^{-x} = 0,$$

and

$$u_1'(-e^{-x}) + v_1'(-xe^{-x} + e^{-x}) = e^x.$$

Then

$$\begin{aligned} u_1' &= \frac{-xe^{-x}e^x}{e^{-x}(-xe^{-x} + e^{-x}) - xe^{-x}(-e^{-x})} \\ &= -\frac{xe^x}{e^{-x}} \\ &= -xe^{2x} \end{aligned}$$

and

$$v_1' = \frac{e^x e^{-x}}{e^{-2x}} = e^{2x}.$$

Therefore,

$$u_1 = -\frac{x}{2}e^{2x} + \frac{e^{2x}}{4} + C_1$$

$$v_1 = \frac{e^{2x}}{2} + C_2.$$

The general solution is

$$y = \left(\frac{-x}{2}e^{2x} + \frac{e^{2x}}{4} + C_1\right)e^{-x} + \left(\frac{e^{2x}}{2} + C_2\right)xe^{-x}$$

The method of solution of a linear differential equation of order n with constant coefficients

$$(D^n + c_{n-1}D^{n-1} + \cdots + c_1D + c_0)y = Q(x)$$

is as for the second order case, except that we must find the roots of an nth degree characteristic equation in order to obtain a general solution for the corresponding homogeneous equation, and the difficulties of finding a particular solution are also increased. The case of a linear equation with nonconstant coefficients is discussed in the next section.

PROBLEMS

Solve the following differential equations:

1. $(y^2 + 1)\,dx + (2xy + 1)\,dy = 0.$
2. $y' + 2y = e^{-x}.$
3. $x\,dy + (y - \sin x)\,dx = 0.$
4. $(D^2 + 4)y = 0.$
5. $(D^2 - 4)y = 0.$
6. $y'' - 2y' - 8y = 0.$
7. $y'' - 3y' + 5y = 0$, when $x = 0, y' = 0, y = 0.$
8. $3y'' - 4y' + 7y = 0$, when $x = 1, y' = 1, y = 0.$
9. $y'' + y = \tan x.$
10. $y'' - 4y = e^x \sin x.$
11. $y'' - 2y' + y = x^{17}e^x.$
12. $y'' - y = \sin x.$
13. $y''' - 3y'' + y' - 3y = 0.$
14. $y'' - 4y' + 4y = 0.$
15. $(D^2 - 9)y = e^x.$
16. $(D^2 - 3D - 4)y = x^2.$
17. $y'' + y' + y = x.$

18. An equation of the form

$$y' + R(x)y = S(x)y^k \qquad (k \neq 0, 1),$$

is called a *Bernoulli equation*. Such an equation can be reduced to a linear equation by multiplying each side by $(1 - k)y^{-k}$ to obtain

$$(1 - k)y^{-k}y' + (1 - k)R(x)y^{1-k} = (1 - k)S(x).$$

Letting $v = y^{1-k}$ so that $v' = (1 - k)y^{-k}y'$, we can write the equation as

$$v' = (1 - k)R(x)v = (1 - k)S(x).$$

Using this method, solve the following differential equations:

a) $y' + (1/x)y = x^4y^3$.
b) $y' + xy = xy^2$.
c) $y' - 2y = e^xy^2$.
d) $xy' + y = x^2e^xy^2$.

14.5 SERIES SOLUTIONS AND NUMERICAL METHODS

Suppose we have a linear differential equation

$$b_0(x)y^{(n)} + b_1(x)y^{(n-1)} + \cdots + b_n(x)y = G(x) \tag{14.12}$$

with polynomial coefficients; then $x = x_0$ is a *singular point* if $b_0(x_0) = 0$. In this section we discuss series solutions of (14.12) valid near points which are not singular. We specialize in second order differential equations and state without proof:

Theorem 3. If $x = 0$ is not a singular point of the linear equation with polynomial coefficients

$$b_0(x)y'' + b_1(x)y' + b_2(x)y = 0 \tag{14.13}$$

then there is a solution

$$y = \sum_{n=0}^{\infty} a_nx^n,$$

with two arbitrary constants a_0 and a_1, which converges inside the interval centered at $x = 0$ and extending to the nearest singular point.

Thus in solving an equation of the form (14.13), our problem is to determine the constants a_n, $n \geq 2$. Since $b_0(x)$ is not zero in the interval with which we are concerned, we can divide through by $b_0(x)$ and replace (14.13) by

$$y'' + p(x)y' + q(x)y = 0, \tag{14.14}$$

$p(x)$, $q(x)$ rational functions of x whose denominators do not equal zero at $x = 0$.

Let us suppose that

$$y = \sum_{n=0}^{\infty} a_nx^n,$$

is a solution to (14.14). When $x = 0$, $y = a_0$, and since

$$y' = \sum_{n=0}^{\infty} na_n x^{n-1},$$

$a_1 = y'(0)$. In fact, by Taylor's formula, we can write

$$y(x) = \sum_{n=0}^{\infty} y^{(n)}(0)\,\frac{x^n}{n!},$$

where the right-hand side converges in an interval about $x = 0$.

We observe that

$$y''(x) = -p(x)y'(x) - q(x)y(x),$$

so that $y''(0)$ can be computed if $y'(0)$ and $y(0)$ are known. Similarly,

$$y'''(x) = -p(x)y''(x) - p'(x)y'(x) - q(x)y'(x) - q'(x)y(x),$$

and thus $y'''(0)$ can be determined once $y''(0)$ is known. This can be continued to find $y^{(n)}$ for any n. In practice, we want to find an easy way to find these values. We note that the solution is completely determined by the values $y(0)$ and $y'(0)$; these are referred to as *boundary values*. If the original differential equation is of degree higher than two, the method of solution is analogous, but more complicated.

Example. Solve

$$(1 - x^2)y'' - 6xy' - 4y = 0. \tag{14.15}$$

Solution. The singular points are ± 1, so there is a solution

$$y = \sum_{n=0}^{\infty} a_n x^n,$$

valid for $|x| < 1$ and a_0, a_1 arbitrary. Substituting this solution in (14.15), we obtain

$$\sum_{n=0}^{\infty} n(n-1)a_n x^{n-2} - \sum_{n=0}^{\infty} n(n-1)a_n x^n - \sum_{n=0}^{\infty} 6na_n x^n - \sum_{n=0}^{\infty} 4a_n x^n = 0,$$

which simplifies to

$$\sum_{n=0}^{\infty} n(n-1)a_n x^{n-2} - \sum_{n=0}^{\infty} (n^2 + 5n + 4)a_n x^n = 0,$$

and then to

$$\sum_{n=0}^{\infty} n(n-1)a_n x^{n-2} - \sum_{n=2}^{\infty} (n-1)(n+2)a_{n-2} x^{n-2} = 0.$$

Now a power series can be equal to zero for all x if and only if each of its coefficients is zero, so

$$n(n - 1)a_n - (n - 1)(n + 2)a_{n-2} = 0,$$

$$a_n = \frac{n + 2}{n} a_{n-2}.$$

By observing the a_n for even n and odd n separately, we see that

$$a_2 = \tfrac{4}{2}a_0 \qquad a_3 = \tfrac{5}{3}a_1$$

$$a_4 = \tfrac{6}{4}a_2 \qquad a_5 = \tfrac{7}{5}a_3$$

$$\vdots \qquad\qquad \vdots$$

$$a_{2k} = \frac{2k + 2}{2k} a_{2k-2}, \qquad a_{2k+1} = \frac{2k + 3}{2k + 1} a_{2k-1}.$$

Now we note that for $k \geq 1$

$$a_2 a_4 \ldots a_{2k} = \frac{4 \cdot 6 \ldots \cdot (2k + 2)}{2 \cdot 4 \ldots \cdot (2k)} a_0 a_2 \ldots a_{2k-2}$$

$$a_3 a_5 \ldots a_{2k+1} = \frac{5 \cdot 7 \ldots \cdot (2k + 3)}{3 \cdot 5 \ldots \cdot (2k + 1)} a_1 a_3 \ldots a_{2k-1}$$

so that

$$a_{2k} = (k + 1)a_0 \quad \text{and} \quad a_{2k+1} = \frac{2k + 3}{3} a_1.$$

This yields

$$y = a_0 + \sum_{k=1}^{\infty} a_{2k}x^{2k} + a_1 x + \sum_{k=1}^{\infty} a_{2k+1}x^{2k+1}$$

$$= a_0 \left[1 + \sum_{k=1}^{\infty} (k + 1)x^{2k}\right] + a_1 \left[x + \sum_{k=1}^{\infty} \frac{2k + 3}{3} x^{2k+1}\right].$$

If it is not so easy to see how the coefficients can be computed in terms of the boundary values, it is possible to use a computer on a recursive formula such as

$$a_n = \frac{n + 2}{n} a_{n-2},$$

or to compute whatever finite number of terms is necessary for whatever degree of accuracy we desire (see Chapter 13).

PROBLEMS

Solve the following differential equations:

1. $\dfrac{d^2y}{dx^2} = xy.$
2. $(x^2 + 1)y'' + xy' + y = 0.$
3. $y' = x + y^2.$
4. $y'' - 3(x + 2)y' - 2y = 0.$
5. $(\sin x)y'' + (\cos x)y' + y = 0.$

14.6 APPLICATIONS

In this section we present in some detail differential equations which occur in several different fields.

1. Assume that as a population P increases its growth rate declines. Suppose P cannot exceed the value M. Assume that the growth rate of P is proportional to the product of P and the difference $M - P$:

$$\frac{dP}{dt} = kP(M - P),$$

k a positive constant. The equation is separable:

$$\frac{dP}{P(M - P)} = k\,dt.$$

Using partial fractions we obtain

$$\frac{1}{MP} + \frac{1}{M(M - P)}\,dP = k\,dt.$$

Integrating yields

$$\frac{1}{M}\log P - \frac{1}{M}\log(M - P) = kt + C$$

$$\frac{1}{M}\log\frac{P}{M - P} = kt + C$$

$$\frac{P}{M - P} = e^{Mkt + MC}.$$

Denoting the population at $t = 0$ by P_0,

$$P_0 = \frac{e^{MC}M}{1 + e^{MC}},$$

$$e^{MC} = \frac{P_0}{M - P_0}.$$

Thus

$$P = \frac{MP_0}{P_0 + (M - P_0)e^{-Mkt}}.$$

This equation is a good approximation to the growth of fruit flies in a jar and to the spread of epidemics in a human population.

2. Another similar problem involves Newton's law of cooling (or heating): the temperature of a body changes at a rate which is proportional to the difference in temperature between the outside medium and the body itself. This is a variation of the growth toward equilibrium situation discussed in Chapter 8.

A thermometer which reads 70°F. is placed inside an oven whose temperature is 250°F. Five minutes later it reads 200°F. How long will it take for it to read 250°F.?

We let T represent the reading of the thermometer at time t. We have

$$t = 0 \qquad t = 5$$

$$T = 70^\circ \qquad T = 200^\circ$$

$$\frac{dT}{dt} = k(250^\circ - T).$$

Solving the differential equation we get

$$T = 250 + Ce^{kt}.$$

From the initial temperature reading, we have

$$70 = 250 + C,$$

so that $C = -180$. Now

$$T = 250 - 180e^{kt}.$$

Using the second reading, we obtain

$$200 = 250 - 180e^{5k}$$

$$\tfrac{5}{18} = e^{5k}$$

$$k = -0.256.$$

Therefore,

$$T = 250 - 180e^{-0.256t}$$

and letting $T = 250$ we can solve for t.

3. What initial velocity must a particle projected in a radial direction have to escape the earth?

We assume that the only force acting on the particle is that of gravity. According to Newton's law of gravitation, the acceleration of the particle is inversely proportional to the square of the distance of the particle from the center of the earth. Let r be that distance, R the radius of the earth, t time, a the acceleration, and k the constant of proportionality (Fig. 14.1). Then

$$a = \frac{dv}{dt} = \frac{k}{r^2}.$$

When $r = R$, $a = -g$, the gravitational constant. (g is 32 ft/sec^2 in the British system.) Therefore,

$$-g = \frac{k}{R^2},$$

and we write

$$a = \frac{-gR^2}{r^2}.$$

Remembering that $v = dr/dt$, we write, using the chain rule,

$$a = \frac{dv}{dt} = \frac{dr}{dt} \cdot \frac{dv}{dr} = v\frac{dv}{dr}.$$

Now we have

$$v\frac{dv}{dr} = \frac{-gR^2}{r^2}.$$

This is a separable differential equation and its solution is given by

$$v^2 = \frac{2gR^2}{r} + C.$$

If the velocity at which the particle leaves the surface of the earth is denoted by v_0,

$$C = v_0^2 - 2gR.$$

This yields

$$v^2 = \frac{2gR^2}{r} + v_0^2 - 2gR.$$

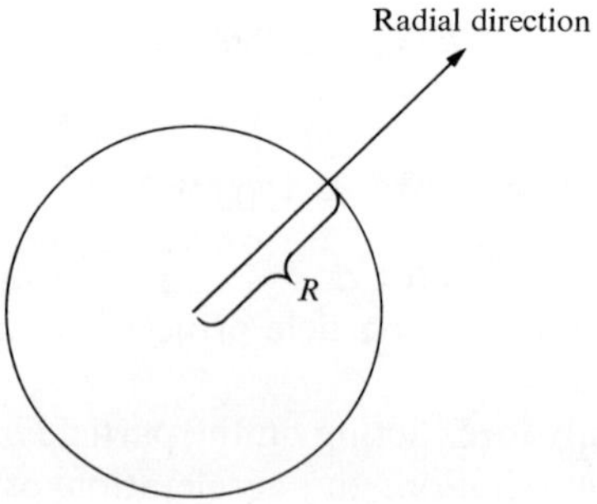

Fig. 14.1

Now, in order for the particle to escape, the velocity must remain positive. In particular, since the function is continuous and positive at $r = R$, it can never be zero. Therefore, we must have

$$v_0^2 - 2gR \geq 0,$$

$$v_0 \geq \sqrt{2gR}.$$

We call v_0 the *escape velocity*.

4. Consider a parasite population x living on a host population y. The rate of change of x with respect to time t is a function of both x and y; moreover, if the parasites affect the hosts, the rate of change of y depends on both x and y. In fact, since the product xy represents the number of ways of pairing a parasite with a host, it is reasonable to assume that

$$dx/dt = c_1x + c_2xy$$

$$dy/dt = c_3y + c_4xy,$$

c_i a constant, $i = 1, \ldots, 4$, $c_2 \neq 0 \neq c_4$.

Suppose we are interested in the effects on the host population of increasing or decreasing the parasite population; that is, we want to find dy/dx. We have

$$\frac{dy}{dx} = \frac{dy/dt}{dx/dt} = \frac{c_3y + c_4xy}{c_1x + c_2xy} \quad \text{if} \quad \frac{dx}{dt} \neq 0.$$

This equation is separable:

$$\left(\frac{c_1}{y} + c_2\right) dy - \left(\frac{c_3}{x} + c_4\right) dx = 0.$$

The solution is given (implicitly) by

$$c_2y + c_1 \log y = c_4x + c_3 \log x + k.$$

This is not very convenient if we want y in terms of x. Hence we "solve" for y by graphic methods.

Consider an equation of the form

$$F(x) + G(y) = C,$$

C a fixed constant. We want to use graphs to show whether this has a solution in the form of a differentiable function $y(x)$ and to obtain an approximation when the solution exists. We write

$$G(y) = -F(x) + C, \tag{14.16}$$

and graph $G(u)$ and $-F(u)$ in a new plane (see Fig. 14.2).

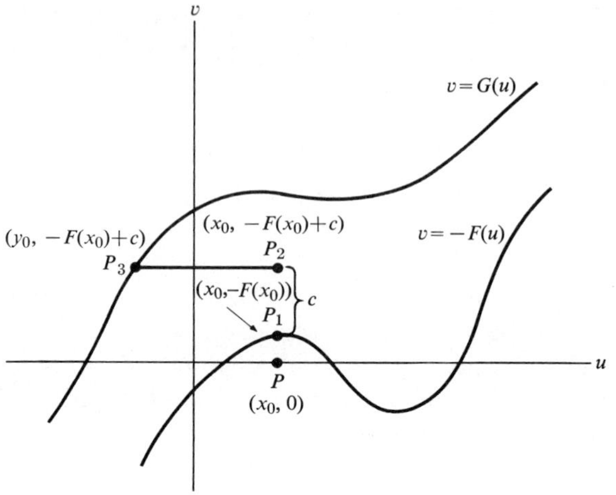

Fig. 14.2

Since C is fixed we now ask whether there are pairs of values of x and y for which (14.16) holds. Let us decide on a fixed x_0 and locate the point $P_0 = (x_0, 0)$ in the new plane. Next we locate the point P_1 on the curve $v = -F(u)$ with coordinates $(x_0, -F(x_0))$. P_2 will be the point with coordinates $(x_0, -F(x_0) + C)$. (As shown, $C > 0$, but the procedure for $C \leq 0$ is analogous.) Finally we draw a horizontal line through P_2, hoping that it intersects the curve $v = G(u)$ at some point P_3. If so, we designate P_3 by $(y_0, -F(x_0) + C)$.

We now have a pair of values (x_0, y_0) for which

$$G(y_0) = -F(x_0) + C.$$

If repeating the procedure gives other pairs (x_i, y_i) which satisfy (14.16), we make a table of values and construct an approximation of a function $y = y(x)$ from these values.

Let us return to the parasite problem and suppose $c_1 = c_2 = c_3 = 1$, $c_4 = k = -1$, so that we have

$$y + \log y = -x + \log x - 1.$$

Then $G(u) = u + \log u$, $-F(u) = u - \log u$; their graphs are shown in Fig. 14.3. We can make a table of values by using the procedure above:

x	y
1	0.5
2	0.7
3	0.95
5	1.8
10	5.1

and approximate $y = y(x)$ as shown by the graph in Fig. 14.4.

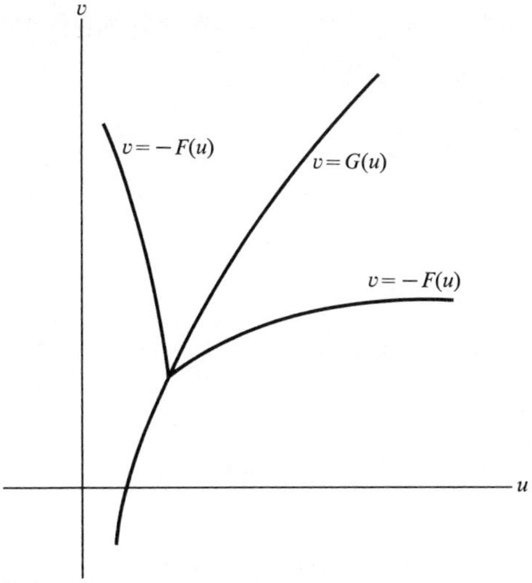

Fig. 14.3

5. Suppose dye is injected in a vein near the heart. The dye is carried to the heart where it is mixed with the blood passing through the heart. At each stroke of the heart a quantity of the mixture of the blood and dye is forced out and replaced by undyed blood. The concentration of dye depends, among other things, upon how much blood has been ejected, which in turn depends on the stroke volume of the heart and the number of strokes per minute.

We assume that the concentration is proportional to the sum of the volume of the blood ejected and the rate at which the amount ejected changes. On the other hand, the change in concentration is proportional to the volume of blood ejected. Thus we have

$$\frac{d^2C}{dt^2} + \frac{dC}{dt} = kC,$$

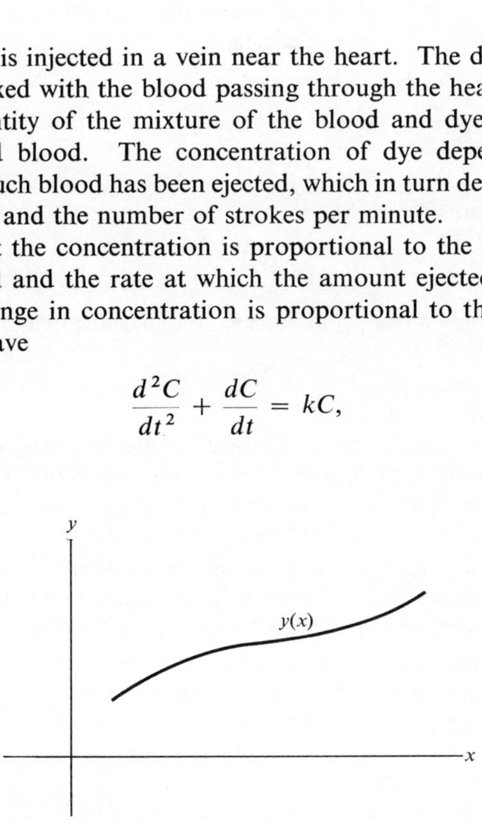

Fig. 14.4

where C is the concentration and k is a negative constant. Writing this linear differential equation as

$$(D^2 + D + k)C = 0,$$

we see that the general solution is

$$C = C_1 e^{(1+\sqrt{1-4k})t/2} + C_2 e^{(-1-\sqrt{1-4k})t/2}.$$

PROBLEMS

1. If $7y + 2 \log y = x^2 + 3 \log x - 5$, define y as a function of x by means of a table.
2. The Deese* equation is

$$\frac{dR}{dN} = k(M - R),$$

where

$$R = \text{response strength},$$
$$N = \text{number of reinforcements},$$
$$M = \text{fixed maximum value of } R,$$
$$k = \text{a constant}.$$

Solve for R if $R = 0$ when $N = 0$.

3. In the Domar model we assume that savings is a function of income:

$$S(t) = \gamma y(t),$$

γ the propensity to save, and that

$$I(t) = \beta \frac{dy}{dt},$$

where I denotes investment, β the accelerator. Assume there is a savings-investment equilibrium

$$S(t) = I(t).$$

Solve for y.

4. A rabbit starts at position $(a, 0)$ and travels along the x-axis in a positive direction. A dog starts at position $(0, b)$ with direction always toward the rabbit. If both move with constant velocity, find the equation of the path of the dog. (*Hint:* use a parametric representation for the path and consider the tangent to the path of the dog.)
5. If a specimen is immersed in a large vat of a solution of some chemical, the chemical penetrates the specimen as time progresses. Assuming that the concentration k in the vat remains constant, find the concentration $c(t)$ in the specimen if
   a) the rate of increase of $c(t)$ is directly proportional to the difference in the concentration in the vat and that in the specimen;
   b) the rate of increase of $c(t)$ is proportional to the concentration in the specimen and inversely proportional to the sum of the concentration of the specimen and the concentration in the vat.

---

* Deese, J., *The Psychology of Learning*, McGraw-Hill, New York, 1952.

6. Soup at a temperature of 190°F. cools in a room whose temperature is 72°F. The rate of cooling is directly proportional to the difference in the temperature of the soup and of the surrounding air. If the soup has cooled to 100°F. in 10 minutes, how long does it take it to cool to 75°F.?

MISCELLANEOUS PROBLEMS

Solve the following differential equations by whatever means is appropriate:

1. $\frac{dy}{dx} = e^{x-y}$.
2. $x\,dy + (y + \cos x)\,dx = 0$.
3. $\arcsin x\,dy + \frac{y}{\sqrt{1 - x^2}}\,dx = 0$.
4. $x\,dy - y\,dx = (x^2 + y^2)\,dx$.
5. $y'' + y' + y = 0$.
6. $(D^2 + 2D + 1)y = e^x$.
7. $y'' + 4y' + 3y = 5$.
8. $\sqrt{1 + x^2}\,dy + \sqrt{y^2 - 1}\,dx = 0$.
9. $(x + y)\,dx = -2x\,dy$.
10. $\frac{d^2y}{dx^2} - y = x$.
11. $2xy\,dy = (x^2 + y^2)\,dx$.
12. $2x^2 + e^x - 2y\frac{dy}{dx} = 0$.
13. $x\,dy + y\,dx = y\,dy$.
14. $\frac{d^2y}{dx^2} - \frac{dy}{dx} - 6y = 0$.
15. $y'' + y = 2 \sin x$.

14.7 DIFFERENCE EQUATIONS

Suppose y is a function of t (usually t is time), but that t takes on only integral values; that is, we evaluate y only at intervals rather than continuously. Then we usually write y_k for $y(t)$, $t = k$, the value of y after k intervals, and y_{k+n} for the value n intervals later.

A *linear difference equation* is an equation of the form

$$y_{k+n} + a_{n-1}y_{k+n-1} + \cdots + a_0 y_k = f(k), \qquad (14.17)$$

(the a_i may be constants or functions of k).

The *order* of a term of a difference equation is the number of intervals. The *order* of a difference equation is the largest order of a term with nonzero coefficient; the order of (14.17) is n. The *degree* of a difference equation is the degree of the term of highest order; in a linear difference equation, all terms are of degree one, and it is with these that we shall be primarily concerned.

If in (14.17), $f(k) = 0$, the difference equation is said to be *homogeneous*.

Examples

1.
$$2y_{k+3} + 3y_{k+2} - 5y_{k+1} + 2y_k = 4.$$

This is third-order, first-degree, nonhomogeneous.

2.
$$2y_{k+2}^3 - 5y_k = 0.$$

This is second-order, third-degree, homogeneous.

The complete solution of a linear nth-order difference equation (14.17) with constant coefficients is of the form

$$y_k = y_{k_g} + y_{k_p},$$

where y_{k_g} is the general solution of the homogeneous linear difference equation with constant coefficients:

$$y_{k+n} + a_{n-1}y_{k+n-1} + \cdots + a_0 y_k = 0, \tag{14.18}$$

and y_{k_p} is a particular solution of (14.17).

The solutions of the analogous homogeneous linear differential equation are of exponential form e^{rk} or $k^n e^{rk}$ for suitable constant r. In the case of difference equations, it is customary to write

$$m = e^r$$

and to try to find solutions of the form

$$y_k = m^k.$$

We use the notation

$$Ey_i = y_{i+1},$$

and define the linear difference operator

$$L = E^n + a_{n-1}E^{n-1} + \cdots + a_1 E + a_0.$$

We observe that

$$\begin{aligned} Em^k &= m^{k+1} = m(m^k) \\ &\vdots \\ E^n m^k &= m^n(m^k), \end{aligned}$$

and that therefore

$$Lm^k = (m^n + a_{n-1}m^{n-1} + \cdots + a_1 m + a_0)m^k.$$

Thus $y_k = m^k$ will satisfy (14.18) if m is a root of the characteristic equation

$$m^n + a_{n-1}m^{n-1} + \cdots + a_1 m + a_0 = 0. \tag{14.19}$$

As in the case of differential equations, it can be shown that if $m_1, \ldots, m_n$ are real distinct roots of (14.19), so that $m_1^k, \ldots, m_n^k$ are roots of (14.18), then any linear combination of these solutions is also a solution of the homogeneous linear difference equation (14.18). Moreover,

$$y_k = c_1 m_1^k + \cdots + c_n m_n^k,$$

is the general solution of (14.18).

Now suppose m_1 is a root of multiplicity j of the characteristic equation (14.19). Then as for differential equations,

$$y_k = m_1^k(c_1 + c_2 k + \cdots + c_j k^{j-1}),$$

is the part of the solution corresponding to this multiple root.

Finally, the part of the solution corresponding to a pair of complex conjugate roots can be written as

$$y_k = c_1 e^{(\alpha+\beta i)k} + c_2 e^{(\alpha-\beta i)k}.$$

Examples

1. Solve $y_{k+2} - 2y_{k+1} - 8y_k = 0.$

Solution. Let $y_k = m^k$. Then

$$(m^2 - 2m - 8)m^k = 0, \quad m = 4, -2.$$

Hence the general solution is

$$y_k = c_1 4^k + c_2(-2)^k.$$

2. Solve $y_{k+3} - 3y_{k+1} - 2y_k = 0.$

Solution. Let $y_k = m^k$. Then

$$(m^3 - 3m - 2)m^k = 0,$$

$m = -1, -1, 2$. Hence the general solution is given by

$$y_k = c_1(-1)^k + c_2 k(-1)^k + c_3 2^k.$$

The methods for finding a particular solution to a nonhomogeneous linear difference equation are not so straightforward. Mostly they are the same as used for differential equations, such as the variation of parameters we have discussed. However, summation is the finite analog of the integration used in the solution of differential equations, so we look at it briefly.

To find the solution to $dy/dx = f(x)$ we integrate

$$y = \int f(x)\, dx \qquad \text{(if possible).}$$

Consider the difference equation

$$y_{k+1} - y_k = f_k. \tag{14.20}$$

Letting $y_0 = c$, we have

$$y_1 = c + f_0,\ y_2 = c + f_0 + f_1, \ldots, y_k = c + \sum_{n=1}^{k} f_{n-1}.$$

In this case c is the general solution to the homogeneous equation

$$y_{k+1} - y_k = 0,$$

and a particular solution to (14.20) is given by

$$y_k = \sum_{n=1}^{k} f_{n-1}.$$

For the general first-order linear difference equation, the method is similar. Suppose

$$y_k = cu_k, \qquad c \text{ an arbitrary constant,}$$

is the general solution to

$$y_{k+1} + ay_k = 0,$$

and we seek a particular solution to

$$y_{k+1} + ay_k = f_k,$$

where a may be a function of k. We assume that such a solution is of the form

$$y_{k_p} = C_k u_k, \qquad C_k \text{ a function of } k.$$

We have

$$y_{k+1} = C_{k+1}u_{k+1} = C_k u_{k+1} + (C_{k+1} - C_k)u_{k+1},$$

so that

$$C_k(u_{k+1} + au_k) + (C_{k+1} - C_k)u_{k+1} = f_k.$$

But $u_{k+1} + au_k = 0$, since u_k is a solution to the homogeneous equation. Thus the condition we need on C_k is

$$C_{k+1} - C_k = \frac{f_k}{u_{k+1}}.$$

But this is a difference equation of the form (14.20) so

$$C_k = \sum_{n=1}^{k} \frac{f_{n-1}}{u_n}.$$

Hence the particular solution is

$$y_{k_p} = u_k \sum_{n=1}^{k} \frac{f_{n-1}}{u_n},$$

and the complete solution is

$$y_k = cu_k + u_k \sum_{n=1}^{k} \frac{f_{n-1}}{u_n}.$$

Examples

1. Solve

$$y_{k+1} - ky_k = k! \qquad (k \geq 1).$$

Solution. The solution to the corresponding homogeneous equation is

$$y_{k_g} = c(k - 1)!;$$

the particular solution is

$$y_k = (k - 1)! \sum_{n=1}^{k} \frac{(n - 1)!}{(n - 1)!} = k(k - 1)! = k!.$$

Therefore

$$y_k = k! + c(k - 1)! \qquad (k \geq 1, \quad c \text{ an arbitrary constant}).$$

2. The Duesenberry* model is a multiplier-accelerator approach to the analysis of income determination. His basic assumptions are that an excess of investment over saving generates income, that the rate of investment is a function of the capital-output ratio, and that investment generates income and expands capital capacity. Thus he has a large number of variables, the most important of which are influenced by the current or earlier values of the others. After various simplifying procedures he arrives at a system of difference equations.

   The income equation

$$Y_k = (\delta + \rho)Y_{k-1} + (\eta + \mu)K_{k-1},$$

and the capital equation

$$K_k = \delta Y_{k-1} + [\eta + (1 - \chi)]K_{k-1},$$

where χ is a constant and there are four parameters described as follows:

δ represents the effect of changes in income on investment.

ρ reflects the influence of capital stock on investment working through both marginal efficiency and profit.

---

* James S. Duesenberry, *Business Cycles and Economic Growth*, New York, 1958.

η is the marginal propensity to consume out of the gross national product.

μ reflects the changes in capital stock on profits through the influence of profits in dividends on consumption.

We can solve for Y_k to get

$$Y_k = [(\delta + \rho) + (\eta + 1 - \chi)]Y_{k-1} + [\delta(\eta + \mu) - (\delta + \rho)(\eta + 1 - \chi)]Y_{k-2},$$

the generalized multiplier-accelerator model. This is a homogeneous second-order linear difference equation, so if we assign numerical values to the parameters and assume two initial conditions, the solution is straightforward.

Difference equations are of interest not only because they accurately represent the variational pattern of some systems, but also because they may be used to approximate solutions to problems involving differential equations. To do this one solves difference equations with ever decreasing intervals, hoping to achieve a good approximation to the continuous case.

We shall not discuss methods of solutions for difference equations of higher degree; the reader is referred to Hildebrand: *Methods of Applied Mathematics.* We have looked at examples of a system of linear difference equation in several variables (Section 11.5); the use of matrix techniques is central to the solution of such systems.

PROBLEMS

1. The growth of money invested at compound interest can be written as a difference equation. If r is the rate of interest compounded annually, P_k the amount after k years, then
$$P_{k+1} = (1 + r)P_k.$$
Solve for P_k.

2. If
$$f_k - f_{k+1} = -3k + 1,$$
find f_k.

3. Solve the following difference equations:
   a) $y_{k+2} - 3y_{k+1} - 4y_k = 0, y_0 = 0, y_1 = 10.$
   b) $y_{k+2} + 2y_{k+1} - y_k = 0, y_0 = 1, y_1 = 2.$
   c) $y_{k+2} - 2y_k = 0.$
   d) $y_{k+1} + 3y_k = 0.$
   e) $y_{k+2} - 5y_{k+1} + 4y_k = 16.$
   f) $y_{k+2} - 2y_{k+1} + 3y_k = 2k^2 + 4k - 1.$

4. One model of economic growth leads to the equation
$$(1 - c/g)y_{k+1} - y_k = 0.$$
Solve for y_k.

5. The Samuelson multiplier-acceleration principle assumes that national income at time t is the sum of estimated consumer demand $c(t)$ and estimated investment demand $I(t)$, with

$$c(t) = ky_{t-1}, \qquad (t \text{ takes on only integral values}),$$

k the multiplier constant, and $I(t) = A[y_{t-1} - y_{t-2}]$, A the acceleration constant. From this we get a "backward" difference equation

$$y_t = (k + A)y_{t-1} - Ay_{t-2}.$$

If $k = 1$ and $A = 7$, convert this to a "forward" difference equation and solve for y_t.

6. (Gambler's Ruin). Two men are tossing coins. A wins if the coins match; B wins if they do not. A has p coins and B has q coins. Find the probability that A eventually wins all of B's coins. (*Hint:* suppose at a certain time in the play A has k coins, $0 < k < p + q$. If u_k is the probability that A eventually wins all of B's coins, then

$$u_k = \tfrac{1}{2}(u_{k-1} + u_{k+1}),$$

where $u_0 = 0$ and $u_{p+q} = 1$.)

APPENDIXES

APPENDIX A. LIMITS

In this appendix we prove several theorems about limits which were stated in Chapter 4.

Theorem 1. If

$$\lim_{x \to x_0} f(x) = A \qquad \text{and} \qquad \lim_{x \to x_0} g(x) = B,$$

then

$$\lim_{x \to x_0} (f(x) + g(x)) = A + B.$$

Proof. Let U be a neighborhood of the real number $A + B$ of radius ε. We want to find a neighborhood V of x_0 such that $x \in V$, $x \neq x_0$ guarantees that $f(x) + g(x) \in U$, that is, that

$$|f(x) + g(x) - (A + B)| < \varepsilon.$$

If $f(x) - A$ and $g(x) - B$ can be made sufficiently small, their sum can also; moreover, we have a way to get bounds on $f(x) - A$ and $g(x) - B$.

Let U_1 be a neighborhood of A of radius $\varepsilon/2$ and U_2 a neighborhood of B of radius $\varepsilon/2$. Since

$$\lim_{x \to x_0} f(x) = A$$

and

$$\lim_{x \to x_0} g(x) = B,$$

there are neighborhoods of x_0, say V_1 of radius δ_1 and V_2 of radius δ_2, such that

$$x \in V_1, x \neq x_0, \text{ implies } f(x) \in U_1, \qquad \text{i.e., } |f(x) - A| < \varepsilon/2,$$

and

$$x \in V_2, x \neq x_0, \text{ implies } g(x) \in U_2, \qquad \text{i.e., } |g(x) - B| < \varepsilon/2.$$

Let δ be the smaller of δ_1 and δ_2 and V a neighborhood of x_0 of radius δ. Then

$$x \in V, x \neq x_0, \text{ implies } |f(x) - A| < \varepsilon/2 \qquad \text{and} \qquad |g(x) - B| < \varepsilon/2,$$

so that

$$|f(x) + g(x) - (A + B)| \leq |f(x) - A| + |g(x) - B| < \varepsilon/2 + \varepsilon/2 = \varepsilon,$$

as required.

Theorem 2. If

$$\lim_{x \to x_0} f(x) = A \quad \text{and} \quad \lim_{x \to x_0} g(x) = B,$$

then

$$\lim_{x \to x_0} f(x)g(x) = AB.$$

Proof. Let U be a neighborhood of AB of radius ε. The situation here is more complicated; we must be more clever in picking the bounds on $f(x) - A$ and $g(x) - B$ as it is not just a simple case of addition. In fact, we also need a bound on $f(x)$ itself.

If $B \neq 0$, let U_1 be a neighborhood of A of radius the smaller of 1 and $\varepsilon/2|B|$ and U_2 a neighborhood of B of radius $\varepsilon/2(|A| + 1)$. Then there are neighborhoods of x_0, say V_1 of radius δ_1 and V_2 of radius δ_2, such that

$$x \in V_1, x \neq x_0 \quad \text{implies that} \quad |f(x) - A| < \varepsilon/2|B| \quad \text{and} \quad |f(x) - A| < 1,$$

and

$$x \in V_2, x \neq x_0 \quad \text{implies that} \quad |g(x) - B| < \varepsilon/2(|A| + 1).$$

Observe that

$$|f(x) - A| < 1 \quad \text{yields} \quad |f(x)| < |A| + 1.$$

Let δ be the smaller of δ_1 and δ_2 and V a neighborhood of x_0 of radius δ. Then for $x \in V$, $x \neq x_0$,

$$\begin{aligned} |f(x)g(x) - AB| &= |f(x)g(x) - f(x)B + f(x)B - AB| \\ &\leq |f(x)g(x) - f(x)B| + |f(x)B - AB| \\ &\leq |f(x)|\,|g(x) - B| + |B|\,|f(x) - A| \\ &< (|A| + 1)\frac{\varepsilon}{2(|A| + 1)} + |B|\frac{\varepsilon}{2|B|} \\ &= \frac{\varepsilon}{2} + \frac{\varepsilon}{2} \\ &= \varepsilon. \end{aligned}$$

If $B = 0$, we simply choose a neighborhood of A of radius $\varepsilon/2$ and the proof goes similarly.

Theorem 3. If

$$\lim_{x \to x_0} f(x) = A \quad \text{and} \quad \lim_{x \to x_0} g(x) = B \neq 0,$$

then

$$\lim_{x \to x_0} \frac{f(x)}{g(x)} = \frac{A}{B}.$$

Proof. We prove that if $\lim_{x \to x_0} g(x) = B \neq 0$, then $\lim_{x \to x_0} 1/g(x) = 1/B$. Combining this with Theorem 2 yields the desired result.

Let U be a neighborhood of $1/B$ of radius ε, U_1 a neighborhood of B of radius the smaller of $|B|/2$ and $\varepsilon|B|^2/2$. Then let V_1 be a neighborhood of x_0 such that

$$x \in V_1,\ x \neq x_0 \quad \text{implies} \quad |f(x) - B| < \frac{|B|}{2},$$

from which

$$\frac{|B|}{2} = |B| - \frac{|B|}{2} < |f(x)|,$$

so that

$$\frac{1}{|f(x)|} < \frac{2}{|B|}.$$

Let V_2 be a neighborhood of x_0 such that

$$x \in V_2,\ x \neq x_0 \quad \text{implies} \quad |f(x) - B| < \frac{\varepsilon|B|^2}{2}.$$

Let V be the smaller of V_1 and V_2. Then for $x \in V$, $x \neq x_0$,

$$\begin{aligned} \left|\frac{1}{f(x)} - \frac{1}{B}\right| &= \frac{|B - f(x)|}{|Bf(x)|} \\ &< \frac{2}{|B||B|}\,|B - f(x)| \\ &< \frac{2}{|B||B|}\,\frac{\varepsilon}{2}\,|B|^2 \\ &= \varepsilon. \end{aligned}$$

Theorem 4. If

$$\lim_{x \to x_0} f(x) = A \qquad \text{and} \qquad \lim_{x \to x_0} f(x) = B,$$

then $A = B$.

Proof. To show that $A = B$, we show that $|A - B|$ is arbitrarily small. Let $\varepsilon > 0$ be given. Let U_1 be a neighborhood of A of radius $\varepsilon/2$ and U_2 a neighborhood of B of radius $\varepsilon/2$. Then there is a neighborhood V of x_0 such that

$$x \in V,\ x \neq x_0 \quad \text{implies} \quad |f(x) - A| < \varepsilon/2 \qquad \text{and} \qquad |f(x) - B| < \varepsilon/2.$$

Then

$$\begin{aligned} |A - B| &= |A - f(x) + f(x) - B| \\ &\leq |A - f(x)| + |f(x) - B| \\ &< \varepsilon/2 + \varepsilon/2 \\ &= \varepsilon. \end{aligned}$$

The last two theorems of this appendix are trivial exercises in selecting the proper neighborhood of x_0, given a neighborhood of the limit.

Theorem 5. If c is a real number and $f(x) = c$ for all x, then

$$\lim_{x \to x_0} f(x) = c.$$

Proof. Let U be a neighborhood of c of radius ε. For all x in *any* neighborhood V of x_0, $f(x) = c$, so

$$|f(x) - c| = 0 < \varepsilon,$$

i.e., $f(x) \in U$. Thus

$$\lim_{x \to x_0} f(x) = c.$$

Theorem 6. If

$$f(x) = x \quad \text{for all} \quad x,$$

then

$$\lim_{x \to x_0} f(x) = x_0.$$

Proof. Let U be a neighborhood of x_0 of radius ε. Then for $x \in V$, a neighborhood of x_0 of radius ε,

$$|f(x) - x_0| = |x - x_0| < \varepsilon,$$

as required.

The proofs in this appendix are manifestations of a common technique, the $\varepsilon - \delta$ process. Once the principle is understood, many proofs depend only upon ingenuity in selecting the right bounds.

APPENDIX B. COMMON LOGARITHMS

This appendix is devoted to the use of common logarithms for computational purposes.

If $a > 0$, $a \neq 1$, is a real number and if $N = a^n$, then $n = \log_a N$, that is, n is the *logarithm* of N to the *base a*. We shall concern ourselves here only with logarithms to the base 10, called *common* logarithms:

$$\text{if} \quad N = 10^n \quad \text{then} \quad n = \log_{10} N.$$

Throughout this appendix we omit the subscript indicating that 10 is the base.

The fundamental laws of logarithms correspond to those for exponents:

$$N = 10^n, \qquad M = 10^m$$

$$10^n 10^m = 10^{n+m} \qquad \log(NM) = \log N + \log M$$

$$\frac{10^n}{10^m} = 10^{n-m} \qquad \log \frac{N}{M} = \log N - \log M$$

$$(10^n)^k = 10^{nk} \qquad \log N^k = k \log N.$$

Since the logarithms of most integers are irrational numbers, tables of logarithms, such as the ones given in this text, give decimal approximations of the logarithms. The approximations for common logarithms consist of two parts: the *mantissa*, given by the table, and the *characteristic*, which is computed as follows.

Any number can be expressed as a number between 1 and 10 multiplied by some power of 10. For example,

$$418 = 4.18 \times 10^2$$

$$15 = 1.5 \times 10^1$$

$$0.018 = 1.8 \times 10^{-2}$$

$$3 = 3 \times 10^0.$$

The exponent on 10 is the number of places the decimal must be moved to the left if positive, to the right if negative. This exponent is the characteristic part of the common logarithm of the number. Suppose we want to find the common logarithm of 418. We observe from the table that the mantissa is 0.6212. The characteristic is 2 so

$$\log 418 = 2.6212.$$

Similarly,

$$\log 15 = 1.1761$$

$$\log 0.018 = -2 + 0.2553$$

$$\log 3 = 0.4771.$$

It is customary to keep the characteristic and mantissa separate in the case of the negative characteristic. Usually we write

$$\log 0.018 = 8.2553 - 10.$$

to aid in computation.

Another problem arises. The tables given here are for 3-digit numbers. What if the logarithm of 5773 is required? In a case like this we must resort to a process known as *interpolation*. Observe that the (mantissa) logarithm of 577 is 0.7612 and that of 578 is 0.7619. We set up a proportion:

$$\frac{3}{10} = \frac{x}{7},$$

where 10 is the difference between 5770 and 5780, 7 is the difference in their logarithms, 3 the difference between 5770 and 5773, and x the approximation of the difference in the logarithms. We find that $x = 2.1$, so adding 2 to 7612 yields

$$\log 5773 = 3.7614.$$

We have not yet looked at the reverse procedure—given a number x, find the number whose logarithm is x. Suppose the given logarithm is 1.5729. A look at the table tells us that 0.5729 is the mantissa of 374, and the characteristic is 1, so

$$1.5729 = \log 3.74 \times 10^1 = \log 37.4.$$

Of course, it is more difficult if the mantissa does not happen to be in the table, but we simply reverse the interpolation process. For example, to find the number whose logarithm is 9.2860 − 10, we observe that the mantissa of 193 is 0.2856 and that of 194 is 0.2878. Setting up a proportion yields

$$\frac{4}{22} = \frac{x}{10}$$

$$x = 1.8.$$

We add 2 to 1930 to get 1932. Therefore,

$$9.2860 - 10 = \log 1.932 \times 10^{-1} = \log 0.1932.$$

One more point to note: if the logarithm is given in a negative form, for example -1.0235, it must be converted to a positive mantissa before continuing. To do this we subtract from 10:

$$\begin{array}{r} 10.0000 \\ -\ 1.0235 \\ \hline 8.9765. \end{array}$$

Thus $-1.0235 = 8.9765 - 10$ and we find

$$\text{mantissa } 947 = 9763$$

$$\text{mantissa } 948 = 9768$$

$$\frac{2}{5} = \frac{x}{10}$$

$$x = 4.$$

Now we have $-1.0235 = 8.9765 - 10 = \log 9.474 \times 10^{-2} = \log 0.09474$.

A word of warning: The logarithm of a negative number is not defined, but numbers between 0 and 1 have negative logarithms.

We give a few sample computations involving logarithms.

1. Find

$$\frac{3.25 \times 6251 \times 430}{0.0921 \times 64}.$$

Solution. Let N be the solution. Then

$$\begin{aligned} \log N &= \log 3.25 + \log 6251 + \log 430 - (\log 0.0921 + \log 64) \\ &= 0.5118 + 3.7960 + 2.6335 - (8.9643 - 10 + 1.8062) \\ &\qquad\qquad \text{(employing interpolation when required)} \\ &= 6.9413 - (10.7705 - 10) \\ &= 6.9413 - 0.7705 \\ &= 6.1708 \end{aligned}$$

$$N = 1.482 \times 10^6 = 1{,}482{,}000.$$

Note that this is an *approximate* solution.

2. Find

$$\frac{\sqrt{492 \times 516}}{(634)^2 \times (0.0911)^3}$$

Solution. Let N be the solution.

$$\begin{aligned}
\log N &= \tfrac{1}{2}(\log 492 + \log 516) - (2 \log 634 + 3 \log 0.0911) \\
&= \tfrac{1}{2}(2.6920 + 2.7126) - \big(2(2.8021) + 3(8.9595 - 10)\big) \\
&= 2.7023 - (5.6042 + 26.8785 - 30) \\
&= 2.7023 - (5.6042 + 6.8785 - 10) \\
&= 2.7023 - (12.4827 - 10) \\
&= 2.7023 - 2.4827 \\
&= 0.2196 \\
N &= 1.658.
\end{aligned}$$

3. Find $(1034 + 143)^{51}$.

Solution. Let N be the solution. Note that logarithms are of no help in addition problems. First perform the addition:

$$1034 + 143 = 1177.$$

Then

$$\begin{aligned}
\log N &= 51 \log 1177 \\
&= 51(3.0708) \\
&= 156.6108 \\
N &= 4.081 \times 10^{156}.
\end{aligned}$$

APPENDIX C. MATRICES AND CRAMER'S RULE

Cramer's Rule

Another method for solving a system

$$\begin{aligned} a_{11}x_1 + \cdots + a_{1n}x_n &= b_1 \\ &\vdots \\ a_{n1}x_1 + \cdots + a_{nn}x_n &= b_n \end{aligned}$$

of linear equations is given by Cramer's rule:

$$x_i = \frac{\det \begin{pmatrix} a_{11} & \cdots & b_1 & \cdots & a_{1n} \\ \vdots & & & & \\ a_{n1} & \cdots & b_n & \cdots & a_{nn} \end{pmatrix}}{\det(a_{ij})},$$

(with $b_1, \ldots, b_n$ in the ith column)

where the numerator is the determinant of the $n \times n$ matrix obtained by replacing the ith column of the coefficient matrix by

$$\begin{pmatrix} b_1 \\ \vdots \\ b_n \end{pmatrix}.$$

Note that if $\det(a_{ij}) = 0$ this method does not work; in this case we know (see Chapter 10) that there is no unique solution to the system. Moreover, Cramer's rule is applicable only to systems in which the number of equations is equal to the number of unknowns.

Example. Solve

$$\begin{aligned} 3x_1 + 2x_2 - 4x_3 &= 2 \\ -x_1 \qquad + 2x_3 &= 3 \\ 2x_1 - 2x_2 + 3x_3 &= 1. \end{aligned}$$

Solution

$$x_1 = \frac{\det\begin{pmatrix} 2 & 2 & -4 \\ 3 & 0 & 2 \\ 1 & -2 & 3 \end{pmatrix}}{\det\begin{pmatrix} 3 & 2 & -4 \\ -1 & 0 & 2 \\ 2 & -2 & 3 \end{pmatrix}} = \frac{18}{18} = 1,$$

$$x_2 = \frac{\det\begin{pmatrix} 3 & 2 & -4 \\ -1 & 3 & 2 \\ 2 & 1 & 3 \end{pmatrix}}{18} = \frac{63}{18} = \frac{7}{2},$$

$$x_3 = \frac{\det\begin{pmatrix} 3 & 2 & 2 \\ -1 & 0 & 3 \\ 2 & -2 & 1 \end{pmatrix}}{18} = \frac{36}{18} = 2.$$

Existence of Solutions

Theorem. The system

$$\begin{aligned} a_{11}x_1 + \cdots + a_{1n}x_n &= k_1 \\ &\vdots \\ a_{m1}x_1 + \cdots + a_{mn}x_n &= k_m \end{aligned} \tag{*}$$

has a solution if and only if the rank of the coefficient matrix $A = (a_{ij})$ is equal to the rank of the augmented matrix $(A \mid K)$.

Proof. By the Gauss process we can transform the system (*) into the form (including possible rearrangement of subscripts)

$$\begin{aligned} b_{11}x_1 + \quad \cdots \quad + b_{1n}x_n &= c_1 \\ b_{22}x_2 + \quad + b_{2n}x_n &= c_2 \\ \vdots \qquad\qquad &\ \ \vdots \\ b_{jj}x_j + \cdots + b_{jn}x_n &= c_j \\ 0x_{j+1} + \cdots + 0x_n &= c_{j+1} \\ \vdots \qquad\qquad & \\ 0x_{j+1} + \cdots + 0x_n &= c_n. \end{aligned}$$

The rank of the coefficient matrix is j. If the rank of the augmented matrix is also j, then $c_i = 0$, $i = j + 1, \ldots, n$, and the system has a solution. On the other hand, if the rank of the augmented matrix is greater than j, at least one of the c_i, $i = j + 1, \ldots, n$ is nonzero and the system is inconsistent.

Inverse

The following gives another method for finding the inverse of an $n \times n$ matrix, if one exists. Recall that if the ith row and the jth column of an $n \times n$ matrix are deleted, we denote the remaining submatrix by M_{ij} and let

$$A_{ij} = (-1)^{i+j} \det M_{ij},$$

calling A_{ij} the *cofactor* of a_{ij}. Then

$$A^{-1} = \frac{1}{\det A} \, {}^t(A_{ij}),$$

where ${}^t(A_{ij})$ is the transpose of the matrix whose (i, j)th entry is the cofactor of a_{ij}. Note that A^{-1} exists if and only if $\det A \neq 0$.

Example. Find the inverse of

$$A = \begin{pmatrix} 2 & 1 & 0 \\ -1 & 5 & 4 \\ 0 & 4 & 5 \end{pmatrix}.$$

Solution

$$\det A = 23.$$

$$A_{11} = \det\begin{pmatrix} 5 & 4 \\ 4 & 5 \end{pmatrix} = 9 \qquad A_{23} = -\det\begin{pmatrix} 2 & 1 \\ 0 & 4 \end{pmatrix} = -8$$

$$A_{12} = -\det\begin{pmatrix} -1 & 4 \\ 0 & 5 \end{pmatrix} = 5 \qquad A_{31} = \det\begin{pmatrix} 1 & 0 \\ 5 & 4 \end{pmatrix} = 4$$

$$A_{13} = \det\begin{pmatrix} -1 & 5 \\ 0 & 4 \end{pmatrix} = -4 \qquad A_{32} = -\det\begin{pmatrix} 2 & 0 \\ -1 & 4 \end{pmatrix} = -8$$

$$A_{21} = -\det\begin{pmatrix} 1 & 0 \\ 4 & 5 \end{pmatrix} = -5 \qquad A_{33} = \det\begin{pmatrix} 2 & 1 \\ -1 & 5 \end{pmatrix} = 11$$

$$A_{22} = \det\begin{pmatrix} 2 & 0 \\ 0 & 5 \end{pmatrix} = 10$$

$$A^{-1} = \frac{1}{23}\begin{pmatrix} 9 & -5 & 4 \\ 5 & 10 & -8 \\ -4 & -8 & 11 \end{pmatrix}.$$

TABLES

Trigonometric Functions

Angle					Angle				
De-gree	Ra-dian	Sine	Co-sine	Tan-gent	De-gree	Ra-dian	Sine	Co-sine	Tan-gent
0°	0.000	0.000	1.000	0.000					
1°	0.017	0.017	1.000	0.017	46°	0.803	0.719	0.695	1.036
2°	0.035	0.035	0.999	0.035	47°	0.820	0.731	0.682	1.072
3°	0.052	0.052	0.999	0.052	48°	0.838	0.743	0.669	1.111
4°	0.070	0.070	0.998	0.070	49°	0.855	0.755	0.656	1.150
5°	0.087	0.087	0.996	0.087	50°	0.873	0.766	0.643	1.192
6°	0.105	0.105	0.995	0.105	51°	0.890	0.777	0.629	1.235
7°	0.122	0.122	0.993	0.123	52°	0.908	0.788	0.616	1.280
8°	0.140	0.139	0.990	0.141	53°	0.925	0.799	0.602	1.327
9°	0.157	0.156	0.988	0.158	54°	0.942	0.809	0.588	1.376
10°	0.175	0.174	0.985	0.176	55°	0.960	0.819	0.574	1.428
11°	0.192	0.191	0.982	0.194	56°	0.977	0.829	0.559	1.483
12°	0.209	0.208	0.978	0.213	57°	0.995	0.839	0.545	1.540
13°	0.227	0.225	0.974	0.231	58°	1.012	0.848	0.530	1.600
14°	0.244	0.242	0.970	0.249	59°	1.030	0.857	0.515	1.664
15°	0.262	0.259	0.966	0.268	60°	1.047	0.866	0.500	1.732
16°	0.279	0.276	0.961	0.287	61°	1.065	0.875	0.485	1.804
17°	0.297	0.292	0.956	0.306	62°	1.082	0.883	0.469	1.881
18°	0.314	0.309	0.951	0.325	63°	1.100	0.891	0.454	1.963
19°	0.332	0.326	0.946	0.344	64°	1.117	0.899	0.438	2.050
20°	0.349	0.342	0.940	0.364	65°	1.134	0.906	0.423	2.145
21°	0.367	0.358	0.934	0.384	66°	1.152	0.914	0.407	2.246
22°	0.384	0.375	0.927	0.404	67°	1.169	0.921	0.391	2.356
23°	0.401	0.391	0.921	0.424	68°	1.187	0.927	0.375	2.475
24°	0.419	0.407	0.914	0.445	69°	1.204	0.934	0.358	2.605
25°	0.436	0.423	0.906	0.466	70°	1.222	0.940	0.342	2.748
26°	0.454	0.438	0.899	0.488	71°	1.239	0.946	0.326	2.904
27°	0.471	0.454	0.891	0.510	72°	1.257	0.951	0.309	3.078
28°	0.489	0.469	0.883	0.532	73°	1.274	0.956	0.292	3.271
29°	0.506	0.485	0.875	0.554	74°	1.292	0.961	0.276	3.487
30°	0.524	0.500	0.866	0.577	75°	1.309	0.966	0.259	3.732
31°	0.541	0.515	0.857	0.601	76°	1.326	0.970	0.242	4.011
32°	0.559	0.530	0.848	0.625	77°	1.344	0.974	0.225	4.332
33°	0.576	0.545	0.839	0.649	78°	1.361	0.978	0.208	4.705
34°	0.593	0.559	0.829	0.675	79°	1.379	0.982	0.191	5.145
35°	0.611	0.574	0.819	0.700	80°	1.396	0.985	0.174	5.671
36°	0.628	0.588	0.809	0.727	81°	1.414	0.988	0.156	6.314
37°	0.646	0.602	0.799	0.754	82°	1.431	0.990	0.139	7.115
38°	0.663	0.616	0.788	0.781	83°	1.449	0.993	0.122	8.144
39°	0.681	0.629	0.777	0.810	84°	1.466	0.995	0.105	9.514
40°	0.698	0.643	0.766	0.839	85°	1.484	0.996	0.087	11.43
41°	0.716	0.656	0.755	0.869	86°	1.501	0.998	0.070	14.30
42°	0.733	0.669	0.743	0.900	87°	1.518	0.999	0.052	19.08
43°	0.750	0.682	0.731	0.933	88°	1.536	0.999	0.035	28.64
44°	0.768	0.695	0.719	0.966	89°	1.553	1.000	0.017	57.29
45°	0.785	0.707	0.707	1.000	90°	1.571	1.000	0.000	

Natural Logs

n	$\log_e n$	n	$\log_e n$	n	$\log_e n$
0.0		4.5	1.5041	9.0	2.1972
0.1	7.6974	4.6	1.5261	9.1	2.2083
0.2	8.3906	4.7	1.5476	9.2	2.2192
0.3	8.7960	4.8	1.5686	9.3	2.2300
0.4	9.0837	4.9	1.5892	9.4	2.2407
0.5	9.3069	5.0	1.6094	9.5	2.2513
0.6	9.4892	5.1	1.6292	9.6	2.2618
0.7	9.6433	5.2	1.6487	9.7	2.2721
0.8	9.7769	5.3	1.6677	9.8	2.2824
0.9	9.8946	5.4	1.6864	9.9	2.2925
1.0	0.0000	5.5	1.7047	10	2.3026
1.1	0.0953	5.6	1.7228	11	2.3979
1.2	0.1823	5.7	1.7405	12	2.4849
1.3	0.2624	5.8	1.7579	13	2.5649
1.4	0.3365	5.9	1.7750	14	2.6391
1.5	0.4055	6.0	1.7918	15	2.7081
1.6	0.4700	6.1	1.8083	16	2.7726
1.7	0.5306	6.2	1.8245	17	2.8332
1.8	0.5878	6.3	1.8405	18	2.8904
1.9	0.6419	6.4	1.8563	19	2.9444
2.0	0.6931	6.5	1.8718	20	2.9957
2.1	0.7419	6.6	1.8871	25	3.2189
2.2	0.7885	6.7	1.9021	30	3.4012
2.3	0.8329	6.8	1.9169	35	3.5553
2.4	0.8755	6.9	1.9315	40	3.6889
2.5	0.9163	7.0	1.9459	45	3.8067
2.6	0.9555	7.1	1.9601	50	3.9120
2.7	0.9933	7.2	1.9741	55	4.0073
2.8	1.0296	7.3	1.9879	60	4.0943
2.9	1.0647	7.4	2.0015	65	4.1744
3.0	1.0986	7.5	2.0149	70	4.2485
3.1	1.1314	7.6	2.0281	75	4.3175
3.2	1.1632	7.7	2.0412	80	4.3820
3.3	1.1939	7.8	2.0541	85	4.4427
3.4	1.2238	7.9	2.0669	90	4.4998
3.5	1.2528	8.0	2.0794	95	4.5539
3.6	1.2809	8.1	2.0919	100	4.6052
3.7	1.3083	8.2	2.1041		
3.8	1.3350	8.3	2.1163		
3.9	1.3610	8.4	2.1282		
4.0	1.3863	8.5	2.1401		
4.1	1.4110	8.6	2.1518		
4.2	1.4351	8.7	2.1633		
4.3	1.4586	8.8	2.1748		
4.4	1.4816	8.9	2.1861		

Exponential Functions

x	e^x	e^{-x}	x	e^x	e^{-x}
0.00	1.0000	1.0000	2.5	12.182	0.0821
0.05	1.0513	0.9512	2.6	13.464	0.0743
0.10	1.1052	0.9048	2.7	14.880	0.0672
0.15	1.1618	0.8607	2.8	16.445	0.0608
0.20	1.2214	0.8187	2.9	18.174	0.0550
0.25	1.2840	0.7788	3.0	20.086	0.0498
0.30	1.3499	0.7408	3.1	22.198	0.0450
0.35	1.4191	0.7047	3.2	24.533	0.0408
0.40	1.4918	0.6703	3.3	27.113	0.0369
0.45	1.5683	0.6376	3.4	29.964	0.0334
0.50	1.6487	0.6065	3.5	33.115	0.0302
0.55	1.7333	0.5769	3.6	36.598	0.0273
0.60	1.8221	0.5488	3.7	40.447	0.0247
0.65	1.9155	0.5220	3.8	44.701	0.0224
0.70	2.0138	0.4966	3.9	49.402	0.0202
0.75	2.1170	0.4724	4.0	54.598	0.0183
0.80	2.2255	0.4493	4.1	60.340	0.0166
0.85	2.3396	0.4274	4.2	66.686	0.0150
0.90	2.4596	0.4066	4.3	73.700	0.0136
0.95	2.5857	0.3867	4.4	81.451	0.0123
1.0	2.7183	0.3679	4.5	90.017	0.0111
1.1	3.0042	0.3329	4.6	99.484	0.0101
1.2	3.3201	0.3012	4.7	109.95	0.0091
1.3	3.6693	0.2725	4.8	121.51	0.0082
1.4	4.0552	0.2466	4.9	134.29	0.0074
1.5	4.4817	0.2231	5	148.41	0.0067
1.6	4.9530	0.2019	6	403.43	0.0025
1.7	5.4739	0.1827	7	1096.6	0.0009
1.8	6.0496	0.1653	8	2981.0	0.0003
1.9	6.6859	0.1496	9	8103.1	0.0001
2.0	7.3891	0.1353	10	22026	0.00005
2.1	8.1662	0.1225			
2.2	9.0250	0.1108			
2.3	9.9742	0.1003			
2.4	11.023	0.0907			

Common Logs

N	0	1	2	3	4	5	6	7	8	9
10	.0000	.0043	.0086	.0128	.0170	.0212	.0253	.0294	.0334	.0374
11	.0414	.0453	.0492	.0531	.0569	.0607	.0645	.0682	.0719	.0755
12	.0792	.0828	.0864	.0899	.0934	.0969	.1004	.1038	.1072	.1106
13	.1139	.1173	.1206	.1239	.1271	.1303	.1335	.1367	.1399	.1430
14	.1461	.1492	.1523	.1553	.1584	.1614	.1644	.1673	.1703	.1732
15	.1761	.1790	.1818	.1847	.1875	.1903	.1931	.1959	.1987	.2014
16	.2041	.2068	.2095	.2122	.2148	.2175	.2201	.2227	.2253	.2279
17	.2304	.2330	.2355	.2380	.2405	.2430	.2455	.2480	.2504	.2529
18	.2553	.2577	.2601	.2625	.2648	.2672	.2695	.2718	.2742	.2765
19	.2788	.2810	.2833	.2856	.2878	.2900	.2923	.2945	.2967	.2989
20	.3010	.3032	.3054	.3075	.3096	.3118	.3139	.3160	.3181	.3201
21	.3222	.3243	.3263	.3284	.3304	.3324	.3345	.3365	.3385	.3404
22	.3424	.3444	.3464	.3483	.3502	.3522	.3541	.3560	.3579	.3598
23	.3617	.3636	.3655	.3674	.3692	.3711	.3729	.3747	.3766	.3784
24	.3802	.3820	.3838	.3856	.3874	.3892	.3909	.3927	.3945	.3962
25	.3979	.3997	.4014	.4031	.4048	.4065	.4082	.4099	.4116	.4133
26	.4150	.4166	.4183	.4200	.4216	.4232	.4249	.4265	.4281	.4298
27	.4314	.4330	.4346	.4362	.4378	.4393	.4409	.4425	.4440	.4456
28	.4472	.4487	.4502	.4518	.4533	.4548	.4564	.4579	.4594	.4609
29	.4624	.4639	.4654	.4669	.4683	.4698	.4713	.4728	.4742	.4757
30	.4771	.4786	.4800	.4814	.4829	.4843	.4857	.4871	.4886	.4900
31	.4914	.4928	.4942	.4955	.4969	.4983	.4997	.5011	.5024	.5038
32	.5051	.5065	.5079	.5092	.5105	.5119	.5132	.5145	.5159	.5172
33	.5185	.5198	.5211	.5224	.5237	.5250	.5263	.5276	.5289	.5302
34	.5315	.5328	.5340	.5353	.5366	.5378	.5391	.5403	.5416	.5428
35	.5441	.5453	.5465	.5478	.5490	.5502	.5514	.5527	.5539	.5551
36	.5563	.5575	.5587	.5599	.5611	.5623	.5635	.5647	.5658	.5670
37	.5682	.5694	.5705	.5717	.5729	.5740	.5752	.5763	.5775	.5786
38	.5798	.5809	.5821	.5832	.5843	.5855	.5866	.5877	.5888	.5899
39	.5911	.5922	.5933	.5944	.5955	.5966	.5977	.5988	.5999	.6010
40	.6021	.6031	.6042	.6053	.6064	.6075	.6085	.6096	.6107	.6117
41	.6128	.6138	.6149	.6160	.6170	.6180	.6191	.6201	.6212	.6222
42	.6232	.6243	.6253	.6263	.6274	.6284	.6294	.6304	.6314	.6325
43	.6335	.6345	.6355	.6365	.6375	.6385	.6395	.6405	.6415	.6425
44	.6435	.6444	.6454	.6464	.6474	.6484	.6493	.6503	.6513	.6522
45	.6532	.6542	.6551	.6561	.6571	.6580	.6590	.6599	.6609	.6618
46	.6628	.6637	.6646	.6656	.6665	.6675	.6684	.6693	.6702	.6712
47	.6721	.6730	.6739	.6749	.6758	.6767	.6776	.6785	.6794	.6803
48	.6812	.6821	.6830	.6839	.6848	.6857	.6866	.6875	.6884	.6893
49	.6902	.6911	.6920	.6928	.6937	.6946	.6955	.6964	.6972	.6981
50	.6990	.6998	.7007	.7016	.7024	.7033	.7042	.7050	.7059	.7067
51	.7076	.7084	.7093	.7101	.7110	.7118	.7126	.7135	.7143	.7152
52	.7160	.7168	.7177	.7185	.7193	.7202	.7210	.7218	.7226	.7235
53	.7243	.7251	.7259	.7267	.7275	.7284	.7292	.7300	.7308	.7316
54	.7324	.7332	.7340	.7348	.7356	.7364	.7372	.7380	.7388	.7396

Common Logs

N	0	1	2	3	4	5	6	7	8	9
55	.7404	.7412	.7419	.7427	.7435	.7443	.7451	.7459	.7466	.7474
56	.7482	.7490	.7497	.7505	.7513	.7520	.7528	.7536	.7543	.7551
57	.7559	.7566	.7574	.7582	.7589	.7597	.7604	.7612	.7619	.7627
58	.7634	.7642	.7649	.7657	.7664	.7672	.7679	.7686	.7694	.7701
59	.7709	.7716	.7723	.7731	.7738	.7745	.7752	.7760	.7767	.7774
60	.7782	.7789	.7796	.7803	.7810	.7818	.7825	.7832	.7839	.7846
61	.7853	.7860	.7868	.7875	.7882	.7889	.7896	.7903	.7910	.7917
62	.7924	.7931	.7938	.7945	.7952	.7959	.7966	.7973	.7980	.7987
63	.7993	.8000	.8007	.8014	.8021	.8028	.8035	.8041	.8048	.8055
64	.8062	.8069	.8075	.8082	.8089	.8096	.8102	.8109	.8116	.8122
65	.8129	.8136	.8142	.8149	.8156	.8162	.8169	.8176	.8182	.8189
66	.8195	.8202	.8209	.8215	.8222	.8228	.8235	.8241	.8248	.8254
67	.8261	.8267	.8274	.8280	.8287	.8293	.8299	.8306	.8312	.8319
68	.8325	.8331	.8338	.8344	.8351	.8357	.8363	.8370	.8376	.8382
69	.8388	.8395	.8401	.8407	.8414	.8420	.8426	.8432	.8439	.8445
70	.8451	.8457	.8463	.8470	.8476	.8482	.8488	.8494	.8500	.8506
71	.8513	.8519	.8525	.8531	.8537	.8543	.8549	.8555	.8561	.8567
72	.8573	.8579	.8585	.8591	.8597	.8603	.8609	.8615	.8621	.8627
73	.8633	.8639	.8645	.8651	.8657	.8663	.8669	.8675	.8681	.8686
74	.8692	.8698	.8704	.8710	.8716	.8722	.8727	.8733	.8739	.8745
75	.8751	.8756	.8762	.8768	.8774	.8779	.8785	.8791	.8797	.8802
76	.8808	.8814	.8820	.8825	.8831	.8837	.8842	.8848	.8854	.8859
77	.8865	.8871	.8876	.8882	.8887	.8893	.8899	.8904	.8910	.8915
78	.8921	.8927	.8932	.8938	.8943	.8949	.8954	.8960	.8965	.8971
79	.8976	.8982	.8987	.8993	.8998	.9004	.9009	.9015	.9020	.9025
80	.9031	.9036	.9042	.9047	.9053	.9058	.9063	.9069	.9074	.9079
81	.9085	.9090	.9096	.9101	.9106	.9112	.9117	.9122	.9128	.9133
82	.9138	.9143	.9149	.9154	.9159	.9165	.9170	.9175	.9180	.9186
83	.9191	.9196	.9201	.9206	.9212	.9217	.9222	.9227	.9232	.9238
84	.9243	.9248	.9253	.9258	.9263	.9269	.9274	.9279	.9284	.9289
85	.9294	.9299	.9304	.9309	.9315	.9320	.9325	.9330	.9335	.9340
86	.9345	.9350	.9355	.9360	.9365	.9370	.9375	.9380	.9385	.9390
87	.9395	.9400	.9405	.9410	.9415	.9420	.9425	.9430	.9435	.9440
88	.9445	.9450	.9455	.9460	.9465	.9469	.9474	.9479	.9484	.9489
89	.9494	.9499	.9504	.9509	.9513	.9518	.9523	.9528	.9533	.9538
90	.9542	.9547	.9552	.9557	.9562	.9566	.9571	.9576	.9581	.9586
91	.9590	.9595	.9600	.9605	.9609	.9614	.9619	.9624	.9628	.9633
92	.9638	.9643	.9647	.9652	.9657	.9661	.9666	.9671	.9675	.9680
93	.9685	.9689	.9694	.9699	.9703	.9708	.9713	.9717	.9722	.9727
94	.9731	.9736	.9741	.9745	.9750	.9754	.9759	.9763	.9768	.9773
95	.9777	.9782	.9786	.9791	.9795	.9800	.9805	.9809	.9814	.9818
96	.9823	.9827	.9832	.9836	.9841	.9845	.9850	.9854	.9859	.9863
97	.9868	.9872	.9877	.9881	.9886	.9890	.9894	.9899	.9903	.9908
98	.9912	.9917	.9921	.9926	.9930	.9934	.9939	.9943	.9948	.9952
99	.9956	.9961	.9965	.9969	.9974	.9978	.9983	.9987	.9991	.9996

ANSWERS TO SELECTED EXERCISES

ANSWERS TO SELECTED EXERCISES

Section 1.1

1. a) {Alaska, Washington, Oregon, California, Hawaii}
c) $\{-5, -4, -3, -2, -1\}$
e) $\{x \mid x \text{ is an integer and } x > 10\}$
g) $\{HH, HT, TH, TT\}$ i) $\emptyset$

2. a) $A \cup B = \{a, b, c, d, e\}$; $A \cap B = \{c, d\}$
c) $A \cup B = \{1, 2, 3, 4, 5\}$; $A \cap B = \{2, 4, 5\}$
e) $A \cup B =$ all real numbers; $A \cap B = \{x \mid 5 < x < 10\}$

3. a) $\emptyset$, {red}, {pink}, {gold}, {red, pink}, {red, gold}, {pink, gold}, {red, pink, gold}

4. a) all nonpositive real numbers c) $\{x \mid x \le 5\}$
e) $\{x \mid x \le a \text{ or } x \ge b\}$ g) $\emptyset$

5. a) $A = \{x \mid x > 5\}$ $B = \{x \mid -3 < x < 10\}$
$A \cap B = \{x \mid 5 < x < 10\}$ $A \cup B = \{x \mid x > -3\}$
$\bar{A} = \{x \mid x \le 5\}$ $\bar{B} = \{x \mid x \le -3 \text{ or } x \ge 10\}$
c) $A = \{x \mid x \le -3\}$ $B = \{x \mid x > -10\}$
$A \cap B = \{x \mid -10 < x \le -3\}$ $A \cup B = \{x \mid -10 < x\}$
$\bar{A} = \{x \mid x > -3\}$ $\bar{B} = \{x \mid x \le -10\}$

6. a)

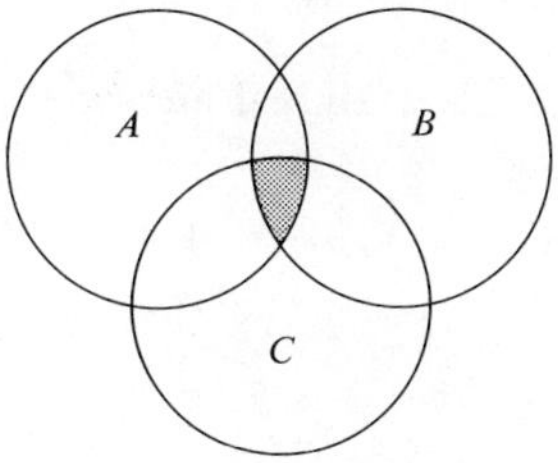

c)

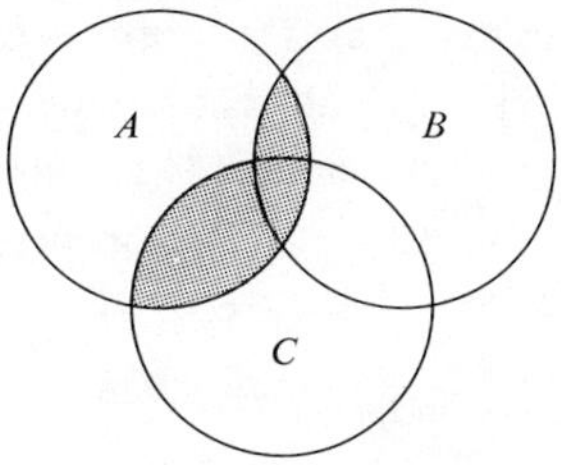

e)

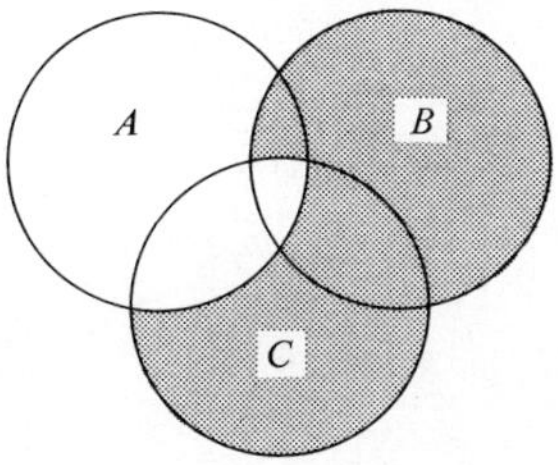

g)

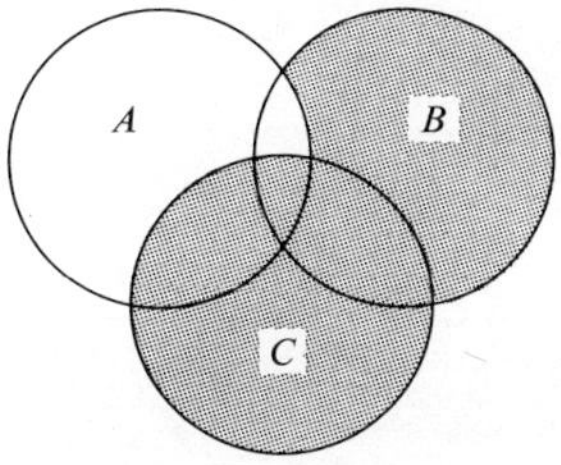

7. 710 **9.** 81; 4

Section 1.2

1. a) $\frac{7}{12}$ c) $\frac{10}{9}$ e) $\frac{3}{10}$ g) x^2y/z^3 i) $\frac{23}{20}$ k) $\frac{1}{36}$
2. a) $[-3, 2]$ c) $[-3, 1)$ e) $\emptyset$ g) $(-1, 5]$
3. a) $\{x \mid 2 < x < 3\}$ c) $\{x \mid -10 \leq x < -5\}$ e) $\{x \mid -2 < x < 3\}$
g) $\{x \mid -4 < x \leq -3\}$
4. a) $A \cap B = [1, \frac{3}{2})$ c) $A \cap B = (-2, -1)$, $A \cup B =$ all real numbers
5. a) any interval (a, b) where $a < 0$ and $b > 0$, e.g., $(-1, 1)$
c) any interval (a, b) where $a < 2$ and $b > 2$, e.g., (3.2, 5.2)
6. $\frac{4}{15}$ **7.** $\frac{4}{15}$ **8.** 40

Section 1.3

1. a) $-3, \frac{1}{3}$ c) $-1/2, 2$ e) $-\pi, 1/\pi$ g) 0, no multiplicative inverse
i) $-2, \frac{1}{2}$ k) $3, -1/3$ m) $-c/d, d/c$ $(d, c \neq 0)$
2. a) $\frac{32}{243}y^{10}$ c) $8a^{9/2}$ e) $169a^{10}$ g) $\dfrac{1}{\sqrt[5]{2}x^{2/5}y^{1/5}}$
3. If $a \neq 0$, multiply by the inverse of a:

$$\frac{1}{a}ab = 0, \qquad b = 0$$

5. a) at most one element c) precisely one element
6. a) $\{x \mid x \geq -1\}$ c) $\{x \mid x \leq 1\}$
7. a) $x < 2$ c) $x < 1$ e) $x \geq 1$ g) $x \geq -5/6$ i) $x \geq 1$
8. Let $c =$ cost per person. Then $c \leq \$7.50$.

Section 1.4

1. a) $(-6, 6)$ c) $\emptyset$ e) $(-2, 3)$
2. a) $[1, 5]$ c) all real numbers e) $x = -6, 4$ g) all real numbers
3. a) $|2y|$ c) impossible to simplify
4. a) 2 c) 10 e) use properties of absolute value g) 7 i) 4 k) 4
5. a) $(-5, 1)$ c) $(-6, -4)$

Section 1.5

1. a) $-1/3$ c) $\pm 2\sqrt{3}$
2. a)
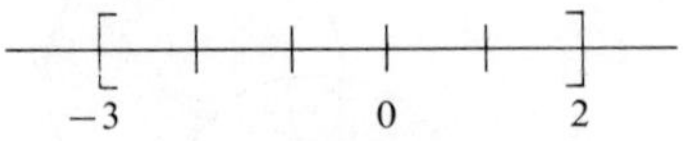

c)
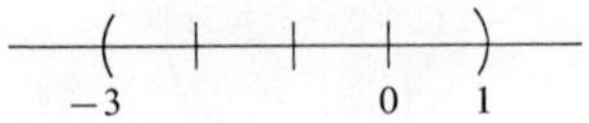

3. a)

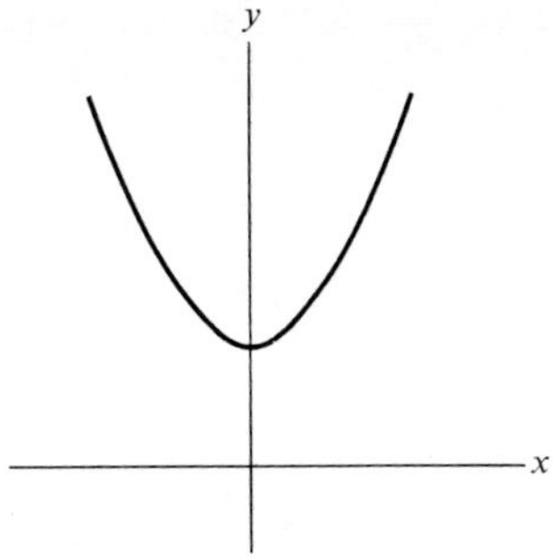
y
x

c)

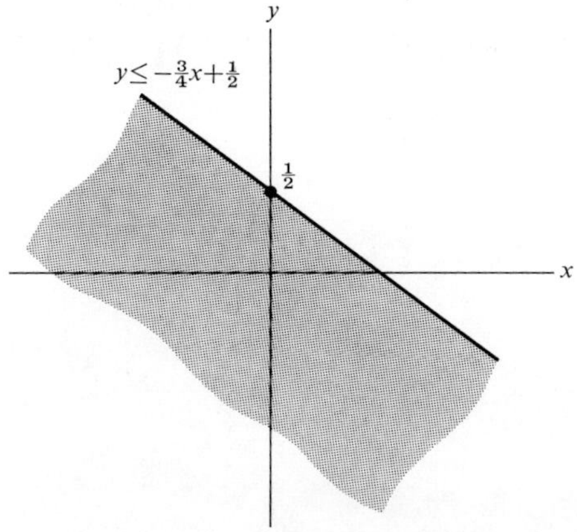
y
$y \leq -\frac{3}{4}x + \frac{1}{2}$
$\frac{1}{2}$
x

e)

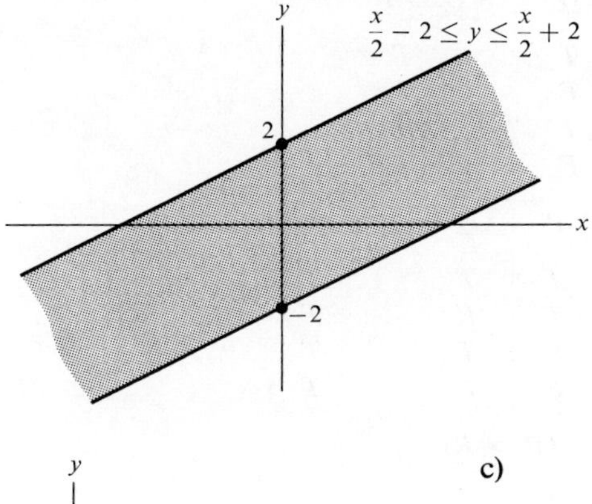
y
$\frac{x}{2} - 2 \leq y \leq \frac{x}{2} + 2$
2
−2
x

4. a)

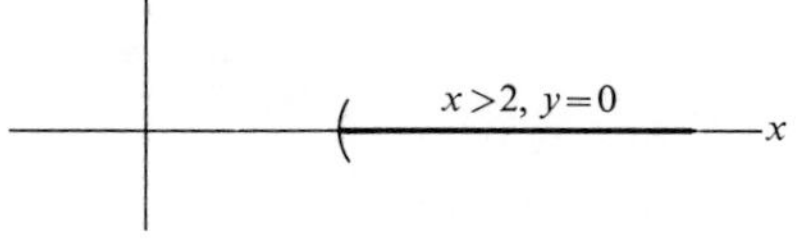
y
$x > 2,\ y = 0$
x

c)

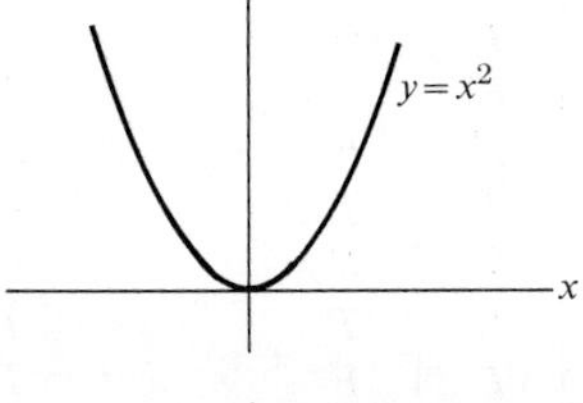
y
$y = x^2$
x

5.

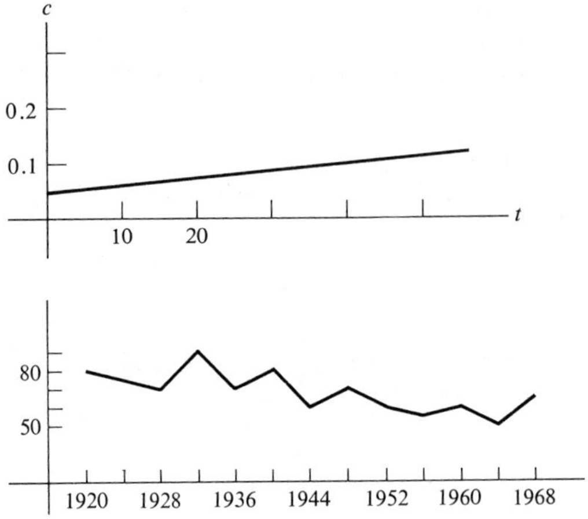

7.

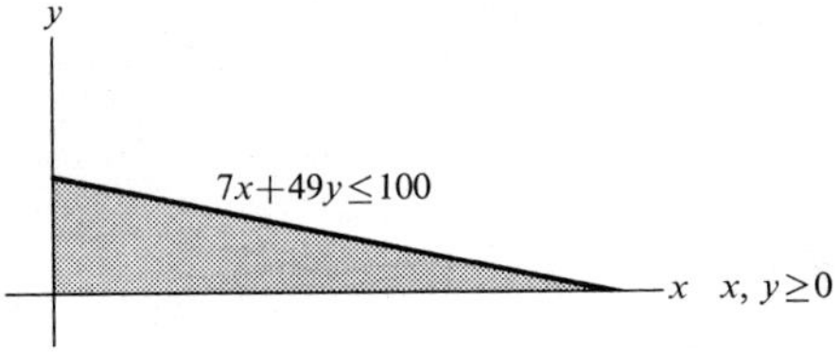

Section 1.6

2. a) no

$P \Rightarrow Q$	P	Q	$\sim Q \cup P$
T	T	T	T
F	T	F	T
T	F	T	F
T	F	F	T

c) yes

$P \Rightarrow (Q \Rightarrow P)$	P	Q	$Q \Rightarrow Q$
T	T	T	T
T	T	F	T
T	F	T	T
T	F	F	T

3.

P	Q	R	$\sim Q \wedge (P \Rightarrow R)$
T	T	T	F
T	T	F	F
T	F	T	T
T	F	F	F
F	T	T	F
F	T	F	F
F	F	T	T
F	F	F	T

4. a) P = it will rain, Q = I shall stay home $\sim P \Rightarrow \sim Q$
c) P = our team will win, Q = Johnson plays $P \Rightarrow Q$
e) P = you play the sonata, Q = the evening is a success $\sim(P \Rightarrow Q) \wedge \sim(Q \Rightarrow P)$

6. a) $\sum_{i=0}^{n} 2^i$ c) $\sum_{i=1}^{5} (i^2 - 2)$ e) $\sum_{i=2}^{n} (2i + 2)$

7. a) 6, 12, 22, 36, 54 c) $-5, -4, 3, 22, 59$ e) $-4, -3, -2, -1, 0$

9. $\sum_{i=1}^{n} (2n - 1) = n^2$

Section 2.1

1. a) $\{\frac{1}{3}\}$ c) $\{2\}$ e) $\{-1/3\}$

3. a) 4

5. a)

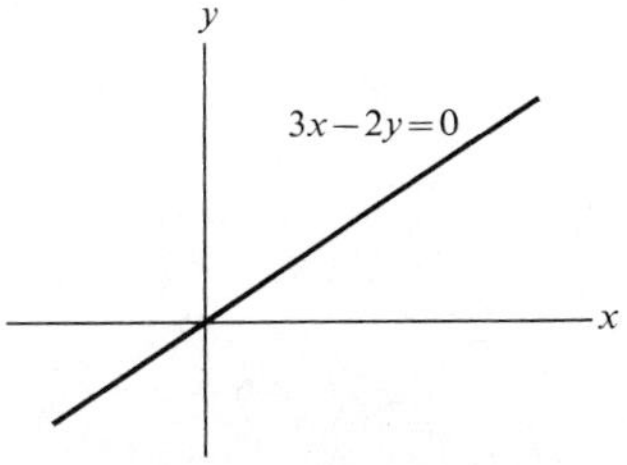

c)

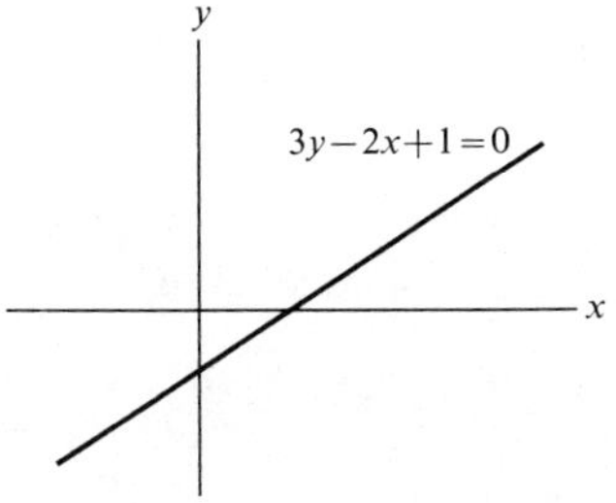

6. a) slope = $\frac{3}{2}$, y-intercept = 0
c) slope = $\frac{2}{3}$, y-intercept = $-1/3$
e) slope not defined, no y-intercept

7. a) $y = \frac{1}{2}x$ c) $y = -2x - 1$ e) $y = 2x - 2$ g) $y = -3x + 4$
i) $y = \frac{1}{2}x + 1$ k) $y = 5x + 22$ m) $y = (-3/2)x + \frac{5}{2}$ o) $y = \frac{3}{2}x - \frac{1}{2}$
q) $y = \frac{1}{3}x - 3$ s) $y = \frac{2}{3}x - \frac{7}{3}$

8. a) $y = (-3/2)x + \frac{1}{2}$ c) $y = \frac{2}{5}x - \frac{1}{5}$

Section 2.2

1. a) $1 + 8i$ c) $4 + i$

2. a) $4 - 6i$ c) $3 - 3i$

3. a) sum = $7 - 3i$, product = $22 - 14i$
c) sum = $\sqrt{2} - 3i$, product = $-3\sqrt{2}i$
e) sum = $-2 + 2i$, product = $-15 - 10i$

4. a) $\frac{3}{13} + \frac{2}{13}i$ c) $\frac{1}{4} - \frac{1}{4}i$ e) $\frac{2}{13} - \frac{3}{13}i$ g) $\frac{1}{10} + \frac{1}{5}i$

5. a) $x = \pm\frac{1}{2}$

6. a) 3 c) 3

Section 2.3

1. a) $\dfrac{2 \pm \sqrt{2}i}{3}$ c) $\dfrac{-1 \pm \sqrt{7}}{2}$ e) $-7, 2$ g) $-1, -3/2$ i) $0, \frac{5}{3}$

2. a) $4, -3$ c) $\frac{1}{4}, \frac{1}{3}$

3. a) 1 c) 0 e) $\dfrac{-2 \pm \sqrt{2}i}{3}$

5. $\dfrac{5 - \sqrt{13}}{2}$

6. a) $1, \dfrac{-1 \pm \sqrt{7}i}{4}$ c) $-1, \frac{1}{2}, \dfrac{-1 \pm \sqrt{3}i}{2}$ e) $-1, 2$

7. a) 0 c) -1111 e) 40

8. a) 0.7 c) 1.3 e) 0.6

9. $t = 2.6$

Section 2.4

1. a) $x - y$ c) $\dfrac{3}{x - y}$

2. a) $\dfrac{x}{x^2 + xy + y^2}$ c) $\dfrac{4x^2 - y^2}{x - y}$

3. a) $x = 3$ c) no solution

5. a) $x = -3 \pm 3\sqrt{2}$ c) $x = \dfrac{7 \pm \sqrt{69}}{10}$

Section 2.5

1. a) $x = \frac{19}{7}, y = -1/7$ c) $x = 4, y = 3; x = -5, y = 0$
e) $x = 1, y = -1, z = 0$ g) $x = 1, y = -1, z = 0$ i) no solution

3. $x = \frac{2}{5}, y = -1/5$

5. approximately 44.4 ml 90% solution and 55.6 ml water

7. 65, 56

Section 2.6

1. $p = 28.5$ $x = 20$ **3.** 6.35 years **5.** 13%

7. 60,000 to wife, 20,000 to each child **9.** 9.8%

11. equilibrium price $= 2\frac{2}{3}$, equilibrium quantity $= 6$

13. 47.5, 159 **15.** 21.4% **17.** 172.50

19. 300 ml of 40%, 700 ml of 90% solution

Section 3.1

1. a) 720 c) $-1079/90$ e) $\dfrac{(n + 2)! - 2n}{n(n + 2)!}$

3. a) $4! = 24$ c) $5! = 120$

4. a) 90 c) 500

5. a) 650 c) 2 **7.** 15,600 **9.** 36 **13.** 10

Section 3.2

1. $x^4 + 12x^3y + 54x^2y^2 + 108xy^3 + 8y^4$

3. a) 126 c) 114,688

5. $27a^3 - 108a^2b + 144ac^2 - 64c^3$

6. a) 448 c) 720

Section 3.3

1. {(1, 1), (1, 2), (1, 3), (1, 4), (1, 5), (1, 6), (2, 1), (2, 2), (2, 3), (2, 4), (2, 5), (2, 6), (3, 1), (3, 2), (3, 3), (3, 4), (3, 5), (3, 6), (4, 1), (4, 2), (4, 3), (4, 4), (4, 5), (4, 6), (5, 1), (5, 2), (5, 3), (5, 4), (5, 5), (5, 6), (6, 1), (6, 2), (6, 3), (6, 4), (6, 5), (6, 6)}

3. a) $\frac{1}{2}$ c) $\frac{1}{4}$

4. a) $\frac{1}{26}$ c) $\frac{10}{13}$

5. a) $\frac{138}{2652}$ c) $\frac{1560}{2652}$

7. $\frac{22}{11!}$

8. a) $\dfrac{95 \cdot 94 \cdot 93}{100 \cdot 99 \cdot 98}$ c) $1 - \dfrac{95 \cdot 94 \cdot 93}{100 \cdot 99 \cdot 98}$

9. a) $\frac{40}{81}$ c) $\frac{4}{81}$ e) $\frac{52}{81}$

11. $\frac{3}{28}$

13. 0.11, 0.0121

14. a) $\frac{3}{8}$ c) $\frac{15}{16}$

15. a) $\frac{156}{2652}$ c) $\frac{2}{2652}$

16. a) $\dfrac{995 \cdot 994 \cdot 993 \cdot 992 \cdot 991 \cdot 990 \cdot 989 \cdot 988 \cdot 987 \cdot 986}{1000 \cdot 999 \cdot 998 \cdot 997 \cdot 996 \cdot 995 \cdot 994 \cdot 993 \cdot 992 \cdot 991}$ c) 0

17. a) $\frac{5}{12}$ c) $\frac{70}{132}$ e) $\frac{9}{132}$

Section 3.4

1. 0.000001 **3.** $\dfrac{960 \cdot 40}{1000 \cdot 999}$

4. a) $\frac{1}{169}$

5. a) $\dfrac{20 \cdot 19 \cdot 18 \cdot 17}{10{,}000 \cdot 9999 \cdot 9998 \cdot 9997}$ c) $\dfrac{80 \cdot 9980 \cdot 9979 \cdot 9978}{10{,}000 \cdot 9999 \cdot 9998 \cdot 9997}$

7. a) $\frac{1}{2}$ c) 0 e) $\frac{31}{45}$ g) $\frac{9}{14}$

8. a) $\frac{1}{64}$ c) $\frac{1}{27}$

9. a) $\frac{99}{663}$ c) $\frac{3}{51}$ e) $\frac{3}{51}$

10. a) 0.1325 c) 0.65

11. a) $\frac{1}{6}$ c) $\frac{1}{6}$

12. a) $\dfrac{25 \cdot 24 \cdot 23 \cdot 22 \cdot 21 \cdot 20 \cdot 19 \cdot 18 \cdot 17 \cdot 16}{2000 \cdot 1999 \cdot 1998 \cdot 1997 \cdot 1996 \cdot 1995 \cdot 1994 \cdot 1993 \cdot 1992 \cdot 1991}$

c) $\dfrac{1975 \cdot \ldots \cdot 1966}{2000 \cdot \ldots \cdot 1991}$

13. a) 0.99 c) $(0.99)^{100}$

14. a) $\frac{3}{8}$ c) $\frac{1}{3}$

15. $\dfrac{100!}{40!\,60!}\left(\dfrac{1}{2}\right)^{100}$

Section 4.1

1. a) yes c) yes e) no g) yes i) yes k) no m) yes o) yes

2. a) polynomial c) neither e) rational g) neither

3.1a) domain = all real numbers, image = all real numbers
1c) domain = $\{1, 2, 3, 4\}$, image = $\{3, 5, 4, -5\}$
1g) domain = all real numbers, image = all integers
1i) domain = all real numbers ≥ -2 image = all non-negative real numbers
1m) domain = image = all real numbers
1o) domain = all real numbers except 2 image = all real numbers except 0
2a) domain = all real numbers, image = $\{x \mid x \geq -17/4\}$
2c) domain = $\{x \mid |x| \geq \sqrt{2}\}$, image = $\{x \mid x \geq 0\}$
2e) domain = $\{x \mid x \neq -1/5\}$, image = all real numbers except those between approximately $-2/3$ and $+1/2$

5. yes; there are two possibilities:
a) The number of deaths from heart disease per 100,000 is a function of the state.
b) The state is a function of the number of deaths from heart disease.
Clearly (a) is a more useful practical notion.

7. a) $f \circ g = 9x^2 + 12x + 3$, domain = all real numbers
$g \circ f = 3x^2 - 1$, domain = all real numbers

c) $f \circ g = \dfrac{8x^2 + 1}{2x - 5}$, domain = $\{x \mid x \neq \frac{5}{2}\}$

$g \circ f = \dfrac{4x^2 + 2}{x - 5}$, domain = $\{x \mid x \neq 5\}$

e) $(f \circ g)(x) = 2\sqrt{x + 1}$, domain = $\{x \mid x \geq -1\}$
$(g \circ f)(x) = \sqrt{2x + 2}$, domain = $\{x \mid x \geq -1\}$

g) $(f \circ g)(x) = \sqrt{6x^2 + 1}$, domain = all real numbers
$(g \circ f)(x) = 6x - 2$, domain = all real numbers

i) $(f \circ g)(x) = |x^2 + 1|$, domain = all real numbers
$(g \circ f)(x) = |x - 1|^2 + 2$, domain = all real numbers

Section 4.2

1. a) 1, yes c) $\lim_{x \to 0} f(x)$ does not exist e) $\sqrt{5}$, yes g) 1, no

2. a) $\lim_{x \to +\infty} f(x)$ does not exist c) $\lim_{x \to \infty} f(x)$ does not exist e) $\lim_{x \to +\infty} f(x) = 2$

3. x = selling price, $f(x)$ = percentage return on investment, $\lim_{x\to\infty} f(x) = 12\%$

4. a) 1 c) 0 e) $\sqrt{15}$ g) $a^2 + 4a - 5$ i) does not exist k) $2\sqrt{3}$
m) does not exist

5. 22

6. a) yes c) yes

Section 4.3

1. a) -6 c) $\frac{13}{2}$ e) $\frac{98}{5}$ g) -1 i) $\dfrac{4x_0^{\,3} + 4x_0^{\,2} - 7}{x_0^{\,2} - x_0 + 1}$

Section 4.4

1. a) yes c) continuous except at $x = 1$ e) yes

2. a) continuous for all values of x c) not continuous at $x = -1$

3. yes

5. yes, -2

7. a) no c) yes e) yes

Section 5.1

1. a) -1 c) 1

3. a) -2 c) 1 e) $10x_0 - 2$

5. a) -1 c) 11 e) $3x_0^{\,2} + 5x_0 - 1$

Section 5.2

1. a) $4x + 2$ c) $-2/x^3$ e) $9x^2 - 3$ g) $15x^2 - 4x + 1$ i) $1/\sqrt{2x + 1}$

3. $f'(x) = 9x^2 - 5, f'(0) = -5, f'(2) = 31, f'(-2) = 31$

5. $y = -x$

7. a) Let $D(t)$ = difficulty of learning, $P(t)$ = amount of practice:

$$\frac{dD}{dt} = c\, D(t)P(t), \qquad c < 0$$

c) Let N = population:

$$\frac{dN}{dt} = cN$$

Section 5.3

1. a) $30x^4 - 20x^3 + 2$ c) $-12x^5 - 16x^3 - 2x + 5$
e) $-5x^4 + 12x^2 + 2x$ g) $72x^3 + 10x - 4$

2. a) $(6x^3 - 2x^2 - x + 5)(28x^3 - 18x^2 + 10x)$
$+ (7x^4 - 6x^3 + 5x^2 + 2)(18x^2 - 4x - 1)$

5. $(\frac{5}{4}, -17/8)$

6. a) 100

7. $\dfrac{dP}{dx} = p - \dfrac{3V}{2k}\left(\dfrac{x}{k}\right)^{1/2}$. When $x = \dfrac{4k^3p^2}{9V^2}$

8. a) $(\frac{5}{4}, -1/8)$ c) $(-2/3, -19/3)$

9. $y = 20x - 20$

Section 5.4

1. a) $y = 77x - 161$ c) $y = \frac{161}{4}x - 57$
e) $y = 25x + 16$ g) $y = 29x - 27$

2. a) $(\frac{5}{2}, -21/4)$ c) $(\frac{1}{2}, \frac{3}{4})$ e) $(115, -415)$ g) $(0, 4)$

3. Rises to 110.25 feet, hits the ground when $t = \frac{5}{4}$.

5. a) $\{y \mid y \le \frac{37}{12}\}$

6. a) $\frac{1}{3}$

7. a) $\frac{7}{8}$ c) -4

Section 5.5

1. a) $\dfrac{(3x^2 - 14)(48x^2 - 10x + 2) - (16x^3 - 5x^2 + 2x - 1)6x}{(3x^2 - 14)^2}$

c) $\dfrac{(6x^2 - 1)(17x^4 - 5x + 5)[(3x^2 - 1)2 + 9x^2(2x - 5)]}{(6x^2 - 1)^2(17x^4 - 5x + 5)^2} - \dfrac{(3x^3 - 1)(2x - 5)[(6x^2 - 1)(68x^3 - 5) + 12x(17x^4 - 5x + 5)]}{(6x^2 - 1)^2(17x^4 - 5x + 5)^2}$

e) $x^{-1/2}$ g) $(2x^{1/3} + 1)((-9/2)x^{-7/4} + 2x^{-3/2}) + \frac{2}{3}x^{-2/3}(6x^{-3/4} - 4x^{-1/2} + 5)$

i) $(-4/7)x^{-2}$

k) $\dfrac{(4x^2 + 15x - 2)(9x^2 - 10x) - (3x^3 - 5x^2 + 1)(8x + 15)}{(4x^2 + 15x - 2)^2}$

m) $\dfrac{(6x^2 - 17x + 1)[9x^2(2x + 3) + 2(3x^3 - 1)] - (3x^3 - 1)(2x + 3)(12x - 17)}{(6x^2 - 17x + 1)^2}$

o) $\dfrac{\frac{3}{2}x^{1/2}}{(x^{3/2} - 5)^2}$

Section 5.6

1. $v = 9t^2 - 14$, $a = 18t$

3. yes, $s = \frac{2}{3}t^3 - \frac{5}{2}t^2 + t + 10$

5. a) $f'(x) = 6x, f''(x) = 6, f'''(x) = 0$
c) $f'(x) = 6x^2 - 4, f''(x) = 12x, f'''(x) = 12$
e) $f'(x) = \dfrac{1}{2\sqrt{x}} + 5 + \dfrac{1}{x^2}$; $f''(x) = -\dfrac{1}{4x^{3/2}} - \dfrac{2}{x^3}$; $f'''(x) = \dfrac{3}{8x^{5/2}} + \dfrac{6}{x^4}$

g) $f'(x) = \dfrac{-4x}{(2x^2+1)^2}$; $f''(x) = \dfrac{-4(2x^2+1)^2 + 32x^2(2x^2+1)}{(2x^2+1)^4}$;

$$f'''(x) = \frac{(2x^2+1)^4[-32x(2x^2+1) + 256x^3 + 64x]}{(2x^2+1)^8} - \frac{16x(2x+1)^3[-4(2x^2+1)^2 + 32x^2(2x^2+1)]}{(2x^2+1)^8}$$

7. $s = t^3 + 2t^2 - t$

Section 5.7

1. a) $3(3x^3 - 4x + 1)(9x^2 - 4)$ c) $-2(7x^5 - 4x - 1)^{-3}(35x^4 - 4)$

e) $\frac{1}{2}(7x^2 - 4x + 2)^{-1/2}(14x - 4)(-4x + 1)^{-1/3} + (7x^2 - 4x + 2)^{1/2}(-1/3)(-4)(-4x + 1)^{-4/3}$

g) $\dfrac{(4x^3 - 5x^2)^{2/3}3(16x^2 - 5x + 1)^2(32x - 5) - (16x^2 - 5x + 1)^3}{(4x^3 - 5x^2)^{4/3}} \times \dfrac{\frac{2}{3}(4x^3 - 5x^2)^{-1/3}(12x^2 - 10x)}{(4x^3 - 5x^2)^{4/3}}$

i) $7(ax^4 - bx^3 - cx + d)(4ax^3 - 3bx^2 - c)$

k) $\frac{1}{2}(6x^2 + \sqrt{2x})^{-1/2}\left(12x + \dfrac{\sqrt{2}}{2}x^{-1/2}\right)$

m) $\dfrac{-3(18x^2 - 5x)}{(6x^3 - 5x + 5)^{-4}}$

o) $\dfrac{(3x^3 - 4x^2)^{1/2}\frac{3}{2}(3x - 1)^{-1/2} - (3x - 1)^{1/2}\frac{1}{2}(3x^3 - 4x^2)^{-1/2}(9x^2 - 8x)}{3x^3 - 4x^2}$

2. a) $v = \frac{1}{2}(t^2 - 2t + 3)^{-1/2}(2t - 2)$

$a = (t^2 - 2t + 3)^{-1/2} + (t - 1)(-1/2)(t^2 - 2t + 3)^{-3/2}(2t - 2)$

c) $v = -2(2t^4 - 5t)^{-3}(8t^3 - 5)$

$a = -2(2t^4 - 5t)^{-3}(24t^2) + 6(2t^4 - 5t)^{-4}(8t^3 - 5)^2$

3. a) $y = 18750x - 21875$ c) $y = 12x + 8$

Section 5.8

1. a) $\frac{1}{6}$ c) $1/3x^2$ e) $1/2x$ g) $1/(2x + 2)$

2. a) $-x/y$ c) $3x^2/2y$ e) $(2x - 1)/2y$ g) $\dfrac{x - y^2x}{x^2y - 2y}$

i) $\dfrac{1 - 2y}{2y + 2x - 1}$ k) $x^2(x^2 + 1)^{-1/2} + (x^2 + 1)^{1/2}$ m) $\dfrac{-4x}{y}$

o) $\dfrac{1 + 3y}{2y + 1 - 3x}$

3. a) tangent line $y = -x + 1/\sqrt{2}$
normal line $y = x + 1/\sqrt{2}$

4. a) $f^{-1}(y) = \dfrac{y^2 - 1}{3}$, $(f^{-1})'(y) = \frac{2}{3}y$

c) $f^{-1}(y) = 1 + \sqrt{9 + y}$, $(f^{-1})'(y) = \frac{1}{2}(9 + y)^{-1/2}$

Section 6.1

1. a) no maximum nor minimum
c) no maximum nor minimum
e) maximum of $-10/7$ at $x = \frac{2}{7}$, no minimum

2. a) 1 c) 20 e) -5

3. a) local maximum of $-11/3$ at $x = +1/3$, no local minimum
c) local minimum of $-157/243$ at $x = \frac{5}{9}$; local maximum of 5 at $x = -1$
e) local minimum of 1 at $x = 0$; no local maximum

4. a) $x = -3/4$ c) $x = 1, -1/3$

5. a) maximum of 4 at $x = 1$; minimum of -1 at $x = 0$
c) maximum of 2 at $x = 0$; minimum of -1 at $x = 1$

Section 6.2

1. a) minimum at $x = \frac{1}{3}$ c) no critical points e) no critical points
g) local maximum at $x = -1$; local minimum at $x = \frac{7}{9}$ i) no critical points

3. maximum of $\frac{643}{243}$ at $x = -5/9$; minimum of -4 at $x = 1$

7. Assuming that a cut must actually be made, a piece $\pi/(4 + \pi)$ mm long should be bent into a circle and the remainder into a square. The absolute maximum area can be obtained by bending the whole piece into a circle.

9. by road for $12 - 3\sqrt{3}$ miles and then a straight line cross-country to the cabin,

11. radius of base $= \sqrt{10/3\pi} =$ height

13. optimal lot size $\frac{6}{5}$ thousands, so reorder 4 times a year in lots of 1250

15. radius of base $= \sqrt[3]{500/\pi}$; height $= (1000/\pi)(\pi/500)^{2/3}$

Section 6.3

1. $c = 2/\sqrt{3}$ **3.** $c = \sqrt{3} - 2$ **5.** $c = 0$

7. a) $\frac{13}{6}$ c) 0

8. a) $\frac{17}{4}$ c) $\frac{1}{2}$

Section 6.4

1. a) vertical asymptote $x = -4$; horizontal asymptote $y = 2$
c) vertical asymptotes $x = -2$, $x = \frac{4}{3}$; horizontal asymptote $y = 0$
e) horizontal asymptote $y = 0$

2. a) symmetric about the x- and y-axes c) symmetric about the line $x = \frac{5}{4}$
e) symmetric about x- and y-axes g) symmetric about $x = 1$

3. a) ellipse c) parabola e) ellipse g) parabola

4. a) the function is constant
c) increasing for $x > -2$, decreasing for $x < -2$
e) decreasing for all $x \neq \frac{1}{3}$

5. a) concave upward for $x < 0$, concave downward for $x > 0$
c) concave upward for $x < -1$, concave downward for $x > -1$
e) concave upward for $x > \frac{4}{3}$, concave downward for $x < \frac{4}{3}$

8. a) increasing for $x > 0$; decreasing for $x < 0$; concave downward everywhere.
c) decreasing everywhere; concave upward for $x > 1$; concave downward for $x < 1$.
e) increasing everywhere; concave upward for $x < -1/2$; concave downward for $x > -1/2$.
g) increasing on $(-4/3, 0)$, decreasing elsewhere; concave upward for $x < -2/3$; concave downward for $x > -2/3$.

Section 6.5

1. $\dfrac{dR}{dL} = 9L^2 + 36L + 34$

3. The profit is larger if none is passed on. If none is passed on, the maximum profit with tax is 688,900 whereas before its imposition the maximum profit is 765,625.

5. Let T = tax as a percentage of income I. Then

$$\frac{dT}{dI} = \frac{519}{I^2}.$$

7. approximately 19.6 mph **9.** $1/2\pi$ ft/min **11.** $\dfrac{dp}{dt} = 14$ lb/in²/sec

13. 60,350 **15.** $-2/\sqrt{221}$ **17.** $-1436/\sqrt{1066}$, yes

Section 7.1

1. $\frac{500}{3}$ **3.** 3 **5.** $\frac{32}{3}$ **7.** 40 **9.** $\frac{32}{3}$ **11.** 44

Section 7.2

1. a) $-2/3$ c) $-185/3$ e) $-1/3$ g) $-110/3$ i) $-215/6$ k) $-2/3$

2. a) $-4/3$ c) $-7/3$ e) $-4/3\sqrt{2}$

Section 7.3

1. a) $x^5/5 - x^4/4 + x^2 - 5x$ c) $\frac{2}{3}t^6 - \frac{5}{4}t^4 + t^2/2 - 4t$
e) $\frac{2}{7}\sqrt{7y}$ g) $\dfrac{4}{3}x^3 - \dfrac{x^4}{2} + \dfrac{x^2}{2} - x$

3. a) $3x^2$ c) $\sqrt{1 + x}$ e) $-(4x - 2)$

5. 1 **7.** $\frac{1}{16}$

8. a) $\frac{7}{3}$ c) $-67/12$ e) 2 g) $\frac{134}{21}$

9. $2x(1 + x^4)^{1/2}$

Section 7.4

1. a) $-8/5$ c) $\frac{1}{3}(5^{3/2} - 1)$ e) $\frac{2}{5}(\sqrt{5b^2 + 2} - \sqrt{5a^2 + 2})$

3. $s = \frac{7}{2}t^2 - 6t + \frac{15}{2}$ **5.** $s = \dfrac{2}{3}t^3 - \dfrac{t^2}{2} + 40t + \dfrac{2759}{6}$ **7.** 1625

10. a) $\frac{32}{3}$ c) $-\sqrt{3} + \frac{17}{3}$

11. $\frac{9}{4}10^{4/3} + 70$ **13.** $\dfrac{39}{2}Q - \dfrac{Q^2}{4}$; 39 **15.** 60

Section 7.5

1. a) $\frac{3}{2}x^2 + 2x + C$ c) $\frac{1}{36}(2x - 5)^{18} + C$ e) $\frac{1}{3}(2x + 3)^{3/2} + C$
g) $x^3/3 - 2\sqrt{x} + C$ i) $(-1/2)(t^2 + 4t - 3)^{-1} + C$
k) $\frac{3}{4}x^4 + \frac{2}{3}x^3 - 4x + C$ m) $(-1/12)(t^3 - 4)^{-4} + C$

2. a) $y = x^3/3 + 2x + C$ c) $y = 2\sqrt{x} + \frac{2}{3}x^{3/2} + C$ e) $(-1/2)y^{-2} = x^2 + C$
g) $\frac{1}{8}(2t + 1)^4 + C = s$ i) $y = (-2/3)(2 - x)^{3/2} + C$
k) $2(1 + y)^{1/2} = x + C$

3. $y = \frac{4}{3}x^3 - \frac{5}{2}x^2 + 3x + 2$ **5.** 10.05

Section 7.6

1. 158.1 **3.** approximately 9940

4. a) convergent, 1 c) convergent, 2 e) divergent g) divergent
i) 1 k) $\frac{1}{4}$

7. $1 - \dfrac{2}{\sqrt[3]{18}}$

Section 8.1

1. a) $\pi/36$ c) $\pi/60$ e) $5\pi/12$ g) $\pi/18$ i) $8\pi/9$

2. a) $180/\pi$ c) 270 e) 22.5 g) 105 i) 165

3. a) $2k\pi$, $\pm\pi/3 + 2k\pi$, k an integer c) $2k\pi$, k an integer
e) $\pi/4 + k\pi/2$, k an integer

5. a) 0.924 c) 0 e) $\frac{1}{2}$

7. a) $3\pi/2$ c) $\pi/6$ e) $-\sqrt{3}/2$ g) $-\pi/6$ i) $-\pi/3$

Section 8.2

1. a) $3 \cos 3x$ c) $-6x \sin (3x^2 + 1)$ e) y/x g) $\frac{2}{3} \sec^2 2x(1 + \tan 2x)^{-2/3}$

i) $\dfrac{-\sin 2x}{\sqrt{2 + \cos 2x}}$ k) $\frac{1}{2} \cos \dfrac{x}{2}$ m) $-\dfrac{1}{2\sqrt{x}} \sin \sqrt{x}$

2. a) $1/\sqrt{4 - x^2}$ c) $-4/\sqrt{1 - 4x^2}$ e) $1/\sqrt{x}(x + 1)$ g) $2x/\sqrt{1 - x^4}$

3. $y = 4x - \frac{4}{3}\pi + \sqrt{3}$

5. a) local maximum at $x = \frac{2}{3}\pi$, maximum at $x = 2\pi$, local minimum at $x = 4\pi/3$ concave downward on $(0, \pi)$, concave upward on $(\pi, 2\pi)$, points of inflection at $x = 0, \pi, 2\pi$.

c) no local maximum nor minimum, concave downward for $-\pi/2 < x < 0$, concave upward for $0 < x < \pi/2$, point of inflection at $x = 0$.

7. 0.075 rad/min **9.** 15; $\frac{15}{4}$

Section 8.3

1. a) 4 c) $2\sqrt{2} - \frac{4}{3}\sqrt{3}$ e) $(1/\sqrt{3}) - 1$ g) $\pi/2$ i) $\sqrt{2} - \sqrt{7}/2$
k) $(3\sqrt{3} - 1)/24$ m) $\frac{1}{3} \arctan \frac{2}{3}$

2. a) $-\frac{1}{4} \cos 4x + C$ c) $\frac{1}{4} \sin^2 2x + C$ e) $-\frac{1}{2}(\sin x + \cos x)^{-2} + C$

g) $\frac{1}{2} \arctan \dfrac{x}{2} + C$ i) $-\frac{1}{2}(1 + x^2)^{-1} + C$ k) $\frac{1}{2} \sec^2 x + C$

3. a) 0 c) $2\pi^2$ e) 1.109 g) π i) $\dfrac{\pi^2}{2}$

Section 8.4

1. a) $\dfrac{2x}{1 + x^2}$ c) $\dfrac{4}{(2x + 2)(x - 1)}$ e) $2e^{2x}$ g) $\dfrac{e^{x/2}}{2}$

i) $e^x \cos e^x$ k) 1 m) $2 \log x + (\log x)^2$ o) $\dfrac{2}{x}$

q) $\dfrac{1}{2x} (\log 2x)^{-1/2}$ s) $1 + \log |x|$

2. a) $e/2$ c) 1 e) 0 g) 0

3. minimum at $x = -1$, concave upward for $x > -2$, concave downward for $x < -2$, point of inflection at $x = -2$.

5. $\dfrac{dH}{dx} = 100ie^{-ix}$

Section 8.5

1. a) $\dfrac{e^{3x}}{3} + C$ c) $\frac{1}{2} \log (1 + 2e^x) + C$ e) $-\frac{1}{2}e^{-2x} + C$

g) $\frac{1}{6} \log (2x^3 - 4) + C$ i) $\frac{1}{4} \log (1 + 2x^2) + C$ k) $x - \log |x + 1| + C$
m) 0

3. approximately 61 lb **5.** approximately 1.024×10^6

6. a) approximately 20.4 c) approximately 0.068

7. approximately 4.67

Section 8.6

1. approximately 4010, approximately 1418 **3.** approximately 12,800
5. approximately 6.7 years
7. approximately 14 years **9.** $1 + 6e$ **11.** $\frac{1}{2} + \frac{3}{2}e^2$
13. a) 3.85 b) 5
15. \$54.88

Section 9.1

1. $\frac{\partial f}{\partial x} = 12x^3 + z^2$; $\frac{\partial f}{\partial y} = -z$; $\frac{\partial f}{\partial z} = -y + 2zx$; $\frac{\partial^3 f}{\partial z^3} = 0$

3. $\frac{\partial f}{\partial x} = \log y$; $\frac{\partial^2 f}{\partial y^2} = \frac{-x}{y^2}$; $\frac{\partial^2 f}{\partial x\, \partial y} = \frac{1}{y}$

5. $\frac{-6}{25}$ **7.** $\frac{\partial f}{\partial x} = -x^2$; $\frac{\partial f}{\partial y} = y^2$ **9.** $\frac{\partial^2 f}{\partial x^2} = 2x$; $\frac{\partial^2 f}{\partial x\, \partial y} = 4y$

11. $\frac{\partial^2 f}{\partial x\, \partial y} = 6xe^y$; $\frac{\partial^2 f}{\partial y^2} = 3x^2e^y$ **13.** -1 **15.** $\frac{\partial f}{\partial x} = x + 2$; $\frac{\partial f}{\partial y} = -y - 2$

Section 9.2

1. a) $(\frac{1}{3}, -5/6)$ c) $(-1, 2k\pi)$, k an integer $(1, (2k + 1)\pi)$, k an integer
e) $(0, 0, 0)$ g) $(0, 0)$ i) $(x, 0, x)$
2. a) none c) open e) open, bounded g) closed, bounded i) bounded
3. a) none c) none e) none g) none

5. $q_1 = \frac{8p_1 - 3p_2}{55}$, $q_2 = \frac{3p_1 - 8p_2}{55}$ **6.** 1

11. $x = \frac{16}{3}$, $y = \frac{8}{3}$ **13.** length $= \frac{100}{3}$, height $=$ width $= \frac{50}{3}$

Section 9.3

1. a) 9 c) $\frac{5}{3}$ e) $\frac{1}{2}$ g) -1 i) $e^3 - e^2 - e + 1$ k) $(\log 2)^2$
2. a) $-65/34$ c) 0 e) $\frac{1}{18}$ g) 0
3. 1

Section 9.4

1. a) $y = \frac{68}{35}x + \frac{38}{21}$ c) $y = (-9/14)x + \frac{37}{14}$
2. a) $y = \frac{25}{84}x^2 - \frac{5}{4}x + \frac{79}{42}$ c) $y = (-27/56)x^2 + \frac{993}{280}x - 31/10$

3. $\frac{13}{64}$ **5.** $\frac{500}{R}$

7. a) $e^2 - 2e + 1$ c) $\frac{3}{2}\log 2$

Section 10.1

1. a) $x = -5/17, y = -8/17$ c) $x = \frac{9}{14}, y = \frac{11}{14}, z = \frac{2}{7}$
e) $\{(x, y, z) \mid x + y + z = 1\}$ g) $x = \frac{3}{5}, y = -2/5$ i) $w = \frac{12}{7}, z = \frac{23}{7}$

3. 5 of truck A, 1 truck B, 1 truck C **5.** $p = \frac{15}{4}, y = \frac{25}{2}$

Section 10.2

1. a) 3 c) 0 e) -1 g) 432 i) 3 k) -6

2. a) 2 c) 2 e) 4 g) 4 i) 3 k) 4

4. a) consistent; $x = 119/11, y = -18/11, z = -23/11$ c) inconsistent
e) $\{(x, y, x, 2 - y) \mid x, y \text{ real numbers}\}$ g) $x = \frac{3}{2}, y = 1, z = \frac{1}{2}$
i) $\{(x, x, 1 - x, 4 - x) \mid x \text{ a real number}\}$

5. 2 of A, 2 of B, 2 of C, $\frac{1}{5}$ of D

Section 10.3

1. a) $\begin{pmatrix} 5 & 1 & 8 \\ 6 & 7 & 17 \\ -2 & -3 & 15 \end{pmatrix}$ c) $\begin{pmatrix} 5 & 10 & -1 \\ 8 & 18 & 3 \\ 3 & 6 & -9 \end{pmatrix}$

e) $\begin{pmatrix} 5 & 3 & 7 \\ 3 & -6 & 3 \\ 5 & 41 & 15 \end{pmatrix}$ g) $\begin{pmatrix} 5 & 3 & 7 \\ 3 & -6 & 3 \\ 5 & 41 & 15 \end{pmatrix}$

2. a) 38

c) $\begin{pmatrix} 0 & 6 & 4 \\ 1 & 11 & -5 \\ 10 & -2 & 2 \end{pmatrix}$ e) $\begin{pmatrix} 17 & 12 & -7 \\ -2 & 16 & 13 \\ 15 & -12 & 23 \end{pmatrix}$

g) -20

3. a) $\begin{pmatrix} 1 & 3 & 2 & 6 \\ 2 & -4 & 1 & 7 \end{pmatrix}$ c) $\begin{pmatrix} 1 & 1 & 0 & 1 \\ 0 & 0 & 1 & 0 \\ 1 & 1 & 0 & 0 \\ 0 & 0 & 1 & 1 \end{pmatrix}$

4. a) $BA = \begin{pmatrix} 78 & 57 & 50 & 30 \\ 14 & 5 & 48 & 9 \end{pmatrix}$ c) $AB = \begin{pmatrix} 11 & 6 & 9 & 4 \\ 6 & 3 & 13 & -3 \\ 7 & 3 & 28 & 8 \\ 55 & 69 & 51 & -18 \end{pmatrix}$

$$BA = \begin{pmatrix} 30 & -1 & 18 & 17 & 19 \\ 10 & -29 & -3 & -19 & 20 \\ 19 & 13 & 8 & 18 & 6 \\ 17 & -1 & 1 & 5 & 8 \\ -2 & -13 & -2 & -12 & 10 \end{pmatrix}$$

e) $AB = \begin{pmatrix} -10 & 1 \\ 5 & 4 \end{pmatrix}$, $BA = \begin{pmatrix} 1 & 2 & -3 \\ -1 & 0 & 7 \\ 3 & 7 & -7 \end{pmatrix}$

g) AB not defined, $BA = \begin{pmatrix} 15 & 7 & 12 & 5 \\ 3 & 3 & 5 & 5 \end{pmatrix}$

5.

	Cost	Transportation cost
Collar and leash	7.56	0.064
Coat	7.17	0.103
Cushion	8.82	0.3

6. a) $\frac{-1}{6}\begin{pmatrix} -2 & 2 & -2 \\ 2 & 1 & -7 \\ -2 & -1 & 1 \end{pmatrix}$ c) $\begin{pmatrix} 13 & 19 & 5 \\ -10 & -15 & -4 \\ 2 & 3 & 1 \end{pmatrix}$

e) $\frac{1}{4}\begin{pmatrix} 2 & -2 \\ -1 & 3 \end{pmatrix}$ g) $\frac{-1}{33}\begin{pmatrix} -7 & 5 & -6 \\ 8 & -1 & -12 \\ -2 & -8 & 3 \end{pmatrix}$

i) no inverse

6. a) inconsistent c) $x = 1, y = -1, z = -1$

7. 19

Section 10.4

1. a) $\lambda = -1, 3$ characteristic vectors $\begin{pmatrix} x \\ 2x \end{pmatrix}, \begin{pmatrix} x \\ -2x \end{pmatrix}$

c) $\lambda = \pm\sqrt{2}$ characteristic vectors $\begin{pmatrix} x \\ x/\sqrt{2} \end{pmatrix}, \begin{pmatrix} x \\ -x/\sqrt{2} \end{pmatrix}$

e) $\lambda = 6, 7$ characteristic vectors $\begin{pmatrix} x \\ x \end{pmatrix}, \begin{pmatrix} x \\ 2x \end{pmatrix}$

g) $\lambda = 2, \pm 3$ characteristic vectors $\begin{pmatrix} 0 \\ y \\ 0 \end{pmatrix}, \begin{pmatrix} x \\ 0 \\ x \end{pmatrix}, \begin{pmatrix} x \\ 0 \\ -x \end{pmatrix}$

i) $\lambda = 7, 2, 3, -1$ characteristic vectors $\begin{pmatrix} x \\ 0 \\ 0 \\ 0 \end{pmatrix}, \begin{pmatrix} 0 \\ y \\ 0 \\ 0 \end{pmatrix}, \begin{pmatrix} 0 \\ 0 \\ z \\ 0 \end{pmatrix}, \begin{pmatrix} 0 \\ 0 \\ 0 \\ w \end{pmatrix}$

k) $1, 5; \begin{pmatrix} x \\ -x/3 \end{pmatrix}, \begin{pmatrix} x \\ x \end{pmatrix}$

m) $\pm\sqrt{3}; \left(\frac{\sqrt{3}-1}{2}x\right), \left(\frac{-\sqrt{3}-1}{2}x\right)$

o) $1, -2, 3; \begin{pmatrix} x \\ -x \\ -x \end{pmatrix}, \begin{pmatrix} x \\ x/11 \\ -\frac{14}{11}x \end{pmatrix}, \begin{pmatrix} x \\ x \\ x \end{pmatrix}$

q) $-1, 1 \pm \sqrt{6}; \begin{pmatrix} x \\ 0 \\ 0 \end{pmatrix}, \begin{pmatrix} 0 \\ y \\ \frac{\sqrt{6}}{2}y \end{pmatrix}, \begin{pmatrix} 0 \\ y \\ \frac{-\sqrt{6}}{2}y \end{pmatrix}$

3. $\begin{pmatrix} x \\ x \\ x \end{pmatrix}$

Section 10.5

1. a)

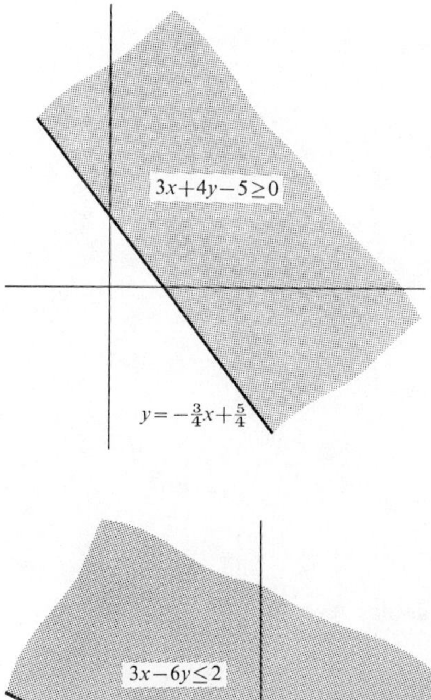

c)

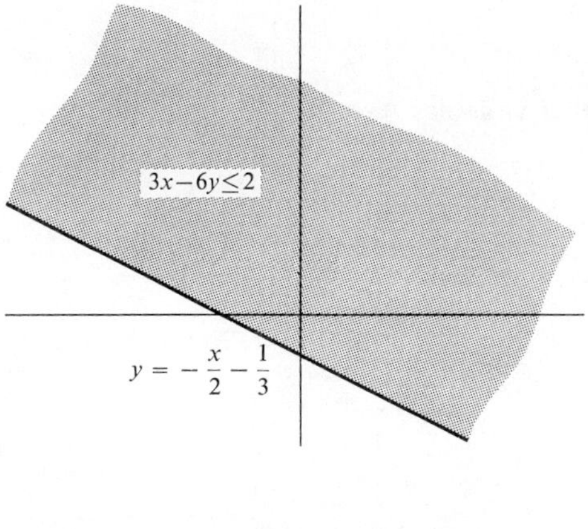

3. a)

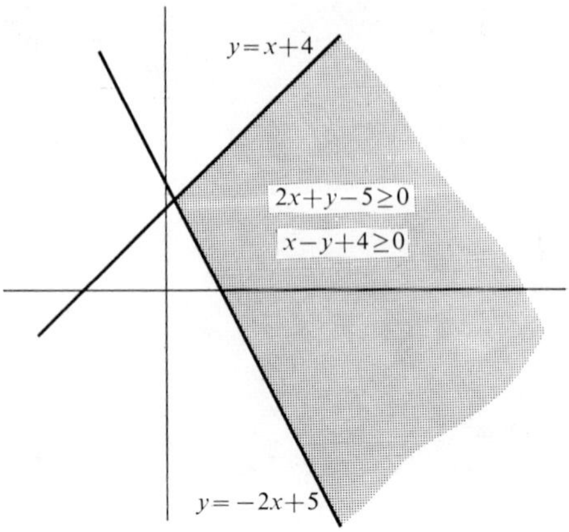

c)

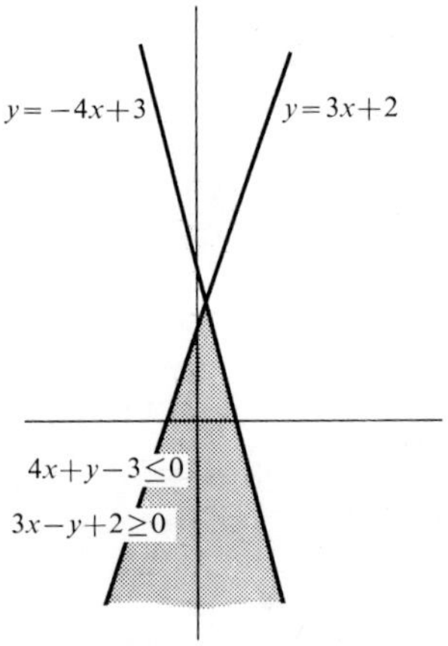

5. $x = \frac{20}{7}$, $y = 0$

7. 2000 Monsters, 500 Super Monsters

Section 10.6

1. $x_1 = 2$, $x_2 = 1$, $x_3 = 3$, all other variables zero

Section 11.1

1. a) $(6, -1, 6, 13)$ c) -7 e) 21 g) $\sqrt{14}$ i) $-14\sqrt{198}$

2. a) $\frac{1}{\sqrt{30}}(2, -1, 3, 4)$ c) $\frac{1}{\sqrt{114}}(2, -5, 9, 2)$

3. a) $(-4, -1, 5, 5)$ c) 3 e) $3\sqrt{46}$

4. a) $\frac{1}{\sqrt{14}}(1, 0, -2, -3)$ c) $\frac{1}{\sqrt{83}}(7, 3, -5, 0)$

5. a) -13 c) 0 e) $\begin{pmatrix} 16 & 48 & 13 \\ -16 & 64 & 31 \\ 28 & 10 & -13 \end{pmatrix}$ g) -3 i) $\begin{pmatrix} 9 & 5 \\ -12 & -5 \\ 12 & 5 \end{pmatrix}$

7. $\begin{matrix} \text{gas} \\ \text{coal} \\ \text{electricity} \end{matrix} \begin{pmatrix} 165 \\ 95 \\ 350 \end{pmatrix}$

Section 11.2

1. a) yes c) yes e) no g) no i) no k) no

2. a) yes c) yes e) yes g) yes

3. a) no c) yes e) no

Section 11.3

1. a) no c) no e) yes g) no i) no k) no

2. e) $2x^2 - 3y^2$

3. a) $\begin{pmatrix} 1 & \frac{3}{2} \\ \frac{3}{2} & 1 \end{pmatrix}$ c) $\begin{pmatrix} 2 & 0 & 0 & 0 \\ 0 & -4 & 0 & 0 \\ 0 & 0 & -5 & 0 \\ 0 & 0 & 0 & 1 \end{pmatrix}$

e) $\begin{matrix} 3 & & & & & & & \\ & 4 & & & 0 & & & \\ & & 5 & & & & & \\ & & & 6 & & & & \\ & & & & 7 & & & \\ & & & & & 8 & & \\ 0 & & & & & & 9 & \\ & & & & & & & 10 \end{matrix}$ g) $\begin{pmatrix} 2 & -\frac{3}{2} & 1 & 0 \\ -\frac{3}{2} & 0 & 0 & 0 \\ 1 & 0 & -1 & 0 \\ 0 & 0 & 0 & 1 \end{pmatrix}$

i) $\begin{pmatrix} 0 & 1 & \frac{3}{2} & 2 \\ 1 & 0 & 3 & 4 \\ \frac{3}{2} & 3 & 0 & 6 \\ 2 & 4 & 6 & 0 \end{pmatrix}$

4. a) $x^2 - 2xz + 6yz + 2y^2 + 4z^2$ c) $2x^2 + 4xz + y^2 + 8yz + 3z^2$

Section 11.4

1. a) linearly independent c) linearly independent e) linearly independent
g) linearly dependent i) linearly dependent k) linearly independent

5. a) 2 c) 4 e) 3 g) 4

7. a) $(-1, 2)$ c) $(0, 1)$

8. a) $(5, -4, 3)$

9. a) $(-1, 3)$ c) $(\frac{2}{3}, \frac{1}{3})$

Section 11.5

1. a) yes c) yes e) no g) yes i) no k) yes

2. a) isomorphism c) surjective

4. a) $\begin{pmatrix} 3 & -1 & 0 \\ 0 & 0 & 1 \end{pmatrix}$ c) $\begin{pmatrix} 1 & 0 & 0 \\ 0 & 0 & 0 \\ 0 & 1 & 0 \end{pmatrix}$ e) $\begin{pmatrix} 1 & 0 \\ 0 & 2 \\ 2 & 0 \end{pmatrix}$

5. $\begin{pmatrix} 11 & 3 & -1 \\ -14 & -5 & 1 \end{pmatrix}$

7. $(6, \frac{8}{3}, 7)$

9. $\begin{pmatrix} \frac{5}{4} & 0 & \frac{5}{2} \\ \frac{3}{2} & 1 & 2 \end{pmatrix}$

10. a) $\begin{pmatrix} 0 & 1 & 0 \\ 0 & 0 & 2 \\ 0 & 0 & 0 \end{pmatrix}$ c) $\begin{pmatrix} 0 & -1 \\ 1 & 0 \end{pmatrix}$

Section 11.6

1. $\frac{23}{96}$ **3.** $\frac{2071}{5184}$ **5.** Eros 50.78333%, Lib 28.57083%, Porno 20.64583%

7.

leather	14831
wool	30858
cedar chips	12360
machine time	18260
labor time	10095
miscellaneous	2263

9. a) $u^2 + 6v^2$ c) $-3u^2 - 4v^2$ e) $u^2 - \frac{1}{2}v^2 - \frac{1}{2}w^2$

g) presently in diagonalized form i) $\dfrac{1 + \sqrt{37}}{2}u^2 + \dfrac{1 - \sqrt{37}}{2}v^2$

10. a) $\begin{pmatrix} 3 & 0 \\ 0 & 0 \end{pmatrix}$ c) $\begin{pmatrix} 4 & 0 \\ 0 & -1 \end{pmatrix}$ e) $\begin{pmatrix} 1 & 0 & 0 \\ 0 & 2 & 0 \\ 0 & 0 & 2 \end{pmatrix}$ g) $\begin{pmatrix} 1 & 0 \\ 0 & 4 \end{pmatrix}$

i) $\begin{pmatrix} 3 + \sqrt{10} & 0 \\ 0 & 3 - \sqrt{10} \end{pmatrix}$

Section 11.7

1. $(2y - 2x, 2x - x^2)$ **3.** $(4z - 3y, -3x + 2z, 4z + 2y)$

5. $(\log yz, x/y, x/z - e^z)$ **7.** $\begin{pmatrix} 2 & 0 \\ 0 & 9 \\ 1 & 1 \end{pmatrix}$ **9.** $(0, 8, 8)$ **11.** $(-\pi, 0)$

12. a) $(2xy - 3y^2)\,dx + (x^2 - 6xy)\,dy$
c) $(2xy^2 - z^2 + 3yz)\,dx + (2x^2y + 3xz)\,dy + (3xy - 2xz)\,dz$

13. -2

15. $\begin{pmatrix} 4x+4z & 0 & 4x+4z \\ 0 & 2y+2z & 2y+2z \\ 1 & 1 & 2 \end{pmatrix}$

17. $(2yz - 2xy^2, 2xz - 2x^2y + 2yz^2, 2xy + 2y^2z)$

19. $(yz \cos xy, xz \cos xy, \sin xy)$ **21.** (e, e) **23.** $\begin{pmatrix} 4 & 0 & 0 \\ 0 & 1 & -1 \end{pmatrix}$

25. $(4, 6)$ **27.** $(3, 0, 3)$

29. a) $(6x - y)\,dx - x\,dy$ c) $(e^y - 2xze^{x^2})\,dx + xe^y\,dy - e^{x^2}\,dz$

31. $\begin{pmatrix} 4 & 0 \\ 2 & 4 \end{pmatrix}$

33. $\dfrac{\partial F}{\partial u} = e^u \sin(v^3e^u - vwe^u) + (v^3e^{2u} - vwe^{2u}) \cos(v^3e^u - vwe^u)$

35. $(1, -2)$ **37.** (e^2, e^2, e^2) **39.** $(-\frac{1}{2}e^{-1/2}, 0, -\frac{1}{4}e^{-1/2})$

Section 11.8

1. 1 **3.** 4 **5.** $\frac{9}{8}$ **7.** $\frac{17}{4}$ **9.** $48 + 5\sqrt{7}$

Section 12.1

1. $\dfrac{x^2}{2} \log|x| - x^2/4 + C$ **3.** $\dfrac{-1}{8} \sin x \cos x + \dfrac{1}{8} x + \dfrac{\sin^3 x \cos x}{4} + C$

5. $x(\log|x|)^2 - 2x \log|x| + 2x + C$ **7.** $(x^2 - 2x + 2)e^x + C$

9. $\dfrac{x}{8}(2x+1)^4 - \dfrac{1}{80}(2x+1)^5 + C$ **11.** $x^2 \sin x + 2x \cos x - 2 \sin x + C$

13. $\dfrac{e^x}{2}(\sin x - \cos x) + C$ **15.** $\dfrac{x}{3} - \dfrac{1}{9} \log|9x + 3| + C$

Section 12.2

1. $\dfrac{1}{4} \log\left|\dfrac{x-3}{x+5}\right| + C$ **3.** $-\frac{1}{4}(2x - 5)^{-2} + C$

5. $\log|x + 1| - 2(x+1)^{-1} + C$ **7.** $\frac{1}{3} \log|(x-5)^8(x+1)| + C$

9. $x + \log\left|\dfrac{(x-3)^9}{(x-2)^4}\right| + C$ **11.** $\frac{1}{4} \log(2x^2 + 1) + C$

13. $\frac{1}{2} \log\left|\dfrac{x-4}{x+2}\right| + C$

15. $\frac{1}{2} \log(x^2 + x + 1) + \dfrac{1}{\sqrt{3}} \arctan\left(\dfrac{2}{\sqrt{3}} x + \dfrac{1}{\sqrt{3}}\right) + C$

17. $\log\left|\dfrac{x-3}{x-1}\right| + (x+1)^{-1} + C$ **19.** $\frac{1}{2} \log(x^2 + 1) + C$

21. $\log \dfrac{P}{a - bP} = t + C$

Section 12.3

1. a) $\dfrac{7x}{2}\sqrt{4 - x^2} + 14 \arcsin \dfrac{x}{2} + C$ c) $x - \frac{2}{3} \log |2 - 3x| + C$

e) $\arcsin (x - 1) + C$ g) $2\sqrt{\log x} + C$ i) $\dfrac{1}{\sqrt{3}} \arctan \sqrt{3}x + C$

k) $\dfrac{-\sqrt{16 - x^2}}{16x} + C$

2. a) $\dfrac{x}{2}(\sin (\log |x|) - \cos (\log |x|)) + C$ c) $x \tan x + \log |\cos x| + C$

e) $5 \arctan (x + 2) + C$ g) $-(9 - x^2)^{1/2} + C$ i) $\arcsin \dfrac{x}{3} + C$

k) $\frac{1}{3} \arcsin 3x + C$ m) $\dfrac{1}{\sqrt{5}} \arcsin \sqrt{\tfrac{5}{2}}x + C$

o) $\frac{1}{2} \arcsin x - \dfrac{x}{2}\sqrt{1 - x^2} + C$ q) $\arcsin \dfrac{x}{2} + C$

s) $-(4 - x^2)^{1/2} + C$ u) $\dfrac{2}{\sqrt{7}} \arctan \left(\dfrac{2}{\sqrt{7}}x + \dfrac{3}{\sqrt{7}}\right) + C$

Section 12.4

1. trapezoid rule: $T_4 = \frac{243}{64}$ midpoint rule: $M_4 = \frac{477}{128}$
Simpson's rule: $S_4 = \frac{15}{4}$

3. trapezoid rule: $T_4 = 0.6035\pi$; midpoint rule: $M_4 = 0.6535\pi$;
Simpson's rule: $S_4 = 0.636\pi$

5. trapezoid rule: $T_4 = \frac{101}{60}$; midpoint rule: $M_4 = \frac{1488}{945}$;
Simpson's rule: $S_4 = \frac{73}{45}$

7. trapezoid rule: $T_4 = \frac{287}{260}$; trapezoid rule: $M_4 = \frac{251668}{226525}$;
trapezoid rule: $S_4 = \frac{431}{390}$

Section 13.1

1. 48,000; 4,140,000; 497

3. a) 8722 c) approximately $\frac{1}{2}$

4. a) converges c) diverges e) converges g) diverges

7. $2^{64} - 1$ **9.** 150,000 **11.** approximately 8450

Section 13.2

1. 100,000,000

3. a) converges c) converges e) converges g) diverges i) diverges
k) converges

4. a) absolutely convergent c) divergent e) absolutely convergent
g) absolutely convergent

5. 3333

Section 13.3

1. a) $-2 < x < 2$ c) $-2 < x < 0$ e) $-1 \leq x \leq 3$
g) $0 < x < \frac{1}{2}$ i) $-3 < x < 3$ k) $-1 < x < 1$

3. 18

5. a) $f(x) = \sum_{n=1}^{\infty} \frac{(-1)^{n+1}}{n} x^n$ c) $f(x) = \sum_{n=0}^{\infty} \frac{e}{n!}(x-1)^n$

e) $\sum_{n=1}^{\infty} \frac{(x-1)^n(-1)^{n-1}}{n}$

g) $\frac{1}{2} + \sum_{n=1}^{\infty} \frac{(-1)^{n-1}(x-2)^n(3\cdot 5\cdots(2n-3))}{2^{2n-1}n!\,2^{n-1}}$

6. a) $1 - x + \frac{x^2}{2} - \frac{x^3}{6}$ c) $1 - x^2 + x^4 - x^6$ e) $1 + x^2 + \frac{x^4}{2} + \frac{x^6}{6}$

7. 0.693

10. a) local maximum at $(\frac{1}{5}, \frac{1}{3})$ c) saddle point at (0, 0)
e) saddle point at (0, 0)

Section 13.4

1. $\frac{3}{2} - \frac{2}{\pi} \sum_{k=1}^{\infty} \frac{\sin(2k-1)x}{2k-1}$

3. $\frac{\pi}{4} - \sum_{k=1}^{\infty} \frac{2\cos(2k-1)x}{(2k-1)^2} + \sum_{k=1}^{\infty} \frac{2x\sin(2k-1)x}{(2k-1)}$

5. $\frac{-4}{\pi} \sum_{k=1}^{\infty} \frac{\sin(2k-1)}{(2k-1)} x$

Section 13.5

1. a) $f(x) = \sum_{n=0}^{\infty} (n+1)x^n$ c) $f(x) = \sum_{n=0}^{\infty} \frac{2x^{2n+1}}{2n+1}$

e) $f(x) = 1 + x - \frac{x^2}{2!} - \frac{x^3}{3!} + \frac{x^4}{4!} + \frac{x^5}{5!} - \frac{x^6}{6!} + \cdots$

2. a) $1 + x - \frac{1}{3}x^3 - \frac{1}{6}x^4 - \frac{1}{30}x^5$ c) $x - \frac{1}{24}x^3 + \frac{1}{24}x^4 - \frac{71}{1920}x^5$

e) $1 - x^2 + \frac{x^4}{3} - \frac{2}{45}x^6 + \frac{x^8}{315}$

Section 14.1

1. a) $e^x + e^{-y} + C = 0$ c) $\log y - \frac{1}{2}(\log x)^2 + C = 0$

e) $\log xy + C = 0$ g) $\log \frac{y}{x} + y - x + C = 0$

i) $\arctan x - \log y + C = 0$

2. a) $\log \frac{y}{2} - 1/2x + 1/2 = 0$ c) $3 \log \sin x + 2 \log y = 0$

3. $-\frac{4}{k} \log \frac{|k/\sqrt{L} - \sqrt{Lr}|}{|\sqrt{Lr}|} = At + \frac{BP_0}{\lambda} e^{\lambda t}$

$$r = \frac{k}{L(e^{Akt/4 + (kBP_0/4\lambda)e^{\lambda t}} + 1)}$$

Section 14.2

1. $\frac{3}{2}x^2 + 8xy - 3x - \frac{5}{2}y^2 + 6y + C = 0$ **3.** $\frac{x^2y^2}{2} + xy + C = 0$

5. $\frac{x^2}{2} + 3yx + C = 0$ **7.** $e^x - x^2 \cos y + C = 0$

9. $y \arctan x + C = 0$

Section 14.3

1. $y = \frac{x}{2} \tan (c/x)^3$ **3.** $y = x \log cx$ **5.** $y = \frac{2 + (cx)^{1/3}}{1 - (cx)^{1/3}}$

7. $y = \frac{\sqrt{3}x}{2} \tan \log (cx)^{\sqrt{3}/2} + x$

Section 14.4

1. $y^2x + x + y + C = 0$ **3.** $y = -\frac{\cos x}{x} + C$

5. $y = C_1e^{2x} + C_2e^{-2x}$ **7.** $y = e^{3x/2}\left(C_1 \cos \frac{\sqrt{11}x}{2} + C_2 \sin \frac{\sqrt{11}x}{2}\right)$

9. $y = (C_1 - \log |\sec x + \tan x|) \cos x + C_2 \sin x$ **11.** $y = \frac{x^{19}e^x}{342}$

13. $y = C_1e^{3x} + C_2 \cdot \sin x + C_3 \cos x$

14. a) $y = \sqrt{3/2}\,(-x)^{5/2}$ c) $y = -3e^{-x}$

Section 14.5

1. $y = a_0\left(1 + \frac{x^3}{2 \cdot 3} \quad \frac{x^6}{2 \cdot 3 \cdot 5 \cdot 6} + \frac{x^9}{2 \cdot 3 \cdot 5 \cdot 6 \cdot 8 \cdot 9} + \cdots\right)$

$$+ a_1\left(x + \frac{x^4}{3 \cdot 4} + \frac{x^7}{3 \cdot 4 \cdot 6 \cdot 7} + \frac{x^{10}}{3 \cdot 4 \cdot 6 \cdot 7 \cdot 9 \cdot 10} + \cdots\right)$$

3. $y = a_0 + {a_0}^2x + \left({a_0}^3 + \frac{1}{2}\right)x^2 - \left(\frac{{a_0}^4}{3} + a_0\right)x^3 - \left(\frac{2{a_0}^5}{3} - \frac{{a_0}^2}{4} + \frac{a_0}{2}\right)x^4 + \cdots$

5. $y = a_0\left(1 - x + \frac{x^2}{4} + \frac{x^3}{18} + \frac{13x^4}{576} + \cdots\right)$

Section 14.6

1.

x	y
$\frac{1}{4}$	1.19
$\frac{1}{2}$	0.95
1	0.6
2	0.25
3	0.025

3. $y = e^{\gamma t/\beta}$ **5.** a) $c(t) = k(1 - e^{At})$

Miscellaneous Problems

1. $e^x + e^{-y} + C = 0$ **3.** $y \arcsin x + C = 0$

5. $y = e^{-1/2}\left(C_1 \cos \frac{\sqrt{3}x}{2} + C_2 \sin \frac{\sqrt{3}x}{2}\right)$ **7.** $= C_1 e^{-x} + (\frac{5}{3} e^{3x} + C_2)e^{-3x}$

9. $y = \frac{1}{3}(Cx^{-1/2} - x)$ **11.** $y^2 = x^2 + Cx$

13. $y^2 - 2yx + C = 0$ **15.** $(C_1 - x) \cos x + C_2 \sin x = y$

Section 14.7

1. $P_k = (1 + r)^k P_0$

3. a) $y_k = 2(4^k - (-1)^k)$ c) $y_k = A(2)^k + B(-2)^k$

e) $y_k = A4^k + B + \frac{11}{2}k$

5. $y_t = A(\frac{7}{3})^t + B(-1/6)^t$

INDEX

INDEX

ABCDEFGH798765432